MICROBIAL PATHOGENICITY
IN MAN AND ANIMALS

Other Publications of the
*Society for General Microbiology**
THE JOURNAL OF GENERAL MICROBIOLOGY
THE JOURNAL OF GENERAL VIROLOGY

SYMPOSIA

* Published by the Cambridge University Press, except for the first Symposium, which was published by Blackwell's Scientific Publications Limited.

MICROBIAL PATHOGENICITY IN MAN AND ANIMALS

TWENTY-SECOND SYMPOSIUM OF THE
SOCIETY FOR GENERAL MICROBIOLOGY
HELD AT
IMPERIAL COLLEGE LONDON
APRIL 1972

CAMBRIDGE
Published for the Society for General Microbiology
AT THE UNIVERSITY PRESS
1972

Published by the Syndics of the Cambridge University Press
Bentley House, 200 Euston Road, London NW1 2DB
American Branch: 32 East 57th Street, New York, N.Y.10022

© Cambridge University Press 1972

Library of Congress Catalogue Card Number: 75–177940

ISBN: 0 521 08430 X

Printed in Great Britain
at the University Printing House, Cambridge
(Brooke Crutchley, University Printer)

CONTRIBUTORS

AUSTWICK, P. K. C., Nuffield Institute of Comparative Medicine, The Zoological Society of London, Regent's Park, London.

BABLANIAN, R., Department of Microbiology and Immunology, State University of New York, Downstate Medical Centre, New York.

BANG, F. B., Department of Pathobiology, The Johns Hopkins University, School of Hygiene and Public Health, Baltimore.

BURROWS, R., Animal Virus Research Institute, Pirbright.

CRAIG, J. P., Department of Microbiology and Immunology, State University of New York, Downstate Medical Centre, New York.

GLYNN, A. A., Department of Bacteriology, Wright–Fleming Institute, London.

HALL, J. G., Department of Experimental Pathology, University of Birmingham.

HIRSCH, J. G., The Rockefeller University, New York.

LOWRIE, D. B., Department of Microbiology, University of Birmingham.

MIMS, C. A., Department of Microbiology, John Curtin School of Medical Research, Canberra.

NEWTON, B. A., The Molteno Institute, University of Cambridge.

PARISH, W. E., The Lister Institute of Preventive Medicine, Elstree.

PEARCE, J. H., Department of Microbiology, University of Birmingham.

SAVAGE, D. C., Department of Microbiology, University of Texas, Austin.

SMITH, H., Department of Microbiology, University of Birmingham.

STONER, H. B., Medical Research Council Toxicology Unit, Carshalton.

WEBB, H. E., St Thomas' Hospital, London.

WHITTLESTONE, P., School of Veterinary Medicine, Cambridge.

CONTENTS

EDITORS' PREFACE

It is customary for Symposium editors to lay claim to the timeliness of their volume in summarising advances or presaging developments in their subject. Not immune in this respect we are inclined to hope that our symposium does a little of both! In designing the programme our aim was to examine the mechanisms of bacterial pathogenicity and, on a similar template, to compare and contrast them with those of viruses, fungi, protozoa and the emergingly important mycoplasmas. Exposure of lack of knowledge was considered as important as description of recent advances. Our contributors have responded nobly to the call.

First, it is apparent from the articles that much remains to be explained. But, in the eight years since the last symposium of the Society which dealt with pathogenicity, there have been significant developments in the subject. Perhaps the most striking progress has occurred in elucidating the cellular and humoral mechanisms of host defence and the pathogenic processes of organisms producing enteric disease. To a lesser extent information has accumulated on microbial anti-host defence mechanisms. The importance of host-immune response in the pathological damage wrought by microbial disease has been increasingly recognised and possibly over-emphasised. New paths are being cut into the related problems of mixed culture and mucosal surface infections which will lead to a better understanding of the early stages of disease. Many parallels between the pathogenic activities of the various microbial types have been demonstrated in the symposium and more may become apparent in the future.

Our thanks are due to the contributors, to our office staff and to the Cambridge University Press.

H. SMITH

J. H. PEARCE

Department of Microbiology
University of Birmingham

THE LITTLE-KNOWN DETERMINANTS OF MICROBIAL PATHOGENICITY

H. SMITH

Department of Microbiology, University of Birmingham,
England

INTRODUCTION

Perhaps no discoveries have had such a dramatic impact on human affairs as the proof that micro-organisms cause disease and the subsequent recognition of the species involved. The resulting public health measures have controlled in many countries the worst effects of infectious disease, especially death from bacterial disease. This has been achieved without the need for sophisticated research on the biochemistry of infectious processes. Even vaccines and drugs, which played a significant but smaller part in controlling infection, have been developed largely by empirical methods requiring little or no knowledge of the microbial or host determinants of pathogenicity. It is not surprising, therefore, that the nature of these determinants is still obscure with a few notable exceptions such as the bacterial toxins.

Our symposium, however, comes at a time of renewed interest in microbial pathogenicity. Despite the advances of the past century, infectious disease remains a major problem in human and veterinary medicine with the economically important nuisance and chronic aspects gaining prominence as the fatal consequences become less frequent. And there is much to be done. Effective chemotherapy of troublesome virus diseases is still lacking. Drug resistance of bacteria, protozoa and fungi is increasing. Many vaccines remain unsatisfactory, either due to incomplete immunogenicity or to the hazard of injecting live organisms. The alarming increase in gonorrhoea is a current reminder of the ability of a once-regulated disease to rebound with changing social conditions. New methods of attacking infectious disease are needed. Undoubtedly, as in the past, empirical screening procedures will provide some new methods. However, after nearly 50 years such procedures may be reaching the limit of their usefulness; so far they have failed to provide an effective therapy for virus diseases. There appears good reason for an inclination towards a more rational approach to the problem, namely that of attempting to recognise and then to neutralise the determinants of pathogenicity. This symposium deals with the state of knowledge on these determinants and what might be done to increase it.

In studies of microbial pathogenicity bacteria have received most attention and although many aspects of their pathogenicity are still unexplained in biochemical terms, far more is known about them than about other pathogenic microbes. Furthermore, effort in the bacterial field has revealed difficulties of investigation and produced broad concepts of pathogenesis which probably apply to studies of other microbes. For although pathogenic viruses, mycoplasmas, fungi and protozoa differ from bacteria in many properties, they share the essential capacity of invading and damaging animal hosts. Hence, work on their pathogenicity might benefit from comments and comparisons arising from what is known or surmised about bacterial pathogenicity. In this introductory chapter, I shall attempt to do this from the small available information – more to promote future research than to seek answers at present. In this context therefore, I hope I shall be forgiven if, at points, I push parallels between bacteria and other microbes too far. I shall discuss first the inherent difficulties of research in this field and then the main requirements for pathogenicity, namely abilities; (1) to enter the host usually by surviving on and penetrating mucous membranes; (2) to multiply *in vivo*; (3) to inhibit or not to stimulate host defence mechanisms; and (4) to damage the host. Inevitably I shall mention briefly some of the points discussed in greater detail by other contributors to the symposium. The term virulence is used with respect to the degree of pathogenicity (Miles, 1955).

THE DIFFICULTIES OF INVESTIGATING THE DETERMINANTS OF PATHOGENICITY

Most of the difficulties in attempts to identify virulence determinants of bacteria have stemmed from two facts which apply equally to other microbes. First, with rare exceptions, virulence is determined not by one factor but by several; and loss of any single factor can result in either partial or almost complete loss of virulence (Smith, 1958). Second, virulence is detectable only *in vivo* and is markedly influenced by changes in growth conditions due to selection of types and to phenotypic change (Meynell, 1961). Thus, genetic information which determines virulence may be expressed only under the conditions of the test for virulence, namely during growth *in vivo*. When microbes are removed from infected animals and grown *in vitro* the change of environment will induce phenotypes different from those found *in vivo*; such phenotypes may lack one or more virulence determinants or even possess apparent virulence factors never found *in vivo* (Smith, 1964). Hence, although micro-organisms grown *in vitro* are conveniently studied, in this field of pathogenicity they can be incomplete and even misleading. Early in bacteriology, loss of virulence on sub-culture *in vitro* and increase during animal passage were recognised; and in the past fifteen years direct evidence has accumulated for many bacterial species that organisms grown in infected animals are different chemically and biologically from those grown *in vitro* (Smith, 1958; Fukui, Delwiché, Mortlock & Surgalla, 1962; Beining & Kennedy, 1963; Smith & Fitzgeorge, 1964; Gordon & Gibbons, 1967; Schwab & Brown, 1968; Bekierkunst, 1968; Ekstedt & Yoshida, 1969; Adlam, Pearce & Smith, 1970*a*, *b*; Ward, Watt & Glynn, 1970). In virology, tissue culture is a boon for isolation, estimation and fundamental studies of replication. For studies of pathogenicity, however, sole reliance on growth in the de-differentiated cells of tissue cultures (Ross, Treadwell & Syverton, 1962; Hoorn & Tyrrell, 1969) is inadvisable for the above reasons that are underlined by abundant examples of virus attenuation (Enders, 1952; Ross & Syverton, 1957) and change of parent cell susceptibility (Kaplan, 1955; Ross & Syverton, 1957; Swim, 1959) occurring in tissue culture. The virulence of mycoplasmas also appears to be reduced by sub-culture in laboratory media (Lipman, Clyde & Denny, 1969). For many fungi, the different influences of growth *in vivo* and *in vitro* are expressed in morphological form (Austwick, this volume), as well as chemically and antigenically; yeast forms predominate *in vivo* and some appear to be more pathogenic and immunogenic than mycelial/arthro-spore forms (Mariat, 1964; Romano, 1966; Kong & Levine, 1967; Landay, Wheat, Conant & Lowe, 1967). Pathogenic protozoa grown *in vitro* have shown changes in morphological form, antigenicity and capacity to infect certain hosts (Weitz, 1964).

For problems of pathogenicity which have so far defied solution by use of cultures *in vitro*, the above remarks suggest a study of microbial behaviour *in vivo*. Aspects of pathogenicity might be recognised so that they can be reproduced later *in vitro* by appropriate changes in cultural conditions. It is not easy to gain information on microbial behaviour *in vivo*. But methods for bacteria described elsewhere (Smith, 1964) could be applied to other microbes. Important aspects of bacterial pathogenicity that have been recognised by application of these methods are as follows:

(*a*) the toxins responsible for the main pathological effects of anthrax, cholera and *Escherichia coli* infections (Smith, Keppie & Stanley, 1955;

De, Ghose & Sen, 1960; Finkelstein, Norris & Datta, 1964; Craig, 1965; Taylor, Wilkins & Payne, 1961; Smith & Halls, 1967) and one which may contribute to death from plague (Stanley & Smith, 1967);

(b) the role of immunological reactions in streptococcal glomerulonephritis (Vosti, Lindberg, Kosek & Raffel, 1970);

(c) the antigens associated with prevention of phagocytosis of *Pasteurella pestis* (Burrows, 1955);

(d) a cell-wall material responsible for resistance of *Brucella abortus* to the intracellular bactericidins of bovine phagocytes (Smith & Fitzgeorge, 1964; Fitzgeorge & Smith, 1966);

(e) the enhanced resistance of staphylococci and gonococci grown *in vivo* to the bactericidins of rabbit neutrophils and human serum respectively (Adlam *et al.* 1970a, b; Ward *et al.* 1970);

(f) the nutritional basis for localisation of *Br. abortus* and *Vibrio fetus* in the placentae of certain domestic animals (Keppie, 1964; Lowrie & Pearce, 1970). These results and others such as the induction by conditions *in vivo* of capsulation and phagocytosis resistance of the fungus, *Cryptococcus neoformans* (Fashi, Bulmer & Tacker, 1970) should encourage similar studies with other micro-organisms.

Mechanisms of pathogenicity such as toxin production can be investigated with a single strain of a particular microbial species, but the existence of stable strains of differing virulence greatly increases the experimental scope and allows studies at greater depth. Comparisons of such strains should reveal properties associated only with virulent strains – virulence markers – some of which may be virulence determinants, i.e. produced during infection and directly concerned with pathogenesis. Some virulence determinants might be recognised by observing the influence of products from a virulent strain on the behaviour of an avirulent strain either in animals or in relevant biological tests *in vitro*. These methods, enhanced lately by the techniques of microbial genetics, have been used to good effect in bacteriology, e.g. in studying the influence of nutrition on virulence (Burrows, 1960; Panos & Ajl, 1963) and in identifying cell-wall and capsular materials which inhibit humoral and cellular defence mechanisms (Burrows, 1960; Keppie, Harris-Smith & Smith, 1963; Smith & Fitzgeorge, 1964; Fitzgeorge & Smith, 1966; Medearis, Camitta & Heath, 1968; Friedberg & Shilo, 1970; Voltonen, 1970). However, their use for microbes other than bacteria depends on whether virulence differences between strains can be detected and measured *in vivo*.

Quantitative comparison of virus strains is difficult. The effect in animals (LD_{50}; lesion size or mean death time for the 'same' dose) must be related to amounts of virus particles indicated by plaque counts or egg infection. The latter detect only a small proportion of the total virus particles and therefore may not measure all the particles (which could vary for different strains) capable of multiplying in experimental animals. For example, infection of suckling mice detected more infectious particles of Semliki Forest virus than plaque counts on chick embryo fibroblasts (Bradish, Allner & Maber, 1971). Hence, in virus work, only strains for which conventional tests have indicated large differences in virulence should be compared in order to recognise virulence markers and determinants. Such strains are rare but comparative experiments with them have been informative on virus virulence (Bang & Luttrell, 1961; Huang & Wong, 1963; Mims, 1964; Waterson, Pennington & Allen, 1967; Bradish *et al.* 1971; Pusztai, Gould & Smith, 1971). A similar procedure should be adopted in work with mycoplasmas; estimations of the relative virulence of mycoplasma strains are only just beginning and may also be hampered by uncertainty regarding viable counts (Lipman & Clyde, 1969; Lipman, Clyde & Denny, 1969). With fungi, quantitative virulence comparisons with mycelia might be based on mycelial weight or chitin content (Ride & Drysdale, 1972) and those based on arthrospore or yeast form counts appear relatively easy. The yeast and arthrospore/mycelial forms of the dimorphic fungi seem particularly suitable for comparison, the former appearing more virulent than the latter. Protozoa are so changeable that there seems little point in comparing the virulence of different strains. Life cycle phase, morphological form and antigenic nature are extremely sensitive to environmental conditions and all influence pathogenicity (Weitz, 1964; Lumsden, 1965; Garnham, 1967; Newton, this volume). Keeping one strain in one form for repeated experiments on a particular problem appears difficult and valid comparisons between corresponding forms of different strains seem almost impossible.

SURVIVAL ON AND PENETRATION OF MUCOUS MEMBRANES

Although some microbes enter the host directly by trauma or vector bite, with the possible exception of protozoal diseases, most microbial infections start on the mucous membranes of the respiratory, alimentary and uro-genital tracts. Hence, pathogenic microbes must at least survive, and better multiply, on these membranes from a small inoculum; and this process must occur, not in pure culture, but in competition with many commensals. Subsequently the microbes penetrate into the tissues to varying extents; even in such a localised disease as dysentery organisms penetrate into epithelial cells (Takeuchi, Formal & Sprinz, 1968). A superior ability of some microbial species in accomplishing these early stages of infection might explain why some diseases (e.g. brucellosis, influenza) are more communicable than others (e.g. anthrax, mumps); and surface interactions of pathogenic species might explain enhancing or repressive effects in mixed infections (Henderson, 1964). Clearly precise information on early microbial attack of mucous membranes would aid control of infectious disease but even in bacteriology such information is scanty. Nevertheless, interest in the subject is quickening, as will be clear from later chapters. There is much recent work *in vivo* and *in vitro* on mixed cultures of gut organisms both commensals and pathogens (Savage, this volume; Pearce & Lowrie, this volume). There has also been some electron microscopy of early bacterial attack of gut epithelium (*Salmonella* and *Shigella* spp.; Takeuchi & Sprinz, 1967; Takeuchi *et al.* 1968). But investigation of the conditions for growth of very small inocula of pathogenic bacteria (e.g. *Pasteurella tularensis*; Halman, Benedict & Mager, 1967) are still too few. In virology, attack of the respiratory tract has received some attention and in whole animal experiments the site of membrane lesions and the preferentially attacked cell types have been noted for example in Newcastle disease of chickens (Bang & Foard, 1969) and influenza of experimental animals and man (Ebisawa, Kitamoto, Takeuchi & Makino, 1969). Nevertheless, there appear to be few deeper investigations for example, on the influence of oxygen tension, temperature, pH, commensals and mucus on virus survival and replication on membranes; and on the mechanisms of entry of various viruses into the particular epithelial cells they select for attack (Burrows, this volume). Organ cultures might be used for some experiments, e.g. in attempts to identify the basis for the close adhesion between influenza virus and the cilia and microvilli of nasal epithelium which has been noted in such cultures (Dourmashkin & Tyrrell, 1970; Gould, Ratcliffe, Basarab & Smith, 1972). Hardly any detailed studies appear to exist on the influence of membrane environment and commensals on the ability of fungi, mycoplasmas and protozoa to survive on and penetrate mucous surfaces; although some interesting observations have been made e.g. fatty acids in teat secretions can be fungistatic (Adams & Rickard, 1963), antibiotics remove bacterial commensals of the gut and increase *Candida* infections (Seelig, 1966), viruses increased growth of mycoplasma in organ culture (Reed, 1970), and some gut bacteria are essential as food for *Entamoeba histolytica* (Hoare & Neal, 1955). Studies similar to those occurring now with bacteria could be conducted with other microbes, especially those involving mixed cultures of relevant organisms with small inocula of the pathogenic components.

MULTIPLICATION *IN VIVO*

Virulent microbes must multiply in the host tissues in order to produce their disease syndrome either by increasing locally or by spreading throughout the host. Two qualities are needed for multiplication. First, an inherent ability to replicate in the biochemical conditions of the host tissue and second, an ability to inactivate or not to stimulate host defence mechanisms which would otherwise kill or remove them (see below). The effects of these two qualities *in vivo* are not easy to separate and often it is difficult to assess their relative importance, in the increase of a single infecting population, in the differential behaviour of virulent and attenuated strains and in the different susceptibilities of tissues or hosts to infection. In this section the first quality – ability to replicate – is discussed.

Some information on the influence of nutritional factors on bacterial pathogenicity has been gained by experiments *in vivo* with nutritionally deficient mutants or appropriate anti-metabolites; and by growth experiments *in vitro* using tissue extracts (and their diffusates). In the extracts, cellular defence mechanisms are absent and any bactericidins in them can often be removed or reduced by ultrafiltration or heating. Such experiments have shown that avirulence of *S. typhi*, *P. pestis* and other species can result from inavailability in the host of required nutrients (Burrows, 1960; Panos & Ajl, 1963) and indicated some requirements for growth and toxin production by *Bacillus anthracis* (Tempest & Smith, 1957). Moreover, they have demonstrated a nutritional basis for some examples of host and tissue specificity (Pearce & Lowrie, this volume) as well as suggesting the probable absence of a nutritional basis for the differential ability of virulent and avirulent strains of *Br. abortus* to survive and grow within bovine phagocytes (Burrin, Keppie & Smith, 1966).

Similar investigations with viruses are complicated by the obligate parasitism involved which not only entails complexity of the factors required for replication but increases the difficulty of distinguishing the influence of their absence from that of host factors (defence mechanisms) which actually destroy virus or inhibit replication. Nevertheless, an attempt to distinguish the two types of factors may aid the recognition of virus products which inactivate or are unaffected by host defence mechanisms (see later). The ability of a virus to replicate in a particular cell depends on inherent features of that cell. These features can be involved in one or more stages of replication; attachment and penetration of virus, uncoating, provision of energy and precursors of low molecular weight, synthesis of viral nucleic acid and proteins, assembly and release (Newton, 1970). The characteristics of the host cell which determine these stages of replication might be called 'replication factors', and seem to be the counterparts in virology of the environmental factors such as low molecular weight nutrients, necessary for bacterial multiplication in host tissues or fluids. It is abundantly clear that in cell culture 'replication factors' vary from cell type to cell type and are influenced by changes in environment of the cell. Thus, by comparison with cells supporting full virus replication certain cell types have been shown to lack 'replication factors' at the receptor stage for poliomyelitis virus (Holland, 1964), at penetration for feline virus (Meyer & Enders, 1969), at uncoating for mouse hepatitis virus (Shif & Bang, 1970), at viral nucleic acid synthesis and later maturation stages for some PK-negative mutants of rabbit pox virus (Fenner, 1970) and at late maturation or envelopment for KB-negative mutants of adenovirus type 12, some DK mutants of herpes simplex virus and Sendai virus (Fenner, 1970). And, virus replication in cell culture is affected by temperature and pH changes (Holland, 1964) by small molecular materials such as arginine (Becker, Olshevsky & Levitt, 1967), leucine (Gabliks, 1969) some fatty acids (Koch, Drén & György, 1968) and by extracts of other cells (Simon & Dömök, 1966; Mathews, 1970; Chakrabortty & Beswick, 1971).

In animal infection, variation of availability of 'replication factors' in particular hosts or tissues and under different conditions will affect virus pathogenicity. Attenuated viruses may have a decreased capacity to use the factors. But few investigations of the influence of such factors comparable in depth to the experiments in tissue culture have been conducted either in animals or organ culture. Nevertheless, there are signs of this influence in the available studies. With regard to receptors, there was some parallel between the ability of homogenates of various primate tissues to bind poliomyelitis virus and their susceptibilities to virus replication and damage in infection and some attenuated strains adhered to susceptible nerve tissues less strongly than virulent strains (Holland, 1964). The well-known effect of temperature on virus virulence (Bang & Luttrell, 1961; Smorodintsev, 1960) probably reflects a temperature sensitivity of the enzymes used in virus synthesis rather than an influence on host defence mechanisms. Although the author is unaware of reports of an enhancement of virulence of adenovirus and herpes virus in animals by injections of arginine, vaccinia virus infection of mice, like that in cell culture, was enhanced by leucine (Gabliks, 1969). In future studies organ cultures may play the same role in virology as those with tissue extracts and diffusates in bacteriology in providing a system removed to some extent from the influence of host defence mechanisms without complete destruction of the *in vivo* character of experiments. In organ cultures, the defence mechanisms of the reticulo-endothelial system and the inflammatory response are largely absent. If therefore, infection patterns in organ culture parallel those in whole animals for example in the replication of influenza virus in various ferret tissues (Basarab & Smith, 1969, 1970) host defence mechanisms (other than those present or induced in the infected cells) may not be as important in pathogenicity *in vivo* as 'replication factors'. On the other hand, when the pattern of infection *in vivo* is not repeated in organ culture, for example, in growth of trachoma – inclusion conjunctivitis agents in baboon and guinea-pig conjunctiva (Pearce & Lowrie, this volume) host defence mechanisms may be more important in pathogenicity *in vivo*. Further, when the patterns of infection *in vivo* are repeated in organ cultures, factors such as temperature, pH, bacterial flora and small molecular constituents in surrounding fluids, which could vary *in vivo* but are standardised in the cultures, probably play no major role in pathogenicity.

Investigations of the influence of nutrition on the pathogenicity of mycoplasmas, fungi and protozoa are sparse but could be increased along the same lines as those described above for bacteria and viruses. The author is unaware of any relevant studies of mycoplasmas (Thomas, 1970). However, more is known about the dimorphic fungi. Although host nutrients which determine the selection of morphological form *in vivo* have not yet been recognised, factors are known which, in culture media, produce a conversion of many fungi from saprophytic to parasitic form (Mariat, 1964; Austwick, this volume). Not only would it be interesting to see if similar factors operate *in vivo* to control morphology and therefore pathogenicity but also whether – as for bacteria – nutritional factors explain some fungal localisations such as *Aspergillus fumigatus* in bovine placenta in mycotic abortion (Hugh-Jones & Austwick, 1967). Nutrition seems important in protozoal pathogenicity. For example, the malarial parasite grows within red cells because it needs haemoglobin to form a haematin-containing pigment, haemazoin; and malarial attacks seem to be stimulated by p-amino benzoic acid and methionine (Maegraith, 1955). Since different forms of protozoa vary in their capacities for infecting different hosts, studies of the nutritional basis for morphological change may yield much information on pathogenicity.

Before leaving this section the rate of multiplication of microbes *in vivo* should be mentioned. It will be affected by the various factors described above. The more rapid it is, the more the likelihood of establishing the infection against the activity of the host defence mechanisms. Thus, in virology, there is often a correlation between vigorous replication in cell culture as evidenced by large plaque sizes and virulence in animals (Holland, 1964; Waterson *et al.* 1967; Schloer & Hanson, 1968; Simizu & Takayama, 1969).

What do we know of multiplication rate *in vivo*? It is relatively simple to count the numbers of viable microbes in the tissues of an infected host at any time after inoculation. But these numbers are only the result of microbial multiplication and destruction or removal. In comparisons between avirulent and virulent strains, large differences in the respective resultants will occur as the infections proceed, although the multiplication rate of the two strains may have been similar and masked by a differential resistance to host defence mechanisms. Only recently and in bacteriology has a method been evolved for measuring microbial division rates *in vivo*. Maw & Meynell (1968) used pathogenic bacteria genetically labelled with recognisable characteristics retained by a known proportion of the progeny at each division. Both division rates and death rates were determined and remarkable results were obtained; in the spleens of mice *Salmonella typhimurium* divided at only 5–10 % of the maximum rate *in vitro* and the death rate was extremely small. This type of approach may be used for measuring true division and death rates of other microbes *in vivo* (Maw & Meynell, 1968).

MICROBIAL RESISTANCE TO HOST DEFENCE MECHANISMS

In addition to possessing a biochemical ability to multiply in the host tissues, virulent microbes must also have the capacity to withstand numerous humoral and cellular host defence mechanisms pre-existing in the tissues and induced or mobilised as a result of microbial attack. This capacity is particularly important in the decisive primary lodgement period (Miles, Miles & Burke, 1957) of infection when the protective reactions of the host are most heavily weighted against the few invading microbes. It is also needed later to spread throughout the host against the protective activities of the reticulo-endothelial system. The interplay of microbial and host reactions can vary from complete destruction of the microbes to complete subjugation of the host including near stalemate in chronic infections; indeed in some latent infections and carrier states parasitism might be considered to approach symbiosis with the microbe receiving sustenance from the host and the latter continually immunised by the microbe against disease.

The ability of virulent bacteria to withstand or inhibit the bactericidins in blood and tissues and to prevent the bactericidal action of phagocytes by resisting contact, ingestion or intracellular digestion is well known; and in some cases the compounds – aggressins – responsible for these effects have been recognised (Wilson & Miles, 1964; Dubos & Hirsch, 1965; Smith, 1968; Glynn, this volume). In contrast, although host defence mechanisms against viruses, mycoplasmas, fungi and protozoa are known, in most cases the methods whereby virulent strains of these microbes resist or avoid the defence mechanisms are still not clear; and even more so the microbial compounds involved. Do viruses, mycoplasmas, fungi and protozoa produce counterparts of the bacterial aggressins?

Non-specific defence against virus infections is stronger in adult than young hosts and includes both humoral and cellular factors (Smorodintsev, 1960; Mims, 1964; Holland, 1964; Gresser & Lang, 1966; Notkins, Mergerhagen & Howard, 1970; Zisman, Hirsch & Allison, 1970; Hirsch, Zisman & Allison, 1970; Jandásek, 1970; Thind & Price, 1970; Veltri & Kirk, 1971). The cellular factors can perhaps be divided into two types. There are those present or induced in the cells of any tissue the virus attacks, such as host nucleases capable of destroying virus nucleic acid after uncoating (Newton, 1970) or interferon. Then, there are those present in the phagocytic and other cells of the protective reticulo-endothelial system which often provide a barrier to invasion of virus into susceptible tissues (Mims, 1964; Johnson & Mims, 1968; Basarab & Smith, 1969). Not much is known of the function of the different cells of the reticulo-endothelial system; macrophages appear more important than neutrophils but the latter have not been investigated thoroughly. (Mims, 1964; Gresser & Lang, 1966; Notkins et al. 1970). Virulent viruses have the ability to inhibit or not to induce these host defence mechanisms (Mims, this volume). Thus, carbon clearance experiments show that the reticulo-endothelial system is depressed in some but not all virus infections (Notkins et al. 1970). Virulent strains of influenza virus resist serum inhibitors more than avirulent strains (Smorodintsev, 1960). Virulent strains of some, but not all, viruses induce less interferon than avirulent strains (Bang & Luttrell, 1961; Holland, 1964; Mirchamsy & Rapp, 1969; Postic, Schleupner, Armstrong & Ho, 1969; Campbell, Buera & Tobias, 1970). Virulent strains of ectromelia infect mouse macrophages more readily than avirulent strains (Mims, 1964). Influenza, mumps and Coxsackie viruses reduce the phagocytic activity against bacteria of guinea-pig, mouse and rat phagocytes (Sawyer, 1969; Notkins *et al.* 1970). But what virion constituents or virus-induced products determine the superior resistance to host defence mechanisms of virulent strains of viruses? Is some peculiarity of the envelope protein responsible for resistance to or inhibition of serum and phagocyte viricidins? Could host nucleases be inhibited by a particular folding of the viral nucleic acid or by an internal protein (Newton, 1970)? What chemical difference determines induction of interferon by an avirulent strain in contrast to a virulent strain? Do viruses produce interferon antagonists? The answer to the last question appears to be yes and different strains of Newcastle disease virus appear to have different abilities to antagonise interferon (Lockart, 1967; Truden, Sigel & Dietrich, 1967; Sheaff & Stewart, 1968). However, the chemical nature of the antagonists is as yet unknown. It should not be long before a viral counterpart of the bacterial aggressins is identified.

Tissue extracts and polymorphonuclear leucocytes kill mycoplasmas (Thomas, 1970; Jones & Hirsch, 1971), but whether virulent strains resist these host defence mechanisms by the production of aggressins is not clear. A little more is known in the fungal field. Fungistatic or fungicidal materials are found in body fluids (Adams & Rickard, 1963; Kozinn, Caroline & Taschdjian, 1964) and phagocytes (Kong & Levine, 1967; Lehrer & Cline, 1969; Lehrer, 1970). Since many fungi are larger than phagocytes, mere size may prevent ingestion and thus rapid growth alone can be an aggressive mechanism. As for pneumococci, a capsular polysaccharide prevents the phagocytic ingestion of virulent strains of *C. neoformans* (Bulmer & Sans, 1968; Fashi *et al.* 1970). Like brucellae and tubercle bacilli, some fungi e.g. *Histoplasma capsulatum*, *Coccidioides immitis* and *Candida albicans* can survive and grow within phagocytes (Kong & Levine, 1967; Stanley & Hurley, 1969; Lehrer, 1970). Inhibition of intracellular killing may be due to fungal aggressins comparable to those produced by brucellae (Smith, 1968). Production of aggressins may be related to the yeast form; the non-invasive mycelial dermatophytes probably lack the powerful aggressins of the yeast-like fungi causing the deeper mycoses (Ainsworth, 1955; Kong & Levine, 1967).

Host defence against protozoal infection involves humoral factors, particularly lysis by antibody and complement in some infections such as trypanosomiasis (Hawking, 1955) and phagocytes in others such as malaria and leishmaniasis (Goble & Singer, 1960; Huff, 1963; Adler, 1964). According to recent reports interferon may contribute to defence against malaria (Schultz, Huang & Gordon, 1968). The mechanisms whereby protozoa avoid or inhibit host defence mechanisms are obscure. Trypanosomes appear to avoid lysis by antibody and complement and malarial parasites destruction by macrophages by antigenic plasticity *in vivo* (Hawking, 1955; Weitz, 1964; Brown, 1971). The high mobility of some protozoa may prevent phagocytosis and entamoebae appear to produce factors which kill leucocytes (Jarumilinta & Kradolfer, 1964) but the author is not aware of a protozoal product which prevents ingestion by phagocytes. *Toxoplasma gondii* produces a factor which promotes its penetration of host cells and, as for bacterial aggressins, enhances its virulence for mice (Lycke, Norrby & Remington, 1968). Since *Leishmania* spp. survive and grow within phagocytes (Adler, 1964) they may produce as yet unidentified aggressins similar to those of brucellae.

Although for all microbial types the cells of the reticulo-endothelial system usually provide protection for the host, there are circumstances when such cells provide protection for microbes together with a means of dissemination throughout the host. This occurs when microbes are able to survive and grow within wandering phagocytes due, either to being well-endowed with aggressins against the killing mechanisms of normal phagocytes, or to being ingested by phagocytes deficient in some or all of the normal killing mechanisms. The latter occurs as a result of the natural heterogeneity of function in phagocytic populations and in pathological states such as chronic granulomatous disease of childhood (Smith, 1968). Within a phagocyte, microbes may be protected against anti-microbial agents either present or introduced into body fluids and against more destructive cellular mechanisms. For example, bacteria normally killed by polymorphonuclear cells could be protected by ingestion by less destructive macrophages. There is strong evidence of this protective and disseminating influence of the reticulo-endothelial cells in bacterial (Smith, 1968), viral (Notkins *et al.* 1970) and fungal (Stanley & Hurley, 1969) diseases, and it probably also occurs in protozoal and mycoplasmal disease.

DAMAGING THE HOST

The mechanisms whereby most microbes damage the host are not clear. There are two main problems in investigating these mechanisms. Specific microbial effects on the host must be distinguished from those occurring as a general response to injury (Stoner, this volume); and the microbial products or processes responsible for the specific effects must be recognised. Only the second problem is discussed here.

Bacteriology shows that host damage may result from the production of microbial poisons or toxins. A few of these toxins are of overriding importance in disease, others are of some but lesser significance and many are either not formed or not liberated to any significant extent *in vivo* and remain merely as products of bacterial growth *in vitro* with little relevance to disease (Smith, 1972). Thus, toxins demonstrated *in vitro* may be responsible for damage in disease, but this is not necessarily the case.

Despite the examples in bacterial disease, toxins responsible for the pathological effects of most viral, mycoplasmal, fungal and protozoal diseases have not been recognised. There are two explanations for an apparent lack of toxins. First, relevant toxins exist but have yet to be demonstrated, possibly – as for the toxic complex of *B. anthracis* and the entero-toxins of *V. cholerae* and *E. coli* – by a closer study of microbial behaviour in more natural environments (Smith, 1968). Second, the host is harmed by means other than direct action of a toxin. In diseases where large microbial populations accumulate, the growing microbes might produce emboli or deplete some tissues of essential nutrients. In intracellular diseases, cells might burst after excessive microbial multiplication. Host damage might also result from the evocation of hypersensitivity reactions by otherwise non-toxic microbial products. In bacteriology only the last explanation appears important and tuberculosis and streptococcal infections are the main examples (Smith, 1972; Parish, this volume). The implication of hypersensitivity reactions in other bacterial diseases is less clear. Skin tests clearly indicate hypersensitive states in many bacterial diseases. But, mere demonstration of a state of hypersensitivity is no proof of its importance in the main pathology of the disease. The systemic and local effects of disease must be simulated by hypersensitivity reactions evoked in a sensitised host by products of the appropriate microbe. The difficulty of obtaining such proof in bacteriology should be borne in mind in studies on viruses, mycoplasmas, fungi and protozoa.

Viruses produce cytopathic effects in individual cells, both in tissue culture and in disease. Such cell damage acting in localised areas of animal hosts, e.g. poliomyelitis virus in the motor neurones of the anterior horn, or more widespread, e.g. in pox virus infections, can account for the many pathological effects of virus infections such as paralysis in poliomyelitis and shock in pox virus infections (Smith, 1963).

As discussed above, the cytopathic effects on cells could result from a passive role on the part of the virus; cells might burst after acting merely as hosts for virus replication or they might die as a result of the profound redirection of metabolic processes required by such replication. Cell damage might also occur as a consequence of the 'cut off' phenomenon, that is, cessation of host cell macromolecular synthesis (Martin & Kerr, 1968). For some viruses there is increasing evidence that cytopathic effects may occur by a direct 'toxic' action of viral products. First there are experiments in whole animals showing that massive doses of some viruses such as influenza virus and pox viruses produce rapid toxic effects (Smith, 1963; Tamm & Horsfall, 1965). Then there are results of experiments in tissue culture on polio virus, influenza virus, mengo virus, Newcastle disease virus and vaccinia virus, indicating that virus replication is not essential for cytopathic effects which require in some cases virus-induced protein synthesis (Bablanian, this volume). Support for the fact that virus replication and cytopathic effect need not be closely linked is provided by the observation that virulent strains of some viruses have greater cytopathic effects than avirulent strains (Bang & Luttrell, 1961; Waterson et al. 1967; Amako & Dales, 1967; Reeve & Poste, 1971; Reeve, Alexander, Pope & Poste, 1971).

Are the virus-induced proteins which accompany these cytopathic effects cytotoxins? At present this is not known and, apart possibly from the penton of adenovirus (Ginsberg, Pereira, Valentine & Wilcox, 1966), a virus cytotoxin has not been characterised. To answer this question, Stephen & Birkbeck (1969) advocated a direct approach compatible to that used in bacteriology, namely to isolate virus-induced components from infected cells, to free them from intact virus and to attempt to produce toxic effects in uninfected cells. This approach requires the design of techniques for the difficult process of introducing potentially cytotoxic viral products into fresh cells. Hence, evidence for production of virus-induced cytotoxins in tissue culture is being sought. However if this evidence is obtained, it will still remain to be proved that such cytotoxins are responsible for the pathological effects of virus disease or even some of the rapid toxic effects produced in animals by massive doses of some viruses (see above). There is increasing evidence that evocation of hypersensitivity effects enters into the pathology of virus infections (Webb & Hall, this volume). Skin reactions indicate that hypersensitive states occur in many virus diseases such as smallpox, mumps, measles and virus encephalitis (Wilson & Miles, 1964). The possibility of 'auto-allergic' phenomena occurring in virus infections is greater than in bacterial disease because many viruses appear to incorporate host components into their structure. Deciding whether evocation of hypersensitivity reactions or direct toxicity is responsible for the main pathological effects of virus disease is not made easier by the present lack of knowledge of virus toxicity. Nevertheless, experimental studies in which pathological effects of virus diseases have been provoked or made worse by introducing antibody or immune cells into an infected host, indicate that evocation of hypersensitivity or 'auto-allergic' reactions are involved in the pathology of rashes (Mims, 1966), lymphocytic choriomeningitis (Mims & Tosolini, 1969; Oldstone & Dixon, 1970a) and viral encephalitides (Webb & Hall, this volume). Cytotoxic factors may be released in some cases by immunospecific reactions (Oldstone & Dixon, 1970b).

Mycoplasmas appear to damage host tissues by similar mechanisms to those described for bacteria and viruses. They may harm cells passively by excessive usage of arginine (Thomas, 1968, 1970). *Mycoplasma gallisepticum* and *Mycoplasma neurolyticum* appear to produce neurotoxins responsible for the effects of disease and hydrogen peroxide production may be toxic to some cells (Thomas, 1968, 1970). Finally, close adherence of mycoplasmas to cells may change surface antigens leading to hypersensitivity or 'auto-immune' effects (Thomas, 1968, 1970).

In some mycotic diseases, mechanical blockage by the large size and mycelial form of some fungi could damage the host. Fungi also produce toxins such as aflatoxin and sporidesmin (Wright, 1968). These toxins are produced in foodstuffs outside the animal host, and may be in infection, but this has yet to be demonstrated unequivocally (Ainsworth, 1955; Wright, 1968). Peptidases (Chattaway, Ellis & Barlow, 1963), collagenase (Rippon & Peck, 1967) and elastase (Rippon, 1967) appear to be involved in dermatophyte toxicity.

Hypersensitivity occurs in many mycoses and probably explains to a large degree the pathology of some fungal skin diseases. The compounds responsible are ill-defined. Hypersensitive states are present in the deep mycoses, but whether or not hypersensitivity phenomena enter into their main pathology is a matter for speculation (Ainsworth, 1955; Kong & Levine, 1967).

The overall toxic effects of protozoa on their hosts have been investigated, especially those of *Plasmodium* spp. – for example red cell destruction, anuria and circulatory failure that occurs in malaria (Maegraith, 1955; Huff, 1963; Garnham, 1967; Newton, this volume). Clearly cell destruction by growth of protozoa within cells could contribute to damage in such diseases as malaria and leishmania. A microbial toxin has not yet been recognised as being unequivocally responsible for the main pathological effects of any protozoal disease. In malaria, increased capillary permeability leading to shock and brain damage appears to be mediated by kallikrein, kinins and adenosine, but a malarial toxin responsible for the release of such host products has not been demonstrated (Onabanjo & Maegraith, 1970*a*, *b*; Onabanjo, Bhobani & Maegraith, 1970). It is possible that no malarial toxin exists and that the whole pathological syndrome follows from destruction of red blood cells. The lytic effects of *Entamoeba* spp. on tissues are known (Hoare & Neal, 1955) but the mechanisms of cell destruction are not clear; it is possible that lysosomal enzymes are transferred from the entamoebae to the host cells by tubules formed between the two cells on contact (Eaton, Meerovitch & Costerton, 1970). Toxic products of trypanosomes (Lumsden, 1965) and toxoplasmas (Jacobs, 1967) are known, but their importance in disease is not clear.

Hypersensitivity and 'auto-allergic' phenomena undoubtedly occur in protozoal infections (Garnham, 1967; Huff, 1963; Adler, 1964) but whether they are responsible for important pathological effects is hard to judge. It is possible in malaria that antibodies to host antigens changed by red blood cell parasitisation may, by opsonisation, promote phagocytosis and destruction of unparasitised red blood cells.

I hope this survey over the field of microbial pathogenicity has provided sufficient parallels between the actual or surmised mechanisms of pathogenicity of the various microbial species to indicate the mutual advantage of discussion between microbiologists interested in the various species.

REFERENCES

ADAMS, E. W. & RICKARD, C. G. (1963). The antistreptococcic activity of bovine teat canal keratin. *American Journal of Veterinary Research*, **24**, 122.

ADLAM, C., PEARCE, J. H. & SMITH, H. (1970*a*). Virulence mechanisms of staphylococci grown *in vivo* and *in vitro*. *Journal of Medical Microbiology*, **3**, 147.

ADLAM, C., PEARCE, J. H. & SMITH, H. (1970*b*). The interaction of staphylococci grown *in vivo* and *in vitro* with polymorpho-nuclear leucocytes. *Journal of Medical Microbiology*, **3**, 157.

ADLER, S. (1964). Leishmania. In *Advances in Parasitology*, ed. B. Dawes, vol. 2, 35. New York: Academic Press.

AINSWORTH, G. C. (1955). Pathogenicity of fungi in man and animals. *Symposium of the Society for General Microbiology*, **5**, 242.

AMAKO, K. & DALES, S. (1967). Cytopathology of mengovirus infection. I. Relationship between cellular disintegration and virulence. *Virology*, **32**, 184.

BANG, B. F. & FOARD, M. A. (1969). Experimentally induced changes in nasal mucous secretory systems and their effect on virus infection of chickens. II. Effects on adsorption of Newcastle Disease virus. *Journal of Experimental Medicine*, **130**, 121.

BANG, F. B. & LUTTRELL, C. N. (1961). Factors in the pathogenesis of virus diseases. In *Advances in Virus Research*, ed. K. M. Smith and M. A. Lauffer, vol. 8, 199. New York: Academic Press.

BASARAB, O. & SMITH, H. (1969). Quantitative studies on the tissue localization of influenza virus in ferrets after intranasal and intravenous or intracardial inoculation. *British Journal of Experimental Pathology*, **50**, 612.

BASARAB, O. & SMITH, H. (1970). Growth patterns of influenza virus in cultures of ferret organs. *British Journal of Experimental Pathology*, **51**, 1.

BECKER, Y., OLSHEVSKY, U. & LEVITT, J. (1967). The role of arginine in the replication of herpes simplex virus. *Journal of General Virology*, **1**, 471.

BEINING, P. R. & KENNEDY, E. R. (1963). Characteristics of a strain of *Staphylococcus aureus* grown *in vivo* and *in vitro*. *Journal of Bacteriology*, **85**, 732.

BEKIERKUNST, A. (1968). Tubercle bacilli grown *in vivo*. *Annals of the New York Academy of Sciences*, **154**, 79.

BRADISH, C. J., ALLNER, K. & MABER, H. B. (1971). The virulence of original and derived strains of Semliki Forest virus for mice, guinea-pigs and rabbits. *Journal of General Virology*, **12**, 141.

BROWN, K. N. (1971). Protective immunity to malaria provides a model for the survival of cells in an immunologically hostile environment. *Nature, London*, **230**, 163.

BULMER, G. S. & SANS, M. D. (1968). *Cryptococcus neoformans*. III. Inhibition of phagocytosis. *Journal of Bacteriology*, **95**, 5.

BURRIN, D. H., KEPPIE, J. & SMITH, H. (1966). The isolation of phagocytes and lymphocytes from bovine blood and the effect of their extracts on the growth of *Brucella abortus*. *British Journal of Experimental Pathology*, **47**, 70.

BURROWS, T. W. (1955). The basis of virulence for mice of *Pasteurella pestis*. *Symposium of the Society for General Microbiology*, **5**, 152.

BURROWS, T. W. (1960). Biochemical properties of virulent and avirulent strains of bacteria: *Salmonella typhosa* and *Pasteurella pestis*. *Annals of the New York Academy of Sciences*, **88**, 1125.

CAMPBELL, J. B., BUERA, J. G. & TOBIAS, F. M. (1970). Influence of blood clearance rates on interferon production and virulence of Mengo virus plaque mutants in mice. *Canadian Journal of Microbiology*, **16**, 821.

CHAKRABORTTY, A. S. & BESWICK, T. S. L. (1971). Morphological changes and resistance to vaccinia virus induced in human amnion cells by yeast extract. *Journal of Medical Microbiology*, **4**, 115.

CHATTAWAY, F. W., ELLIS, D. A. & BARLOW, A. J. E. (1963). Peptidases of dermatophytes. *Journal of Investigative Dermatology*, **41**, 31.

CRAIG, J. P. (1965). A permeability factor (toxin) found in cholera stools and culture filtrates and its neutralization by convalescent cholera sera. *Nature, London*, **207**, 614.

DE, S. N., GHOSE, M. L. & SEN, A. (1960). Activities of bacteria-free preparations from *Vibrio cholerae*. *Journal of Pathology and Bacteriology*, **79**, 373.

DOURMASHKIN, R. R. & TYRRELL, D. A. J. (1970). Attachment of two myxoviruses to ciliated epithelial cells. *Journal of General Virology*, **9**, 77.

DUBOS, R. J. & HIRSCH, J. G. (1965). *Bacterial and Mycotic Infections of Man*, 4th edition. Philadelphia: J. B. Lippincott Co.

EATON, R. D., MEEROVITCH, E. & COSTERTON, J. W. (1970). The functional morphology of pathogenicity in *Entamoeba histolytica*. *Annals of Tropical Medicine and Parasitology*, **64**, 299.

EBISAWA, I. T., KITAMOTA, O., TAKEUCHI, Y. & MAKINO, M. (1969). Immuno-cytolytic study of epithelial cells in influenza. *American Review of Respiratory Diseases*, **99**, 507.

EKSTEDT, R. D. & YOSHIDA, K. (1969). Immunity to staphylococcal infection in mice. Effect of living versus killed vaccine. Role of circulating antibody and induction of protection-inducing antigen(s) *in vitro*. *Journal of Bacteriology*, **100**, 745.

ENDERS, J. F. (1952). In *Viral and Rickettsial Infections of Man*, eds. T. M. Rivers and F. L. Horsfall, 126. Philadelphia: J. B. Lippincott Co.

FASHI, F., BULMER, G. S. & TACKER, J. R. (1970). *Crytococcus neoformans*, IV. The not-so-encapsulated yeast. *Infection and Immunity*, **1**, 526.

FENNER, F. (1970). The genetics of animal viruses. *Annual Reviews of Microbiology*, **24**, 297.

FINKELSTEIN, R. A., NORRIS, H. T. & DATTA, N. K. (1964). Pathogenesis of experimental cholera in infant rabbits. I. Observations of the intraintestinal infection and experimental cholera produced with cell-free products. *Journal of Infectious Diseases*, **114**, 203.

FITZGEORGE, R. B. & SMITH, H. (1966). The chemical basis of the virulence of *Brucella abortus*. VII. The production *in vitro* of organisms with an enhanced capacity to survive intracellularly in bovine phagocytes. *British Journal of Experimental Pathology*, **47**, 558.

FRIEDBERG, D. & SHILO, M. (1970). Role of cell wall structure of *Salmonella* in the interaction with phagocytes. *Infection and Immunity*, **2**, 279.

FUKUI, G. M., DELWICHÉ, E. A., MORTLOCK, R. P. & SURGALLA, M. J. (1962) Oxidative metabolism of *Pasteurella pestis* grown *in vivo* and *in vitro*. *Journal of Infectious Diseases*, **110**, 143.

GABLIKS, J. (1969). Interaction of leucine vaccinia virus infection in mice and cell cultures. *Journal of Infectious Diseases*, **120**, 679.

GARNHAM, P. C. C. (1967). Malaria in mammals excluding man. In *Advances in Parasitology*, ed. B. Dawes, vol. 5, 139. New York: Academic Press.

GINSBERG, H. S., PEREIRA, H. G., VALENTINE, R. C. & WILCOX, W. C. (1966). A proposed terminology for the adenovirus antigens and virion morphological sub-units. *Virology*, **28**, 782.

GOBLE, F. C. & SINGER, I. (1960). The reticuloendothelial system in experimental malaria and trypanosomiasis. *Annals of the New York Academy of Sciences*, **88**, 149.

GORDON, D. F., JR & GIBBONS, R. J. (1967). Glycolytic activity of *Streptococcus mitis* grown *in vitro* and in gnotobiotic animals. *Journal of Bacteriology*, **93**, 1735.

GOULD, E. A., RATCLIFFE, N. A., BASARAB, O. & SMITH, H. (1972). Studies of the basis of localization of influenza virus in ferret organ cultures. *British Journal of Experimental Pathology*, In press.

GRESSER, I. & LANG, D. J. (1966). Relationships between viruses and leucocytes. *Progress in Medical Virology*, **8**, 62.

HAWKING, F. (1955). The pathogenicity of protozoal and other parasites: general considerations. *Symposium of the Society for General Microbiology*, **5**, 176.

HALMANN, M., BENEDICT, M. & MAGER, J. (1967). Nutritional requirements of *Pasteurella tularensis* for growth from small inocula. *Journal of General Microbiology*, **49**, 451.

HENDERSON, D. W. (1964). Mixed populations *in vivo* and *in vitro*. *Symposium of the Society for General Microbiology*, **14**, 241.

HIRSCH, M. S., ZISMAN, B. & ALLISON, A. C. (1970). Macrophages and age-dependent resistance to herpes simplex virus in mice. *Journal of Immunology*, **104**, 1160.

HOARE, C. A. & NEAL, R. A. (1955). Host-parasite relations and pathogenesis in infections with *Entamoeba histolytica*. *Symposium of the Society for General Microbiology*, **5**, 230.

HOLLAND, J. J. (1964). Viruses in animals and in cell culture. *Symposium of the Society for General Microbiology*, **14**, 257.

HOORN, B. & TYRRELL, D. A. J. (1969). Organ cultures in virology. *Progress in Medical Virology*, **11**, 408.

HUANG, C. H. & WONG, C. (1963). Relation of the peripheral multiplication of Japanese B Encephalitis virus to the pathogenesis of the infection in mice. *Acta Virologica*, **7**, 322.

HUFF, C. G. (1963). Experimental research in avian malaria. In *Advances in Parasitology*, ed. B. Dawes, vol. 1, 1. New York: Academic Press.

HUGH-JONES, M. E. & AUSTWICK, P. K. C. (1967). Epidemiological studies in mycotic abortion. I. The effect of climate on incidence. *Veterinary Record*, **81**, 273.

JACOBS, L. (1967). Toxoplasma and Toxoplasmosis. In *Advances in Parasitology*, ed. B. Dawes, vol. 5, 1. New York: Academic Press.

JANDÁSEK, L. (1970). Influence of antileukocyte serum on intra-peritoneal vaccinia virus infection of rats. *Acta Virologica*, **14**, 467.

JARUMILINTA, R. & KRADOLFER, F. (1964). The toxic effect of *Entamoeba histolytica* on leukocytes. *Annals of Tropical Medicine*, **58**, 375.

JOHNSON, R. T. & MIMS, C. A. (1968). Pathogenesis of viral infections of the nervous system. *New England Journal of Medicine*, **278**, 23 and 84.

JONES, T. C. & HIRSCH, J. G. (1971). The interaction *in vitro* of *Mycoplasma pulmonis* with mouse peritoneal macrophages and L-cells. *Journal of Experimental Medicine*, **133**, 231.

KAPLAN, A. S. (1955). The susceptibility of monkey kidney cells to polio virus *in vivo* and *in vitro*. *Virology*, **1**, 377.

KEPPIE, J. (1964). Host and tissue specificity. *Symposium of the Society for General Microbiology*, **14**, 44.

KEPPIE, J., HARRIS-SMITH, P. W. & SMITH, H. (1963). The chemical basis of the virulence of *Bacillus anthracis*. IX. Its agressins and their mode of action. *British Journal of Experimental Pathology*, **44**, 446.

KOCH, A., DRÉN, C. S. & GYÖRGY, E. (1968). Saturated fatty acids in poliovirus host cell interaction. I. Stimulation and inhibition of viron intake. *Acta Microbiologica Academiae Scientiarum Hungaricae*, **15**, 77.

KONG, Y. M. & LEVINE, H. B. (1967). Experimentally induced immunity in the mycoses. *Bacteriological Reviews*, **31**, 35.

KOZINN, P. J., CAROLINE, L. & TASCHDJIAN, C. L. (1964). Conjunctiva contains factor inhibiting growth of *Candida albicans*. *Science*, **146**, 1479.

LANDAY, M. E., WHEAT, R. W., CONANT, N. F. & LOWE, E. P. (1967). Serological comparison of spherules and arthrospores of *Coccidioides immitis*. *Journal of Bacteriology*, **94**, 1400.

LEHRER, R. I. (1970). Measurement of candidacidal activity of specific leukocyte types in mixed cell populations. I. Normal, myeloperoxidase-deficient and chronic granulomatous disease neutrophils. *Infection and Immunity*, **2**, 42.

LEHRER, R. I. & CLINE, M. J. (1969). Interactions of *Candida albicans* with human leucocytes and serum. *Journal of Bacteriology*, **98**, 996.

LIPMAN, R. P. & CLYDE, W. A. (1969). The interrelationship of virulence, cytadsorption and peroxide formation in *Mycoplasma pneumoniae*. *Proceedings of the Society for Experimental Biology and Medicine*, **131**, 1163.

LIPMAN, R. P., CLYDE, W. A. & DENNY, F. W. (1969). Characteristics of virulent, attenuated and avirulent *Mycoplasma pneumoniae* strains. *Journal of Bacteriology*, **100**, 1037.

LOCKART, R. Z., JR (1967). Recent progress in research on interferons. *Progress in Medical Virology*, **9**, 451.

LOWRIE, D. B. & PEARCE, J. H. (1970). Placental localization of *Vibrio fetus*. *Journal of Medical Microbiology*, **3**, 607.

LUMSDEN, W. H. R. (1965). Biological aspects of trypanosomiasis research. In *Advances in Parasitology*, ed. B. Dawes, vol. 3, 1. New York: Academic Press.

LYCKE, E., NORRBY, R. & REMINGTON, J. (1968). Penetration enhancing factor extracted from *Toxoplasma gondii* which increases its virulence for mice. *Journal of Bacteriology*, **96**, 785.

MAEGRAITH, B. G. (1955). The pathogenicity of plasmodia and entamoebae. *Symposium of the Society for General Microbiology*, **5**, 207.

MARIAT, F. (1964). Saprophytic and parasitic morphology of pathogenic fungi. *Symposium of the Society for General Microbiology*, **14**, 85.

MARTIN, E. M. & KERR, I. M. (1968). Virus-induced changes in host-cell macromolecular synthesis. *Symposium of the Society for General Microbiology*, **18**, 15.

MATHEWS, M. B. (1970). Tissue specific factor required for the translation of a mammalian viral R.N.A. *Nature, London*, **228**, 661.

MAW, J. & MEYNELL, G. G. (1968). The true division and death rates of *Salmonella typhimurium* in the mouse spleen determined with superinfecting phage P 22. *British Journal of Experimental Pathology*, **40**, 597.

MEDEARIS, D. N., JR, CAMITTA, B. M. & HEATH, E. C. (1968). Cell wall composition and virulence in *Escherichia coli*. *Journal of Experimental Medicine*, **128**, 399.

MEYER, P. T. & ENDERS, J. F. (1969). Feline Herpes virus infection in fused cultures of naturally resistant human cells. *Journal of Virology*, **3**, 469.

MEYNELL, G. G. (1961). Phenotypic variation and bacterial infection. *Symposium of the Society for General Microbiology*, **11**, 174.

MILES, A. A. (1955). The meaning of pathogenicity. *Symposium of the Society for General Microbiology*, **5**, 1.

MILES, A. A., MILES, E. M. & BURKE, J. (1957). The value and duration of defence reaction of the skin to the primary lodgement of bacteria. *British Journal of Experimental Pathology*, **38**, 79.

MIMS, C. A. (1964). Aspects of the pathogenesis of virus diseases. *Bacteriological Reviews*, **28**, 30.

MIMS, C. A. (1966). Pathogenesis of rashes in virus diseases. *Bacteriological Reviews*, **30**, 739.

MIMS, C. A. & TOSOLINI, F. A. (1969). Pathogenesis of lesions in lymphoid tissue of mice infected with lymphocytic choriomeningitis (LCM) virus. *British Journal of Experimental Pathology*, **50**, 584.

MIRCHAMSY, H. & RAPP, F. (1969). Role of interferon in replication of virulent and attenuated strains of measles virus. *Journal of General Virology*, **4**, 513.

NEWTON, A. A. (1970). The requirements of a virus. *Symposium of the Society for General Microbiology*, **20**, 323.

NOTKINS, A. L., MERGERHAGEN, S. E. & HOWARD, R. H. (1970). Effect of virus infections on the function of the immune system. *Annual Reviews of Microbiology*, **24**, 525.

OLDSTONE, M. B. A. & DIXON, F. J. (1970*a*). Pathogenesis of chronic disease associated with persistant lymphocytic choriomeningitis viral infection. II. Relationship in the anti-lymphocytic choriomeningitis immune response to tissue injury in chronic lymphocytic choriomeningitis disease. *Journal of Experimental Medicine*, **131**, 1.

OLDSTONE, M. B. A. & DIXON, F. J. (1970*b*). Tissue injury in lymphocytic choriomeningitis viral infection; virus induced immunologically specific release of a cytotoxic factor from immune lymphoid cells. *Virology*, **42**, 805.

ONABANJO, A. O. & MAEGRAITH, B. G. (1970*a*). Kallikrein as a pathogenic agent in *Plasmodium knowlesi* infection in *Macaca mullatta*. *British Journal of Experimental Pathology*, **51**, 523.

ONABANJO, A. O. & MAEGRAITH, B. G. (1970*b*). The probable pathogenic role of adenosine in malaria. *British Journal of Experimental Pathology*, **51**, 581.

ONABANJO, A. O., BHOBANI, A. R. & MAEGRAITH, B. G. (1970). The significance of kinin-destroying enzymes in *Plasmodium knowlesi* malarial infection. *British Journal of Experimental Pathology*, **51**, 534.

PANOS, C. & AJL, S. J. (1963). Metabolism of microorganisms as related to their pathogenicity. *Annual Reviews of Microbiology*, **17**, 297.

POSTIC, B., SCHLEUPNER, C. J., ARMSTRONG, J. A. & HO, M. (1969). Two variants of Sindbis virus which differ in interferon induction and serum clearance. I. The phenomenon. *Journal of Infectious Diseases*, **120**, 339.

PUSZTAI, R., GOULD, E. A. & SMITH, H. (1971). Infection patterns in mice of a virulent and an avirulent strain of Semliki Forest virus. *British Journal of Experimental Pathology*, In press.

REED, S. E. (1970). The growth of mycoplasmas in organ culture and its enhancement by concurrent virus infection. *Journal of Medical Microbiology*, **3**, xi.

REEVE, P., ALEXANDER, D. J., POPE, G. & POSTE, G. (1971). Studies on the cytopathic effects of Newcastle disease virus: metabolic requirements. *Journal of General Virology*, **11**, 25.

REEVE, P. & POSTE, G. (1971). Studies on the cytopathogenicity of Newcastle disease virus: relation between virulence, polykaryocytosis and plaque size. *Journal of General Virology*, **11**, 17.

RIDE, J. P. & DRYSDALE, R. B. (1972). A rapid chemical estimation of filamentous fungi in plants. *Physiological Plant Pathology*, In press.

RIPPON, J. W. (1967). Elastase production by ringworm fungi. *Science*, **157**, 947.

RIPPON, J. W. & PECK, G. L. (1967). Experimental infection with *Streptomyces madurae* as a function of collagenase. *Journal of Investigative Dermatology*, **49**, 371.

ROMANO, A. H. (1966). *Dimorphism In The Fungi – An Advanced Treatise*, vol. 2, 181. New York: Academic Press.

ROSS, J. D. & SYVERTON, J. T. (1957). Use of tissue cultures in virus research. *Annual Reviews of Microbiology*, **11**, 459.

Ross, J. D., Treadwell, P. E. & Syverton, J. T. (1962). Cultural characterization of animal cells. *Annual Reviews of Microbiology*, **16**, 141.

Sawyer, W. D. (1969). Interaction of influenza virus with leukocytes and its effect on phagocytosis. *Journal of Infectious Diseases*, **119**, 541.

Schloer, G. M. & Hanson, R. P. (1968). Relationship of plaque size and virulence for chickens of 14 representative Newcastle disease virus strains. *Journal of Virology*, **2**, 40.

Schultz, W. W., Huang Kun-Yen & Gordon, C. F. B. (1968). Role of interferon in experimental mouse malaria. *Nature, London*, **220**, 709.

Schwab, J. H. & Brown, R. R. (1968). Modification of antigenic structure *in vivo*: quantitative studies on the processing of streptococcal cell wall antigens in mice. *Journal of Immunology*, **101**, 930.

Seelig, M. S. (1966). Mechanisms by which antibiotics increase the incidence and severity of candidiasis and alter the immunological defences. *Bacteriological Reviews*, **30**, 442.

Sheaff, E. T. & Stewart, R. B. (1968). Substances which inhibit interferon; a substance which enhances virus growth and is antagonistic to interferon action. *Canadian Journal of Microbiology*, **14**, 965.

Shif, I. & Bang, F. B. (1970). *In vitro* interaction of mouse hepatitis virus and macrophages from genetically resistant mice. I. Adsorption of virus and growth curves. *Journal of Experimental Medicine*, **131**, 843.

Simizu, B. & Takayama, N. (1969). Isolation of two plaque mutants of Western equine encephalitis virus differing in virulence for mice. *Journal of Virology*, **4**, 799.

Simon, M. & Dömök, I. (1966). Enhancing effect of human erythrocyte extracts on the susceptibility of monkey kidney cells to certain enteroviruses. *Acta Microbiologica Academiae Scientiarum Hungaricae*, **13**, 229.

Smith, H. (1958). The use of bacteria grown *in vivo* for studies on the basis of their pathogenicity. *Annual Reviews of Microbiology*, **12**, 77.

Smith, H. (1964). Microbial behaviour in natural and artificial environments. *Symposium of the Society for General Microbiology*, **14**, 1.

Smith, H. (1968). Biochemical challenge of microbial pathogenicity. *Bacteriological Reviews*, **32**, 164.

Smith, H. (1972). The role of toxins in the pathogenesis of microbial disease. *Israel Journal of Medical Sciences*, In press.

Smith, H. & Fitzgeorge, R. B. (1964). The chemical basis of the virulence of *Brucella abortus*. V. The basis of intracellular survival and growth in bovine phagocytes. *British Journal of Experimental Pathology*, **45**, 174.

Smith, H., Keppie, J. & Stanley, J. L. (1955). The chemical basis of virulence of *Bacillus anthracis*. V. The specific toxin produced by *B. anthracis in vivo*. *British Journal of Experimental Pathology*, **36**, 460.

Smith, H. W. & Halls, S. (1967). Studies on *Escherichia coli* enterotoxin. *Journal of Pathology and Bacteriology*, **93**, 531.

Smith, Wilson. (1963). *Mechanisms of Virus Infection*. London and New York: Academic Press.

Smorodintsev, A. A. (1960). Basic mechanisms of non-specific resistance to viruses in animals and man. *Advances in Virus Research*, **7**, 327.

Stanley, J. L. & Smith, H. (1967). The chemical basis of the virulence of *Pasteurella pestis*. IV. The components of the guinea pig toxin. *British Journal of Experimental Pathology*, **48**, 124.

Stanley, V. C. & Hurley, R. (1969). The growth of Candida species in cultures of mouse peritoneal macrophages. *Journal of Pathology*, **97**, 357.

Stephen, J. & Birkbeck, T. H. (1969). The biochemistry of virus cytotoxicity. *Journal of General Microbiology*, **59**, xvi.

SWIM, H. E. (1959). Microbiological aspects of tissue culture. *Annual Reviews of Microbiology*, **13**, 141.

TAMM, I. & HORSFALL, F. L. (1965). *Viral and rickettsial infections of man*, 4th edition, ed. I. Tamm. Philadelphia: Lippincott.

TAKEUCHI, A. & SPRINZ, H. (1967). Electron microscopy studies of experimental *Salmonella* infections in the preconditioned guinea pig. II. Response of intestinal mucosae to invasion by *Salmonella typhimurium*. *American Journal of Pathology*, **51**, 137.

TAKEUCHI, A., FORMAL, S. B. & SPRINZ, H. (1968). Experimental acute colitis in the rhesus monkey following peroral infection with *Shigella flexneri*: an electron microscope study. *American Journal of Pathology*, **52**, 503.

TAYLOR, J., WILKINS, M. P. & PAYNE, J. M. (1961). Relation of rabbit gut reaction to enteropathogenic *Escherichia coli*. *British Journal of Experimental Pathology*, **42**, 43.

TEMPEST, D. W. & SMITH, H. (1957). The effect of metabolite analogues on growth of *Bacillus anthracis* in the guinea pig and on the formation of virulence-determining factors. *Journal of General Microbiology*, **17**, 739.

THIND, I. S. & PRICE, W. H. (1970). The role of serum protective factor and neutralizing antibody on pathogenesis of experimental infection with Langat virus in mice. *Journal of Infectious Disease*, **21**, 378.

THOMAS, L. H. (1968). Mycoplasmas as pathogens. *Yale Journal of Biology and Medicine*, **40**, 44.

THOMAS, L. H. (1970). Mycoplasmas as infectious agents. *Annual Reviews of Medicine*, **21**, 179.

TRUDEN, J. L., SIGEL, M. M. & DIETRICH, L. S. (1967). An interferon antagonist; its effect on interferon action in Mengo-infected Ehrlich ascites tumour cells. *Virology*, **33**, 95.

VELTRI, R. W. & KIRK, B. E. (1971). An antiviral substance in the tissues of mice acutely infected with lymphocytic choriomeningitis virus. *Journal of General Virology*, **10**, 17.

VOLTONEN, V. (1970). Mouse virulence of Salmonella strains; the effect of different smooth-type O side-chains. *Journal of General Microbiology*, **64**, 255.

VOSTI, K. L., LINDBERG, L. H., KOSEK, J. & RAFFEL, S. (1970). Experimental streptococcal glomerulonephritis: longterm study of a laboratory model ressembling human acute post-streptococcal glomerulonephritis. *Journal of Infectious Diseases*, **122**, 249.

WARD, M. E., WATT, P. J. & GLYNN, A. A. (1970). Gonococci in urethral exudates possess a virulence factor lost on subculture. *Nature, London*, **227**, 382.

WATERSON, A. P., PENNINGTON, T. H. & ALLAN, W. H. (1967). Virulence in New-castle disease virus. A preliminary study. *British Medical Bulletin*, **23**, 138.

WEITZ, B. (1964). The reaction of trypanosomes to their environment. *Symposium of the Society for General Microbiology*, **14**, 112.

WILSON, G. S. & MILES, A. A. (1964). *Topley and Wilson's Principles of Bacteriology and Immunity*. London: Edward Arnold.

WRIGHT, D. E. (1968). Toxins produced by fungi. *Annual Reviews of Microbiology*, **22**, 269.

ZISMAN, B., HIRSCH, M. S. & ALLISON, A. C. (1970). Selective effect of antimacrophage serum, silica and antilymphocyte serum on pathogenesis of herpes virus infection of young adult mice. *Journal of Immunology*, **104**, 1155.

SURVIVAL ON MUCOSAL EPITHELIA, EPITHELIAL PENETRATION AND GROWTH IN TISSUES OF PATHOGENIC BACTERIA

D. C. SAVAGE

Department of Microbiology, University of Texas,
Austin, Texas, U.S.A.

INTRODUCTION

Many bacterial pathogens induce disease in susceptible mammalian hosts by infecting mucosal membranes. To be successful pathogens, these micro-organisms must be able to survive on mucosal epithelia, frequently in competition with indigenous microbial populations. They then must be able to penetrate into the mucosae or at least produce toxic products that penetrate into susceptible subepithelial tissues. Unfortunately, little is known about the biochemical mechanisms by which such organisms infect susceptible mucosae (Smith, 1968). In fact, so little is known about these processes that significant gaps exist in the knowledge of the pathogenesis of diseases caused by several types of bacteria.

These gaps in understanding seem particularly serious from the point of view of public health. Bacterial infections of mucosal surfaces are among the most significant problems faced in medicine today. Diarrhoeal diseases, frequently of bacterial aetiology and synergistic with malnutrition, are usually the largest factors in morbidity and mortality in children in underdeveloped countries (Gordon & Scrimshaw, 1970), and even among certain population groups in so-called developed countries (Reller & Spector, 1970). Likewise, respiratory diseases of bacterial aetiology are common in all populations (Johanson, Pierce & Sanford, 1969). Such infections involving drug-resistant organisms often follow therapy with antibacterial drugs (Tillotson & Finland, 1969).

In the discussion to follow, therefore, I shall examine the earliest events in the pathogenesis of some diseases of the upper respiratory and gastrointestinal tracts. The intent is not to review in depth what is known about such diseases but to develop some generalisations about their pathogenesis. Experimental findings supporting these generalisations derive from studies of the indigenous mammalian microbiota as well as from studies of specific bacterial diseases. The generalisations can be outlined as follows:

(1) Most bacterial pathogens infecting mucosal epithelia in animals attach preferentially to certain epithelia as a first step in inducing disease. The organisms attach to the epithelia through specific surface–surface interactions and may grow rapidly while on the epithelial surfaces.

(2) Some such pathogens initiate disease while attached to particular epithelia; others initiate disease only after having penetrated into epithelia or into sub-epithelial tissues. Many bacterial types that penetrate into sub-epithelial tissues also grow preferentially in certain tissues.

(3) The point at which pathogens gain access to mucosal epithelia is critical in the development of infection and thus disease. An important process limiting access of bacterial pathogens to mucosal epithelia is that of microbial interference between the pathogens and indigenous micro-organisms. This particular interference operates primarily by mechanisms regulating the population levels and localisation of indigenous microbes.

ATTACHMENT OF BACTERIA TO MAMMALIAN MUCOSAL EPITHELIA

Bacteria infecting mucosal epithelia must assume initially an intimate relationship with the epithelial cells. Some such organisms attach directly to the cell membranes. Particular bacterial types attach selectively to certain epithelia. These attachments undoubtedly are mediated by particular surface characteristics of the bacteria and are influenced by the physiological state of the epithelial cells.

Selectivity of bacterial attachment

Selective attachment to only certain epithelia is particularly common among the bacterial members of the mammalian indigenous microbiota. Certain members of the human oral microbiota, *Streptococcus salivarius* and *Streptococcus sanguis*, adhere readily to the surfaces of the tongues and cheeks of humans and to epithelial cells in cheek scrapings obtained from humans, hamsters and germ-free rats (Gibbons & van Houte, 1971). These two organisms do not adhere well, however, to teeth. In contrast, another member of the oral microbiota, *Streptococcus mutans*, adheres only poorly to cheek and tongue cells but attaches readily to teeth (Gibbons & van Houte, 1971).

Indigenous bacteria also probably attach selectively to the mucosal epithelium in the upper respiratory tract. Viridans streptococci may disappear from the pharyngeal epithelium during penicillin therapy but will return shortly after such treatment is ended (Sprunt & Redman, 1968).

Likewise, certain microbial types in the gastro-intestinal microbiota selectively colonise the mucosal epithelia. Lactobacilli colonise in thick layers the keratinising stratified squamous epithelium of the non-secreting portion of the stomachs of all normal mice (Pl. 1, Figs. 1 and 2) (Savage, Dubos & Schaedler, 1968; Savage, 1970), rats (Brownlee & Moss, 1961) and swine (Dubos, Schaedler, Costello & Hoet, 1965; Tannock & Smith, 1970).

In mice and rats raised under conventional conditions, the stomachs are populated also by yeasts of the genus *Torulopsis* (Brownlee & Moss, 1961; Savage & Dubos, 1967). These organisms colonise the mucin on the columnar epithelium of the secreting portion of the stomach (Pl. 2, Figs. 1 and 2). They appear to be adapted well to growth in the stomach mucin.

Similarly, in both rats (Savage, 1969a) and mice (Savage, unpublished observations) from certain conventional colonies the villous epithelium in the ileum is populated by long chains of short rods or cocci (Pl. 3, Figs. 1 and 2). In this case, however, these bacteria may be attached securely to ileal epithelial cells. In the ilea of normal mice, bacterial cells of similar morphology attached to epithelial cells have been seen in preparations examined in the electron microscope (Hampton & Rosario, 1965, 1966).

Finally, in this respect, again in all normal mice (Savage et al. 1968) and rats (Savage, 1970), the large bowels contain bacteria that associate intimately with the mucosal epithelia (Pl. 3, Fig. 3). These organisms are anaerobic fusiform (Pl. 3, Fig. 4) and spiral-shaped bacteria and spirochaetes (Savage, McAllister & Davis, 1971) that colonise the mucus investing the mucosal epithelium of the caecum and colon. Layers of these bacteria are particularly prominent on the epithelium in the colon (Savage et al. 1968; Savage et al. 1971).

Under normal circumstances, these micro-organisms colonise only the particular mucosal epithelium described in each case. If removed from the epithelium by treatment with antibacterial drugs, they will recolonise the same epithelium after the therapy is discontinued (Savage & Dubos, 1968; Savage, 1969b). Thus, they appear to associate quite selectively with their particular epithelium. These selective associations may be important in the regulation of the composition of the indigenous microbiota (see p. 40). Likewise, selective attachment to certain epithelia may be important in the pathogenesis of some bacterial diseases of mucosal membranes.

Role of the bacterial cell surface in epithelial attachment

Particular surface properties of bacterial cells undoubtedly mediate their attachment to mucosal epithelia. Unfortunately, virtually nothing is known concerning such properties. Bacterial pili (fimbriae) (Brinton, 1965) could possibly serve to secure some microbes to epithelial membranes. Certain strains of *Escherichia coli* adhere through pili to mammalian erythrocyte membranes (Brinton, 1965). Some types of bacteria lack pili, however, and may still attach to mucosal epithelia (Lankford, 1960).

Some pathogenic bacteria produce erythrocyte agglutinins (Keogh, North & Warburton, 1947; Lankford, 1960) that could serve to attach the micro-organisms to epithelia. These agglutinins bind to membranes of red blood cells and thereby aggregate the cells into clumps. Particular agglutinins usually act only on erythrocytes from certain animal types. This property could explain the apparent specificity of bacterial attachment to epithelial cell membranes assuming the latter to resemble blood cell membranes. Again, however, not all bacterial types known to infect mucosae may produce such agglutinins.

Other surface properties of microbes more similar to capsules or envelopes may mediate the attachments in some cases. In the rodent stomach, indigenous lactobacilli attach to keratinised squamous cells (Savage *et al.* 1968) and to each other through a substance on their surfaces that may be a mucopolysaccharide (A. Takeuchi, personal communication). This attachment seems to be quite specific. The indigenous lactobacilli attach only to the keratinised epithelium in the stomachs or guts of monocontaminated germ-free mice. In contrast, non-indigenous lactobacilli fail to attach to any epithelium in such animals (Table 1) (Savage, Schaedler & Dubos, 1967). Consequently, the mucopolysaccharide on the surface of the indigenous bacteria may have considerable specificity for attaching to keratinised epithelium. Similar substances could mediate the specific attachment of pathogenic bacterial cells to other mucosal epithelia. Undoubtedly, however, the mechanisms of such attachments vary with the bacterial type and epithelial surface involved.

Possible effects of epithelial physiology on bacterial attachment

The physiological state of epithelia also may influence the attachment and penetration of bacteria. In humans, upper respiratory passages are lined with epithelia of the non-keratinising stratified squamous, pseudostratified columnar ciliated or stratified columnar types (Ham, 1965). In general, cells in such epithelia arise from multiplication in lower strata, rise to the surface of the epithelium, and desquamate at the surface. As these cells rise to the surface, they undergo physiological changes that can be marked in certain epithelia. Surface cells can be physiologically uniform, however, in respiratory epithelia (Ham, 1965). In contrast, the physiology of surface cells can vary considerably in

Table 1. *Colonisation of the digestive tracts of germ-free mice by lactic acid bacteria*

Savage, Schaedler & Dubos, 1967

Bacterial type*	Culture results		Bacterial layer in stomach
	Stomach	Caecum	
Indigenous			
Lactobacillus (Rhizoid†)	+ + + +‡	+ + + +	Complete
Lactobacillus (Compact†)	+ + + +	+ + + +	Slight
Streptococcus (Group N)	+ + +	+ + +	Slight
Non-indigenous			
Lactobacillus lactis	+ +	+ + + +	None
Lactobacillus acidophilus	+ +	+ + + +	None

* Germ-free mice were monocontaminated with each bacterial type.
† Refers to colonial form. See Dubos & Schaedler (1962).
‡ + + + + = > 10^9 bacteria per g of organ homogenate.
    + + = < 10^5 bacteria per g of organ homogenate.

gastro-intestinal epithelia. Gastro-intestinal epithelium is composed of columnar cells in most areas and of stratified squamous epithelium in portions of the stomachs of certain animal species (Ham, 1965; Hummel, Richardson & Fekete, 1966). The columnar cells multiply only in the crypts of Lieberkuhn (Leblond & Stevens, 1948). They then migrate up the surface to the tips of villi in the small intestine or to the lumenal surface in the stomach and large bowel. The cells are extruded into the lumen at the tips of the villi or in certain extrusion zones on the mucosal surface (Leblond & Stevens, 1948).

As these epithelial cells migrate out of the crypts and up to the extrusion zones, they undergo physiological changes detectable as alterations in activity levels of certain enzymes. For example, in the small bowel of conventional rodents, cells nearest the villous tips have highest activities of dipeptidases, disaccharidases and alkaline phosphatases (Nordström, Dahlquist & Josefsson, 1967; Moog & Grey, 1968). Apparently, these enzyme activities increase as the cells move up the villous surface. Presumably then, if the rates of migration of these cells changed, the activities of these enzymes also would be affected in some way.

The migration rate of the epithelial cells is much slower in the intestines of germ-free rodents than in those of conventional animals (Abrams, Bauer & Sprinz, 1963; Lesher, Walburg & Sacher, 1964). This finding undoubtedly explains the high alkaline phosphatase activity we have detected in duodenal epithelia of germ-free mice (Yolton, Stanley & Savage, 1971). Such activity is three- to four-fold higher in germfree duodenums than in those of mice with a conventional indigenous microflora. The alkaline phosphatase activity is high in the germ-free intestines, in all probability because the activity is higher than normal in cells on the villous bases. The activity could rise to high levels in cells low on the villi because the cells mature physiologically at normal rates but move up the villi more slowly than normal.

Enzymes such as alkaline phosphatases, disaccharidases and dipeptidases are involved in membrane function on the lumenal border of intestinal epithelial cells (Moog & Grey, 1968). Changes in the activities of these enzymes cculd alter considerably the membrane physiology of these cells. In particular, changes in the normal physiology of cell membranes could involve alterations in the ability of the cells to accept attaching bacteria. In certain intestinal infections, the bacterial pathogens attach more readily to epithelia near villous tips than at villous bases (Drees & Waxler, 1970; Staley, Jones & Corley, 1969). In other infections, the bacteria attach more readily to the base of the villi (Freter, 1969). The sites of attachment of these organisms may be dictated, at least in part, by the physiological state of the cell membranes. Thus, the altered physiology of the epithelial cells of germ-free animals may be partly responsible for the enhanced susceptibility of these animals to certain infectious diseases (Gordon, 1965).

MODES OF BACTERIAL INFECTION OF MUCOSAL EPITHELIUM

Up until now, three general points have been made concerning attachments of bacteria to mucosal epithelia. First, the attachment may be selective. Second, the attachments may be mediated through particular surface properties of the bacteria. Third, the physiological state of epithelial cells themselves may influence such attachments. I shall now

Table 2. *Modes of induction of disease in humans by bacterial pathogens of mucosal membranes*

Mode	Representative diseases	Etiological agent	Area of body infected*	References
Bacteria attach to mucosal epithelium; rarely penetrate into epithelium or submucosa; induce disease by multiplying on epithelium and producing exotoxin	Diphtheria Pertussis Cholera Neonatal enteritis	*Corynebacterium diphtheriae* *Bordetella pertussis* *Vibrio cholerae* *Escherichia coli*	Upper respiratory Upper respiratory Small intestine Small intestine	Pappenheimer, 1965 Brown & Brenn, 1931; Buddingh, 1970 Freter, 1969; Finkelstein, 1969 Smith, 1971
Bacteria attach to and penetrate into mucosal epithelia; rarely penetrate into sub-epithelial tissues; induce disease by multiplying within and killing epithelial cells. Some types may produce exotoxins	Bacillary dysentery Neonatal enteritis	*Shigella* spp. *Escherichia coli*	Large intestine Large intestine	Takeuchi *et al.* 1968 Staley *et al.* 1970*a, b*
Bacteria attach to and penetrate through mucosal epithelia into sub-epithelial tissues; induce disease by multiplying in submucosal tissues; some grow intracellularly in macrophages	Streptococcal sore throat Typhoid fever *Salmonella* enteritis Neonatal enteritis	*Streptococcus pyogenes* *Salmonella typhi* *Salmonella* spp. *Escherichia coli*	Upper respiratory Small intestine Small intestine Small intestine	McCarty, 1965 Sprinz *et al.* 1966 Takeuchi & Sprinz, 1967 Smith & Halls, 1968; Smith, 1971

* Area of body commonly infected.

note, where possible, the applicability of these phenomena in the pathogenesis of infectious diseases of mucosal epithelia. In addition, I shall point out that after associating closely with epithelia, bacterial pathogens may induce disease through at least three processes (Table 2) (Takeuchi, 1967; Sprinz, 1969). In the first the pathogen attaches to mucosal epithelia but rarely penetrates beyond that point. Such organisms induce disease by multiplying on the epithelium and elaborating diffusible toxins. In the second, the pathogens attach to and enter mucosal epithelial cells but rarely penetrate beyond those cells into the sub-mucosa. Such organisms induce disease by multiplying within epithelial cells. They may destroy the cells, thus eroding and ulcerating the mucosal epithelium. Some types produce diffusible exotoxins. In the third process, the pathogens attach to epithelial cells, then penetrate through and perhaps around the cells deeply into the submucosal tissues. Such organisms induce disease while multiplying, often intracellularly in macrophages, within the normally sterile submucosal tissues.

Attachment without penetration

Some bacteria induce disease by attaching to mucosal epithelia but as a rule do not penetrate into the epithelia. These organisms multiply on the epithelia and elaborate exotoxins that diffuse into the epithelium and submucosal tissues. The damage to host tissues characteristic of the disease is due to the action of the toxins (Table 2).

The aetiological agent of whooping cough, *Bordetella pertussis*, is the best described example of such bacterial pathogens infecting the upper respiratory tract. This bacterium localises in the muco-ciliary blanket of the bronchi in humans (Brown & Brenn, 1931). This localisation has proved to be quite selective. If inoculated into 15-day-old embryonated eggs, *B. pertussis* eventually localises on the ciliated epithelium of the bronchial mucosa of the embryo. In contrast, if inoculated into 12- to 13-day-old embryos, the organism fails to localise on the non-ciliated epithelium found on the bronchial mucosa during that period of development of the embryo (Buddingh, 1952; Buddingh, 1970).

The mechanism is unknown by which *B. pertussis* attaches to the ciliated epithelial cells. Freshly isolated strains of the organism produce agglutinins for human and fowl red blood cells (Keogh *et al.* 1947). These agglutinins conceivably could mediate the attachment. Unfortunately, no direct evidence exists implicating any substance produced by the bacteria in the attachment mechanism. However it attaches to ciliated epithelium, once attached, *B. pertussis* rarely penetrates into the epithelium (Brown & Brenn, 1931). Subepithelial necrosis and inflammation are characteristics of the disease, however, and may be due to action of a heat labile exotoxin (Munoz, 1971).

Similarly, *Corynebacterium diphtheriae* produces a potent exotoxin while multiplying in close association with the epithelium of the nasopharynx. The exotoxin diffuses into the submucosa and thus into the blood where it is carried throughout the body. All major symptoms of the disease can be attributed to activities of this toxin (Pappenheimer, 1965). Unfortunately for the thesis of this report, however, I can only infer that *C. diphtheriae* attaches specifically to the epithelium of the upper respiratory tract. Experimental findings on this point are lacking.

Some bacteria infecting the gastro-intestinal canal also induce disease without penetrating beyond superficial attachment to mucosal epithelia. *Vibrio cholerae* attaches primarily to epithelial cells in small intestines (Taylor, Maltby & Payne, 1958; Lankford, 1960; Freter, Smith & Sweeney, 1961; Freter, 1969; Freter, 1970). The attachment to small intestinal epithelium seems to be relatively specific; the large bowel is rarely affected (Lankford, 1960). The organisms attach, at least in an experimental model, mainly at the base of the villi and rarely penetrate into or through the intestinal epithelium (Freter, 1969). The bacterial surface properties that mediate the attachment are unknown. *V. cholerae* produces erythrocyte agglutinins. But, as with *B. pertussis*, no direct evidence links the agglutinins with epithelial attachment (Lankford, 1960).

Gross damage to the intestinal epithelium is not detectable in cholera by conventional histological methods (Lankford, 1960). Alterations in the epithelial cells are, however, detectable with the electron microscope (Patnaik & Ghosh, 1966). These ultrastructural changes may be due to an exotoxin produced by the organism (Finkelstein, 1969; Burrows & Musteikis, 1966; Pierce, Greenough & Carpenter, 1971). This entero-toxin enters the mucosa and affects vascular permeability (Finkelstein, Nye, Atthasampunna & Charunmethee, 1966; Finkelstein, 1969) or induces hypersecretion of isotonic fluids by intestinal epithelial cells (Norris & Majno, 1968; Greenough & Carpenter, 1969; Pierce *et al.* 1971). The result is an outpouring of fluids from the blood into the intestinal lumen.

Certain strains of *E. coli* that cause diarrhoeal disease in calves and neonatal swine also adhere by some unknown process to the epithelium of the small bowel (Arbuckle, 1970; Smith & Halls, 1968; Smith, 1971). Moreover, such strains produce enterotoxins (Smith & Halls, 1967, 1968). In appropriate hosts, these exotoxins can by themselves induce watery diarrhoea similar to that seen in experimental cholera (Smith & Halls, 1967; Smith, 1971). Consequently, these *E. coli* strains appear to behave pathogenically as does *V. cholerae*. Interestingly, such a cholera-like disease in humans was reported recently to have been caused by *E. coli* strains able to produce enterotoxins (Sack *et al.* 1971).

Micro-organisms such as *B. pertussis* (Munoz, 1971), *C. diphtheriae* (Pappenheimer, 1965), *V. cholerae* (Finkelstein, 1969) and the cyto-toxigenic strains of *E. coli* (Smith & Halls, 1967, Sack *et al.* 1971) produce their exotoxins *in vitro* in proper media. Thus, these micro-organisms can produce the toxins whether or not they are attached to mucosal epithelia. When present in the upper respiratory or the intestinal tracts, however, they may produce sufficient toxin to be harmful to the host only while sticking to the epithelium and multiplying to considerable population levels. Thus, the ability of these organisms to adhere selectively to certain epithelia and multiply in those areas may be essential in the pathogenesis of these particular diseases.

Attachment with penetration into epithelial cells

Some bacteria attach to mucosal epithelia, then penetrate into the epithelial cells. These organisms multiply within the epithelial cells, kill the cells, and thus induce disease by eroding and ulcerating the mucosal epithelium (Table 2).

In the intestinal canal of humans and primates, *Shigella* organisms attach by some unknown mechanism to colonic epithelial cells, penetrate into the cells, and then multiply within them (La Brec, Schneider, Magnani & Formal, 1964; Ogawa *et al.* 1966; Takeuchi, Formal & Sprinz, 1968). These organisms rarely penetrate beyond the epithelium into the lamina propria (Takeuchi *et al.* 1968). They multiply extensively within the epithelial cells and pass through the lateral plasmalemma into adjacent cells. The bacteria seem almost to punch their way into the adjacent cells moving extremely vigorously inside infected cells (Ogawa, 1970). A mechanism for such movement is as yet unknown. The infected epithelial cells die in patches spreading over the mucosa. The patches manifest themselves histologically as eroded spots in the colonic epithelium (Takeuchi *et al.* 1968). Red blood cells and inflammatory cells pass through these ulcers into the intestinal lumen, thus producing the bloody diarrhoea of classical bacillary dysentery.

Attachment of shigellae to colonic mucosal epithelial cells again can only be inferred to be selective. The organism can attach to and penetrate into HeLa cells in tissue culture (Ogawa, Yoshikura, Nakamura & Nakaya, 1967; Ogawa, Nakamura & Nakaya, 1968) and guinea-pig corneal cells (Piechaud, Szturm-Rubenstein & Piechaud, 1958; Mackel, Langley & Venice, 1961). The organism appears to be reasonably selective for colonic epithelium, however; it rarely infects the small bowel in the natural disease in humans and monkeys (Takeuchi *et al.* 1968). Whether or not the attachment is selective, these organisms probably must penetrate into mucosal epithelia to be able to cause disease. Strains of shigellae lacking this ability are avirulent (Formal, Gemski, Baron & La Brec, 1971).

As noted some enteropathogenic strains of *E. coli* may cause diarrhoeal diseases in calves and neonatal swine similarly to the way *V. cholera* produces disease in man (Smith, 1971). Some such strains appear to behave, however, more similarly to shigellae. Certain *E. coli* strains attach to, penetrate into and probably multiply within mucosal epithelial cells in the large intestines of neonatal pigs (Staley, Corley & Jones, 1970*a*). These bacteria penetrate occasionally into the sub-epithelial tissues (Staley *et al.* 1970*a, b*), although such penetration is probably not a major factor in the infection. These diseases also involve erosion of the mucosal epithelium especially in later stages (Staley *et al.* 1970*a*). Such erosion could be due to multiplication of the bacteria within the epithelial cells or to toxins produced by the bacteria. Some of these strains produce enterotoxins that induce diarrhoea in experimental animals (Moon, Whipp, Engstron & Baetz, 1970).

Attachment with penetration into sub-epithelial tissues

Some types of bacterial pathogens penetrate more deeply into the sub-epithelial tissues than the organisms discussed so far (Table 2). In the upper respiratory tract, *Streptococcus pyogenes* frequently penetrates beyond the mucosal epithelium into the submucosal tissue (McCarty, 1965). This organism is particularly invasive into the lymphoid tissue in the pharynx. The mechanism is not known by which *S. pyogenes* adheres to the pharyngeal epithelium. This organism produces a variety of enzymic and toxic aggressins (McCarty, 1965), however, that may mediate its penetration through the epithelium and into the submucosa.

In the intestinal canal, some bacterial pathogens also penetrate beyond the epithelium into the sub-epithelial tissues. Such infections are characteristic of members of the genus *Salmonella*. These bacteria penetrate through epithelial cells and the basement membrane into the lamina propria, particularly in the small intestine (Takeuchi & Sprinz, 1967; Sprinz *et al.* 1966). Here they are ingested by macrophages (Takeuchi & Sprinz, 1967). These organisms apparently multiply readily and preferentially within macrophages (Mackaness, 1964).

In the early stages of salmonellosis, damage to the intestinal epithelium is slight. The microvilli degenerate at the site of bacterial penetration but this damage appears to be transient and reversible (Takeuchi & Sprinz, 1967). In later stages of the disease, however, epithelial degeneration can be as severe as that seen in shigellosis.

Some strains of *E. coli* that cause enteritis in calves may behave pathogenically similarly to the salmonellae (Smith & Halls, 1968). That is, such strains penetrate into sub-epithelial tissues in the small intestine. They attach most readily at the tips but penetrate most vigorously at the bases of villi (Staley *et al.* 1969). These strains may not produce enterotoxin (Smith, 1971). Therefore, they differ considerably from the strains that induce a disease similar to cholera and to some degree from those that induce diseases similar to shigellosis.

Thus *E. coli* strains exist that behave pathogenically in appropriate hosts very similarly to the behaviour of each of the three major bacterial pathogens of the mammalian intestinal canal. Such *E. coli* strains share with *V. cholera*, *Shigella* species or *Salmonella* species the ability to attach to and sometimes to penetrate into intestinal mucosae.

The ability of *Shig. flexneri* to penetrate into epithelial cells is under genetical control of a single locus on the bacterial chromosome (Formal *et al.* 1971). The mapping of this chromosomal locus was accomplished in part by intergeneric conjugation between various *E. coli* strains and *Shig. flexneri* 2A strains (Formal *et al.* 1971). Thus, the genetical information for such attachment and penetration can be transmitted by conjugation between certain strains of *E. coli* and *Shig. flexneri*. Likewise, the genetical information for production of enterotoxins by *E. coli* is also readily transmissable by conjugation (Smith & Halls, 1967; Smith, 1971). Intergeneric exchange of genetical information for the particular virulence factors discussed in this paper may be common among enteric bacteria. It is possible, therefore, that virulent strains of such organisms may be created relatively commonly from avirulent strains in the mammalian bowel.

Mechanisms of epithelial penetration by bacterial pathogens

As discussed, the biochemical mechanisms mediating the attachments of these bacterial pathogens to mucosal epithelia are unknown. Likewise, the processes by which some of the organisms penetrate into and through epithelia cannot be described in biochemical terms. Studies with the electron microscope have been made, however, of shigellae, salmonellae and *E. coli* penetrating into intestinal epithelia (Takeuchi, 1967; Takeuchi *et al.* 1968; Staley *et al.* 1969).

These organisms apparently pass into epithelial cells by a process very similar to phagocytosis (Staley *et al.* 1969). As the bacteria approach the lumenal (apical) surface of the intestinal epithelial cells, the microvilli (brush border) begin to degenerate in the immediate region of the bacteria. This degeneration begins when salmonellae are within 350 nm (Takeuchi, 1967) and *E. coli* are within 250 nm (Staley *et al.* 1969) of the brush border. As the bacteria pass through the degenerating brush border and into the cells, they are surrounded by inverted cytoplasmic membranes. Therefore, bacteria within the cell cytoplasm are enclosed in membrane-bound vacuoles similar to phagocytic vacuoles seen in macrophages and other phagocytic cells (Takeuchi, 1967; Staley *et al.* 1969). Digestion of the bacteria within such vacuoles has not been observed.

The salmonellae (Takeuchi, 1967) and some strains of *E. coli* (Staley *et al.* 1970*b*) surrounded by the vacuole membranes pass entirely through the epithelial cells and into the lamina propria. Salmonellae also penetrate into the lamina propria by passing through the intercellular junctional complex between epithelial cells (Takeuchi, 1967). This process seems less frequent, however, than passage of the envacuolated bacteria through the cells.

If pathogenic bacteria penetrate into intestinal epithelial cells by a phagocytic process, then the biochemistry of this process possibly could be studied with epithelial cells in tissue culture. A particularly important aspect of such a study would be the mechanism that induces the epithelial cells to ingest the bacteria. Not all bacteria capable of attaching to epithelia are able to penetrate into the cells (Hampton & Rosario, 1965). Therefore, pathogens able to penetrate into epithelial cells must elaborate substances that induce phagocytic activity in such cells. Such substances could be attached to the surfaces of the bacterial cells or could be excreted into the surrounding environment. They would be extremely important factors in the pathogenesis of these bacterial diseases of mucosal membranes.

BACTERIAL GROWTH RATES ON EPITHELIA AND WITHIN STERILE TISSUES

Some bacterial pathogens that penetrate into subepithelial tissues grow best only within particular tissues (Keppie, 1964; Smith, 1968). As has been noted, such pathogens often are selective initially as well in attaching to a particular epithelium. Little is known concerning the biochemistry of the epithelial attachments and growth of bacteria on such epithelia. Likewise, in only a few cases, is the biochemical basis known for the localisation and growth of bacterial pathogens in normally sterile tissues (Smith, 1968).

Research on such problems has progressed slowly, at least in part because methods have not been available for measuring the true growth rates of bacteria on epithelial surfaces and within tissues. Recently, however, ingenious methods have been developed that permit such rates to be estimated in certain bacteria–host systems (Meynell, 1959). These methods will be described in detail because they may prove useful in solving problems of bacterial growth *in vivo*.

Estimation of bacterial growth rates within tissues

Growth rates are difficult to determine for organisms growing in any natural setting (Brock, 1971) but are particularly difficult to estimate for bacteria growing on the various epithelia or within the normally sterile tissues of mammals. Such rates cannot be determined from estimates of populations of viable organisms; viable counts reveal only the net result of multiplication and death of bacteria in tissues and are even less revealing for bacteria on epithelia. Organisms on epithelia are not only multiplying and dying but also are being removed from the membrane by excretory processes.

Meynell's method for estimating bacterial growth rates *in vivo* gives the true division rate and depends upon the disposition in dividing bacteria of certain genetical elements (Meynell, 1959; Meynell & Subbaiah, 1963; Maw & Meynell, 1968). Some such genetic markers, when introduced into a recipient bacterium, do not integrate into the chromosome and do not replicate in the cytoplasm. Consequently, when the recipient bacterium divides, the marker is distributed to only one of the daughter cells. Thus, in each succeeding generation derived from a single recipient cell, only one cell will contain the non-replicating marker. In a bacterial population including a known percentage of recipients containing such a non-replicating marker, the percentage of the population containing the marker theoretically halves with each succeeding generation (Meynell, 1959). The proportion of a bacterial population containing such non-replicating markers can be determined at the beginning and end of a known period. The number of bacterial generations occurring within the period can be calculated from such data.

This principle was first used in efforts to estimate the growth rate *in vivo* of *E. coli* injected intravenously into mice and later recovered from the spleen. The mice were injected with *E. coli* K 12, lysogenised with bacteriophage λb and superinfected with λhc. In this system, λhc, a mutant of λb, can enter a cell lysogenised with λb, but cannot replicate within such a cell. Thus, λhc served as the non-replicating marker in the population of *E. coli* injected into the mice. The proportion of super-infected bacteria in the spleens did not usually change in 30 h following inoculation. Apparently, *E. coli* K 12 could not produce viable progeny in the spleen (Meynell, 1959).

The principle was used later to estimate the growth rate of *Salmonella typhimurium* also injected intravenously into mice and later recovered from the spleen (Maw & Meynell, 1968). In these experiments, the mice were injected with *S. typhimurium* SR 120 lysogenised with a mutant of bacteriophage P 22 (here designated X) and superinfected with another mutant of P 22 (here designated as C). The superinfecting phage C served as the non-replicating marker. Growth rates again were derived from estimates of the proportions of the bacterial populations containing the superinfecting phage in the mouse spleens at the beginning and end of known periods. Such rates were compared with rates derived by the same method from the superinfected bacteria growing in culture media. The division rate of the organisms in the spleen proved to be only 5–10 % of the maximum observed *in vitro*. Interestingly, however, the death rate of the bacteria in the spleen also was extremely small. Thus, the bacteria presumably could persist in the spleens for long periods.

Estimation of bacterial growth rates on mucosal epithelia

In the experiments described above, the investigators estimated growth rates of bacteria within tissues of an animal; they had to account only for multiplication and death of the bacteria. Methods for estimating growth rates of bacteria on mammalian epithelial surfaces must account, however, not only for multiplication and death of the organisms but also for loss of bacterial cells from the site. Meynell & Subbaiah (1963) developed techniques to account for such excreted bacteria when they estimated the division rate of *S. typhimurium* growing in the intestinal canals of mice. Again they employed a bacterial population containing a non-replicating genetic marker but they used a system involving abortive transduction, rather than a superinfecting phage. In transduction, a fragment of the genome of a donor bacterium is transferred by a bacteriophage to a recipient bacterium. Normally, such a genetic fragment is incorporated into the recipient genome and subsequently replicates as a part of that genome. In abortive transduction, however, the genetic fragment fails to incorporate into the recipient genome and remains in the cytoplasm as a non-replicating genetic element. Such elements behave during cell division just as do superinfecting bacteriophages.

Meynell & Subbaiah passed *S. typhimurium* containing an abortively transduced genetic marker into the stomachs of mice and then estimated the division rate, excretion rate and death rate of the bacteria. The division rate was estimated by successive measurements of the proportion of abortive transductants in the intestines of the animals. The excretion rate was estimated from comparison of the dry weight of the caecal and colonic contents with the dry weight of voided faeces. The death rate was calculated from these rates and estimates of the total populations of viable bacteria in the guts. Growth rates were determined for streptomycin-resistant *S. typhimurium* given to untreated mice and mice pretreated with streptomycin.

In untreated normal mice, the bacteria multiplied slowly if at all and were eliminated from the gut in the faeces and by a weak bactericidal mechanism. In contrast, in streptomycin-treated mice, the organism grew as well in the gut as in broth media. The streptomycin obviously abolished the bactericidal mechanism of the gut. The authors concluded that the bactericidal mechanism was due to microbial interference from indigenous micro-organisms.

Such a finding indicates the potential value of these methods of estimating the growth rates of bacteria *in vivo*. Surely, if applied appropriately, these and similar methods could yield information of great value on many of the problems discussed in this report. Studies of microbial interference limiting bacterial infections of mucosal membranes may be particularly susceptible of advancement through application of Meynell's methods.

MICROBIAL INTERFERENCE AND BACTERIAL INFECTIONS OF MUCOSAL MEMBRANES

Indigenous micro-organisms are known to interfere with bacterial infections of mammalian mucosal epithelia (Savage & McAllister, 1970). In the upper respiratory tract, indigenous micro-organisms can limit infections by staphylococci (Boris, 1968), group A streptococci (Sanders, 1969), and gram-negative coliform bacteria (Sprunt & Redman, 1968; Sprunt, Leidy & Redman, 1971). Likewise, in the gastro-intestinal canal, the indigenous microbiota inhibits infections by *V. cholerae* (Freter, 1956), shigellae (Freter, 1956; Hentges, 1969; Hentges & Maier, 1970), salmonellae (Miller & Bohnhoff, 1963; Maw & Meynell, 1968;

Meynell, 1963; Abrams & Bishop, 1966) and *Candida* (Huppert, Cazin & Smith, 1955; Nishikawa *et al.* 1969).

Infection of mucosal epithelia by these pathogens is restricted in all likelihood by mechanisms that serve primarily to regulate the population levels and localisation of the various types of indigenous micro-organisms (Savage & McAllister, 1970). This thesis has been developed from recent findings on the development and quality of the indigenous microbiota.

Microbial interference and the indigenous microbiota

The words 'indigenous microbiota' are used to describe the microbial populations resident in animals. Systematic studies spanning several years have been made of the indigenous microbiota in the gastro-intestinal canals of laboratory rodents and led to the proposal that the indigenous microbiota is composed of the 'normal' and the 'autochthonous' biotas (Dubos *et al.* 1965). The 'normal biota' is thought to be composed of micro-organisms that can be found only in an individual community of a certain type of mammal. Such micro-organisms are ubiquitous in the community and establish themselves in all of its members. The normal biota can vary from community to community. In contrast, the 'autochthonous biota' is thought to be uniform in all communities of a certain type of mammal. Autochthonous micro-organisms seem to be involved in a special relationship with their mammalian hosts. The micro-organisms and the animal appear to have evolved together.

The concept of autochthony in indigenous microbes was drawn primarily from the following findings:

(1) Populations of certain indigenous micro-organisms colonise in a constant pattern the mammalian alimentary canal early in life (Table 3), rise to high levels soon after initial colonisation and remain at those levels throughout the lives of healthy animals (Schaedler, Dubos & Costello, 1965*a*, Dubos *et al.* 1965, Savage *et al.* 1968; Lee *et al.* 1968; Savage & McAllister, 1971).

(2) The various types of indigenous micro-organisms are not just randomly distributed in the alimentary canal but are localised and spatially arranged in micro-environments in various regions (Savage, 1970). Particular microbial populations always colonise the same general areas of the gastro-intestinal tract (Schaedler *et al.* 1965*b*). Some types attach specifically to the mucosal epithelium in particular regions of the stomachs or intestines (Table 4).

(3) Microbial populations that predominate in the mammalian

Table 3. *Development of the caecal flora in specific pathogen-free mice*

Schaedler *et al.* 1965*a*; Savage *et al.* 1968; Lee *et al.* 1971; Savage & McAllister 1971*

Bacteria	Age of mice in days										
	3	5	7	9	11	13	15	17	19	21	30
Lactobacilli	N†	4	9	9	9	9	9	9	9	9	9
Enterococci	N	N	±	6	6	8	8	5	4	4	4
Coliforms	N	N	N	N	±	4	8	7	6	6	4
Bacteroides	N	N	N	N	N	±	8	10	10	10	10
Fusiform-shaped anaerobes	N	N	N	N	N	±	10	11	11	11	11
Spiral-shaped anaerobes‡	(−)	(−)	(−)	(−)	(−)	(±)	(+)	(+)	(+)	(+)	(+)

* Specific pathogen-free mice were raised in protected environments that prevent transmission of murine pathogens.
† Data recorded as the average of the $\log_{10}$ of the numbers of bacteria per g of fresh tissue; ±, bacteria cultured infrequently in low numbers; N, no organisms cultured. Parentheses indicate that results are based upon microscopic observation of organisms in caecal homogenates or tissue sections; +, observed; −, not observed.
‡ Spirochaetes and spiral-shaped bacteria.

Table 4. *Population levels and localisation of indigenous micro-organisms that associate with mucosal epithelia in the gastro-intestinal tracts of adult mammals*

Micro-organism	Animal	Population level*	Epithelium on which localised	Reference
Lactobacilli	Mouse	10^9	Non-secreting, stomach	Savage et al. 1968
	Rat	?	Non-secreting, stomach	Brownlee & Moss, 1961
Yeasts (Torulopsis)	Mouse	10^7–10^8	Secreting, stomach	Savage & Dubos, 1967
	Rat	?	Secreting, stomach	Brownlee & Moss, 1961
Anaerobic streptococci	Mouse	10^9	Non-secreting, stomach	Savage et al. 1968
	Mouse	?	Villous, ileum	Savage, personal observation
	Rat	?	Villous, ileum	Savage, 1969a
	Human	10^8–10^{10}	Mucosal, colon	Nelson & Mata, 1970
Spirochaetes	Mouse	10^9–10^{10}	Mucosal, caecum and colon	Savage et al. 1971; Gordon & Dubos, 1970
	Rat	?	Mucosal, caecum and colon	C. P. Davis, (personal communication)
	Monkey	?	Mucosal, colon	A. Takeuchi, (personal communication)
	Human†	?	Rectum	Harland & Lee, 1967
Spiral-shaped bacteria	Mouse	10^9	Mucosal, caecum and colon	Savage et al. 1971
Fusiform-shaped bacteria	Mouse	10^{11}	Mucosal, caecum and colon	Savage et al. 1968; Gordon & Dubos, 1970
	Rat	?	Mucosal, caecum and colon	Savage, unreported observation
	Human†	?	Villous, ileum	O'Brien, 1968

* Usually reported as number of micro-organisms per g of whole bowel. See individual citations for culture methods. Question mark (?) indicates that quantitative cultures have not yet been reported. In these cases the evidence is based upon histological observations only.
† Patients with bowel diseases.

alimentary canal are composed of anaerobic bacteria in adult animals (Table 5) (Howard, 1967) and humans (Draser, Shiner & McLeod, 1969). In fact, almost all types of micro-organisms commonly found in the mammalian bowel are capable of anaerobic growth. *E. coli*, for example, is facultative and thus can grow anaerobically (Stanier, Doudoroff & Adelberg, 1970). Similarly, microaerophilic lactobacilli and streptococci, capable of growing while incubated in an air atmos-phere, are incapable of utilising oxygen as a terminal electron acceptor and obtain energy from fermentative metabolism (Stanier *et al.* 1970). The population levels of these facultative and microaerophilic organisms are relatively low in most animals, however; the predominating popula-tions by far are oxygen-intolerant anaerobic bacteria (Howard, 1967; Savage *et al.* 1968; Drasar *et al.* 1969; Lee *et al.* 1968; Gordon & Dubos, 1970). In mice, the latter organisms are fusiform-shaped rods of the genera *Fusobacterium, Eubacterium,* and *Clostridium*; spirochaetes; and spiral-shaped bacteria. The fusiform-shaped bacteria outnumber all others by a ratio of from 10 to 1000 to 1 (Savage *et al.* 1968; Gordon & Dubos, 1970).

Thus, the indigenous microflora in the gastro-intestinal canal of animals and also of man is composed predominantly of anaerobic bacteria most of which are oxygen-intolerant. These bacteria colonise the animals early in life; some of them grow in microcolonies – some-times in microlayers – in intimate association with certain mucosal epithelia. Many types of such microbes are considered to be auto-chthonous members of the indigenous microbiota (Dubos *et al.* 1965). The autochthonous microbes may be particularly active in interfering with infections by bacterial pathogens.

Microbial interference is known to play a role in regulating the localisation and population levels of certain of these indigenous micro-organisms (Savage & McAllister, 1970). As noted, both lactobacilli and yeast can colonise the mouse stomach; the lactobacilli colonise the non-secreting epithelium (Pl. 1, Figs. 1 and 2); the yeast colonise the secreting surface (Pl. 2, Figs. 1 and 2) (Savage *et al.* 1968; Savage & Dubos, 1967). In mice given penicillin at appropriate dosage levels in their drinking water the lactobacilli disappear from the keratinised epithelium (Savage, 1969*b*). Under these circumstances, the yeast colonises and forms a layer on the non-secreting epithelium (Table 6). As long as the penicillin is administered, the yeast remains on both the secreting and the non-secreting epithelium. When the penicillin treatment is stopped, the lactobacilli recolonise the keratinised tissues. Thereafter, yeast layers can be seen again only on the secreting tissue.

Table 5. *Oxygen tolerance of indigenous bacterial flora of the gastro-intestinal tracts of adult specific pathogen-free mice*

Bacterial type	Oxygen tolerance*	Area of tract	Population level†	Reference
Lactobacilli	NA	All	10^8–10^9	Schaedler et al. 1965a; Savage & McAllister, 1971
Coliforms	NA	Caecum, colon	10^4–10^6	Schaedler et al. 1965a
Enterococci	NA	Caecum, colon	10^4–10^6	Schaedler et al. 1965a
Anaerobic streptococci	T	All	10^9	Schaedler et al. 1965a
Spiral-shaped‡	T and I	Caecum, colon	10^9–10^{10}	Lee et al. 1968; Gordon & Dubos, 1970; Savage et al. 1971
Bacteroides	T and I	Caecum, colon	10^8–10^{10}	Schaedler et al. 1965a; Savage & McAllister, 1971
Fusiform-shaped bacteria§	I	Caecum, colon	10^{10}–10^{11}	Lee et al. 1968; Gordon & Dubos, 1970; Savage et al. 1971

* Oxygen tolerance is indication of whether or not micro-organisms are killed by oxygen; NA means not applicable; organisms grow in an atmosphere containing oxygen; T means tolerant, organisms grow only anaerobically but do not die during short-term exposure to oxygen; I means intolerant, organisms grow only anaerobically on pre-reduced media and die if exposed to oxygen.

† No. of bacteria per g of whole bowel. See references for culture methods.

‡ Spirochaetes and spiral-shaped bacteria.

§ Includes bacteria of the genera *Fusobacterium*, *Eubacterium* and *Clostridium*.

The lactobacilli apparently are better adapted than are the yeast for growth on the keratinised cells and can displace the yeast and re-establish on the tissue.

Thus in the normal animal, the keratinised cells, the lactobacilli and the immediate environment form an ecosystem that appears to be quite stable. As long as the lactobacilli are in their niche, other microbes seem unable to establish in it. The observed microbial interference, then, is the consequence of the interfering microbes (the lactobacilli) acting in a particular environment upon a particular tissue. The tissue and environment are as much involved in the interference phenomenon as are the lactobacilli themselves.

Table 6. *Microbial interference between indigenous yeast and lactobacilli in the stomachs of laboratory mice*

Savage, 1969b

Drug treatment	No. mice	Stomach tissue cultured	Micro-organisms cultured from the tissues*		Stomach layers present†	
			Yeast	Lactobacilli	Yeast	Lactobacilli
None	50	Secreting	8(7–8)	7(5–8)	14/15	0/15
		Non-secreting	5(4–6)	9(8–9)	0/15	15/15
Penicillin‡	50	Secreting	8(7–9)	NC	21/21	0/21
		Non-secreting	8(6–9)	NC	20/21	0/21
Penicillin then none§	20	Secreting	ND	ND	ND	ND
		Non-secreting	5(4–7)	9(7–9)	ND	ND

* Culture results recorded as the average of the $\log_{10}$ of the numbers of bacteria per g of fresh columnar (secreting) or keratinised squamous (non-secreting) tissue; range in parentheses; NC, no bacteria cultured; ND, not done.

† Layers are observed microscopically in gram-stained frozen sections of stomach. Results are expressed as the ratio of animals with layers to the total number examined.

‡ Aqueous penicillin (0·3 g/l) given as drinking water for up to 8 days; mice sacrificed immediately thereafter.

§ Aqueous penicillin (0·3 g/l) given as drinking water for 5 days; followed by water for 6 days; mice sacrificed thereafter.

A similar displacement phenomenon apparently takes place as the indigenous microflora colonises the infant mouse gut (Table 3). Coliforms and enterococci colonise the large bowel within one week after birth (Schaedler *et al.* 1965a). The populations of these micro-organisms usually increase quickly to quite high levels in the infants, remain at those levels for a few days to a week, then decline to the lower levels characteristic of adult animals. When their population levels are high, coliforms and enterococci can be seen in microcolonies in the mucous layers on the colonic mucosal epithelium (Savage *et al.* 1968; Savage & McAllister, 1971, in preparation). At the end of the second week after birth of the animals these microcolonies are surrounded in the mucin layers by fusiform-shaped bacteria (Savage *et al.* 1968) and spiral-shaped micro-organisms (Savage *et al.* 1971). The tapered and spiral-shaped rods continue to increase in numbers thereafter, until they form distinct layers on the mucosal epithelium by about the fifteenth day after birth (Pl. 3, Figs. 3 and 4). At about this time the populations of coliforms and enterococci drop to adult levels. These bacteria may be displaced from the mucous layer by the anaerobic tapered and spiral-shaped bacteria.

Similarly to the lactobacillus–yeast interaction, this interference between the facultative and anaerobic microbes takes place on a particular epithelium in a particular environment. Again, the epithelium and the environment must be considered along with the anaerobic bacteria as components of the interference system.

A similar hypothesis may explain why the populations of coliform and enterococcal bacteria increase to such high levels in the large bowels of mice given penicillin and certain other antimicrobial drugs (Dubos, Schaedler & Stephens, 1963). In such animals, the layers of fusiform-shaped bacteria disappear from the caecal and colonic epithelium (Savage & Dubos, 1968; Savage & McAllister, 1971). The intact layers of these anaerobes may be necessary in adult animals to control the low population levels of coliforms and enterococci.

Mechanisms of microbial interference regulating the indigenous microbiota

Microbial antagonism regulating the indigenous microflora on any body surface is undoubtedly complex mechanistically. Some mechanisms may involve more or less direct interactions between various types of indigenous microbes. For example, interfering organisms could elaborate antimicrobial substances such as antibiotics or bacteriocins that could kill or inhibit the growth of alien microbes.

Bacteriocins are produced by bacteria of several types indigenous to mammalian body surfaces (Nomura, 1967). Some bacteriocins have been used experimentally to inhibit the growth of sensitive bacterial pathogens in certain well-localised infections (Braude & Siemienski, 1968; Streelman, Snyder & Six, 1970). In other experimental systems, however, these substances have failed to limit the growth of bacteriocin-sensitive bacteria under conditions closely approximating those found in natural populations of indigenous micro-organisms (Ikari, Kenton & Young, 1969). Undoubtedly, the major factor limiting the activity *in vivo* of such antimicrobial substances is the spatial distribution in micro-environments of the microbes of the indigenous biota. Bacterio-cins produced by microbes in one micro-environment probably fail to reach sensitive organisms located in another environment.

Direct antagonism among members of the indigenous biota also could be due to microbial competition. In many natural systems, a particular microbial type may so effectively utilise an essential nutrient that other organisms requiring the nutrient are unable to grow in the immediate environment. This process has been defined as true microbial competi-tion (Brock, 1966). Such a mechanism could operate in micro-environ-ments on epithelial surfaces of mammalian mucosae. For example, indigenous bacteria and yeasts growing in mucous layers investing various epithelial surfaces may metabolise certain constituents of the mucin as carbon and energy sources (Savage & McAllister, 1970). The mucin constituents may be used so efficiently by the indigenous microbes that they are rendered unavailable as nutrients for other micro-organisms. So far this hypothesis has proved to be difficult to confirm experimentally.

Another hypothesis seems more likely, however, to explain how microbes growing in mucous layers may prevent other bacteria from growing in the layers. It was proposed to explain why *S. typhimurium* fails to grow well in the alimentary canals of mice unless the alimentary microflora of such animals is disrupted (Meynell & Subbaiah, 1963). The interference was due to toxicity of *S. typhimurium* to volatile organic acids in a reducing environment. The volatile acids were thought to be produced by anaerobic indigenous micro-organisms particularly fusiform-shaped bacteria (Meynell, 1963). Although the mechanism was proposed to explain resistance to a particular infection, it would undoubtedly operate as well between members of the indigenous microbiota.

As noted, the fusiform-shaped bacteria and also spiral-shaped microbes grow in enormous populations in mucous layers investing the epithelium of the large bowels of rodents (Savage, 1970). Some of these organisms probably use mucin as a source of carbon and energy (Savage & McAllister, 1970) and produce volatile organic acids as a consequence of that metabolism (Meynell, 1963). Those volatile acids are toxic to various bacteria in a reducing environment (Hentges, 1969; Hentges & Maier, 1970). A reducing environment must prevail in the mucous layers where the oxygen-intolerant indigenous anaerobes grow. Consequently, any organism susceptible to the toxicity of the acids would be excluded vigorously from the anaerobic environment in the mucin layers.

Indigenous microbes also may interfere directly with other micro-organisms through their specificity for attachment to a certain epithelium. The antagonism of indigenous lactobacilli for indigenous yeasts may involve such specific attachment (Savage, 1969b; Savage & McAllister, 1970).

Indigenous micro-organisms may also affect host properties that then may *in turn* indirectly influence the localisation and size of the indigenous microbial populations. For example, indigenous microbes are known to influence bowel motility (Abrams & Bishop, 1967). Rapid bowel motility undoubtedly affects populations of indigenous microbes in the intestinal canal especially in the small bowel. Intestinal stasis in the small bowel invariably results in overgrowth, in the static area, of micro-organisms usually found predominantly in the colon (Donaldson, 1967).

Microbial interference between indigenous and pathogenic
bacteria

All these mechanisms, direct and indirect, may operate between indigenous microbes and pathogens as well as between various types of indigenous micro-organisms. By forming volatile organic acids in a reducing environment, the indigenous microbiota interferes directly with bowel infections of mice with *S. typhimurium* (Meynell, 1963) and *Shig. flexneri* (Hentges & Maier, 1970). By stimulating bowel motility, the indigenous microbiota interferes indirectly with alimentary infections of mice by *S. typhimurium*. In the latter case, the bacteria presumably are propelled so rapidly through the bowel, they cannot attach to the epithelial cells to initiate infection (Abrams & Bishop, 1966). Unfortunately, however, in most cases the mechanisms are not known by which the indigenous microbiota inhibits specific bacterial infections.

By the same token, the mechanisms are unknown by which specific pathogens overcome interference from indigenous microbes. Pathogenic bacteria succeed in infecting mucosal epithelia either because they can suppress or bypass interference or because they enter hosts in which that interference is operating inefficiently. Evidence for the first alternative is virtually nonexistent. Some pathogens can invade mucosal epithelia, however, only after the interference activities of the indigenous microbiota are disrupted by malnutrition or by other as yet unknown factors.

Significant alterations in diet can effect alterations in the population levels and types of micro-organisms in the alimentary indigenous flora (Dubos & Schaedler, 1962; Lee, Gordon, Lee & Dubos, 1971) Undoubtedly such radical changes also affect sharply the interference activities of the biota. Inadequate diets are known to predispose animals to diarrhoeal and other diseases of mucosal epithelia (Gordon & Scrimshaw, 1970). Moreover, dietary manipulation has been used in several experimental settings to alter susceptibility of animals to certain infectious diseases (Dubos & Schaedler, 1962; Dubos, Savage & Schaedler, 1966). A bacterial pathogen entering a malnourished host may meet little or no interference from indigenous micro-organisms.

Many environmental extremes also can predispose mammals to infectious diseases. Little is known, however, of the changes in the indigenous microbiota effected by changes in environment. Research on these problems may be stimulated by recent advances in methods for studying the indigenous biota.

SUMMARY

Many bacterial pathogens induce disease in their mammalian hosts after having attached selectively to particular mucosal epithelia. Some such organisms rarely penetrate into the mucosae but multiply on the epithelia and produce exotoxins. These toxins induce characteristic symptoms of the diseases. Other such pathogens penetrate into and multiply within mucosal epithelial cells. These organisms destroy the epithelium by multiplying to high population levels within the cells and often also by producing exotoxins. Still other bacterial pathogens attach to mucosal epithelia and then penetrate into sub-epithelial tissues. These bacteria multiply in submucosal tissues, often within macrophages, and may not produce exotoxins. They multiply in these normally sterile tissues at much slower rates than in culture media.

Some pathogens may fail to attach to or grow on mucosal surfaces because some of the surfaces are populated by indigenous micro-organisms. The access of these pathogenic bacteria to mucosal epithelia may be limited by microbial interference regulating the localisation and population levels of the indigenous microbes themselves.

The point at which this interference operates may be critical in the pathogenesis of these bacterial diseases. Such diseases may be enhanced in hosts in which the indigenous microbiota is disrupted. The biota may be disrupted in animals treated with antibiotics, suffering malnutrition or exposed to other less well-defined adverse conditions. Under such circumstances, certain pathogens may multiply on mucosal surfaces as rapidly as they multiply *in vitro* in culture media. Thus, unrestricted access of these bacterial pathogens to mucosal epithelia undoubtedly enhances development of these diseases. Indigenous microbial populations on those epithelia restricting the access of these pathogens to mucosal surfaces must be important in the resistance of mammals to such diseases.

REFERENCES

ABRAMS, G. D. & BISHOP, J. E. (1966). Effect of the normal microbial flora on the resistance of the small intestine to infection. *Journal of Bacteriology*, **92**, 1604–8.

ABRAMS, G. D. & BISHOP, J. E. (1967). Effect of the normal microbial flora on gastrointestinal motility. *Proceedings of the Society for Experimental Biology and Medicine*, **126**, 301–4.

ABRAMS, G. D., BAUER, H. & SPRINZ, H. (1963). Influence of the normal flora on mucosal morphology and cellular renewal in the ileum. *Laboratory Investigation*, **12**, 355–64.

ARBUCKLE, J. B. R. (1970). The location of *Escherichia coli* in the pig intestine. *Journal of Medical Microbiology*, **3**, 333–40.

BORIS, M. (1968). Bacterial interference: Protection against staphylococcal disease. *Bulletin of the New York Academy of Medicine*, **44**, 1212–21.

BRAUDE, A. E. & SIEMIENSKI, J. S. (1968). The influence of bacteriocins on resistance to infection by gram-negative bacteria. *Journal of Clinical Investigation*, **47**, 1763–73.

BRINTON, C. C. (1965). The structure, function, synthesis and genetic control of bacterial pili and a molecular model for DNA and RNA transport in gram negative bacteria. *Transactions of the New York Academy of Sciences*, **27**, 1003–54.

BROCK, T. D. (1966). In *Microbial ecology*, 126–7. Englewood Cliffs: Prentice-Hall.

BROCK, T. D. (1971). Microbial growth rates in nature. *Bacteriological Reviews*, **35**, 39–58.

BROWN, J. H. & BRENN, L. (1931). A method for the differential staining of gram-positive and gram-negative bacteria in tissue sections. *Bulletin of Johns Hopkins Hospital*, **4**, 69–73.

BROWNLEE, A. & MOSS, W. (1961). The influence of diet on lactobacilli in the stomach of the rat. *Journal of Pathology and Bacteriology*, **82**, 513–16.

BUDDINGH, G. J. (1952). Bacterial and mycotic infections of the chick embryo. *Annals of the New York Academy of Sciences*, **55**, 282–7.

BUDDINGH, G. J. (1970). The chick embryo for the study of infection and immunity. *Journal of Infectious Diseases*, **121**, 660–3.

BURROWS, W. & MUSTEIKIS, G. M. (1966). Cholera infection and toxin in the rabbit ileal loop. *Journal of Infectious Diseases*, **116**, 183–90.

DONALDSON, R. M., JR (1967). Role of enteric microorganisms in malabsorption. *Federation Proceedings*, **26**, 1426–31.

DRASAR, B. S., SHINER, M. & McLEOD, G. M. (1969). Studies on the intestinal flora. I. The bacterial flora of the gastrointestinal tract in healthy and achlorhydric persons. *Gastroenterology*, **56**, 71–9.

DREES, D. T. & WAXLER, G. L. (1970). Enteric colibacillosis in gnotobiotic swine: a fluorescence microscopic study. *American Journal of Veterinary Research*, **31**, 1147–57.

DUBOS, R. J., SAVAGE, D. C. & SCHAEDLER, R. W. (1966). Biological Freudianism, Lasting effects of early environmental influences. *Pediatrics*, **38**, 789–800.

DUBOS, R. J. & SCHAEDLER, R. W. (1962). The effect of diet on the fecal bacteria flora of mice and on their resistance to infection. *The Journal of Experimental Medicine*, **115**, 1161–72.

Dubos, R. J., Schaedler, R. W. & Stephens, M. (1963). The effect of antibacterial drugs on the fecal flora of mice. *The Journal of Experimental Medicine*, **117**, 231–43.

Dubos, R. J., Schaedler, R. W., Costello, R. & Hoet, P. (1965). Indigenous, normal, and autochthonous flora of the gastrointestinal tract. *The Journal of Experimental Medicine*, **122**, 67–76.

Finkelstein, R. A. (1969). The role of choleragen in the pathogenesis and immunology of cholera. *Texas Reports on Biology and Medicine*, **27**, Suppl. 1, 181–201.

Finkelstein, R. A., Nye, S. W., Atthasampunna, P. & Charunmethee, P. (1966). Pathogenesis of experimental cholera: effect of choleragen on vascular permeability. *Laboratory Investigation*, **15**, 1601–9.

Formal, S. B., Gemski, P., Baron, L. S. & LaBrec, E. H. (1971). A chromosomal locus which controls the ability of *Shigella flexneri* to evoke keratoconjunctivitis. *Infection and Immunity*, **3**, 73–9.

Freter, R. (1956). Experimental enteric shigella and vibrio infections in mice and guinea pigs. *The Journal of Experimental Medicine*, **104**, 411–18.

Freter, R. (1969). Studies on the mechanism of action of intestinal antibody in experimental cholera. *Texas Reports on Biology and Medicine*, **27**, Suppl. 1, 299–316.

Freter, R. (1970). Mechanism of action of intestinal antibody in experimental cholera. II. Antibody-mediated antibacterial reaction at the mucosal surface. *Infection and Immunity*, **2**, 556–62.

Freter, R., Smith, H. L., Jr & Sweeney, F. J., Jr (1961). An evaluation of intestinal fluids in the pathogenesis of cholera. *Journal of Infectious Diseases*, **109**, 35–42.

Gibbons, R. J. & van Houte, J. (1971). Selective bacterial adherence to oral epithelial surfaces and its role as an ecological determinant. *Infection and Immunity*, **3**, 567–73.

Gordon, H. A. (1965). Germ-free animals in research: An extension of the pure culture concept. *Triangle*, **7**, 108–121.

Gordon, J. E. & Scrimshaw, N. S. (1970). Infectious disease in the malnourished. *Medical Clinics of North America*, **54**, 1495–1508.

Gordon, J. H. & Dubos, R. J. (1970). The anaerobic bacterial flora of the mouse cecum. *The Journal of Experimental Medicine*, **132**, 251–60.

Greenough, W. B. III & Carpenter, C. C. J. (1969). Fluid loss in cholera. A current perspective. *Texas Reports on Biology and Medicine*, **27**, Suppl. 1. 203–12.

Ham, A. (1965). In *Histology*, 5th ed., pp. 170, 673, 692, 751. Philadelphia: Lippincott.

Hampton, J. C. & Rosario, B. (1965). The attachment of microorganisms to epithelial cells in the distal ileum of the mouse. *Laboratory Investigation*, **14**, 1464–81.

Hampton, J. C. & Rosario, B. (1966). A variation in the plasma membrane of intestinal epithelial cells. *The Anatomical Record*, **156**, 369–82.

Harland, W. A. & Lee, F. D. (1967). Intestinal spirochetosis. *British Medical Journal*, **3**, 718–19.

Hentges, D. J. (1969). Inhibition of *Shigella flexneri* by the normal intestinal flora. II. Mechanisms of inhibition by coliform organisms. *Journal of Bacteriology*, **97**, 513–17.

Hentges, D. J. & Maier, B. R. (1970). Inhibition of *Shigella flexneri* by the normal intestinal flora. III. Interactions with *Bacteroides fragilis* strains *in vitro*. *Infection and Immunity*, **2**, 364–70.

Howard, B. H. (1967). Intestinal microorganisms of ruminants and other vertebrates. In *Symbiosis*, ed. S. M. Henry, vol. II, 317–85. New York: Academic Press.

HUMMEL, K. P., RICHARDSON, F. L. & FEKETE, E. (1966). Anatomy. In *Biology of the Laboratory Mouse*, ed. E. L. Green, 2nd edition, 247–308. New York: McGraw-Hill.
HUPPERT, M., CAZIN, J., JR & SMITH, H., JR (1955). Pathogenesis of *Candida albicans* infection following antibiotic therapy. II. The effect of antibiotics on the incidence of *Candida albicans* in the intestinal tract of mice. *Journal of Bacteriology*, **70**, 440–7.
IKARI, N. S., KENTON, D. M. & YOUNG, V. M. (1969). Interaction in the germfree mouse intestine of colicinogenic and colicin-sensitive microorganisms. *Proceedings of the Society for Experimental Biology and Medicine*, **130**, 1280–4.
JOHANSON, W. G., PIERCE, A. K. & SANFORD, J. P. (1969). Changing pharyngeal bacterial flora of hospitalized patients. Emergence of gram-negative bacilli. *New England Journal of Medicine*, **281**, 1137–40.
KEOGH, E. V., NORTH, E. A. & WARBURTON, M. F. (1947). Haemagglutinins of the *Hemophilus* group. *Nature, London*, **160**, 63.
KEPPIE, J. (1964). Host and tissue specificity. In *Microbial behaviour in vivo and in vitro*, ed. H. Smith and J. Taylor, 44–63. London: Cambridge University Press.
LA BREC, E. H., SCHNEIDER, H., MAGNANI, T. J. & FORMAL, S. B. (1964). Epithelial cell penetration as an essential step in the pathogenesis of bacillary dysentery. *Journal of Bacteriology*, **88**, 1503–18.
LANKFORD, C. E. (1960). Factors of virulence of *Vibrio cholerae*. *Annals of The New York Academy of Sciences*, **88**, 1203–12.
LEBLOND, C. P. & STEVENS, C. E. (1948). The constant renewal of the intestinal epithelium in the albino rat. *The Anatomical Record*, **100**, 357–78.
LEE, A., GORDON, J. & DUBOS, R. J. (1968). Enumeration of the oxygen-sensitive bacteria usually present in the intestine of healthy mice. *Nature, London*, **220**, 1137–9.
LEE, A., GORDON, J., LEE, C. J. & DUBOS, R. J. (1971). The mouse intestinal microflora with emphasis on the strict anaerobes. *The Journal of Experimental Medicine*, **133**, 339–52.
LESHER, S., WALBURG, H. E. & SACHER, G. A. (1964). Generation cycle in the duodenal crypt cells of germ-free and conventional mice. *Nature, London*, **202**, 884–6.
MACKANESS, G. B. (1964). The behavior of microbial parasites in relation to phagocytic cells *in vitro* and *in vivo*. In *Microbial behaviour in vivo and in vitro*, ed. H. Smith and J. Taylor, 213–40. London: Cambridge University Press.
MACKEL, D. C., LANGLEY, L. F. & VENICE, L. A. (1961). The use of the guinea pig conjunctivae as an experimental model for the study of virulence of *Shigella* organisms. *American Journal of Hygiene*, **73**, 219–23.
MAW, J. & MEYNELL, G. G. (1968). The true division and death rates of *Salmonella typhimurium* in the mouse spleen determined with superinfecting phage P_{22}. *The British Journal of Experimental Pathology*, **49**, 597–613.
MCCARTY, M. (1965). The hemolytic streptococci. In *Bacterial and Mycotic Infections of Man*, ed. R. J. Dubos and J. Hirsch. 4th edition, 356–90. Philadelphia: Lippincott.
MEYNELL, R. G. (1959). Use of superinfecting phage for estimating the division rate of lysogenic bacteria in infected animals. *Journal of General Microbiology*, **21**, 421–37.
MEYNELL, G. G. (1963). Antibacterial mechanisms of the mouse gut. II. The role of Eh and volatile fatty acids in the normal gut. *British Journal of Experimental Pathology*, **44**, 209–19.

MEYNELL, G. G. & SUBBAIAH, T. V. (1963). Antibacterial mechanisms of the mouse gut. I. Kinetics of infection by *Salmonella typhimurium* in normal and streptomycin-treated mice studied with abortive transductants. *British Journal of Experimental Pathology*, **44**, 197–208.

MILLER, C. P. & BOHNHOFF, M. (1963). Changes in the mouse's microflora associated with enhanced susceptibility to salmonella infection following streptomycin treatment. *Journal of Infectious Diseases*, **113**, 59–66.

MOOG, F. & GREY, R. D. (1968). Alkaline phosphatase isozymes in the duodenum of the mouse: Attainment of pattern of spatial distribution in normal development and under the influence of cortisone or actinomycin D. *Developmental Biology*, **18**, 481–500.

MOON, H. W., WHIPP, S. C., ENGSTROM, G. W. & BAETZ, A. L. (1970). Response of rabbit ileal loop to cell-free products from *Escherichia coli* enteropathogenic for swine. *Journal of Infectious Diseases*, **121**, 182–7.

MUNOZ, J. (1971). Protein toxins from *Bordetella pertussis*. In *Microbial Toxins*, ed. S. Kadis, T. C. Montie and S. J. Ajl, vol. 2a, 271–300. New York: Academic Press.

NELSON, D. P. & MATA, L. J. (1970). Bacterial flora associated with the human gastrointestinal mucosa. *Gastroenterology*, **58**, 56–61.

NISHIKAWA, T., HATANO, H., OHNISHI, N., SASAKI, S. & NOMURA, T. (1969). Establishment of *Candida albicans* in the alimentary tract of the germ-free mice and antagonism with *Escherichia coli* after oral inoculation. *Japanese Journal of Microbiology*, **13**, 263–76.

NOMURA, M. (1967). Colicins and related bacteriocins. *Annual Reviews of Microbiology*, **21**, 257–84.

NORDSTRÖM, C., DAHLQUIST, A. & JOSEFSSON, L. (1967). Quantitative determination of enzymes in different parts of the villi and crypts of rat small intestine. Comparison of alkaline phosphatase, disaccharidases, and dipeptidases. *Journal of Histochemistry and Cytochemistry*, **15**, 713–21.

NORRIS, H. T. & MAJNO, G. (1968). On the role of the ileal epithelium in the pathogenesis of experimental cholera. *The American Journal of Pathology*, **53**, 263–79.

O'BRIEN, W. (1968). Acute miliary tropical sprue in southeast Asia. *The American Journal of Clinical Nutrition*, **21**, 1007–12.

OGAWA, H. (1970). Experimental approach in studies on pathogenesis of bacillary dystentery – with special references to the invasion of bacilli into intestinal mucosa. *Acta Pathologica Japonica*, **20**, 261–77.

OGAWA, H., HONJO, S., TAKASAKA, M., FUJIWARA, T. & IMAIZUMI, K. (1966). Shigellosis in cynomologus monkeys (*Macaca irus*). IV. Bacteriological and histo-pathological observations on the earlier stage of experimental infection with *Shigella flexneri* 2A. *Japanese Journal of Medical Science and Biology*, **19**, 23–32.

OGAWA, H., NAKAMURA, A. & NAKAYA, R. (1968). Cine-micrographic study of tissue cell cultures infected with *Shigella flexneri*. *Japanese Journal of Medical Science and Biology*, **21**, 259–73.

OGAWA, H., YOSHIKURA, H., NAKAMURA, A. & NAKAYA, R. (1967). Susceptibility of cell cultures from various mammalian tissues to the infection by *Shigella*. *Japanese Journal of Medical Science and Biology*, **20**, 329–39.

PAPPENHEIMER, A. M., JR (1965). The diphtheria bacillus. In *Bacterial and Mycotic Infections of Man*, ed. R. J. Dubos and J. H. Hirsch, 4th edition, 468–89. Philadelphia: Lippincott.

PATNAIK, B. K. & GHOSH, H. K. (1966). Histopathological studies on experimental cholera. *The British Journal of Experimental Pathology*, **47**, 210–14.

PIECHAUD, M., SZTURM-RUBENSTEIN, S. & PIECHAUD, D. (1958). Evolution histologique de la kerato-conjonctivite a bacillus dysenteriques du cobaye. *Annals Institute Pasteur*, **94**, 298–300.

PIERCE, N. F., GREENOUGH, W. G. & CARPENTER, C. C. J. (1971). *Vibrio cholerae* enterotoxin and its mode of action. *Bacteriological Reviews*, **35**, 1–13.

RELLER, L. B. & SPECTOR, M. I. (1970). Shigellosis among Indians. *Journal of Infectious Diseases*, **121**, 355–7.

SACK, R. B., GORBACH, S. L., BANWELL, J. G., JACOBS, B., CHATTERJEE, B. D. & MITRA, R. C. (1971). Enterotoxigenic *Escherichia coli* isolated from patients with severe cholera-like disease. *Journal of Infectious Diseases*, **123**, 378–85.

SANDERS, E. (1969). Bacterial interference. I. Its occurrence among the respiratory tract flora and characterization of inhibition of Group A streptococci by viridans streptococci. *Journal of Infectious Diseases*, **120**, 698–707.

SAVAGE, D. C. (1969a). Localization of certain indigenous microorganisms on the ileal villi of rats. *Journal of Bacteriology*, **97**, 1505–6.

SAVAGE, D. C. (1969b). Microbial interference between indigenous yeast and lactobacilli in the rodent stomach. *Journal of Bacteriology*, **98**, 1278–83.

SAVAGE, D. C. (1970). Associations of indigenous microorganisms with gastrointestinal mucosal epithelia. *The American Journal of Clinical Nutrition*, **23**, 1495–501.

SAVAGE, D. C. & DUBOS, R. J. (1967). Localization of indigenous yeast in the murine stomach. *Journal of Bacteriology*, **94**, 1811–16.

SAVAGE, D. C. & DUBOS, R. J. (1968). Alterations in the mouse cecum and its flora produced by anti-bacterial drugs. *The Journal of Experimental Medicine*, **128**, 97–110.

SAVAGE, D. C. & MCALLISTER, J. S. (1970). Microbial interactions at body surfaces and resistance to infectious diseases. In *Resistance to Infectious Disease*, ed. R. H. Dunlop and H. W. Moon, 113–127. Saskatoon: Modern Press.

SAVAGE, D. C. & MCALLISTER, J. S. (1971). Cecal enlargement and microbial flora in suckling mice given antibacterial drugs. *Infection and Immunity*, **3**, 342–9.

SAVAGE, D. C., DUBOS, R. J. & SCHAEDLER, R. W. (1968). The gastrointestinal epithelium and its autochthonous bacterial flora. *The Journal of Experimental Medicine*, **127**, 67–76.

SAVAGE, D. C., MCALLISTER, J. S. & DAVIS, C. P. (1971). Anaerobic bacteria on the mucosal epithelium of the murine large bowel. *Infection and Immunity*, **4**, 492–502.

SAVAGE, D. C., SCHAEDLER, R. W. & DUBOS, R. (1967). Autochthonous microflora of the mouse stomach. *Bacteriological Proceedings*, **67**, 67.

SCHAEDLER, R. W., DUBOS, R. J. & COSTELLO, R. (1965a). The development of the bacterial flora in the gastrointestinal tract of mice. *The Journal of Experimental Medicine*, **122**, 59–66.

SCHAEDLER, R. W., DUBOS, R. J. & COSTELLO, R. (1965b). Association of germfree mice with bacteria isolated from normal mice. *The Journal of Experimental Medicine*, **122**, 77–82.

SMITH, H. (1968). Biochemical challenge of microbial pathogenicity. *Bacteriological Reviews*, **32**, 164–84.

SMITH, H. W. (1971). The bacteriology of the alimentary tract of domestic animals suffering from *Escherichia coli* infection. *Annals of The New York Academy of Sciences*, **176**, 110–25.

SMITH, H. W. & HALLS, S. (1967). Studies on *Escherichia coli* enterotoxin. *Journal of Pathology and Bacteriology*, **93**, 531–43.

SMITH, H. W. & HALLS, S. (1968). The production of oedema disease and diarrhoea in weaned pigs by the oral administration of *Escherichia coli*: factors that influence the course of the experimental disease. *Journal of Medical Microbiology*, 1, 45–59.

SPRINZ, H. (1969). Pathogenesis of intestinal infections. *Archives of Pathology*, 87, 556–62.

SPRINZ, H., GANGAROSA, E. J., WILLIAMS, M., HORNICK, R. B. & WOODWARD, T. E. (1966). Histopathology of the upper small intestines in typhoid fever. Biopsy study of experimental disease in man. *The American Journal of Digestive Diseases*, 11, 615–24.

SPRUNT, K. & REDMAN, W. (1968). Evidence suggesting importance of role of interbacterial inhibition in maintaining balance of normal flora. *Annals of Internal Medicine*, 68, 579–90.

SPRUNT, K., LEIDY, G. A. & REDMAN, W. (1971). Prevention of bacterial overgrowth. *Journal of Infectious Diseases*, 123, 1–10.

STALEY, T. E., CORLEY, L. D. & JONES, E. W. (1970a). Early pathogenesis of colitis in neonatal pigs monocontaminated with *Escherichia coli*. Fine structural changes in the colonic epithelium. *The American Journal of Digestive Diseases*, 15, 923–35.

STALEY, T. E., CORLEY, L. D. & JONES, E. W. (1970b). Early pathogenesis of colitis in neonatal pigs monocontaminated with *Escherichia coli*. Fine structural changes in the circulatory compartments of the lamina propria and submucosa. *The American Journal of Digestive Diseases*, 15, 937–51.

STALEY, T. E., JONES, E. W. & CORLEY, L. D. (1969). Attachment and penetration of *Escherichia coli* into intestinal epithelium of the ileum in new-born pigs. *The American Journal of Pathology*, 56, 371–92.

STANIER, R. Y., DOUDOROFF, M. & ADELBERG, E. A. (1970). *The Microbial World*, 3rd edition, 580, 663. Englewood Cliffs: Prentice-Hall.

STREELMAN, A. J., SNYDER, I. S. & SIX, E. W. (1970). Modifying effect of colicin on experimental *Shigella* keratoconjunctivitis. *Infection and Immunity*, 2, 15–23.

TAKEUCHI, A. (1967). Electron microscope studies of experimental *Salmonella* infection. I. Penetration into the intestinal epithelium by *Salmonella typhimurium*. *The American Journal of Pathology*, 50, 109–36.

TAKEUCHI, A. & SPRINZ, H. (1967). Electron microscope studies of experimental salmonella infection in the preconditioned guinea pig. II. Response of the intestinal mucosa to the invasion by *Salmonella typhimurium*. *The American Journal of Pathology*, 51, 137–61.

TAKEUCHI, A., FORMAL, S. B. & SPRINZ, H. (1968). Experimental acute colitis in the rhesus monkey following peroral infection with *Shigella flexneri*. *The American Journal of Pathology*, 52, 503–29.

TANNOCK, G. W. & SMITH, J. M. B. (1970). The microflora of the pig stomach and its possible relationship to ulceration of the pars oesophagea. *Journal of Comparative Pathology*, 80, 359–67.

TAYLOR, J. M., MALTBY, P. & PAYNE, J. M. (1958). Factors influencing the response of ligated rabbit-gut segments to infected *Escherichia coli*. *Journal of Pathology and Bacteriology*, 76, 491–9.

TILLOTSON, J. R. & FINLAND, M. (1969). Bacterial colonization and clinical super-infection of the respiratory tract complicating antibiotic treatment of pneumonia. *Journal of Infectious Diseases*, 119, 597–624.

YOLTON, D., STANLEY, C. & SAVAGE, D. C. (1971). Influence of the indigenous gastrointestinal microbial flora on duodenal alkaline phosphatase activity in mice. *Infection and Immunity*, 3, 768–73.

PLATE 1

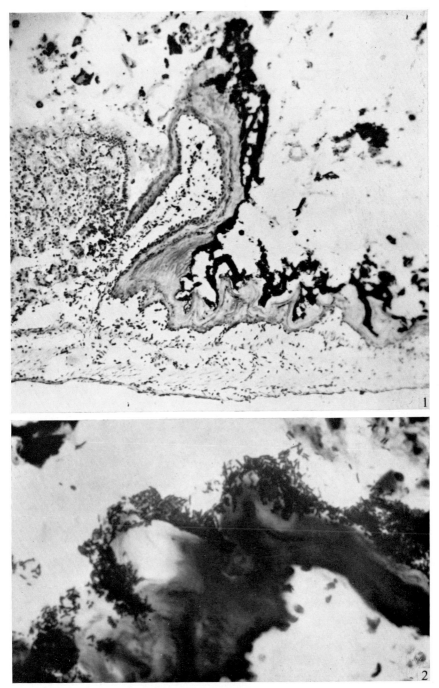

PLATE 2

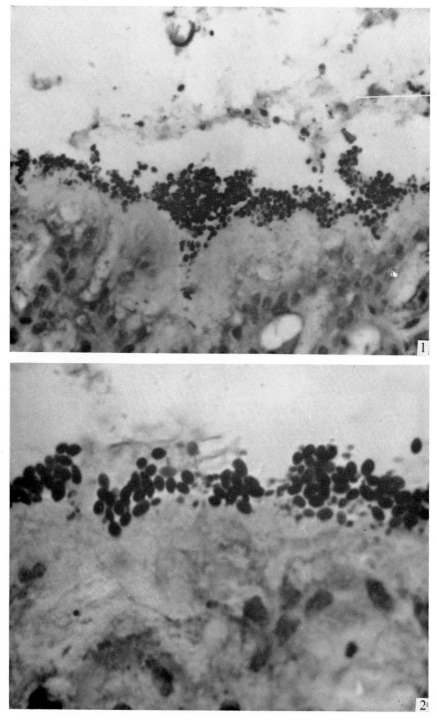

PLATE 3

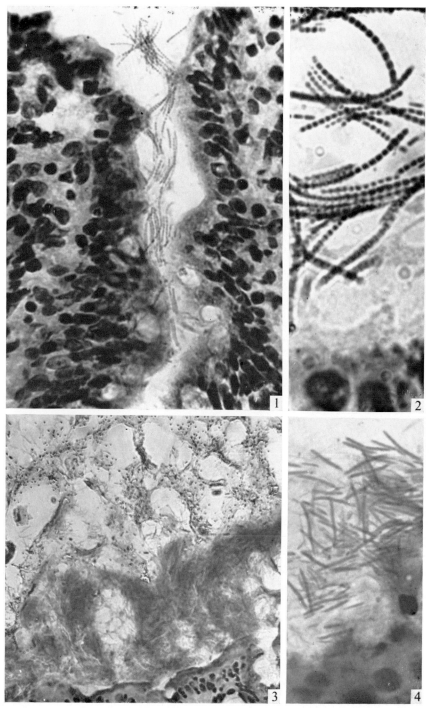

EXPLANATION OF PLATES

PLATE 1

Gram-stained tissue sections showing indigenous Gram-positive bacteria in the stomachs of mice. The bacteria form layers on the stratified squamous epithelium of the non-glandular mucosa. These bacterial layers stop abruptly at the cardiac antrum of the stomachs. The sections are cut with a microtome-cryostat from stomachs frozen with their contents intact (Savage *et al.* 1968. Courtesy of The Rockefeller University Press). Fig. 1. Stomach, adult mouse; the bacteria are the most darkly stained material to the right; the glandular mucosa is to the left; the cardiac antrum is the large fold of tissue in the centre ($\times 40$). Fig. 2. Stomach, adult mouse; the bacteria are the darkly stained rods lying in close association with the keratinised epithelium ($\times 600$).

PLATE 2

Gram-stained tissue sections showing indigenous yeast in the stomachs of mice. The yeast form layers on the columnar secreting epithelium of the glandular mucosa. The sections are prepared as explained for Plate 1, Figs. 1 and 3 (Savage & Dubos, 1967. Courtesy of The Williams and Wilkins Co.). Fig. 1. Stomach, adult mouse; the yeasts are the most darkly stained material intimately associated with the mucosa at the bottom ($\times 500$). Fig. 2. Same as *a* ($\times 1000$).

PLATE 3

Gram-stained tissue sections showing indigenous Gram-variable bacteria in the ileums and colons of laboratory rodents. The sections are prepared as explained for Plate 1, Figs. 1 and 2. (Fig. 1 and 2 courtesy of the Williams and Wilkins Co.; Fig. 3 courtesy of The Rockefeller University Press.) Fig. 1. Ileum, adult rat; the bacteria are the lightly stained chains in the centre between two villi ($\times 1000$). (Savage, 1969*a*). Fig. 2. Same as Fig. 1 ($\times 2475$). Fig. 3. Colon, adult mouse; the bacteria are in the most darkly stained layer between the mucosal epithelium at the extreme bottom and the lumen occupying the upper half of the photo ($\times 775$). (Savage *et al.* 1968). Fig. 4. Same as Fig. 3, showing fusiform shape of bacteria in layers on epithelium ($\times 1140$).

THE PHAGOCYTIC DEFENCE SYSTEM

J. G. HIRSCH

The Rockefeller University, New York, N.Y. 10021

INTRODUCTION

The infectious process begins when bacteria residing on skin or mucous membranes penetrate these barriers or are carried through them and enter the body. The organisms are confronted then with a different micro-environment, namely the extracellular fluid of various tissues or organs. Initially the fate of bacteria must be determined by the presence or absence in this fluid of nutrients and physicochemical conditions favourable for their growth, as well as by anti-bacterial humoral factors. Actually relatively little is known about the suitability of various tissue fluids for microbial survival and growth but it seems likely that these fluids constitute, in many instances, a good culture medium. Certainly ascitic or pleural fluids support good growth of many bacterial species. Most of the studies that have been made on humoral anti-microbial factors have employed as a test material serum obtained from clotted blood. This test material is somewhat inappropriate since the confrontation between bacterial invaders and host-resistance agencies only rarely takes place in the blood stream; usually the crucial early encounters are in tissue fluids or in the lymphatic system. Furthermore, serum obtained from clotted blood differs from the plasma normally circulating in the blood stream and some of the anti-bacterial components of serum are known to be released from formed elements of the blood during clotting (Gengou, 1901; Hirsch, 1960; Jago & Jacox, 1961).

If we assume for purposes of discussion that some serum constituents are present, albeit in lower concentrations, in the various tissue fluids, then one or more of the several anti-bacterial factors of serum (e.g. complement, antibody, lysozyme, transferrin) may influence the fate of invading micro-organisms. The variety and the possible roles of these humoral bactericidal factors are considered elsewhere (Müller-Eberhard, 1965; Coombs & Smith, 1968) and their interaction with bacteria later in this volume (Glynn, this volume); suffice it to say here that they apparently act as the primary host defence only in certain cases and much remains to be learned about them.

In the event of invading bacteria finding the tissue juices a favourable medium for their growth, host resistance depends mainly on phagocytic cells and this aspect of host resistance will be discussed in this review. No attempt will be made at comprehensive coverage of the field but rather, emphasis will be placed on several recent advances in knowledge of cell structure, life history and function in relation to the overall efficiency of the phagocytic system. The discussion will deal with two cell types – the polymorphonuclear neutrophil or heterophil leucocytes and the mononuclear monocytes or macrophages – which account for phagocytic defence against microbial invaders in vertebrates.

ULTRASTRUCTURE OF PHAGOCYTIC CELLS

An electron micrograph of a human neutrophil is shown in Plate 1, Fig. 1. Nuclear lobes (N) show abundant heterochromatic material at the periphery; nucleoli are not seen. The predominant cytoplasmic organelles are membrane-bounded granules (G), which show considerable variation in size and electron density. Other cytoplasmic structures are notably sparse or even absent in polymorphonuclear leucocytes. The centrioles (C) are surrounded by a few vesicles but no other Golgi components are present. Mitochondria are few in number and small in size. Only rare short strips of endoplasmic reticulum are seen. In specimens fixed and processed in other ways, large cytoplasmic deposits of glycogen may be present.

In the rabbit polymorphonuclear leucocyte the granules are of two types, distinguished by their size and electron density (Pl. 1, Fig. 2), by their histochemical reactivity (Wetzel, Spicer & Horn, 1967; Bainton & Farquhar, 1968) and by the time and precise site of their formation in the marrow myelocytes (Bainton & Farquhar, 1966). Recently these two granule populations have been separated from heterophil homogenates (Baggiolini, Hirsch & de Duve, 1969, 1970). The larger, more electron-opaque granules, which form by budding from the concave face of the Golgi apparatus in promyelocytes, qualify as lysosomes (de Duve & Wattiaux, 1966) as they contain numerous digestive enzymes capable of degrading various macromolecules at an acid pH. These larger granules, called primary because of their early formation, or azurophil because of their staining properties, also contain myeloperoxidase in abundance and several anti-bacterial cationic peptides (Zeya & Spitznagel, 1971). The smaller, less dense granules (secondary or specific granules), which form from the convex face of the Golgi apparatus in myelocytes, are not lysosomes since they do not contain acid hydrolases; in them are localised alkaline phosphatase, lysozyme, and a large amount of lactoferrin, an iron-binding protein with anti-bacterial properties (Baggiolini, de Duve, Masson & Heremans, 1970; Masson, 1970).

Plate 2 illustrates the ultrastructure of a human blood monocyte. The nucleus (N) is horseshoe-shaped. In the vicinity of the centriole (C) are numerous small vesicular and saccular elements of the Golgi apparatus, and a number of electron-dense granules (G), similar to those seen in polymorphonuclear leucocytes. The peripheral monocyte cytoplasm contains mitochondria (M), numerous small vesicles varying in size and electron opacity and scattered strips of endoplasmic reticulum.

Plate 3 shows a rabbit alveolar macrophage as viewed by electron microscopy. The nucleus (N) is oval. There is a prominent centrosomal region (C) containing multicentric stacks of flattened Golgi saccules and an assortment of small vesicles. Arranged in rosette fashion around the centrosomal region are mitochondria (M) and numerous very large, membrane-bound granules (G) containing a homogeneous electron-dense material. The peripheral cytoplasm contains rough endoplasmic reticulum (R). Note the highly irregular surface contour.

Thus mononuclear phagocytes possess a rich assortment of cytoplasmic organelles. Several types of structures exhibit lysosomal properties in these cells. Monocytes and bone marrow precursors of monocytes contain a few peroxidase-rich cytoplasmic granules (van Furth, Hirsch & Fedorko, 1970) apparently quite similar in structure and in genesis to the azurophil granules of neutrophils. Macrophages in tissues as a rule possess none of these peroxidase-positive granules but often show large numbers of dense granules (Plate 3) which give a negative cytochemical reaction for peroxidase. These large granules do contain acid hydrolases and thus are lysosomes; they develop from pinocytic vacuoles which acquire digestive enzymes by fusion with Golgi vesicles (Cohn, Fedorko & Hirsch, 1966; Hirsch, Fedorko & Cohn, 1968; Cohn, 1968).

LIFE HISTORY OF PHAGOCYTIC CELLS

Obviously the first requirement for an efficient phagocytic defence is production of these cells in adequate numbers and a delivery system providing for an abundant supply anywhere in the body. In the past decade our knowledge of the production, distribution and fate of phagocytes has expanded considerably.

Several recent reviews are available on life history of the neutrophil (Perry, 1970; Cronkite & Vincent, 1970.) In the adult mammal these cells are produced mainly in bone marrow. They arise from a population of stem cells that have not yet been identified with certainty but probably resemble lymphocytes in morphology. The stem cells appear to be pleuripotential, i.e. capable of giving rise to granulocytic, monocytic, or erythroid cell lines. At any given time a significant proportion of the stem cell population is in a non-replicating 'G$_0$' state. Maturation proceeds through the well-recognised myeloblast, promyelocyte, myelocyte and metamyelocyte stages to the mature marrow neutrophil granulocyte. The proliferative stages (myeloblast, promyelocyte, myelocytes) involve four or more mitotic divisions; kinetic data suggest the likelihood that rate of production of granulocytes can be regulated by changes in the cell cycle time and number of mitoses in the myelocyte stage, as well as by the rate of stem cell proliferation. The maturation phase in the marrow (metamyelocyte through juvenile or stab cell to mature polymorphonuclear leucocyte) involves changes in cell structure and function without cell division. Overall marrow maturation time, from stem cell to mature cell, is approximately fourteen days under steady state conditions. The bone marrow contains a large reserve pool of mature neutrophils (approximately 30×10^{11} in man) which can be rapidly delivered into the blood stream in times of need.

Neutrophils from the marrow enter the blood stream and are carried to all parts of the body. In the blood stream of man there are approximately 60×10^{10} neutrophils, half of which are marginated and half of which are circulating at any given time. The neutrophils leave the blood and move into tissues in random fashion with a mean half time in the circulation of 6–7 h. Emigration into tissues is accomplished by sticking and crawling between endothelial cells (Marchesi & Florey, 1960). Nothing is known of the mechanism by which neutrophils separate the overlapping adherent endothelial cells or penetrate the underlying basement lamina.

The size of the tissue pool of neutrophils is not known accurately. Neutrophils survive in tissues for a few days at most; they are end cells incapable of reproduction or rejuvenation. In order that the system be kept in balance approximately 10^{11} neutrophils must disappear from a man's tissues each day. There is no known leucocyte graveyard to handle this 100 ml of packed cells daily. Most of these disappearing neutrophils probably move into the outside world through various mucous membranes, especially in the intestine.

Mononuclear phagocyte life history is similar in several regards, but there are also important differences from the life history of neutrophils. Blood monocytes are derived from the bone marrow. A proliferating stage of marrow promonocyte has recently been identified (van Furth & Diesselhoff-Den Dulk, 1970) but little else is known about the maturation process; no large reserve pool exists of mature monocytes in the marrow. Marrow monocytes move into the blood stream and spend approximately 32 h there on the average, before emigrating into tissues (van Furth & Cohn, 1968). In tissues these cells are capable of considerable modulation and maturation which takes a different form depending on the particular site. Evidence indicates that skin macrophages (Volkman & Gowens, 1965), alveolar macrophages (Pinkett, Cowdrey & Nowell, 1966), Kupffer cells of the liver (Howard, 1970) and peritoneal macrophages (Balner, 1963; Volkman, 1966) are normally replenished by marrow derived blood monocytes rather than by local proliferation of macrophages or a connective tissue precursor. Macrophages in tissues appear to have a very long life span and ordinarily they do not recirculate (Roser, 1970). Under normal conditions macrophages do not synthesise new DNA nor divide, but under some special circumstances they can be stimulated to do so (Virolainen & Defendi, 1967; Virolainen, 1968; North, 1969). The tissue macrophage turnover rate is quite slow, about 40 days for the mouse peritoneal macrophage (van Furth & Cohn, 1968). Mature macrophages are dispersed throughout the body. They can be placed in two categories: (a) those wandering in tissues and body spaces (e.g. alveolar and peritoneal macrophages, skin macrophages, histiocytes), and (b) those fixed to vascular endothelium forming a sort of a baffle system in blood or lymph (e.g. Kupffer cells of the liver, fixed macrophages of spleen and lymph nodes).

MECHANISMS FOR FACILITATING CONTACT BETWEEN INVADING MICRO-ORGANISMS AND PHAGOCYTIC CELLS

Contact between micro-organisms and phagocytic cells may be considered as a random phenomenon, the frequency of hits being influenced by such variables as rate and direction of movement of the components and their concentrations. In the early phases of an infection, frequently the rate of contact is the limiting factor which determines the rate of removal of micro-organisms. If this rate of removal exceeds the rate of increase in microbial population due to continued penetration from the outside and due to multiplication, then the infection is aborted; if not, the increase in microbial population leads to increasing local pathology and to dissemination of the infection.

Contact with micro-organisms is facilitated by general tissue mechanisms such as the inflammatory reaction, as well as by physiologic properties of the individual cells such as locomotion and chemotaxis. Inflammation has two important effects in terms of the cellular defence mechanism: (*a*) the margination and emigration of blood stream leucocytes in the inflamed region results in marked elevation of the concentration of these cells in the adjacent tissues, thus increasing the opportunity for contact between one of them and bacteria; and (*b*) the exudation of fluid and increased lymphatic drainage may well serve to carry bacteria away from the primary tissue battleground.

In the tissues, leucocytes move in a random or in an oriented fashion to encounter stationary micro-organisms. In some instances the bacteria may also exhibit motility. In the lymphatic or blood streams bacteria are transported passively with the fluid and do not commonly come into contact with phagocytic cells in suspension which also flow passively in the fluid. These micro-organisms circulating in blood or lymph are removed when they come into contact with phagocytic cells fixed to the vascular endothelium in certain filtering organs such as lymph nodes, liver and spleen.

Polymorphonuclear leucocytes exhibit impressive locomotion at rates up to 40 μm/min *in vitro* (Dittrich, 1962), whereas mononuclear phagocytes show motility (change in shape and surface activity) but move from place to place slowly or not at all. Both of these cell types apparently respond with directed movement (chemotaxis) towards various factors in their environment. These chemotactic agents include substances produced by certain bacteria, complement components and factors released by cells. (Ward, Cochrane & Müller-Eberhard, 1965; Ward & Hill, 1970; Sorkin, Borel & Stecher, 1970). As observed under the microscope chemotaxis is limited to small distances, up to 50–100 μm. Most of the recent studies on chemotaxis have utilised the Boyden millipore chamber; with appropriate controls this technique serves nicely for quantitation of chemotactic response. Strictly speaking chemotaxis should refer only to the phenomenon of directed locomotion. A collection of cells in tissues or on a millipore filter does not always indicate chemotaxis, for such accumulations can occur as a result of random locomotion and immobilisation at a specific site. Obviously chemotaxis, even if limited to a short range, can serve as an important physiologic mechanism favouring contact between host cell and parasite. Although there is no solid evidence that chemotaxis operates *in vivo*, many studies *in vitro* establish clearly that leucocytes have the ability to detect certain objects in their environment, presumably by sensing chemical gradients, and to move in a direct line towards these objects.

DETERMINANTS OF PHAGOCYTOSIS

Plate 4, Fig. 1 presents a series of prints from motion pictures of phagocytosis of bacilli by a human polymorphonuclear leucocyte, illustrating the morphologic features and speed of the ingestion process as observed with the phase contrast microscope.

Once the phagocyte and bacterium have come into contact, ingestion rapidly occurs provided the bacterium does not have a surface structure allowing it to escape this fate. Most of the studies done in the first half of the twentieth century on host–parasite relationships were in fact directed towards this particular aspect; namely the recognition of polysaccharide or protein coats on the bacteria which enabled them to resist phagocytosis and neutralisation of this bacterial resistance mechanism by opsonisation, i.e. overcoating with heat-labile factors or with specific antibody.

The more recent studies on phagocytosis have demonstrated that it is a process which has at least two distinct stages, attachment and ingestion (Rabinovitch, 1967a). The determinants for these two stages are different, one from the other, and also differ depending on the nature of the particle (Rabinovitch, 1970). Little is understood of the precise nature of the physicochemical forces responsible for the attachment between bacterium and phagocyte. In the special case of bacteria coated with immunoglobulin, the attachment may well be mediated by the protruding F_c piece of the antibody; both polymorphonuclear (Messner & Jelinek, 1970) and mononuclear (Abramson, LoBuglio, Jandl & Cotran, 1970) phagocytes have an F_c receptor site on their surface.

The ingestion process operates in most phagocytic cells on energy derived from glycolysis (Karnovsky, 1962; Cohn, 1970) although polymorphonuclear leucocytes of some animal species (Karnovsky, Simmons, Noseworthy & Glass, 1970a) and macrophages of some types (Karnovsky, Simmons, Glass, Shafer & D'Arcy Hart, 1970b) appear to depend at least to some measure on oxidative metabolism to provide energy for ingestion. At the ultrastructural level, ingestion appears to be accomplished by an evagination process in which the phagocyte sends out microprojections to encircle the attached microbe, as is shown in Plate 4, Fig. 2 (Jones & Hirsch, 1971). These micropseudopods fuse with one another to enclose the microbe in a pouch, the wall of this pouch being inverted plasma membrane. Nothing is known about the cytoplasmic agencies responsible for the micropseudopod formation or the determinants of the membrane fusion by which the bacterium is translocated inside of the cell.

Neutrophil leucocytes and macrophages are both classed as 'professional' phagocytes (Rabinovitch, 1967b), as based on the high rate at which they ingest bacteria and other particles under suitable conditions, and on the fact that they find particles coated with immunoglobulin particularly appetising (perhaps a reflection of their surface F_c receptor). Few differences between neutrophils and macrophages are recognised in terms of their phagocytic behaviour towards various types of micro-organisms, but these cells need to be compared in this regard more thoroughly, since they do exhibit different appetites – for example macrophages ingest old red cells whereas polymorphonuclear leucocytes do not.

KILLING AND DIGESTION OF BACTERIA IN THE PHAGOCYTIC VACUOLE

Following phagocytosis, certain morphologic and biochemical events play important roles in relation to killing and digestion of the microbes. As pointed out above, engulfed material lies within a cellular stomach, the phagocytic vacuole, lined by internalised plasma membrane. Most bacteria are killed in vacuoles of polymorphonuclear or mononuclear phagocytes in 5–10 min (Wilson, Wiley & Bruno, 1957; Wood, 1960). The few bacterial species that survive in phagocytic vacuoles do so because they are resistant to the bactericidal conditions which develop in this micro-environment or because they alter the physiology of the cell so as to prevent delivery of bactericidal agents to the vacuole.

Metabolic alterations occur in polymorphonuclear leucocytes following a phagocytic stimulus (Karnovsky, 1962; Karnovsky et al. 1970a). Two substances of possible importance in bactericidal action are produced in the course of these metabolic events. Accompanying phagocytosis there occurs marked stimulation of glycolysis, resulting in enhanced production of lactic acid. This organic acid presumably enters and accumulates in the phagocytic vacuole, accounting at least in part for the acidity of this locus. Studies employing indicator dyes coupled to bacteria indicate a vacuolar pH of 4·7 to 5·5 (Sprick, 1956). The low pH and presence of lactic acid are undoubtedly lethal conditions for many micro-organisms (Dubos, 1953) and they are, furthermore, conditions suitable for the action of the leucocyte azurophil granule lysosomal and bactericidal substances. Also associated with phagocytosis is activation of a cytoplasmic NADH oxidase and stimulation of the glutathione cycle and the hexose monophosphate shunt (Karnovsky et al. 1970a). One of the products of these activities is hydrogen peroxide, which may diffuse into the phagocytic vacuole and exert bactericidal action *per se* in some situations (McRipley & Sbarra, 1967), or probably more often cooperate with myeloperoxidase and halides to halogenate and kill microbes (Klebanoff, 1967, 1968).

During and shortly after phagocytosis the granules of polymorpho-nuclear leucocytes coalesce with the phagocytic vacuole and discharge their contents into it (Hirsch, 1962; Zucker-Franklin & Hirsch, 1964). As discussed in the section on structure above, these granules are of two types or classes, azurophil (or primary) and specific (or secondary). The azurophil granules are lysosomal structures containing a rich assortment of enzymes capable of degrading macromolecules at an acid pH, and containing also abundant peroxidase and several cationic proteins with bactericidal potential. The specific granules do not contain lysosomal hydrolases but they are rich in an alkaline phosphatase of unknown function and they also contain lactoferrin, an iron-binding protein with antimicrobial properties and lysozyme, an enzyme capable of degrading exposed bacterial muramic acid.

Thus phagocytosis by polymorphonuclear leucocytes leads to meta-bolic changes and the degranulation response which exposes the ingested bacteria to a veritable barrage of antimicrobial factors, including low pH, lactic acid, hydrogen peroxide, the peroxidase–peroxide mediated halogenation system, cationic proteins, lysozyme and lactoferrin. It is impossible in most situations to determine which of these agents is in the main responsible for the killing. Studies on correlation between specific cellular deficiencies and loss of ability to kill certain microbes are needed to clarify this complex situation.

Polymorphonuclear leucocytes are essentially incapable of altering their structure or their biosynthetic activities. In contrast the macro-phages respond to environmental stimuli with striking changes in appearance and in synthetic activities; in certain circumstances they may even be stimulated to divide (Virolainen & Defendi, 1967; North, 1969). These physiologic differences are correlated with the ability, or the lack of it, of these cells to develop ˝acquired immunity (see section below).

Mononuclear phagocytes vary in their metabolic patterns (Karnovsky *et al.* 1970*b*). Many of these cells subsist primarily on glycolysis as do the polymorphonuclear leucocytes, whereas others, such as the alveolar macrophages, can also use energy derived from respiration for phagocytosis and other functions. During phagocytosis the glycolytic rate increases in macrophages as it does in neutrophils. Information is incomplete on hydrogen peroxide production by macrophages but many of them generate little or no peroxide during phagocytosis (Karnovsky, et al. 1970b).

As mentioned earlier, the lysosomes of mononuclear phagocytes differ in several ways from those of polymorphonuclear phagocytes. Bone marrow promonocytes and circulating monocytes, the 'adolescent' cells of this line, have in their centrosomal region a small number of dense granules which resemble the azurophil granules of the poly-morphonuclear cells, i.e. they contain acid hydrolases and peroxidase. The mature mononuclear phagocytes of tissues, the macrophages, appear in most instances to be devoid of these peroxidase-positive granules. Macrophages stimulated in vivo or in vitro often show large numbers of dense granules containing acid hydrolases but not peroxi-dase. The genesis of these granules has been studied carefully; they are secondary lysosomes formed from endocytic vacuoles. The primary lysosomes of macrophages appear to be small vesicles associated with the Golgi apparatus. Macrophage lysosomes fuse with phagocytic vacuoles but none of these various macrophage lysosomal structures, primary or secondary, has been demonstrated to contain cationic peptides or lactoferrin and many of them have no lysozyme. Thus the number of anti-microbial factors known to occur in macrophages is limited indeed – in some cases only lactic acid – yet these cells seem to kill various bacteria as efficiently as do granulocytes. Almost certainly important bactericidal agents are yet to be discovered in macrophages.

The digestion of bacteria inside leucocytes has been followed by observing alteration in their staining properties and their morphology and it has also been studied in terms of rate of breakdown of macro-molecular material of bacteria labelled with radioisotopes (Cohn, 1963). Digestion in most cases is well along in 30–60 min. The products of bacterial breakdown probably include indigestible components as well as small molecules. The small molecules may move from the phagocytic vacuole into the cell cytoplasm and then into the tissues. Larger molecules and indigestible materials are likely retained within the cell (Cohn, 1970), to be released into the tissues upon death of the cell, or to be carted into the outside world via the intestinal or respiratory tract or the skin. Amoebae are able to eliminate indigestible material by egestion – reverse phagocytosis. Although egestion has been reported also in phagocytic cells (Wilson, 1953), it appears to occur rarely or not at all in most situations (Cohn & Benson, 1965; Cohn & Fedorko, 1969).

ACQUIRED CELLULAR IMMUNITY

Animals which have recovered from an infectious disease will exhibit for some time or forever, an increased ability to handle a second challenge with the same infectious agent. This phenomenon is one of the cornerstones of the science of immunology. In many instances, of course, the acquired immunity is due in large part or even entirely to the presence of antibodies which opsonise or kill the bacteria or detoxify their products. In other instances, the classical example being tuberculosis, the acquired immunity is unrelated to humoral factors, depending instead on enhanced antibacterial activity of the host phagocytic cells (Lurie, 1965).

Acquired immunity has never been demonstrated convincingly in polymorphonuclear leucocytes. In many instances enhanced activity of these cells in immunised animals has been suspected but on closer examination all such activity has turned out to be due to antibodies, not to a change in the cells *per se*. This result is in fact not surprising since polymorphonuclear leucocytes are a short-lived population of cells largely incapable of modulating their structure or functional properties.

Thus acquired cellular immunity is due to changes in the mononuclear cells. The past decade has seen important advances in our understanding of this mononuclear cell response, thanks to the work of Mackaness and his colleagues at the Trudeau Institute (Mackaness, 1970, 1971). In the several systems studied acquired cellular immunity is correlated well in time and in intensity with the development of delayed-type skin hypersensitivity. The early, inductive phase of these responses involves a population of lymphoid cells; the result of their activities, and the mediator mechanism of the immunity, is the appearance of large numbers of 'activated' mononuclear phagocytes with enhanced anti-bacterial activity. The acquired cellular immunity can be transferred to an inexperienced animal by injection of lymphoid cells from a donor animal at the appropriate stage of immune stimulation. The cells capable of this adoptive transfer are short-lived, non-recirculating lymphocytes that move into areas of inflammation (McGregor, Koster & Mackaness, 1971; Koster, McGregor & Mackaness, 1971).

The specificity of acquired cellular immunity has two different aspects and these differences have led to confusion in the minds of some. After primary immunisation, or after transfer of immunity by lymphoid cells, the recall of cellular immunity in an animal long after its initial exposure is highly specific in an immunological sense, since the only organism or antigen which will cause an *accelerated* immune response is the organism or antigen used for stimulating the donor of the lymphoid cells or the animal exposed some time earlier. The response in the case of exposure to this organism or antigen is the acquisition of delayed-type hypersensitivity and of activated macrophages much more rapidly than would be the case in an inexperienced animal. This more rapid development of the immune state can be a crucial factor in limiting the growth of bacteria at an early phase of the infection, resulting in prompt recovery of the animal.

The second aspect of acquired cellular immunity, the expression of the immunity by 'activated' macrophages, appears to be totally lacking in specificity. In other words, at the height of the immune response, whether primary or recall, the macrophages in the lesions are more capable than normal cells of handling not only the organism responsible for the immune response but unrelated organisms as well. No information is yet available on the precise nature of the changes in the activated macrophages which endow them with heightened ability to take in bacteria and suppress their multiplication.

CONCLUDING REMARKS

The biology of phagocytic cells was first studied extensively by Elie Metchnikoff in the late 1800s. He laid a broad and firm foundation of knowledge on these cells. After half a century, there has been in the past 10–15 years a re-awakening of interest in phagocytes. Many of the structures and functions of these cells studied by Metchnikoff have been re-investigated using the tools and approach of modern cell biology. This effort, carried out in many laboratories throughout the world, has shed much new light on many aspects of phagocytic cells – their life history, ultrastructure, metabolism, digestive tract and lysosomal granules and their anti-bacterial systems. It seems safe to predict that the years ahead will continue to be productive ones in terms of new knowledge on these cells, for many basic aspects of their biology remain mysteries. Our knowledge is still superficial, for example, in relation to the sub-cellular mechanisms responsible for locomotion and chemotaxis, the precise physicochemical determinants of phagocytosis and pinocytosis, the function of specific granules of polymorphonuclear leucocytes and the mechanisms of bactericidal action within macrophages.

REFERENCES

ABRAMSON, N., LOBUGLIO, A. F., JANDL, J. H. & COTRAN, R. S. (1970). The interaction between human monocytes and red cells. *Journal of Experimental Medicine*, **132**, 1191.

BAGGIOLINI, M., HIRSCH, J. G. & DE DUVE, C. (1969). Resolution of granules from rabbit heterophil leukocytes into distinct populations by zonal sedimentation. *Journal of Cell Biology*, **40**, 529.

BAGGIOLINI, N. DE DUVE, C., MASSON, P. & HEREMANS, J. (1970). Association of lactoferrin with specific granules in rabbit heterophil leukocytes. *Journal of Experimental Medicine*, **131**, 559.

BAGGIOLINI, M., HIRSCH, J. G. & DE DUVE, C. (1970). Further biochemical and morphological studies of granule fractions from rabbit heterophil leukocytes. *Journal of Cell Biology*, **45**, 586.

BAINTON, D. F. & FARQUHAR, M. G. (1966). Origin of granules in polymorpho-nuclear leukocytes. *Journal of Cell Biology*, **28**, 277.

BAINTON, D. F. & FARQUHAR, M. G. (1968). Differences in enzyme content of azurophil and specific granules of polymorphonuclear leukocytes. *Journal of Cell Biology*, **39**, 286.

BALNER, H. (1963). Identification of peritoneal macrophages in mouse radiation chimeras. *Transplantation*, **1**, 217.

COHN, Z. A. (1963). The fate of bacteria within phagocytic cells. I. The degradation of isotopically labeled bacteria by polymorphonuclear leucocytes and macro-phages. *Journal of Experimental Medicine*, **117**, 27.

COHN, Z. A. (1968). The structure and function of monocytes and macrophages. *Advances in Immunology*, **9**, 163.

COHN, Z. A. (1970). Endocytosis and intracellular digestion. In *Mononuclear Phagocytes*, ed. R. van Furth, 121. Oxford: Blackwell.

COHN, Z. A. & BENSON, B. (1965). The *in vitro* differentiation of mononuclear phagocytes. III. The reversibility of granule and hydrolytic enzyme formation and the turnover of granule constituents. *Journal of Experimental Medicine*, **122**, 455.

COHN, Z. A., FEDORKO, M. E. & HIRSCH, J. G. (1966). The *in vitro* differentiation of mononuclear phagocytes. V. The formation of macrophage lysosomes. *Journal of Experimental Medicine*, **123**, 757.

COHN, Z. A. & FEDORKO, M. E. (1969). The formation and fate of lysosomes. In *Lysosomes in Biology and Pathology*, ed. J. T. Dingle and H. B. Fell, 43. Amsterdam and London: North-Holland Co.

COOMBS, R. R. A. & SMITH, H. (1968). The allergic response to immunity. In *Clinical Aspects of Immunology*, ed. R. R. A. Coombs and P. G. H. Gell, 423. Oxford: Blackwell Scientific Publications Ltd.

CRONKITE, E. P. & VINCENT, P. C. (1970). Granulocytopoiesis. In *Hemopoietic Cellular Proliferation*, ed. F. Stohlman Jr, 211. New York and London: Grune & Stratton.

DE DUVE, C. & WATTIAUX, R. (1966). Functions of lysosomes. *Annual Review o, Physiology*, **28**, 435.

DUBOS, R. J. (1953). Effect of ketone bodies and other metabolites on the survival and multiplication of staphylococci and tubercle bacilli. *Journal of Experimental Medicine*, **98**, 145.

DITTRICH, H. (1962). Physiology of neutrophils. In *The Physiology and Pathology of Leukocytes*, ed. F. Braunsteiner and D. Zucker-Franklin, 130. New York: Grune & Stratton.

GENGOU, O. (1901). De l'origine de l'alexine des sérums normaux. *Annales de L'Institut Pasteur*, **15**, 232.

HIRSCH, J. G. (1960). Comparative bactericidal activities of blood serum and plasma serum. *Journal of Experimental Medicine*, **112**, 15.

HIRSCH, J. G. (1962). Cinemicrophotographic observations on granule lysis in polymorphonuclear leucocytes during phagocytosis. *Journal of Experimental Medicine*, **116**, 827.

HIRSCH, J. G., FEDORKO, M. E. & COHN, Z. A. (1968). Vesicle fusion and formation at the surface of pinocytic vacuoles in macrophages. *Journal of Cell Biology*, **38**, 629.

HOWARD, J. G. (1970). The origin and immunological significance of Kupffer cells. In *Mononuclear Phagocytes*, ed. R. van Furth, 178. Oxford: Blackwell.

JAGO, R. & JACOX, R. F. (1961). Cellular source and character of a heat-stable bactericidal property associated with rabbit and rat platelets. *Journal of Experimental Medicine*, **113**, 701.

JONES, T. C. & HIRSCH, J. G. (1971). The interaction *in vitro* of *Mycoplasma pulmonis* with mouse peritoneal macrophages and L-cells. *Journal of Experimental Medicine*, **133**, 231.

KARNOVSKY, M. L. (1962). Metabolic basis of phagocytic activity. *Physiological Reviews*, **42**, 143.

KARNOVSKY, M. L., SIMMONS, S., NOSEWORTHY, J. & GLASS, E. A. (1970*a*). Metabolic patterns that control the functions of leukocytes. In *Formation and Destruction of Blood Cells*, ed. T. J. Greenwalt and G. A. Jamieson, 207. Philadelphia and Toronto: Lippincott Co.

KARNOVSKY, M. L., SIMMONS, S., GLASS, E. A., SHAFER, A. W. & D'ARCY HART, P. (1970*b*). In *Mononuclear Phagocytes*, ed. R. van Furth, 103. Oxford: Blackwell.

KLEBANOFF, S. J. (1967). Iodination of bacteria; a bactericidal mechanism. *Journal of Experimental Medicine*, **126**, 1063.

KLEBANOFF, S. J. (1968). Myeloperoxidase–halide–hydrogen peroxide antibacterial system. *Journal of Bacteriology*, **95**, 2131.

KOSTER, F. T., McGREGGOR, D. D. & MACKANESS, G. B. (1971). The mediator of cellular immunity. II. Migration of immunologically committed lymphocytes into inflammatory exudates. *Journal of Experimental Medicine*, **133**, 400.

LURIE, M. B. (1965). *Resistance to Tuberculosis*. Cambridge, Mass.: Harvard University Press.

MACKANESS, G. B. (1970). Cellular immunity. In *Mononuclear Phagocytes*, ed. R. van Furth, 461. Oxford: Blackwell.

MACKANESS, G. B. (1971). Resistance to intracellular infection. *Journal of Infectious Diseases*, **123**, 439.

MARCHESI, V. T. & FLOREY, H. W. (1960). Electron micrographic observations on the emigration of leucocytes. *Quarterly Journal of Experimental Physiology*, **45**, 343.

MASSON, P. (1970). *La Lactoferrine, protéine des sécrétions externes et des leucocytes neutrophiles*. Thesis, University Catholique de Louvain, Bruxelles: Éditions Arscia S.A.

McGREGOR, D. D., KOSTER, F. T. & MACKANESS, G. B. (1971). The mediator of cellular immunity. I. The life span and circulations dynamics of the immunologically committed lymphocyte. *Journal of Experimental Medicine*, **133**, 389.

McRIPLEY, R. J. & SBARRA, A. J. (1967). Role of the phagocyte in host-parasite interactions. XI. Relation between stimulated oxidative metabolism and hydrogen peroxide formation and intracellular killing. *Journal of Bacteriology*, **94**, 1417.

MESSNER, R. P. & JELINEK, J. (1970). Receptors for human gamma G globulin on human neutrophils. *Journal of Clinical Investigation*, **49**, 2165.

MÜLLER-EBERHARD, H. J. (1965). The role of antibody, complement and other fractions in host resistance to infections. In *Bacterial and Mycotic Diseases of Man*, ed. R. J. Dubos and J. G. Hirsch, 181. Philadelphia: Lippincott.

NORTH, R. J. (1969). Cellular kinetics associated with the development of acquired cellular resistance. *Journal of Experimental Medicine*, **130**, 299.

PERRY, S. (1970). The formation and destruction of leukocytes. In *Formation and Destruction of Blood Cells*, ed. T. J. Greenwalt and G. A. Jamieson, 194. Philadelphia and Toronto: Lippincott Co.

PINKETT, M. O., COWDREY, C. M. & NOWELL, P. C. (1966). Mixed hemopoietic and pulmonary origin of 'alveolar macrophages' as demonstrated by chromosome markers. *American Journal of Pathology*, **48**, 859.

RABINOVITCH, M. (1967a). The dissociation of the attachment and the ingestion phases of phagocytosis by macrophages. *Experimental Cell Research*, **46**, 19.

RABINOVITCH, M. (1967b). 'Non-professional' and 'Professional' phagocytosis. Particle uptake by L-cells and macrophages. *Journal of Cell Biology*, **35**, 108A.

RABINOVITCH, M. (1970). Phagocytic recognition. In *Mononuclear Phagocytes*, ed. R. van Furth, 299. Oxford: Blackwell.

ROSER, B. (1970). The migration of macrophages *in vivo*. In *Mononuclear Phagocytes*, ed. R. van Furth, 166. Oxford: Blackwell.

SORKIN, E., BOREL, J. F. & STECHER, V. J. (1970). Chemotaxis of mononuclear and polymorphonuclear phagocytes. In *Mononuclear Phagocytes*, ed. R. van Furth, 397. Oxford: Blackwell.

SPRICK, M. G. (1956). Phagocytosis of *M. tuberculosis* and *M. smegmatis* stained with indicator dyes. *American Review of Tuberculosis and Pulmonary Diseases*, **74**, 552.

VAN FURTH, R. & COHN, Z. A. (1968). The origin and kinetics of mononuclear phagocytes. *Journal of Experimantal Medicine*, **128**, 415.

VAN FURTH, R. & DIESSELHOFF-DEN DULK, M. M. C. (1970). The kinetics of promonocytes and monocytes in the bone marrow. *Journal of Experimental Medicine*, **132**, 813.

VAN FURTH, R., HIRSCH, J. G. & FEDORKO, M. E. (1970). Morphology and peroxidase cytochemistry of mouse promonocytes, monocytes and macrophages. *Journal of Experimental Medicine*, **132**, 794.

VIROLAINEN, M. (1968). Hematopoietic origin of macrophages as studied by chromosome markers in mice. *Journal of Experimental Medicine*, **127**, 943.

VIROLAINEN, M. & DEFENDI, V. (1967). Dependence of macrophage growth *in vitro* upon interaction with other cell types. In *The Wistar Institute Symposium Monograph*, no. 7, ed. V. Defendi and M. Stoker, 67. Philadelphia: The Wistar Press.

VOLKMAN, A. (1966). The origin and turnover of mononuclear cells in peritoneal exudates in rats. *Journal of Experimental Medicine*, **124**, 241.

VOLKMAN, A. & GOWANS, J. L. (1965). The origin of macrophages from bone marrow in the rat. *British Journal of Experimental Pathology*, **46**, 62.

WARD, P. A., COCHRANE, C. G. & MÜLLER-EBERHARD, H. J. (1965). The role of serum complement in chemotaxis of leukocytes *in vitro*. *Journal of Experimental Medicine*, **122**, 327.

WARD, P. A. & HILL, J. H. (1970). C 5 chemotactic fragments produced by an enzyme in lysosomal granules of neutrophils. *Journal of Immunology*, **104**, 535.

WETZEL, B. K., SPICER, S. S. & HORN, R. G. (1967). Fine structural localization of acid and alkaline phosphatases in cells of rabbit blood and bone marrow. *Journal of Histochemistry and Cytochemistry*, **15**, 311.

WILSON, A. T. (1953). The egestion of phagocytized particles by leukocytes. *Journal of Experimental Medicine*, **98**, 305.

WILSON, A. T., WILEY, G. G. & BRUNO, P. (1957). Fate of non-virulent group A streptococci phagocytozed by human and mouse neutrophils. *Journal of Experimental Medicine*, **106**, 777.

WOOD, W. B., JR (1960). Phagocytosis, with particular reference to encapsulated bacteria. *Bacteriological Reviews*, **24**, 41.

ZEYA, H. I. & SPITZNAGEL, J. K. (1971). Characterization of cationic protein-bearing granules of polymorphonuclear leukocytes. *Laboratory Investigation*, **24**, 229.

ZUCKER-FRANKLIN, D. & HIRSCH, J. G. (1964). Electron microscope studies on the degranulation of rabbit peritoneal leukocytes during phagocytosis. *Journal of Experimental Medicine*, **120**, 569.

EXPLANATION OF PLATES

PLATE 1

Fig. 1. Electron micrograph of a normal human neutrophil leucocyte. N, nuclear lobes; C, centrioles; G, cytoplasmic granules: × 15 000.

Fig. 2. Electron micrograph of a rabbit polymorphonuclear leucocyte, illustrating the two types of granules. A, large, electron dense azurophil or primary granules; B, smaller, less dense specific or secondary granules: × 15 000.

PLATE 2

Electron micrograph of a normal human blood monocyte, showing a horseshoe-shaped nucleus (N), centrosomal region (C) with Golgi saccules and vesicles, granules (G), and mitochondria (M): × 20 000.

PLATE 3

Electron micrograph of a rabbit alveolar macrophage. N, nucleus; C, centrosomal Golgi region; G, granules; M, mitochondria; R, rough endoplasmic reticulum: × 18 400.

PLATE 4

Fig. 1. Phagocytosis sequence from a motion picture of human polymorphonuclear leucocytes engulfing *Bacillus megaterium*. Phase contrast: approximately × 1200.

Fig. 2. Electron micrographs showing phagocytosis of *Mycoplasma pulmonis* by a mouse peritoneal macrophage 5 min after addition of antimycoplasma antibody. Note the micro-pseudopods which surround the mycoplasmas on the cell surface: × 21 000.

PLATE 1

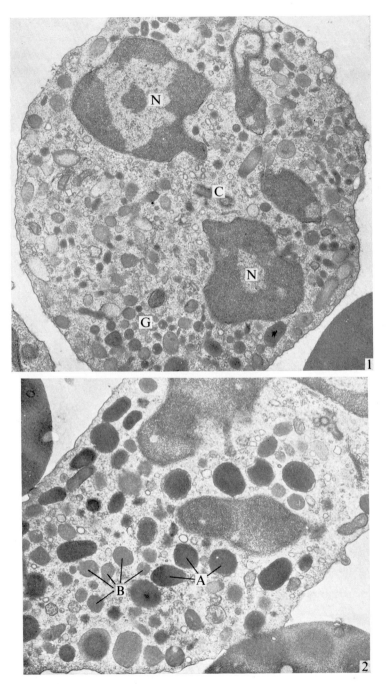

PLATE 2

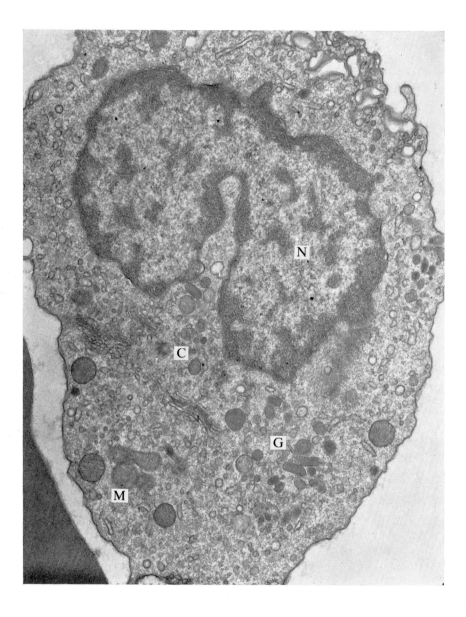

PLATE 3

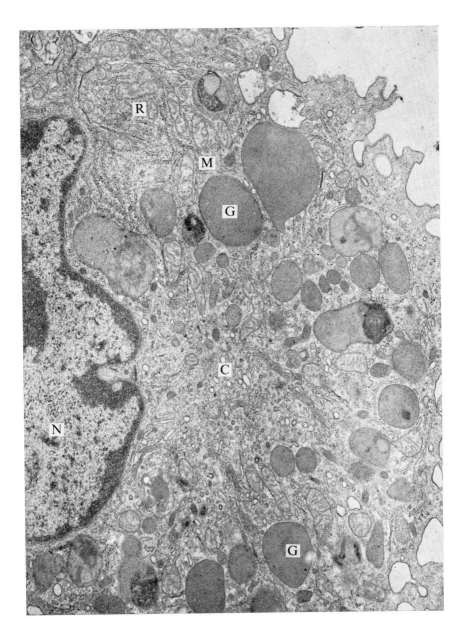

PLATE 4

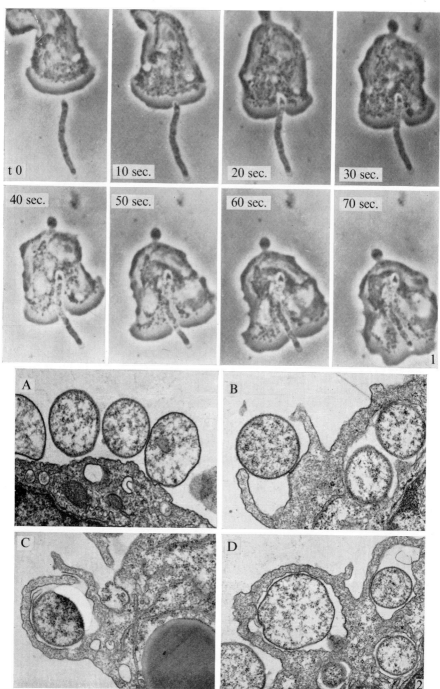

t 0 10 sec. 20 sec. 30 sec.

40 sec. 50 sec. 60 sec. 70 sec.

1

A B

C D

2

BACTERIAL FACTORS INHIBITING HOST DEFENCE MECHANISMS

A. A. GLYNN

Bacteriology Department, St Mary's Hospital
Medical School, London

INTRODUCTION

At the beginning of this century Bail and his colleagues introduced the term aggressin to indicate a new class of bacterial products. Their essential feature was that though themselves non-toxic they were able, when given together with appropriate bacteria, to exacerbate the resulting infection by enabling the bacteria to grow rapidly in the host tissue. They did not stimulate bacterial growth *in vitro* and in any case such a nutritive function would not have justified a new name. Bail thought that the aggressins impeded host defence mechanisms and were not usually produced to any extent *in vitro*.

A variety of properties insisted on by Bail and summarised by Wilson & Miles (1964) included such things as sensitivity to heat and the ability to induce immunity. These are clearly not essential to the basic concept but were derived from particular examples under study. Perhaps the arbitrary nature of such additions increased the polemical nature of the discussions which followed. Since then the term aggressin has not always been widely used and has often been loosely applied to almost any sort of virulence factor even including toxins. In a wider field 'aggression' is usually described, by those who commit it, as self-defence, and the activity of bacterial aggressins would be a better example of this than many. The anthropomorphic use of military analogies has long been a feature of medical microbiology, and provided that it does not interfere with a proper analysis, it is a valuable reminder that the ultimate object is the control of infection in man.

It would be convenient to have a collective noun for all those compounds which are not blatantly toxic, but which can inhibit a variety of host defences. However, taken literally, 'aggressin' gives the impression of something which actively damages the host. A more appropriate term for what are often called aggressins would be 'impedin', first used by Torikata in 1917 to describe a bacterial factor which hindered the combination of an antigen and an antibody (Scott, 1931). Torikata's factor was thermolabile, but again this is not essential for use of the term. Now may be the moment to resurrect the term 'impedin' in a more general sense. Thus 'impedin' could be used for a bacterial factor capable of inhibiting any defence mechanism of multicellular organisms against infection. A pragmatic approach based on function is preferable to a formal definition which would eventually need to be modified. Unlike such chemically homogeneous groups as the bacteriocins, 'impedins' include proteins, polysaccharides, lipids and other compounds. Their chemical heterogeneity is in fact a reason for adopting a collective term. The antibiotics form a good precedent.

Many defence mechanisms, and still more inhibitors, have been described, especially those capsular and cell-wall components which interfere with ingestion and digestion of bacteria by phagocytes (Wilson & Miles, 1964; Dubos & Hirsch, 1965; Smith, 1968). Only a few interactions between 'impedins' and defence mechanisms can be dealt with here. The first sections are devoted to 'impedins' from *Staphylococcus aureus*, *Neisseria gonorrhoeae*, *Mycobacteria* and *Escherichia coli*. Next is a survey of the role of iron in a variety of infections. The last section deals briefly with factors affecting the host's ability to mount an immune response. Throughout, an attempt has been made not only to describe the detailed mechanisms involved but also to assess their significance in the infectious process. Only too often the evidence required for such a judgement does not exist.

THE ANTI-INFLAMMATORY FACTOR IN *STAPHYLOCOCCUS AUREUS*

It is likely that the development of the common staphylococcal skin lesions involves different virulence factors to those of wound infections or systematic staphylococcal lesions. Non-specific factors such as fatigue, psychological stress, nutritional and hormone disturbances increase the incidence of human boils. Their mode of action, however, is unknown and will remain so until the ways in which staphylococci establish themselves in, and cause damage to the tissues and the ways in which they react with host defences are better understood. While in this respect particular immunological states of the host such as delayed hypersensitivity appear to be important (Johnson, Cluff & Goshi, 1961; Johanovsky, 1958; Panton & Valentine, 1929), the effects of variations in the staphylococcus should not be ignored.

It is difficult in animals to produce a local skin lesion corresponding to a boil in man. Elek & Conen (1957), taking post-operative stitch abscesses as their model, demonstrated that the number of staphylococci required to produce a local lesion in man could be reduced ten thousand times if the organisms were implanted on silk thread. This method also worked in mice (James & McLeod, 1961). More conveniently, Noble (1965) injected staphylococci mixed with cotton dust. Agarwal (1967a, b) used Noble's method to study the relationship between staphylococcal virulence and the development of the local inflammatory response. Virulence was judged by the severity of the lesion produced, inflammation by the amount of fluid and leucocytes entering the tissues. After injection of three virulent strains of *Staph. aureus* there was a delay in the onset of local inflammation, rapid early bacterial growth and eventually a severe lesion. Injection of five other strains of *Staph. aureus* and two of *Staphylococcus albus* produced milder lesions; the organisms grew more slowly and there was early onset of fluid exudate and leucocyte emigration. Another organism, *Micrococcus hyicus*, grew slowly and produced only a mild lesion but was associated with a delayed onset of both the fluid and cellular aspects of inflammation.

The protective effect of inflammation in cutaneous infections of guinea-pigs had been shown by Miles and his colleagues (Miles, Miles & Burke, 1957; Burke & Miles, 1958). For a given dose larger lesions were produced if the early stages of inflammation had been suppressed by adrenalin. Agarwal (1967b) suggested that some factor in virulent staphylococci suppressed the initial inflammatory response so enabling the bacteria to multiply more rapidly and produce a severe lesion. He realised that this factor by itself was not enough to determine virulence; for lesion production the organism must also be able to damage the tissues. The factor involved was not α-haemolysin but was associated with the bacterial cells. Heat-killed virulent staphylococci (strain PS 80) injected with cotton dust suppressed the small inflammation usually produced by the cotton alone.

Hill (1968) characterised the factor further using as criteria of activity the ability of extracts to enhance the formation of the lesions by sub-threshold doses of *Staph. aureus* (strain PS 80) and the ability to suppress the early inflammatory oedema which follows infection with non-virulent staphylococci on cotton dust. The factor could be isolated in quantity only from virulent strains in the logarithmic growth phase. Old cultures or non-virulent strains yielded little or no activity. The 'impedin' activity was found in the residue (DOCR) left when cell walls from mechanically disintegrated staphylococci were extracted with desoxycholate. The desoxycholate extract had no activity. DOCR was almost entirely mucopeptide and protein. Both its lesion-enhancing and oedema-inhibiting properties remained after treatment with periodate, formaldehyde, trypsin, or heat at 100° for 10 min. The active factor was made soluble by mild treatment with lysozyme, but prolonged treatment destroyed it. Thus the activity depends more on the mucopeptide than the protein. Hill's factor appears similar to that described by Fisher (1963), despite some differences in the degree of sensitivity to heat and trypsin.

DOCR from virulent staphylococci inhibited chemotaxis of polymorphonuclear leucocytes measured in a Boyden chamber (Weksler & Hill, 1969). The chemotaxis-inhibiting activity differed somewhat from the lesion-enhancing and oedema-inhibiting activities in that it was less easily destroyed by lysozyme and was found in older cultures.

Further evidence of the significance of the DOCR factor in staphylococcal infections of mice is given by Hill (1969). Mice immunised with DOCR were protected against subcutaneous challenge by heterologous as well as homologous strains of staphylococci. In immune mice the lesions were smaller, early inflammatory oedema was not suppressed but was sometimes increased, and the bacteria grew slowly. The significance of DOCR factor in human infections is not yet clear. Sera from patients with chronic staphylococcal infections gave no passive protection to mice. Antibodies to the DOCR may have been absent but rabbit immune serum to DOCR also had only limited protective ability.

PROTEIN A OF *STAPHYLOCOCCUS AUREUS*

Jensen (1958) described an antigen in a crude extract of *Staph. aureus* which appeared to react serologically with all human sera tested. Löfkvist & Sjöquist (1962) showed it to be protein and recommended that it be known as 'protein A' to avoid confusion with polysaccharide A. A preliminary chemical analysis was made by Grov, Myklestad & Oeding (1964); its haemagglutinating ability was destroyed by carboxypeptidase, suggesting that it depended on the presence of an open peptide chain with a free C terminal (Grov, 1965). Di-nitrofluorobenzene and phenylisothiocyanate analysis showed an N-terminal alanine possibly related to the precipitinogen site.

Interest in 'protein A' increased when Forsgren & Sjöquist (1966) found that its reaction with γ-globulin was not that of antigen with antibody. 'Protein A' reacted with 45 % of the γ-globulin present in a pool of human sera, far more than could be accounted for by any one antibody. It also reacted with γG from two myeloma sera which were known to be antibodies to streptolysin O (Kronvall, 1967). The reaction was with H chains and more particularly the Fc portion of γG, but not with L chains or Fab or $F(ab)_2$ portions of the molecule, i.e. not with the antigen binding sites. This held for human (Forsgren & Sjöquist, 1966) rabbit (Forsgren & Sjöquist, 1967) and guinea-pig (Forsgren, 1968a, b) γ-globulins.

Only γG, not γA, γM, γD or γE, globulins react with 'protein A' and even with γG there are differences between species and subclasses. Rabbit γG reacts relatively feebly. Both γ-1 and γ-2 globulins of guinea-pigs react. In the mouse γ-2a, γ-2b and γ-3 but not γ-1 react (Forsgren, 1968a). Of human γG subgroups γG-1, 2 and 4 react but not γG-3. Kronvall & Williams (1969) found in subclasses 1, 2 and 4 of 68 γG myeloma globulins, about a third which gave a good precipition line with 'protein A'; the rest did not precipitate but competitively inhibited precipitation. The amino-acid sequences in the Fc fragment which are needed for the 'protein A' reaction are not yet known but from the subclass distribution described they clearly differ from the requirements for other functions of Fc. Thus complement fixation can occur with γ-3 an dγM but not γ-4 (Ishizaka, Ishizaka, Salmon & Fudenberg, 1967). Reversed passive cutaneous anaphylaxis is not mediated by γ-2 (Terry, 1964). γ-1 and γ-3 but not γ-2 and γ-4 may be cytophilic antibodies (Huber & Fudenberg, 1968). γ-3 but not γ-2 crossed the placenta (Brambell, Hemmings, Oakley & Porter, 1960).

In spite of different affinities of the γG subclasses the addition of 'protein A' to serum may significantly affect the biological function of Fc. Perhaps the most striking is its antiphagocytic effect (Dossett, Kronvall & Williams, 1969). In a system of leucocytes, opsonins and either staphylococci or E. coli 'protein A' decreased the amount of phagocytosis even if added after the organism had been sensitised. The 'protein A' had no effect on the leucocytes and, presumably, acted by attaching itself to the Fc part of the opsonins preventing direct opsonic action and possibly affecting the opsonic role of complement. Its effect on the latter is not altogether clear since, like other aggregating agents such as heat or antigens, 'protein A' will, when combined with γG, enable the latter to fix complement (Sjöquist & Ståhlenheim, 1969). While two strains of Staph. aureus rich in 'protein A' were poorly phagocytosed, Dossett et al. (1969) could not find any clear correlation in eight further strains between 'protein A' content and susceptibility to phagocytosis. A strain of Cowan type I was more easily phagocytosed when grown on mannitol–salt agar, which largely suppresses 'protein A' formation, than on normal media.

Presumably as a result of its ability to enable γG to fix complement, 'protein A' may also give rise to immediate hypersensitivity reactions.

Thus subcutaneous injection of 'protein A' may produce an Arthus reaction in non-immune guinea-pigs (Gustafson, Ståhlenheim, Forsgren & Sjöquist, 1968) but not in rabbits and rats probably because 'protein A' reacts weakly with the γG of the species. However, if human γG is aggregated by 'protein A' the mixture produces an Arthus-like reaction on intradermal injection of rabbits. A similar effect can be produced by giving the 'protein A' intradermally and human γG intravenously (Gustafson, Sjöquist & Ståhlenheim, 1967).

Fascinating though these reactions of protein A with γG are, the role of 'protein A' in the pathogenesis of staphylococcal infection is not yet clear. The protein is found in almost all coagulase-positive staphylococci, i.e. in *Staph. aureus*, but not in *Staph. albus* (Kronvall, Dossett, Quie & Williams, 1971). Whether it helps to determine virulence, or perhaps, like mannitol fermentation, is only associated with it, is undecided. However, all work based on the widespread staphylococcal 'antibodies', much of which was really dealing with 'protein A', will now have to be re-assessed.

NEISSERIA GONORRHOEAE

Experimental infections and clonal types

The gonococcus is nutritionally exacting and relatively fragile under laboratory conditions. Yet it readily infects man and may persist as a chronic disease, particularly in women. This infectivity does not apply to animals and it has proved impossible to set up a satisfactory model infection until recently. Lucas, Chandler, Martin & Schmale (1971) have now successfully induced a gonococcal urethritis in chimpanzees using as inoculum fresh exudates from human patients. But there is still no animal infection induced by cultured organisms. Nor can chimpanzees be regarded as convenient laboratory animals. The ready multiplication of gonococci in the anterior chamber of the rabbit's eye or in embryonated eggs appears to be more in the nature of growth in rich media isolated from host defences than infection of living reactive tissue. The claim by Walsh, Brown, Brown & Pirkle (1963) that the allantoic cavity of 8-day-old chick embryos suppresses non-virulent and selects virulent gonococci could not be confirmed by Ward (1970), who found that recently isolated and old laboratory strains grew equally well in such eggs.

In man, successful experimental infection was achieved by Mahoney, van Slyke, Cutler & Blum (1946) using large doses of recently isolated strains. Infection could more readily be produced by fresh urethral exudates. Kellogg *et al.* (1963) described four morphologically distinct clonal types in cultures of gonococci. Type 1 made up 90 % and type 2 10 % of the colonies seen on primary isolation but disappeared rapidly on unselected subcultures to be replaced by types 3 and 4. By careful selection on suitable media individual clones could be maintained through many transfers over a long period. The interest in these types arises from the claim that type 1, and to a lesser extent type 2, are related to virulence for humans (Kellogg *et al.* 1963, 1968). Starting with a primary isolate of one particular strain (F 62), human infection could still be produced by type 1 organisms even after 770 selected transfers *in vitro* over 35 months. Types 3 and 4 infected 4 of 4 volunteers after 38 passages *in vitro*, but none of 4 after 69 passages. Type 2 was still infective after 440 passages. Type 1 strains were agglutinated by saline and fermented glucose more slowly than types 2, 3 and 4 strains; no other significant differences have been described and the properties giving rise to the apparent virulence of type 1 are unknown.

It is unfortunate that only organisms derived from a single isolate F 62, have so far been used in experimental human infections. Given that virulence and colonial type 1 are associated in strain F 62, it would be useful to test this association more widely. Also the minimal infective dose has not been determined. The actual doses used are said to be 'several billion' presumably over 10^9. The number of organisms concerned in a natural infection is likely to be much less than this, since infection may be transferred by women from whom it is difficult to isolate gonococci. Some 50 000 organisms per ml are required before tubercle bacilli can be seen on a stained smear (Cruickshank, 1965), and the number of gonococci required could be of the same order. Cultural methods are more sensitive. However, caution in judging the natural infecting dose from these criteria is necessary since the increased mucous secretion occurring during sexual intercourse may increase the number of bacteria present at the critical time. Nevertheless in the experimental infection, an infecting dose of over 10^9 organisms could mean that virulent bacteria form such a small percentage of the population that those characteristics which enable them to set up an infection may be hidden from experimental analysis.

The sensitivity of gonococci to complement

Glynn & Ward (1970) examined 60 strains of gonococci for their sensitivity to human complement in the presence of natural human or immune rabbit antibodies. The 60 strains, all but two of which had been recently isolated, could be divided into four main groups ranging from those killed by normal human serum alone to those resistant to complement plus homologous antibody. Bactericidal antibody could be absorbed by lipopolysaccharide extracted from strains of gonococci. The distribution of specific lipopolysaccharides among gonococci was wider than indicated by the bactericidal reaction since antibodies would react in haemagglutination tests with lipopolysaccharides from strains of gonococci resistant to killing. It is possible, though unproven, that the intact bacteria possessed surface blocking factors analogous to the K antigens of *E. coli*. However, the presence of blocking antigens would not explain all the reactions found. Thus strain G36 was killed by complement in the presence of antibodies to strains G37 or G50 but not G4. Yet red cells coated with G36 lipopolysaccharide were agglutinated by all three antisera. Why only anti-G4 should be blocked on intact bacteria G36 is not clear. It could be that anti-G4 reacted with a minor antigenic determinant present only in small amounts and so more readily covered.

Most people have serum antibodies capable of mediating complement killing of cultured gonococci (Spink & Keefer, 1937; Gordon & Johnstone, 1940; Glynn & Ward, 1970). This, taken with the need for a larger inoculum to produce an experimental human infection with cultured gonococci compared with those in urethral exudates suggested that the organisms *in vivo* possess a virulence factor lost on culture. The sensitivity to serum of gonococci in fresh exudate was compared with that of the same strains after subculture (Ward, Watt & Glynn, 1970). The exudate gonococci were resistant to complement, even when pooled rabbit antisera to a variety of strains was added to the system. The majority of strains became sensitive to complement on culture and sensitivity developed after few generations. Exudates were not themselves anti-complementary since a serum-sensitive strain G11, distinguished by its resistance to streptomycin, remained serum sensitive when mixed with exudate and tested in parallel with the indigenous strains. Moreover, gonococci grown on a medium containing prostate extract retained this resistance to complement and recently isolated strains, which had become sensitive, reverted to resistance when grown on prostate medium, whereas old sensitive strains remained sensitive (Watt, Ward & Glynn, to be published). Nor could sensitive strains of *E. coli* be made resistant in this way. The nature of the prostate factor involved is being investigated. The results bear out the dictum of Smith (1968) that the behaviour of pathogens *in vitro* does not necessarily reflect their capabilities *in vivo*. It is tempting to believe that the resistance factor is related to virulence, but direct evidence is lacking. Cultured strains of clonal type 1 were all sensitive to complement. A strain of F62 clonal type 1 obtained from Dr D. S. Kellogg was highly sensitive to killing by antibody and complement and did not become more resistant when grown on prostate medium.

MYCOBACTERIA

The waxy nature of the cell wall of mycobacteria has long been thought to protect them against noxious agents *in vitro*. *In vivo* the system may be even more elaborate. Electron microscopic examination of *Mycobacterium lepraemurium* situated inside phagocytic vacuoles has shown the organisms to be surrounded by an electron transparent space (Yamamoto, Nishiura, Harada & Imaeda, 1958; Chapman, Hanks & Wallace, 1959). The space was not present at first but developed in the third month or so of infection (Allen, Brieger & Rees, 1965). Draper & Rees (1970) have shown by means of platinum shadowing that the space contains tape-like fibres or sheets surrounding and loosely attached to the bacteria. The fibres closely resembled those found round *Mycobacterium tuberculosis* by Imaeda *et al.* (1969) consisting of wax Ds. Since wax Ds is composed of lipopolysaccharide and mycolic acid the bacterial origin of the electron transparent space is clear. Similar spaces have been described round *Mycobacterium leprae* in human material (Imaeda, 1965).

Although such an inert barrier may be protective, it is still unproven, largely because the exact nature of the anti-bacterial mechanism is unknown. Some non-ionic surface active agents may stimulate or suppress tuberculous infection in mice or the growth of *M. tuberculosis* in macrophages. Those agents which increased the growth of tubercle bacilli also produced a greater increase in lysosomal enzyme activity than those which were antituberculous. Similarly, in rat fibroblasts, a medium containing 50 % human cord serum favoured both the growth of tubercle bacilli and the production of lysosomes while a medium containing 10 % calf serum favoured neither; the fibroblasts grew in either medium. Brown, Draper & D'Arcy Hart (1969) suggest that the electron-transparent layer around the mycobacteria impedes the access of large molecular weight nutrients. Lysosomal enzymes by degrading such substances may provide a source of nutrient material small enough to diffuse through. Whether this is so or not, the lysosomal enzymes do not appear to have any direct deleterious action on the mycobacteria. It is interesting to note in this connection that opsonised bacteria are broken down more slowly than unopsonised, presumably by lysosomal enzymes, (Cohn, 1963) yet opsonised organisms are killed more quickly (Jenkin, 1963). D'Arcy Hart, Gordon & Jacques (1969) prepared rat liver lysosomes from animals which had been treated previously with either Triton WR 1339 or Dextran. Triton WR 1339 inhibits and Dextran stimulates the growth of intracellular tubercle bacilli. Triton lysosome juice inhibited the growth of mycobacteria *in vitro*. Dextran lysosome juice was neither inhibitory not stimulatory. In further experiments Triton stimulated the production in lysosomes of an antibacterial lipid. It would be interesting to know the solubility or diffusibility of this lipid in wax Ds which makes up the bacterial protective barrier.

K ANTIGENS OF *ESCHERICHIA COLI*

The inhibitory nature of K antigens is implicit in the manner of their discovery. They were described by Kauffman (1943) as surface antigens rendering *E. coli* inagglutinable by specific anti-O sera and having a relation with pathogenicity in urinary infection (Sjöstedt, 1946) and in appendicitis (Vahlne, 1945). The ability of K antigens to increase the resistance of *E. coli* to complement killing and phagocytosis, the way in which this is brought about and the relation of K antigens to virulence in experimental and human infections will be discussed in turn. K antigens are similar in many ways to the Vi antigen of *Salmonella typhi*, a similarity not without irony in view of the disputes as to the significance of the latter between Kauffmann and the discoverer of Vi, Felix.

K antigens and the serum bactericidal reaction

Complement resistant strains of *S. typhi* are inagglutinable by anti-O serum and are resistant to complement (Muschel, Chamberlain & Osawa, 1958). However, resistant strains become both O-agglutinable and sensitive to complement if grown at 14° or 45°. At these temperatures the production of Vi antigen is reduced (Jude & Nicolle, 1952). Similar relations between O-inagglutinability, complement resistance and growth temperature were found in *E. coli* by Muschel (1960) and confirmed by Glynn & Howard (1970).

Strains of *E. coli* may have a K antigen and still be sensitive to complement, indeed sensitive and resistant strains of the same K (and O) serotype are not uncommon (Milne, 1966). Although some K antigens are more effective than others this does not depend on their serotype. The amount of K antigen is important. It is difficult to measure the absolute amount of K antigen but within one serotype it is quite easy to assay the relative amounts of K antigen in crude extracts of strains by radial immuno diffusion using specific anti-K serum. The results suggest a direct relation between the amount of K antigen and complement resistance (Glynn & Howard, 1970). Immunodiffusion does not allow quantitative comparison between different serotypes, but indirect assays such as the amount of polycation bound by the organism and the effect of temperature suggest the same general relationship.

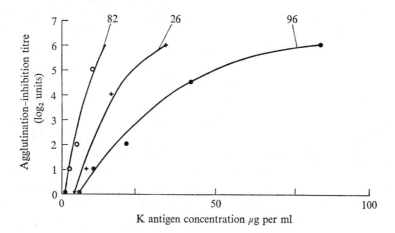

Fig. 1. Inhibition of sheep red-cell agglutination by purified K antigens from *E. coli* strains WF 82, WF 26 and WF 96. Inhibition of agglutination is directly related to complement resistance. Agglutination-inhibition titre ($\log_2$ units); see Howard & Glynn (1971*b*).

A simple and useful measure of the inhibitory activity of K antigens is given by their effect on unrelated agglutination systems. They behave like Vi antigen which, when coated on red blood cells, suppresses their agglutination by specific anti-red cell serum (Ceppellini & Landy, 1963). The agglutination inhibiting activity of comparable extracts of *E. coli*, whether of different strains or the same strain grown at different temperatures, is directly related to complement resistance. However, the activity of purified K antigen from a complement-resistant strain was more than that from a sensitive one. (Fig. 1). Because the amount and nature of K antigen present on red blood cells can be controlled, such a system is more convenient than *E. coli* for investigating the mode of action of K antigens. By using ^{125}I-labelled haemolysin, K antigen was shown to reduce the amount of antibody attaching to red cells so decreasing subsequent lysis by complement. Both γG and γM antibodies were affected. Further, with cells sensitised by a given amount of antibody the subsequent addition of K antigen again reduced complement lysis. This was not due to removal of antibody but to an effect on some or all of the subsequent steps in the lytic chain; that is on antibody activation of C1 or on the reaction of C1–9.

At first sight these results seem to contradict Cepellini & Landy (1963) who concluded from purely serological estimations that Vi did not affect the attachment of red cell antibody but stopped agglutination by impairing subsequent lattice formation. However, the differences are partly due to the different sensitivities of the methods. Thus 25 μg of a K antigen from *E. coli* strain WF 82 decreased agglutination by 50 %, although 2·5 mg would have been needed to reduce antibody binding by the same amount. It is likely, therefore that inhibition of lattice formation was also involved. Similarly agglutination is affected more than lysis in which lattice formation is not relevant.

Under optimal conditions 30 molecules of γM antibody are sufficient to allow complement killing of a gram-negative bacterium (Rowley & Turner, 1968). 2 or 3 γM or 2000 to 3000 γG molecules will permit lysis of red cells (Humphrey & Dourmashkin, 1965). While K antigens impair both killing and haemolysis they do not reduce the amount of bound antibody to this level when tested on red cells. Resistant bacteria contain more K antigen than treated red cells and it is likely that on a bacterium the K antigen is distributed in a more effective way.

K antigens and phagocytosis

O-inagglutinable organisms are difficult to phagocytose (Howard & Glynn, 1971a). The phagocytic index of strains of *E. coli* measured by the method of Biozzi, Benacerraf, Stiffel & Halpern (1954) decreased as the K antigen present increased. Extra O antibody had no effect but anti-K serum was markedly opsonic. The anti-K serum was opsonic *per se* and was not simply covering up K antigen, so allowing better access by anti-O. Although anti-K serum had this marked effect on phagocytosis it caused a barely detectable increase in serum killing of K-rich strains. Vi antigen also inhibits phagocytosis (Bhatnagar, 1935) while anti-Vi, though opsonic, (Felix & Bhatnagar, 1935) has little effect on serum killing of *S. typhi* (Nagington, 1956; Muschel & Treffers, 1956).

There are several possible reasons why K antibodies are opsonic but not bactericidal. Direct comparison is difficult but the results of Rowley & Turner (1966) and Robbins, Kenny & Suter (1965) suggest that killing may need around ten times as much antibody as opsonisation. However, in experiments by Howard & Glynn (1971a) there was no shortage of antibody. Information on the spatial distribution of K antigens on the bacterial surface is badly lacking, but it is likely that they are superficial. Antibody to such a superficial antigen may mediate phagocytosis but not killing since complement might be activated too far from its substrate in the cell wall. Even O antigens may project 150 nm from the surface (Shands, 1965) and it would be interesting to know whether O antibody attached far out is bactericidal. Antibodies to artificially-attached, superficial antigens can be opsonic (Rowley & Turner, 1968), though antibodies to H antigens are not (Felix & Olitzki, 1926). The effect of K antigens on complement may also be important. All nine complement components are needed for killing but C1423 may be enough for phagocytosis. Moreover, though complement accelerates phagocytosis, it is not essential (Ward & Enders, 1933). It would be useful to know the effect of K antigen on individual complement reactions. Red cells passively coated with K antigen and tested with anti-K serum plus complement form a relatively poor lytic system although enough antigen is present to give good agglutination.

The nature of K antigens

With one exception K antigens are acid polysaccharides of varying degrees of polymerisation, depending for their antigenic specificity on relatively common simple sugars (Hungerer et al. 1967; Jann et al. 1968; Luderitz, Jann & Wheat, 1968). Exotic sugars such as are found in O antigens have not so far been described. The acidic groups are hexuronic, hexosaminuronic and neuraminic acids and phosphate. Vi antigen is similar being a polymer of N-acetyl D-galactosamine uronic acid.

The ability of K antigens to impede antibody and complement reactions is the basis of the resistance of many strains of E. coli to serum killing and to phagocytosis. The parallel agglutination-inhibiting activity of K antigen preparations could, therefore, be used as a convenient measure of biological activity in experiments to determine the molecular basis of such activity (Howard & Glynn, 1971b).

The most obvious property calling for investigation was acidity. Immuno-electrophoretic comparison of the K antigen from a complement sensitive strain (WF 96) with that from a resistant strain (WF 82) surprisingly showed that the former moved more rapidly towards the anode. Because of the possible effects in these experiments of molecular size on diffusion through agar, the two antigens together with another from the complement resistant strain WF 26 were compared by chromatography on DEAE-cellulose. More highly charged anions need a buffer of greater molarity for elution. However, there was no consistent relation between the molarity of the eluting buffer and the agglutination-inhibiting activity of the K antigen.

When the antigens were run on columns of Sepharose 6B, 4B and 2B, it became clear that agglutination-inhibiting activity was related to molecular weight. The antigen with the weakest activity (from strain WF 96) was the smallest with a molecular weight of less than 1×10^6; the most active antigen (from strain WF 82) had a molecular weight of

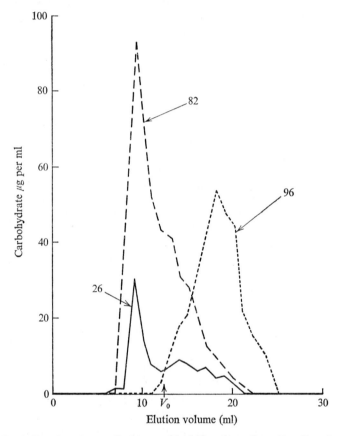

Fig. 2. Fractionation on Sepharose 6B of K antigens from *E. coli* strains WF 82, WF 26 and WF 96. V_0, void volume of column.

over 2×10^7 while the K from strain WF 26 came in between (Figs. 2 and 3). The K antigen from strain WF 82 proved to be heterogeneous when run on Sepharose 2B (Fig. 4). Material from the first peak with a molecular weight of over 2×10^7 had about 20 times the activity of material from the broad second peak with molecular weights between 1×10^6 and 2×10^7. Partial degradation of the WF 82 and WF 26 antigens by weak alkali gave smaller molecules with less activity. Agglutination-inhibiting activity could disappear though antigenic specificity remained.

It is instructive to compare these results with those of Ceppellini & Landy (1963) who found that Vi antigen prepared by electrophoresis was more active and more viscous than that prepared chemically (the more viscous material probably had the higher molecular weight), and with those of Taylor & Walton (1957) who found that the ability of sulphated dextrans to inhibit complement in a haemolytic reaction increased with the molecular weight of the dextran and with the degree of sulphation.

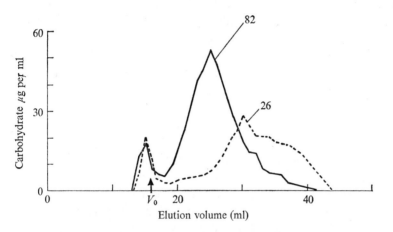

Fig. 3. Fractionation on Sepharose 4B of K antigens from *E. coli* strains WF 82 and WF 26. V_0, void volume of column.

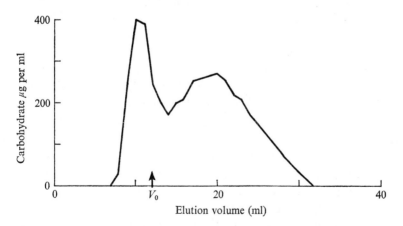

Fig. 4. Fractionation on Sepharose 2B of K antigen from *E. coli* strain WF 82. V_0, void volume of column.

K antigens and bacterial infections

The ability of K antigens to inhibit phagocytosis and complement dependent killing brings them into the category of 'impedins'. To assess their importance in determining pathogenicity we need to consider some infections, experimental and natural.

Rowley (1954) found that complement resistance in strains of *E. coli* was related to their virulence for mice if injected intraperitoneally with mucin. The complement resistance can probably be taken as a measure of the activity of the K antigens present. Rowley used guinea-pig complement to test his strains since mouse complement is not bactericidal *in vitro*. Subsequent work has shown that mouse complement can be effective *in vivo* (Medhurst & Glynn, 1970; Medhurst, Glynn & Dourmashkin, 1971) and that K antigen-rich strains were less readily killed. Roantree & Pappas (1960) found that complement resistant strains of *E. coli* were less readily dealt with than sensitive strains on intravenous injection in rabbits.

Strains of *E. coli* causing bacteraemia in man were more often resistant to complement than strains from the gut. (Roantree & Rantz, 1960). One of the commonest forms of human *E. coli* infection is that of the urinary tract, particularly in pregnant women. The bladder becomes infected with those strains of *E. coli* which happen to be numerous in the gut, although some workers hold that certain O-types are particularly liable to infect. Glynn, Brumfitt & Howard (1971) found no difference in activity between the K antigens of faecal and bladder-infecting strains but showed that K antigen-rich strains were more likely to involve the kidneys.

The role attributed to K antigens does not, of course, exclude the likelihood of other factors affecting pathogenicity. Loss of virulence for mice in strains of *E. coli* with genetic deficiencies in their O antigens has been reported by Medearis, Camitta & Heath (1968). It was not clear, however, whether or not there were concomitant changes in K antigens. Ørskov & Ørskov (1970) have shown by immuno-electrophoresis that enteropathogenic strains of *E. coli* do not possess a separate K antigen but they may have O antigens which move towards the anode suggesting that they contain acidic groups; most O antigens are electrically neutral, the slight displacement towards the cathode being explained by endosmosis. It looks from these results as if K antigens are important in systemic but not in enteric infections with *E. coli*. However, Ørskov & Ørskov (1970) were examining reference laboratory test strains and, while their serotypes are unimpeachable, the latters' behaviour might not always accurately reflect that of different strains, even of the same serotype *in vivo*. For example, strains of serotype 06 K13 have been described which contain different amounts of K antigen markedly affecting their biological activity (Howard & Glynn, 1971*a*). Furthermore, Ingram (1964) reported that K antibodies in colostrum were important in the protection of calves against *E. coli* septicaemia; in animals given colostrum, any deaths from *E. coli* were not septicaemic.

The relation between complement sensitivity and K antigens was established on strains predominantly not enteropathogenic (Glynn & Howard, 1970). It is clearly necessary to examine enteropathogenic strains further. A preliminary investigation of 18 strains (Glynn & Howard, unpublished) showed that though some were complement resistant, acidic polysaccharides with agglutination-inhibiting activity could not be found in any.

IRON AND BACTERIAL INFECTIONS

Introduction

Metal ions are closely concerned with the activity of so many enzymes, vitamins and hormones, that it is not surprising that they can affect the relationship between host and parasite (Weinberg, 1966). Of many metals with biological activity, only iron will be considered. Iron itself can hardly be regarded as either a bacterial agent or a host defence but the ability of a host to deny, or a bacterium to acquire such an essential element results in 'a battle of chelating agents' which may legitimately be discussed here. This presupposes that iron is required by a micro-organism for growth or for the elaboration of specific virulence factors and it may, on occasion, also impair host defence mechanisms such as complement.

A shortage of iron for the bacterium does not necessarily benefit the host. *Corynebacterium diphtheriae* only produces toxin when the environmental iron is below a certain threshold. Similarly, excess of iron suppresses the formation of the toxin of *Clostridium welchii*, tetanus toxin and the α-toxin and enterotoxin of *Staph. aureus*. As well as affecting their production, incubation with iron is said to reduce the toxicity of diphtheria toxin and of *Cl. welchii* toxin. The effect may be largely reversed if the iron is removed by versene. Iron also reduces the mortality and degree of tissue necrosis produced by *E. coli* endotoxin but does not affect its pyrogenic activity (Janoff & Zweifach, 1960).

Iron increased the proportion of rough forms in cultures of *Brucella suis* (Waring, Elberg, Schneider & Green, 1953); the infective dose for mice was the same for organisms grown with or without iron but the latter produced more lesions containing greater numbers of bacteria. A clue to a possible mechanism is that the site of inoculation with iron-deficient bacteria showed little inflammation with few polymorpho-nuclear cells and delayed macrophage accumulation. *In vitro* such organisms inhibited the migration of leucocytes from guinea-pig blood clot recalling the ability of virulent staphylococci to suppress inflammation (Agarwal, 1967*b*).

Schade & Caroline (1944) found that the inhibition of growth of *Shigella dysenteriae, Staph. aureus, E. coli* and *Saccharomyces cerevisiae* by raw egg white could be overcome by iron. Later (Schade & Caroline, 1946) they demonstrated an iron-binding protein in serum which inhibited *Shig. dysenteriae*. Since then there have been numerous studies of the effect of iron on infected animals and on the growth or killing of bacteria in serum. The results with specific bacteria are described first followed by consideration of the possible mechanisms involved.

The action of iron on specific bacteria

Miscellaneous infections

The virulence of *Klebsiella* strains for rats and mice was increased by simultaneous administration of iron ammonium citrate (Martin, Jandl & Finland, 1963). Intraperitoneal infections were most affected by intraperitoneal iron and intravenous infections by intramuscular iron. Some protection against infection was obtained by administering transferrin, but not if iron was given as well. Chandlee & Fukui (1965) also found that iron increased the virulence for guinea-pigs and mice of *Klebsiella* and *Salmonella typhimurium* infections but not of *S. typhi*, *Pseudomonus aeruginosa* or of two strains of streptococci.

M. tuberculosis grows poorly in serum unless a growth factor contained in an ether extract of tubercle bacilli is added (Marks, 1954). Marks' factor appears to be closely related to mycobactin (Francis, Macturk, Madinaveita & Snow, 1953), a member of a class of lipid-soluble iron-chelating agents found in mycobacteria. The tuberculostatic activity of serum was overcome if sufficient iron was added to saturate the transferrin (Kochan, 1969). Copper and zinc were ineffective. Tubercle bacilli (BCG strain) grown in an iron-deficient medium for a fortnight, liberated mycobactin into the culture fluid. Such 'spent' medium would also neutralise serum tuberculostatic activity presumably by allowing the organisms to make use of what iron there was (Kochan, Pellis & Golden, 1971). Guinea-pigs given endotoxin or BCG developed respectively a short- or long-lived'increase in serum tuberculo-static activity which was paralleled by a fall in serum iron and could be overcome by small amounts of iron (Kochan, Golden & Bukovic, 1969). The significance of such a mechanism in the development of resistance to tuberculosis is not yet clear but iron should be taken into account when non-specific effects of cellular immunity are considered.

The serum of patients with listeria infection contains an increased amount of transferrin (Sword, 1966a) and variations in iron metabolism significantly affect listeria infections in mice (Sword, 1966b). Following the administration of iron to mice the LD_{50} of virulent strains of *Listeria monocytogenes* was reduced a hundredfold, while the number of viable bacteria found in the liver early in the infection rose even more markedly. The iron-chelating agent desferrioxamine β-methane sulphate increased the LD_{50} considerably unless adequate amounts of iron were given at the same time. Iron did not exacerbate infection in previously immunised animals.

Pasteurella pestis

When iron is given to an animal at the same time or soon after an experimental infection with *P. pestis* it usually increases the severity of infection produced.

Burrows (1962) has shown that the virulence of *P. pestis* is related to at least four genes, controlling purine synthesis $(Pu+)$, virulence surface antigens (VW), a capsular antigen (Fl) and the ability to produce pigment when grown on a synthetic medium containing haem $(P+)$. Non-pigmented mutants M 7 of a virulent strain M 3 were much less lethal for mice, but their virulence was restored to normal if iron, whether as ferrous sulphate or as haem, was given with the challenge injection (Jackson & Burrows, 1956). Mice dying with infection of a non-pigmented strain showed a marked reduction in the number of tissue bacteria (10^6 compared with 10^{10} from a pigmented strain), suggesting a nutritional limitation. The M 3 and M 7 strains had the same iron requirements *in vitro*, and pigment production might indicate the ability to accumulate sufficient iron to grow *in vivo*. Grown in serum, both M 3 and M 7 strains grew poorly unless sufficient iron was added to saturate the serum iron-binding capacity (Jackson & Morris, 1961). The possible significance of this iron dependency is indicated by the vaccine strain EV 76 whose virulence could also be restored by iron. Non-pigmented mutants of avirulent pigmented strains remained avirulent even when given with iron.

Brubaker, Beesley & Surgalla (1965) have described non-pigmented avirulent mutants which fail to produce pesticin I unless given with iron when both pesticin I and virulence return. Pesticin I is a bacteriocin against *Pasteurella pseudotuberculosis*. Mutants lacking it fail to produce coagulase and fibrinolytic factors but it is not clear whether the two latter are restored by iron. The loss of virulence of mutants lacking pesticin I is well shown when they are injected intraperitoneally or subcutaneously; but only slightly after intravenous injection, suggesting that the mutants have difficulty in establishing the primary lodgment. Although in *P. pestis* infection the effect of iron seems to be largely on the bacterium, Brubaker and his colleagues do not rule out the possibility that the strains not producing pesticin I are also sensitive to a normal component of serum which can be inactivated by ferrous iron. A possible factor is the 7 S gamma globulin described by Bornside, Merritt & Weil (1964) as inhibiting the respiration of *Bacillus subtilis* unless iron was present. Transferrin was not involved.

Clostridium welchii

Specific antitoxin prevented the growth of *Cl. welchii* in live, but not in dead, eggs (Bullen, Dobson & Wilson, 1964). Experiments with a defined culture medium showed that the antitoxin was only effective if the pH and Eh were kept above certain limits. Rogers (1967) found that under suitable conditions of pH and Eh, both normal and anti-clostridial serum at concentrations of 6–12 % would inhibit clostridial growth after a lag period of about 3 h. This inhibition could be overcome by iron, but while ferric salts were only effective if added initially, ferrous salts were still active if added after 3 h. The slight differences between normal and immune sera were attributed to variations in their degree of iron saturation. Purification of the inhibitory serum factors showed that transferrin alone was inadequate and that β 2 and γ-globulins were also necessary. This was confirmed by Rogers, Bullen & Cushnie (1970), who also found that transferrin was bacteriostatic in the presence of 0·24 % horse albumin. However, unlike whole serum or transferrin plus β 2 and γ-globulins, the bacteriostatic effect of trans-ferrin plus albumin could be suppressed by ferric as well as ferrous ions added 3 h after the bacteria. After heating at 56° for 30 min both normal and immune serum were still bacteriostatic at an Eh of 0 mV and when ferric iron was added. However, at high Eh (110 mV) in the presence of excess ferric iron, a heat-labile serum component was also necessary for bacteriostases.

It is suggested that *Cl. welchii* can obtain iron from 83 % saturated transferrin unless β 2 or γ-globulins are also present. Tetanus antitoxin was as effective *in vitro* as *Cl. welchii* α-antitoxin. Since the exact way in which *Cl. welchii* takes up iron is unknown, it is uncertain whether the heat-labile factor required when transferrin-free iron is present acts on the same path as transferrin. Heat lability alone is suggestive but not sufficient evidence to implicate complement.

The serum inhibitory mechanism denies bacteria the use of ferric iron and may be overcome by ferrous iron, or by a fall in Eh to below − 140 mV when ferric ions convert to ferrous. Haem also bypasses the mechanism and even traces (10 μM) may allow bacterial growth.

In vivo the passive protection afforded to guinea-pigs by specific antiserum was overcome if iron was given within 6 h of the start of the infection (Bullen, Cushnie & Rogers, 1967). The growth curve of bacteria in passively immunised guinea-pigs was similar to that in serum *in vitro*, rising at first then falling after 6 h, while in guinea-pigs given iron as well as antibody this descent was abolished. If large doses of antitoxin were given (Rogers *et al.* 1970), then even animals given iron did not die although they developed a severe local infection. Increase in the size of the inoculum again caused death. Iron given to guinea-pigs did not affect the blood level of antitoxin or the extent of phagocytosis. Even in fatal infections there was an excess of antitoxin and death was attributed to bacterial invasion not toxaemia.

While *Cl. welchii* did not grow in normal serum *in vitro*, serum unlike specific antitoxin, did not protect guinea-pigs. Similarly, tetanus antitoxin was ineffective *in vivo* though active in making transferrin bacteriostatic *in vitro*. This taken with the importance of Eh and pH suggests that the major role of antitoxin *in vivo* is to prevent widespread tissue necrosis which would provide ideal conditions for bacterial multiplication. However, Rogers *et al.* (1970) suggest that large doses of α-antitoxin may also have a direct effect on *Cl. welchii*.

Pasteurella septica

Specific antiserum protected mice given several thousand lethal doses of *Pasteurella septica* intraperitoneally, but death occurred if iron ammonium citrate, haematin or crystalline haemoglobin were given at the time of the injection. The iron had less effect if given over 4 h later (Bullen, Wilson, Cushnie & Rogers, 1968).

P. septica grew rapidly in fresh normal rabbit serum but specific antibody produced a fall of about 1 log unit over 1 h, followed by some slow growth. Antibody did not limit growth in heated serum. Iron added to fresh serum and antibody with a P_{O_2} of 80–90 mm did not prevent the initial killing but allowed good growth of bacteria about 6 h later. At low oxygen tensions there was still the initial fall, but regrowth occurred sooner (Bullen & Rogers, 1969; Bullen, Rogers & Lewin, 1971). Even at normal oxygen tension lysed red cells and haematin produced rapid growth starting immediately after the initial killing.

Escherichia coli

E. coli given intraperitoneally does not infect mice, even in the absence of immune serum. Whether the natural antibodies to E. coli have anything to do with this resistance is an open question. Two strains of E. coli became lethal for mice when injected together with iron (Bullen & Rogers, 1969). A similar effect on one of the strains in guinea-pigs (Bullen, Leigh & Rogers, 1968) was accompanied by a massive growth of E. coli in the peritoneal cavity. Enhancement of E. coli peritoneal infection in mice by iron or haemoglobin was also described by Bornside, Bouis & Cohn (1968).

In rats given an intravenous injection of E. coli, intravenous injection of iron on the following days increased the number of bacteria found in the kidneys and frequently resulted in renal abscesses (Fletcher & Goldstein, 1970). Only small molecular weight iron compounds such as iron ammonium citrate and iron sorbitol were effective. Compounds such as iron-dextran or iron-dextrin incapable of passing the glomerulus were not. Iron deposits in the renal tubules from the previous administration of iron did not exacerbate the infection. Similar results were found with iron and Staph. albus and Mycobacterium fortuitum infections in rats and mice.

One strain of E. coli described by Bullen & Rogers (1969) was killed by fresh normal rabbit serum, the other was relatively resistant. Both, however, grew well if iron was added to the system. Heated rabbit serum had a less marked bacteriostatic effect which, too, was readily overcome by iron. Fletcher (1971) found that both iron ammonium citrate and haemoglobin stimulated the growth of E. coli in normal human serum heated to inactive complement. The iron ammonium citrate, but not the haemoglobin, also allowed the growth of E. coli in fresh serum. A curious feature of these experiments was the lag phase of 1 h in which no change in viable count occurred, whether the conditions were arranged to allow subsequent growth or killing.

Possible mechanisms of iron action

Iron in small amounts in the right place can obviously exacerbate a variety of bacterial infections. It may ensure bacteria sufficient iron in spite of competition from transferrin and perhaps other serum factors.

Although a lot remains to be learned about bacterial iron-trapping mechanisms this purely nutritional story is attractive. However, Fletcher (1970) pointed out that the affinity of transferrin for iron is so great that there is normally no free iron in the plasma. Moreover, free iron is highly toxic and transferrin hands on its iron directly to receptors on tissue cells. Possibly bacteria possess functionally similar receptors. Even if free iron is absent after iron administration a greater proportion of transferrin molecules will have two instead of one iron atoms attached and the second is more easily lost.

The relation of complement and antibody to the iron-depriving system, and the effects of iron on other defence mechanisms are less clear. Iron enables *P. septica* and *E. coli* to grow in the presence of fresh serum and antibody which would otherwise have been bactericidal. Both organisms are killed by the complement–antibody system and the question is whether iron inhibits this system, makes the organisms resistant to it or simply swamps it by stimulating rapid bacterial growth. The last possibility is unlikely since haemoglobin, which is as good a stimulant as iron salts, had less effect than the latter on fresh serum suggesting that iron had an additional action (Fletcher, 1971). Fletcher (1971) could not detect any inhibiting effect of iron on the haemolytic action of complement or on lysozyme activity. Bullen *et al.* (1971) found that iron did not affect complement fixation by *P. septica*. It is possible, therefore, that iron acts directly on the bacteria increasing their resistance to complement. Fletcher suggested that iron might fix protein to the bacterial surface so hindering complement action. This seems unlikely, unless the protein concerned had some particular ligand-impairing activity like many K antigens, or if it was present in such quantity as to form a continuous surface layer. Either possibility should be capable of experimental verification. The failure to detect an effect of iron on complement fixation is also against the surface protein theory.

Since the detection of bactericidal activity is more sensitive than the methods used to detect a direct action of iron on complement, the latter cannot be ruled out entirely. Hadding & Müller-Eberhard (1967) using purified complement components found that C 9 was sensitive to iron. Although the haemolytic and bactericidal activities of complement are similar (Rother *et al.* 1964) and all nine complement components are needed to kill bacteria (Inoue, Yonemasu, Takamizawa & Amano, 1968) there are differences. Lysozyme is irrelevant to haemolysis but, though not essential, can markedly accelerate killing and lysis of *E. coli*. Bentonite absorbs from serum a factor, other than lysozyme, which can increase the killing of *E. coli* (Glynn & Milne, 1965, 1967). It is difficult to relate complement to the heat-labile factor required to restrain the growth of the Gram-positive organism *Cl. welchii* at high Eh in the presence of ferric iron.

Iron does not affect the phagocytosis of *P. pestis* (Burrows, 1962), *Cl. welchii* (Bullen *et al.* 1967), *E. coli* (Fletcher, 1971) or staphylococci (Gladstone & Walton, 1970). However, Gladstone & Walton (1970) found that iron would prevent the killing of strains of *Bacillus anthracis*, *Streptococcus faecalis*, *L. monocytogenes*, *Ps. aeruginosa* and *E. coli* by basic proteins extracted from rabbit polymorphonuclear leucocyte lysosomes. Haematin was also effective. The iron appeared to combine with the protein but did not interfere with lysozyme or with the myeloperoxidase system. Iron also protected staphylococci from being killed inside intact polymorphonuclear leucocytes, probably by combining with cationic proteins within the phagosomes (Gladstone & Walton, to be published). Iron-binding lactoferrin was described in polymorphonuclear cells by Masson, Heremans & Schonne (1969), but Gladstone & Walton (1970) found that the slight bactericidal activity of lactoferrin from milk was not reduced by iron.

There is nothing mutually exclusive about the various ways in which iron may overcome anti-bacterial defences and it is likely that the importance of a particular mechanism varies according to the nature of the infecting organism.

The clinical significance of iron in infections

The anaemia of infection is due to a disturbance of iron metabolism. Effects in the reverse direction i.e. of iron on bacterial infections of man are less well-known.

The experiments mentioned on iron and pyelonephritis in rats were stimulated by the finding (Briggs, Kennedy & Goldberg, 1963) that a single injection of iron sorbitol citrate caused an increase in the number of white cells in the urine of patients with chronic pyelonephritis. There was, however, no clinical evidence of exacerbation of infection, and only one urine had a significant number of bacteria.

Patients with acute haemolytic anaemia due to sickle cell disease, malaria or acute bartonellosis are more susceptible to salmonella infections (Kaye, Gill & Hook, 1967). Macrophages *in vitro* were less effectively bactericidally if they had phagocytosed numerous red cells. Similarly, mice were more susceptible to infection with *S. typhimurium*, *E. coli* and *Staph. aureus* following haemolysis due to anti-red-cell serum or to phenylhydrazine. On the other hand, anaemia produced by bleeding had no effect (Kaye & Hook, 1963).

In children suffering from kwashiorkor and treated with anti-malarial drugs, milk, vitamins and iron some deaths occurred from acute bacterial infection. A poor prognosis was related to low serum trans-ferrin levels which, together with the iron given, may have been an important factor (Macfarlane *et al.* 1970). There may be a case for delaying iron in such patients. All the patients had negative tuberculin tests suggesting that cell-mediated immunity was also impaired.

Patients with agammaglobulinaemia may have increased serum anti-bacterial activity (Martin, Gordon, Felts & McCullough, 1957). Martin (1962) showed that the activity against *Bacillus subtilis* was heat stable and could be reversed by the addition of iron or by reducing the pH to 5·5, at which level transferrin binds little or no iron (Martin, 1962). In 4 patients the anti-bacterial activity of serum was correlated with the unsaturated iron-binding capacity.

It seems at present unlikely that excess iron plays a serious part in many human infections but, as the possibility of it doing so becomes more widely known, further examples will no doubt be described and the role of transferrin made more clear.

BACTERIAL FACTORS INHIBITING IMMUNE RESPONSES

Before recent work on transplant and cancer immunology and Burnet's ideas on surveillance, micro-organisms were regarded as the major initiators of immune reactions. Although inhibition of delayed hyper-sensitivity to tuberculin by measles was described by von Pirquet in 1908, active interest in microbial inhibition of immune responses is relatively recent and is particularly associated with viruses (Salaman, 1970). Less is known of such effects due to bacteria but sufficient have been described to make it likely that more exist.

Endotoxin

The biphasic response in resistance to many types of infection which follows the injection of endotoxin received much attention in the 1950s (Rowley, 1964). The preliminary negative phase described by Almroth Wright in relation to typhoid vaccine and emphasised later by Shaw (1911) was shown to follow the injection of *E. coli* cell-walls (Rowley, 1955). Condie, Zak & Good (1955) made rabbits temporarily more susceptible to streptococcal and pneumococcal infection by a single injection of meningococcal endotoxin. Repeated injection increased resistance but depressed antibody formation. Endotoxin inhibited the formation of antibody to actinophage (Bradley & Watson, 1964) and reduced the number of antibody-forming plaques detectable after a primary stimulus by sheep red cells (Franzl & McMaster, 1968; Finger, Fresenius & Angerer, 1971). The inhibition was abolished if loosely bound lipid was removed from the endotoxin. The immunogenicity of Kunins 'common antigen' of *E. coli* was impaired by endotoxin only if the two were injected at the same site and lipid A was present in the endotoxin (Whang & Neter, 1967). As with other immunosuppressive agents the timing is important. A small dose of endotoxin given before or a large dose given with the antigen usually results in inhibition. Small doses given with antigen have an adjuvant effect (Johnson, Gaines & Landy, 1956). Endotoxin also damages the thymus (Landy, Sanderson, Bernstein & Lerner, 1965).

Other bacterial factors

Group A streptococci contain a non-toxic cytoplasmic factor which can suppress both primary and secondary responses to sheep red cells in mice (Malakian & Schwab, 1967). The material was ineffective if given two or three days after the antigen and did not affect immunological memory. *E. coli* contains a cytoplasmic factor which suppressed the antibody response to sheep red cells and, though less effectively, to Vi antigen. Penicillin spheroplasts of *E. coli* contained the same factor. With a suitable injection schedule, the life of skin allografts in rats and mice was prolonged 1·5–2·5 times (Kirpatovski & Stanislavski, 1971). A ribonuclease isolated from *Bacillus* spp. was also an immunosuppressant (Carpenter, Milton & Mowbray, personal communication).

L-Asparaginase from guinea-pig serum or *E. coli* inhibited the growth of some mouse leukaemia cells. The cells were thought to be asparagine dependent because unlike normal cells they lacked asparagine synthetase. However, rapidly growing normal cells such as the proliferating lymphocytes of an immune response *in vitro* (Berenbaum, Ginsburg & Gilbert, 1970) and normal bone marrow cells in tissue culture (Harris, 1969), can also be affected. Asparaginase inhibits the lymphocyte blastogenesis induced by phytohaemagglutinin *in vitro* (Astaldi *et al.* 1969) and *in vivo* (McElwain & Hayward, 1969). Asparaginase prevented the appearance of plaque-forming cells making antibody to sheep erythrocytes (Schwartz, 1969; Müller-Bérat, 1969; Chaknabarty & Friedman, 1970); heat inactivation of the enzyme or the simultaneous administration of much asparagine abolished the effect. However,

Deodhar (1971) found asparagine ineffective in preventing the enhancement of mouse tumour metastases by asparaginase. A single dose of asparaginase suppressed plaque-forming cells most if given one or two days after immunisation (Friedman & Chakrabarty, 1971). Similarly, rejection of skin grafts was least when asparaginase was given after grafting. This would fit with an effect on dividing immunocytes but asparaginase was ineffective if given more than 3 h after immunisation (Berenbaum, 1970).

Cell-mediated immune responses

Mycoplasma hominis inhibited the transformation of human lymphocytes by phytohaemagglutinin (Copperman & Morton, 1966). The mechanism may have been the same as that found in killed *Mycoplasma arthritidis* which inhibited the synthesis of DNA and RNA underlying the transformation reaction (Spitler, Cochrum & Fudenberg, 1968). Other mycoplasmas may not behave similarly and the pathogenetic significance of the reaction is uncertain. Homografts lasted a long time in 7 patients with burns severely infected with *Ps. aeruginosa* (Stone, Given & Martin, 1967); somatic extracts from one of the strains involved could prolong the life of skin grafts in rats. In contrast to this depressant activity streptolysin O and filtrates of staphylococci stimulated lymphocytes to blast formation (Ling, 1968). Specific sensitivity is not involved. Nor probably is a reaction between 'protein A' and cell-bound γ-globulin since rabbit lymphocytes are readily stimulated.

Delayed hypersensitivity to tuberculin may be reduced in measles, tuberculosis, whooping cough (Fanconi & Wallgren, 1952) and leprosy. Floersheim (1965, 1966) diminished the tuberculin reaction in guinea-pigs sensitised with BCG by giving pertussis vaccine or polysaccharide extracts of pertussis before challenge. Pertussis vaccine may stop the proliferation of spleen cells induced by complete Freund's adjuvant and may stimulate the growth of lymphomas in mice (Hirano, Sinkovics, Shullenberger & Howe, 1967). The latter was attributed to pre-commitment of lymphocytes but an active depressant effect is more likely.

In lepromatous forms of leprosy characterised by granulomatous deposits and large numbers of intracellular *M. leprae*, mycobacterial antibodies are high but cell-mediated immune responses low (Turk, 1970). Delayed hypersensitivity is diminished to lepromin, tuberculin and candida extracts (Bullock, 1968). The peripheral lymphocytes respond poorly to phytohaemagglutinin (Dierks & Shephard, 1968). Di-nitrochlorobenzene failed to sensitise many such patients (Waldorf,

Sheagren, Trautman & Block, 1966) but a more potent sensitiser, keyhole limpet haemocyanin, usually succeeded (Turk & Waters, 1969). Although lepromatous patients may have some innate weakness in their cell-mediated immunity (Jamison & Vollum, 1968), most of the depression must be due to the disease, since delayed hypersensitivity reactions return in successfully treated patients. Moreover, a similar depression of cellular immunity is produced in mice infected with *M. lepraemurium* (Ptak, Gaugas, Rees & Allison, 1970).

Leprosy patients with impaired cellular immunity show replacement of most of the paracortical (thymus-dependent) areas of their lymph nodes by phagocytic histiocytes (Turk & Waters, 1968). A purely mechanical infiltration should not just affect the area associated with cellular responses, although this is where the macrophages arrive from the periphery. Whether *M. leprae* produces an inhibitory factor is unknown. Serum from patients with leprosy can inhibit normal lymphocytic transformation due to streptolysin O (Bullock & Fasal, 1968) but auto-antibodies have not been excluded. Dead mycobacteria can suppress cellular immunity in guinea-pigs (Jankovic, 1962).

A factor preventing lymphocyte transformation by phytohaem-agglutinin was found in the serum of patients with secondary syphilis (Levene, Turk, Wright & Grimble, 1969) but again might have been either of treponemal origin or an auto-antibody. Small lymphocytes disappear from neonatal rabbits infected with *Treponema pallidum* (Festenstein, Abrahams & Bokkenheuser, 1967).

SUMMARY

The numerous bacterial factors which can impede host defence mechanisms without being obviously toxic are often called 'aggressins'. Because this implies a noxious rather than a blocking action, 'impedin' is suggested as a more suitable term. On the assumption that pathogenicity always depends on several factors, an 'impedin' might play either a necessary or a purely adjuvant role. However, it is easier to find information concerning the mode of action of individual 'impedins' than on their significance in the interaction of host and parasite.

Since the many host defences may each be interfered with in several ways it is not surprising that 'impedins' are chemically heterogeneous. The inflammation suppressing desoxycholate residue of staphylococcal cell walls is either mucopeptide or closely associated with it. It is certainly associated with staphylococcal virulence for mice and perhaps more generally. Another staphylococcal product, 'protein A', reacts with some types of γG but its significance for virulence is unknown. At present the importance of the unknown component, which confers complement resistance on gonococci found *in vivo* or grown on prostate medium, cannot be assessed. The component is not the same as that responsible for particular colonial types of gonococcus which may differ in virulence. Mycobacterial lipids probably act as impedins by being largely impermeable layers. The acid polysaccharide, K antigens of *E. coli* block a variety of macromolecular reactions including those of antigen with antibody and of complement. K antigens appear to be important in systemic or renal but not in mucosal infections with *E. coli*.

Iron modifies bacterial infections but the mechanisms are still uncertain, as is the clinical significance of the effects described.

Classical immunology was largely devoted to immune responses to infectious agents. It is fitting that the current emphasis on immuno-suppression should have shown that bacteria can do that too.

REFERENCES

AGARWAL, D. S. (1967*a*). Subcutaneous staphylococcal infection in mice. I. The role of cotton dust in enhancing infection. *British Journal of Experimental Pathology*, 48, 436–49.

AGARWAL, D. S. (1967*b*). Subcutaneous staphylococcal infection in mice. II. The inflammatory response to different strains of staphylococci and micrococci. *British Journal of Experimental Pathology*, 48, 468–82.

ALLEN, J. M., BRIEGER, E. M. & REES, R. J. W. (1965). Electron microscopy of the host–cell parasite relation in murine leprosy. *Journal of Pathology and Bacteriology*, 89, 301–6.

ASTALDI, G., BURGIO, G. R., KRC, J., GENOVA, R. & ASTALDI, A. A. (1969). L-asparaginase and blastogenesis. *The Lancet*, i, 423.

BERENBAUM, M. C. (1970). Immunosuppression by L-asparaginase. *Nature, London*, 225, 550–2.

BERENBAUM, M. C., GINSBURG, H. & GILBERT, D. M. (1970). Effects of L-aspara-ginase on lymphocyte target cell reactions *in vitro*. *Nature, London*, 227, 1147–8.

BHATNAGAR, S. S. (1935). Phagocytosis of *B. typhosus* in relation to its antigenic structure and to the antibody components of the sensitising serum. *The British Journal of Experimental Pathology*, 16, 375–84.

BIOZZI, G., BENACERRAF, B., STIFFEL, C. & HALPERN, B. N. (1954). Etude quantita-tive de l'activité granulopexique du système réticuloendothélial chez la souris. *Comptes rendus des séances de la Société de Biologie*, 148, 431–5.

BORNSIDE, G. H., BOUIS, P. J. & COHN, I. (1968). Haemoglobin and *Escherichia coli*, a lethal intraperitoneal combination. *Journal of Bacteriology*, 95, 1567–71.

BORNSIDE, G. H., MERRITT, C. B. & WEIL, A. C. (1964). Reversal by ferric iron of serum inhibition of respiration and growth of *Bacillus subtilis*. *Journal of Bacteriology*, 87, 1443–51.

BRADLEY, S. G. & WATSON, D. W. (1964). Suppression by endotoxin of the immune response to actinophage in the mouse. *Proceedings of the Society for Experi-mental Biology and Medicine*, 117, 570–2.

BRAMBELL, F. W. R., HEMMINGS, W. A., OAKLEY, C. L. & PORTER, R. R. (1960). The relative transmission of the fractions of papain hydrolyzed homologous γ-globulin from the uterine cavity to the foetal circulation in the rabbit. *Proceedings of the Royal Society, Series* B, **151**, 478–482.

BRIGGS, J. D., KENNEDY, A. C. & GOLDBERG, A. (1963). Urinary white cell secretion after iron sorbitol citric acid. *British Medical Journal*, **2**, 352–4.

BROWN, C. A., DRAPER, P. & D'ARCY HART, P. (1969). Mycobacteria and lysosomes: a paradox. *Nature, London*, **221**, 658–60.

BRUBAKER, R. R., BEESLEY, E. D. & SURGALLA, M. J. (1965). *Pasteurella pestis*: role of pesticin I and iron in experimental plague. *Science*, **149**, 422–4.

BULLEN, J. J., CUSHNIE, G. H. & ROGERS, H. J. (1967). The abolition of the protective effect of *Clostridium welchii* type A antiserum by ferric iron. *Immunology*, **12**, 303–12.

BULLEN, J. J., DOBSON, A. & WILSON, A. B. (1964). Bacteriostatic effects of specific antiserum on *Clostridium welchii* type A. The role of Eh and pH of the medium. *Journal of General Microbiology*, **35**, 175–82.

BULLEN, J. J., LEIGH, L. C. & ROGERS, H. J. (1968). The effect of iron compounds on the virulence of *Escherichia coli* for guinea pigs. *Immunology*, **15**, 581–8.

BULLEN, J. J. & ROGERS, H. J. (1969). Bacterial iron metabolism and immunity to *Pasteurella septica* and *Escherichia coli*. *Nature, London*, **224**, 380–2.

BULLEN, J. J., ROGERS, H. J. & LEWIN, J. E. (1971). The bacteriostatic effect of serum on *Pasteurella septica* and its abolition by iron compounds. *Immunology*, **20**, 391–406.

BULLEN, J. J., WILSON, A. B., CUSHNIE, G. H. & ROGERS, H. J. (1968). The abolition of the protective effect of *Pasteurella septica* antiserum by iron compounds. *Immunology*, **14**, 889–98.

BULLOCK, W. E. (1968). Status of immune mechanisms in leprosy. I. Depression of delayed allergic response to skin test antigens. *New England Journal of Medicine*, **278**, 298–304.

BULLOCK, W. E. & FASAL, P. (1968). Impairment of phytohemagglutinin (PHA) and antigen-induced DNA synthesis in leucocytes cultured from patients with leprosy. *International Journal of Leprosy*, **36**, 608.

BURKE, J. & MILES, A. A. (1958). The sequence of vascular events in early infective inflammation. *Journal of Pathology and Bacteriology*, **76**, 1–19.

BURROWS, T. W. (1962). Genetics of virulence in bacteria. *British Medical Bulletin*, **18**, 69–73.

CEPPELLINI, R. & LANDY, M. (1963). Suppression of blood group agglutinability of human erythrocytes by certain bacterial polysaccharides. *Journal of Experimental Medicine*, **117**, 321–38.

CHAKRABARTY, A. K. & FRIEDMAN, H. (1970). L-asparaginase induced immunosuppression: effects on antibody-forming cells and serum titers. *Science*, **167**, 869–70.

CHANDLEE, G. C. & FUKUI, G. M. (1965). The role of iron in endotoxin induced non-specific protection. *Bacteriological Proceedings*, 45.

CHAPMAN, G. B., HANKS, J. H. & WALLACE, J. H. (1959). An electron microscope study of the disposition and fine structure of *Mycobacterium lepraemurium* in mouse spleen. *Journal of Bacteriology*, **77**, 205–11.

COHN, Z. A. (1963). The fate of bacteria within phagocytic cells. II. The modification of intracellular degradation. *Journal of Experimental Medicine*, **117**, 43–53.

CONDIE, R. M., ZAK, S. J. & GOOD, R. A. (1955). Effect of meningococcal endotoxin on resistance to bacterial infection and the immune response of rabbits. *Federation Proceedings*, **14**, 459–60.

COPPERMAN, R. & MORTON, H. E. (1966). Reversible inhibition of mitosis in lymphocyte cultures by non-viable mycoplasma. *Proceedings of the Society for Experimental Biology and Medicine*, **123**, 790–5.

CRUICKSHANK, R. (1965). *Medical Microbiology*, 11th edition, 202. Edinburgh and London: Livingstone.

D'ARCY HART, P., GORDON, A. H. & JACQUES, P. J. (1969). Suggested role of lysosomal lipid in the contrasting effects of Triton WR-1339 and Dextran on tuberculous infection. *Nature, London*, **222**, 672–3.

DEODHAR, S. D. (1971). Enhancement of metastases by L-asparaginase in a mouse tumour system. *Nature, London*, **231**, 319–21.

DIERKS, R. E. & SHEPHARD, C. C. (1968). Effect of phytohemagglutinin and various mycobacterial antigens on lymphocyte cultures from leprosy patients. *Proceedings of the Society for Experimental Biology and Medicine*, **127**, 391–5.

DOSSETT, J. H., KRONVALL, G., WILLIAMS, R. C., JR & QUIE, P. G. (1969). Antiphagocytic effects of staphylococcal protein A. *Journal of Immunology*, **103**, 1405–10.

DRAPER, P. & REES, R. J. W. (1970). Electron transparent zone of mycobacteria may be a defence mechanism. *Nature, London*, **228**, 860–1.

DUBOS, R. J. & HIRSCH, J. G. (1965). *Bacterial and mycotic infections of man*, 4th edition. Philadelphia: Lippincott.

ELEK, S. D. & CONEN, P. E. (1957). The virulence of *Staphylococcus pyogenes* for man. A study of the problems of wound infection. *British Journal of Experimental Pathology*, **38**, 573–86.

FANCONI, G. & WALLGREN, A. (1952). *Textbook of Paediatrics*, 445. London: Heinemann.

FELIX, A. & BHATNAGAR, S. S. (1935). Further observations on the properties of the Vi antigen of *B. typhosus* and its corresponding antibody. *The British Journal of Experimental Pathology*, **16**, 422–34.

FELIX, A. & OLITZKI, I. (1926). The qualitative receptor analysis. II. Bactericidal serum action and qualitative receptor analysis. *Journal of Immunology*, **11**, 31–80.

FESTENSTEIN, H., ABRAHAMS, C. & BOKKENHEUSER, V. (1967). Runting syndrome in neonatal rabbits infected with *Treponema pallidum*. *Clinical and Experimental Immunology*, **2**, 311–20.

FINGER, H., FRESENIUS, H. & ANGERER, M. (1971). Bacterial endotoxins as immunosuppressive agents. *Experientia*, **27**, 456–8.

FISHER, S. (1963). Experimental staphylococcal infection of the subcutaneous tissue of the mouse. *Journal of Infectious Diseases*, **113**, 213–18.

FLETCHER, J. (1970). Iron transport in the blood. *Proceedings of the Royal Society of Medicine*, **63**, 1216–18.

FLETCHER, J. (1971). The effect of iron and transferrin on the killing of *Escherichia coli* in fresh serum. *Immunology*, **20**, 493–500.

FLETCHER, J. & GOLDSTEIN, E. (1970). The effect of parenteral iron preparations on experimental pyelonephritis. *British Journal of Experimental Pathology*, **51**, 280–5.

FLOERSHEIM, G. L. (1965). Effect of pertussis vaccine on the tuberculin reaction. *International Archives of Allergy*, **26**, 340–4.

FLOERSHEIM, G. L. (1966). Further studies pertaining to a *B. pertussis* factor inhibiting the tuberculin reaction. *Experientia*, **22**, 219–20.

FORSGREN, A. (1968a). 'Protein A' from *Staphylococcus aureus*. V. Reaction with guinea pig gamma-globulins. *Journal of Immunology*, **100**, 921–6.

FORSGREN, A. (1968b). 'Protein A' from *Staphylococcus aureus*. VI. Reaction with subunits from guinea pig gamma-1- and gamma-2-globulin. *Journal of Immunology*, **100**, 927–30.

FORSGREN, A. & SJÖQUIST, J. (1966). 'Protein A' from *Staphylococcus aureus*. I. Pseudo-immune reaction with human gamma-globulin. *Journal of Immunology*, **97**, 822–7.

FORSGREN, A. & SJÖQUIST, J. (1967). 'Protein A' from *Staphylococcus aureus*. III. Reaction with rabbit gamma-globulin. *Journal of Immunology*, **99**, 19–24.

FRANCIS, J., MACTURK, H. M., MADINAVEITIA, J. & SNOW, G. A. (1953). Mycobactin, a growth factor for *Mycobacterium johnei*. I. Isolation from *Mycobacterium phlei*. *Biochemical Journal*, **55**, 596–607.

FRANZL, R. E. & MCMASTER, P. D. (1968). The primary immune response in mice. I. The enhancement and suppression of hemolysin production by a bacterial endotoxin. *Journal of Experimental Medicine*, **127**, 1087–107.

FRIEDMAN, H. & CHAKRABARTY, A. K. (1971). L-asparaginase induced inhibition of transplantation immunity: prolongation of skin allografts in enzyme treated mice. *Transplantation Proceedings*, **3**, 826–30.

GLADSTONE, G. P. & WALTON, E. (1970). Effect of iron on the bactericidal proteins from rabbit polymorphonuclear leucocytes. *Nature, London*, **227**, 849–51.

GLYNN, A. A., BRUMFITT, W. & HOWARD, J. C. (1971). K antigens of *Escherichia coli* and renal involvement in urinary tract infections. *The Lancet*, **i**, 514–16.

GLYNN, A. A. & HOWARD, C. J. (1970). The sensitivity to complement of strains of *Escherichia coli* related to their K antigens. *Immunology*, **18**, 331–46.

GLYNN, A. A. & MILNE, C. M. (1965). Lysozyme and immune bacteriolysis. *Nature, London*, **207**, 1309–10.

GLYNN, A. A. & MILNE, C. M. (1967). A kinetic study of the bacteriolytic and bactericidal action of human serum. *Immunology*, **12**, 639–53.

GLYNN, A. A. & WARD, M. E. (1970). Nature and heterogeneity of the antigens of *Neisseria gonorrhoeae* involved in the serum bactericidal reaction. *Infection and Immunity*, **2**, 162–8.

GORDON, J. & JOHNSTONE, K. I. (1940). The bactericidal action of normal sera. *Journal of Pathology and Bacteriology*, **50**, 483–90.

GROV, A. (1965). Studies on antigen preparations from *Staphylococcus aureus*. 2. The influence of modification of functional groups and enzymic digestion on the serological activity of protein A. *Acta Pathologica et Microbiologica Scandinavica*, **65**, 600–6.

GROV, A., MYKLESTAD, B. & OEDING, P. (1964). Immunochemical studies on antigen preparations from *Staphylococcus aureus*. I. Isolation and chemical characterization of antigen A. *Acta Pathologica et Microbiologica Scandinavica*, **61**, 588–96.

GUSTAFSON, G. T., SJÖQUIST, J. & STÅHLENHEIM, G. (1967). Protein A from *Staphylococcus aureus*. II. Arthus-like reaction produced in rabbits by interaction of protein A and human γ-globulin. *Journal of Immunology*, **98**, 1178–81.

GUSTAFSON, G. T., STÅHLENHEIM, G., FORSGREN, A. & SJÖQUIST, J. (1968). Production of anaphylaxis-like cutaneous and systemic reactions in non-immunized guinea-pigs. *Journal of Immunology*, **100**, 530–4.

HADDING, U., MÜLLER-EBERHARD, H. J. (1967). Complement: substitution of the terminal component in immune haemolysin by 1,10-phenanthroline. *Science*, **157**, 442–3.

HARRIS, J. E. (1969). Effect of L-asparaginase on the ability of normal mouse bone-marrow to form soft agar colonies. *Nature, London*, **223**, 850–1.

HILL, M. J. (1968). A staphylococcal aggressin. *Journal of Medical Microbiology*, **1**, 33–43.

HILL, M. J. (1969). Protection of mice against infection by *Staphylococcus aureus*. *Journal of Medical Microbiology*, **2**, 1–7.

HIRANO, M., SINKOVICS, J. G., SHULLENBERGER, C. C. & HOWE, C. D. (1967). Murine lymphoma: augmented growth in mice with pertussis vaccine induced lymphocytosis. *Science*, **158**, 1061–4.

HOWARD, C. J. & GLYNN, A. A. (1971a). The virulence for mice of strains of *Escherichia coli* related to the effect of K antigens on their resistance to phagocytosis and killing by complement. *Immunology*, **20**, 767–77.

HOWARD, C. J. & GLYNN, A. A. (1971b). Some physical properties of K antigens of *Escherichia coli* related to their biological activity. *Infection and Immunity*, **4**, 6–11.

HUBER, H. & FUDENBERG, H. H. (1968). Receptor sites of human monocytes for Ig G. *International Archives of Allergy*, **34**, 18–31.

HUMPHREY, J. H. & DOURMASHKIN, R. R. (1969). The lesions in cell membranes caused by complement. *Advances in Immunology*, **11**, 75–115.

HUNGERER, D., JANN, K., JANN, B., Ørskov, F. & Ørskov, I. (1967). Immunochemistry of K antigens of *Escherichia coli*. IV. The K antigen of *E. coli* 09:K30:H12. *European Journal of Biochemistry*, **2**, 115–26.

IMAEDA, T. (1965). Electron microscopy. Approach to leprosy research. *International Journal of Leprosy*, **33**, 669–88.

IMAEDA, T., KANETSUNA, F. RIEBER, M., GALINDO, B. & CESARI, I. M. (1969). Ultrastructural characteristics of mycobacterial growth. *The Journal of Medical Microbiology*, **2**, 181–6.

INGRAM, P. L. (1964). Some factors influencing the response of young domesticated animals to *Escherichia coli*. In *Microbial behaviour in vivo and in vitro*. *Symposium of the Society for General Microbiology*, **14**, 122–44.

INOUE, K. YONEMASU, K., TAKAMIZAWA, A. & AMANO, T. (1968). Studies on the immune bacteriolysis. XIV. Requirements of all nine components of complement for immune bacteriolysis. *Biken's Journal*, **11**, 203–6.

ISHIZAKA, T., ISHIZAKA, K., SALMON, S. & FUDENBERG, H. (1967). Biologic activities of aggregated gamma-globulin. 8. Aggregated immunoglobulins of different classes. *Journal of Immunology*, **99**, 82–91.

JACKSON, S. & BURROWS, T. W. (1956). The virulence enhancing effect of iron on non-pigmented mutants of virulent strains of *Pasteurella pestis*. *British Journal of Experimental Pathology*, **37**, 577–83.

JACKSON, S. & MORRIS, B. C. (1961). Enhancement of growth of *Pasteurella pestis* and other bacteria in serum by addition of iron. *British Journal of Experimental Pathology*, **42**, 363–8.

JAMES, R. C. & MCLEOD, C. M. (1961). Induction of staphylococcal infections in mice with small inocula introduced on sutures. *British Journal of Experimental Pathology*, **42**, 266–77.

JAMISON, D. G. & VOLLUM, R. L. (1968). Tuberculin conversion in leprous families in Northern Nigeria. *The Lancet*, **ii**, 1271–2.

JANKOVIC, B. D. (1962). Impairment of immunological reactivity in guinea pigs by prior injection of adjuvant. *Nature, London*, **193**, 789–90.

JANN, K., JANN, B., SCHNEIDER, K. F., ØRSKOV, F. & ØRSKOV, I. (1968). Immunochemistry of K antigens of *Escherichia coli*. V. The K antigen of *E. coli* 08:K27(A):H-. *European Journal of Biochemistry*, **5**, 456–65.

JANOFF, A. & ZWEIFACH, B. W. (1960). Inactivation of bacterial exotoxins and endotoxins by iron. *Journal of Experimental Medicine*, **112**, 23–34.

JENKIN, C. R. (1963). The effect of opsonins on the intracellular survival of bacteria. *The British Journal of Experimental Pathology*, **44**, 47–57.

JENSEN, K. (1958). A normally occurring staphylococcus antibody in human serum. *Acta Pathologica et Microbiologica Scandinavica*, **44**, 421–8.

JOHANOVSKY, J. (1958). Role of hypersensitivity in experimental staphylococcal infections. *Nature, London*, **182**, 1454.

JOHNSON, A. G., GAINES, S. & LANDY, M. (1956). Studies on the O antigen of *Salmonella typhosa*. V. Enhancement of antibody response to protein antigens by the purified lipopolysaccharide. *Journal of Experimental Medicine*, **103**, 225–46.

JOHNSON, J. E., CLUFF, L. E. & GOSHI, K. (1961). Studies on the pathogenesis of staphylococcal infection. I. The effect of repeated skin infections. *Journal of Experimental Medicine*, **113**, 235–48.

JUDE, A. & NICOLLE, P. (1952). Persistance à l'état potential de la capacité d'elaborer l'antigene Vi chez le bacille typhique cultivé en série à basse temperature. *Compte rendu hebdomadaire des séances de l'Académie des Sciences, Paris*, **234**, 1718–20.

KAUFFMANN, F. (1943). Über neue thermolabile Körperentigene der Coli-bakterien. *Acta Pathologica et Microbiologica Scandinavica*, **20**, 21–44.

KAYE, D., GILL, F. A. & HOOK, E. W. (1967). Factors influencing host resistance to salmonella infections: the effects of hemolysis and erythrophagocytosis. *American Journal of the Medical Sciences*, **254**, 205–15.

KAYE, D. & HOOK, E. W. (1963). The influence of hemolysis or blood loss on susceptibility to infection. *Journal of Immunology*, **91**, 65–75.

KELLOGG, D. S., COHEN, I. R., NORINS, L. C., SCHROETER, A. L. & REISING, G. (1968). *Neisseria gonorrhoeae*. II. Colonial variation and pathogenicity during 35 months *in vitro*. *Journal of Bacteriology*, **96**, 596–605.

KELLOGG, D. S., PEACOCK, W. L., DEACON, W. E., BROWN, L. & PIRKLE, C. I. (1963). *Neisseria gonorrhoeae*. I. Virulence genetically linked to clonal variation. *Journal of Bacteriology*, **85**, 1274–9.

KIRPATOVSKI, I. D. & STANISLAVSKI, E. S. (1971). Immunosuppressive effects of cell free extracts from *Escherichia coli*. *Transplantation Proceedings*, **3**, 831–4.

KOCHAN, I. (1969). Mechanism of tuberculostasis in mammalian serum. I. Role of transferrin in human serum tuberculostasis. *Journal of Infectious Diseases*, **119**, 11–18.

KOCHAN, I., GOLDEN, C. A. & BUKOVIC, J. A. (1969). Mechanism of tuberculostasis in mammalian serum. II. Induction of serum tuberculostasis in guinea pigs. *Journal of Bacteriology*, **100**, 64–70.

KOCHAN, I., PELLIS, N. R. & GOLDEN, C. A. (1971). Mechanism of tuberculostasis in mammalian serum. III. Neutralisation of serum tuberculostasis by myco-bactin. *Infection and Immunity*, **3**, 553–8.

KRONVALL, G. (1967). Ligand-binding sites for streptolysin O and staphylococcal protein A on different parts of the same myeloma globulin. *Acta Pathologica et Microbiologica Scandinavica*, **69**, 619–21.

KRONVALL, G., DOSSETT, J. H., QUIE, P. Q. & WILLIAMS, R. C. (1971). Occurrence of protein A in staphylococcal strains: quantitative aspects and correlation to antigenic and bacteriophage types. *Infection and Immunity*, **3**, 10–15.

KRONVALL, G. & WILLIAMS, R. C. (1969). Differences in anti-Protein A activity among Ig G subgroups. *Journal of Immunology*, **103**, 828–33.

LANDY, M., SANDERSON, R. P., BERNSTEIN, M. T. & LERNER, E. M. (1965). Involvement of thymus in immune response of rabbits to somatic polysaccharides of gram-negative bacteria. *Science*, **147**, 1591–2.

LEVENE, G. M., TURK, J. L., WRIGHT, D. J. M. & GRIMBLE, A. G. S. (1969). Reduced lymphocyte transformation due to a plasma factor in patients with active syphilis. *The Lancet*, **ii**, 246–7.

LING, N. R. (1968). In *Lymphocyte stimulation*, 152, 154. Amsterdam: North-Holland.

LÖFKVIST, T. & SJÖQUIST, J. (1962). Chemical and serological analysis of antigen preparations from *Staphylococcus aureus*. *Acta Pathologica et Microbiologica Scandinavica*, **56**, 295–310.

LUCAS, C. T., CHANDLER, F., MARTIN, J. E. & SCHMALE, J. D. (1971). Transfer of gonococcal urethritis from man to chimpanzee, an animal model for gonorrhoea. *World Health Organisation* (*WHO/VDT/RES/GON/*71.49).

LÜDERITZ, O., JANN, K. & WHEAT, R. (1968). Somatic and capsular antigens of gram-negative bacteria. In *Comprehensive biochemistry*, ed. M. Florkin and E. H. Stotz, vol. 26*A*, 105–228. Amsterdam: Elsevier.

MACFARLANE, H., REDDY, S., ADCOCK, K. J., ADESHINA, H., COOKE, A. R. & AKENE, J. (1970). Immunity, transferrin and survival in kwashiorkor. *British Medical Journal*, **4**, 268–70.

MAHONEY, J. F., VAN SLYKE, C. J., CUTLER, J. C. & BLUM, H. L. (1946). Experimental gonococcic urethritis in human volunteers. *American Journal of Syphilis, Gonorrhoea and Veneral Diseases*, **30**, 1–39.

MALAKIAN, A. & SCHWAB, J. (1967). Immunosuppressant from group A streptococci. *Science*, **159**, 880–1.

MARKS, J. (1954). The *Mycobacterium tuberculosis* growth factor. *Journal of Pathology and Bacteriology*, **67**, 254–6.

MARTIN, C. M. (1962). Relation of transferrin to serum bacteriostatic activity in agammaglobulinaemic and other patients. *American Journal of Medical Sciences*, **244**, 334–6.

MARTIN, C. M., GORDON, R. S., FELTS, W. R. & MCCULLOUGH, N. B. (1957). Studies on gamma globulin. I. Distribution and metabolism of antibodies and gamma globulin in hypogammaglobulinemic patients. *Journal of Laboratory and Clinical Medicine*, **49**, 607–16.

MARTIN, C. M., JANDL, J. H. & FINLAND, M. (1963). Enhancement of acute bacterial infections in rats and mice by iron and their inhibition by human transferrin. *Journal of Infectious Diseases*, **112**, 158–63.

MASSON, P. L., HEREMANS, J. F. & SCHONNE, E. (1969). Lactoferrin – an iron binding protein in neutrophil leucocytes. *Journal of Experimental Medicine*, **130**, 643–56.

MCELWAIN, T. J. & HAYWARD, S. K. (1969). L-asparaginase and blastogenesis. *The Lancet*, **i**, 527.

MEDEARIS, D. N., CAMITTA, B. M. & HEATH, E. C. (1968). Cell-wall composition and virulence in *Escherichia coli*. *Journal of Experimental Medicine*, **128**, 399–414.

MEDHURST, F. A. & GLYNN, A. A. (1970). *In vivo* bactericidal activity of mouse complement against *Escherichia coli*. *British Journal of Experimental Pathology*, **51**, 498–506.

MEDHURST, F. A., GLYNN, A. A. & DOURMASHKIN, R. R. (1971). Lesions in *Escherichia coli* cell walls caused by the action of mouse complement. *Immunology*, **20**, 441–50.

MILES, A. A., MILES, E. M. & BURKE, J. (1957). The value and duration of defence reactions of the skin to the primary lodgement of bacteria. *British Journal of Experimental Pathology*, **38**, 79–96.

MILNE, C. M. (1966). *Lysozyme and immune bacteriolysis*. M.Sc. Thesis. University of Bristol.

MULLER-BÉRAT, C. N. (1969). Immunosuppressive action of L-asparaginase studied by means of the localised haemolysis in gel assay (L.H.G.A.). *Acta Pathologica et Microbiologica Scandinavica*, **77**, 750–2.

MUSCHEL, L. H. (1960). Bactericidal activity of normal serum against bacterial cultures. II. Activity against *Escherichia coli* strains. *Proceedings of the Society for Experimental Biology and Medicine*, **103**, 632–6.

MUSCHEL, L. H., CHAMBERLAIN, R. H. & OSAWA, E. (1958). Bactericidal activity of normal serum against bacterial cultures. I. Activity against *Salmonella typhi* strains. *Proceedings of the Society for Experimental Biology and Medicine*, **97**, 376–82.

MUSCHEL, L. H. & TREFFERS, H. P. (1956). Quantitative studies on the bactericidal actions of serum and complement. III. *Journal of Immunology*, **76**, 20–7.

NAGINGTON, J. (1956). The sensitivity of *Salmonella typhi* to the bactericidal action of antibody. *The British Journal of Experimental Pathology*, **37**, 397–405.

NOBLE, W. C. (1965). The production of subcutaneous staphylococcal skin lesions in mice. *British Journal of Experimental Pathology*, **46**, 254–62.

ØRSKOV, F. & ØRSKOV, I. (1970). Immunoelectrophoretic patterns of extracts from all *Escherichia coli* O and K antigen test strains. *Acta Pathologica et Microbiologica Scandinavica*, **78B**, 263–4.

PANTON, P. N. & VALENTINE, F. C. O. (1929). Staphylococcal infection and reinfection. *British Journal of Experimental Pathology* **10**, 257–62.

PTAK, W., GAUGAS, J. M., REES, R. J. W. & ALLISON, A. C. (1970). Immune responses in mice with murine leprosy. *Clinical and Experimental Immunology*, **6**, 117–24.

ROANTREE, R. J. & PAPPAS, N. C. (1960). The survival of strains of enteric bacilli in the blood stream as related to their sensitivity to the bactericidal effect of serum. *Journal of Clinical Investigation*, **39**, 82–8.

ROANTREE, R. J. & RANTZ, L. A. (1960). A study of the relationship of the normal bactericidal activity of human serum to bacterial infection. *Journal of Clinical Investigation*, **39**, 72–81.

ROBBINS, J. B., KENNY, K. & SUTER, E. (1965). The isolation and biological activities of rabbit γM and γG anti-*Salmonella typhimurium* antibodies. *Journal of Experimental Medicine*, **122**, 385–402.

ROGERS, H. J. (1967). Bacteriostatic effects of horse sera and serum fractions on *Clostridium welchii* type A, and the abolition of bacteriostasis by iron salts. *Immunology*, **12**, 285–301.

ROGERS, H. J., BULLEN, J. J. & CUSHNIE, G. H. (1970). Iron compounds and resistance to infection. Further experiments with *Clostridium welchii* type A *in vivo* and *in vitro*. *Immunology*, **19**, 521–38.

ROTHER, K., ROTHER, U., PETERSEN, K. F., GEMSA, D. & MITZE, F. (1964). Immune bactericidal action of complement. *Journal of Immunology*, **93**, 319–30.

ROWLEY, D. (1954). The virulence of strains of *Bacterium coli* for mice. *British Journal of Experimental Pathology*, **35**, 528–38.

ROWLEY, D. (1955). Stimulation of natural immunity to *Escherichia coli* infections. *The Lancet*, **i**, 232–4.

ROWLEY, D. (1964). Endotoxin induced changes in susceptibility to infections. In *Bacterial endotoxins*, ed. M. Landy and W. Braun, 359–72. New Brunswick: Rutgers, The State University.

ROWLEY, D. & TURNER, K. J. (1966). Number of molecules of antibody required to promote phagocytosis of one bacterium. *Nature, London*, **210**, 496–8.

ROWLEY, D. & TURNER, K. J. (1968). Passive sensitisation of *Salmonella adelaide* to the bactericidal action of antibody and complement. *Nature, London*, **217**, 657–8.

SALAMAN, M. H. (1970). Immunodepression by mammalian viruses and plasmodia. *Proceedings of the Royal Society of Medicine*, **63**, 11–15.

SCHADE, A. L. & CAROLINE, L. (1944). Raw hen egg white and the role of iron in growth inhibition of *Shigella dysenteriae*, *Staphylococcus aureus*, *Escherichia coli* and *Saccharomyces cerevisiae*. *Science*, **100**, 14–15.

SCHADE, A. L. & CAROLINE, L. (1946). An iron binding component in human blood plasma. *Science*, **104**, 340–1.

SCHWARTZ, R. W. (1969). Immunosuppression by L-asparaginase. *Nature, London*, **224**, 275–6.

SHANDS, J. W. (1965). Localisation of somatic antigen on Gram-negative bacteria by electron microscopy. *Journal of Bacteriology*, **90**, 266–70.

SHAW, G. B. (1911). *The Doctor's Dilemma*. London: Constable.

SJÖQUIST, J. & STÅHLENHEIM, G. (1969). Protein A from *Staphylococcus aureus*. IX. Complement fixing activity of protein A–Ig G complexes. *Journal of Immunology*, **103**, 467–73.

SJÖSTEDT, S. (1946). Pathogenicity of certain serological types of *B. coli*. *Acta Pathologica et Microbiologica Scandinavica*, **63** Suppl., 1–148.

SMITH, H. (1968). Biochemical challenge of microbial pathogenicity. *Bacteriological Reviews*, **32**, 164–84.

SPINK, W. W. & KEEFER, C. S. (1937). Studies of gonococcal infection. I. A study of the mode of destruction of the gonococcus *in vitro*. *Journal of Clinical Investigation*, **16**, 169–76.

SPITLER, L., COCHRUM, K. & FUDENBERG, H. H. (1968). Mycoplasma inhibition of phytohemagglutinin stimulation of lymphocytes. *Science*, **161**, 1148–9.

STONE, H. H., GIVEN, K. S. & MARTIN, J. D. (1967). Delayed rejection of skin homografts in pseudomonas sepsis. *Surgery, Gynaecology and Obstetrics*, **124**, 1067–70.

SWORD, C. P. (1966a). Serum protein alterations induced by *Listeria monocytogenes* infections. *Journal of Immunology*, **96**, 790–6.

SWORD, C. P. (1966b). Mechanisms of pathogenesis in *Listeria monocytogenes* infection. I. Influence of iron. *Journal of Bacteriology*, **92**, 536–42.

TAYLOR, C. E. D. & WALTON, K. W. (1957). The molecular characteristics determining the anticomplementary activity of dextran sulphates. *British Journal of Experimental Pathology*, **38**, 248–55.

TERRY, W. D. (1964). Subclasses of human Ig G molecules differing in heterologous skin sensitising properties. *Proceedings of the Society for Experimental Biology and Medicine*, **117**, 901–4.

TORIKATA, R. (1917). In J. P. Scott, (1931). Aggressins. An outline of the development of the theory and notes on the use of these products. *Journal of Bacteriology*, **22**, 323–37.

TURK, J. L. (1970). Immunological aspects of clinical leprosy. *Proceedings of the Royal Society of Medicine*, **63**, 1053–6.

TURK, J. L. & WATERS, M. F. R. (1968). Immunological basis for depression of cellular immunity and the delayed allergic response in patients with lepromatous leprosy. *The Lancet*, **ii**, 436–8.

TURK, J. L. & WATERS, M. F. R. (1969). Cell mediated immunity in patients with leprosy. *The Lancet*, **ii**, 243–6.

VAHLNE, G. (1945). Serological typing of the colon-bacteria. *Acta Pathologica et Microbiologica Scandinavica*, **62** Suppl., 1–127.

WALDORF, D. S., SHEAGREN, J. N., TRAUTMAN, J. R. & BLOCK, J. B. (1966). Impaired delayed hypersensitivity in patients with lepromatous leprosy. *The Lancet*, **ii**, 773–6.

WALSH, M. J., BROWN, B. C., BROWN, L. & PIRKLE, C. I. (1963). Use of the chick embryo in maintaining and restoring virulence of *Neisseria gonorrhoeae*. *Journal of Bacteriology*, **86**, 478–81.

WARD, H. K. & ENDERS, J. F. (1933). An analysis of the opsonic and tropic action of normal and immune sera based on experiments with the pneumococcus. *Journal of Experimental Medicine*, **57**, 527–47.

112   A. A. GLYNN

WARD, M. E. (1970). *The antigenic structure and virulence of Neisseria gonorrhoeae.* Ph.D Thesis, University of London.

WARD, M. E., WATT, P. J. & GLYNN, A. A. (1970). Gonococci in urethral exudates possess a virulence factor lost on subculture. *Nature, London,* **227**, 382–4.

WARING, W. S., ELBERG, S. S., SCHNEIDER, P. & GREEN, W. (1953). The role of iron in the biology of *Brucella suis*. I. Growth and virulence. *Journal of Bacteriology,* **66**, 82–91.

WEINBERG, E. D. (1966). Roles of metallic ions in host–parasite interactions. *Bacteriological Reviews,* **30**, 136–51.

WEKSLER, B. B. & HILL, M. J. (1969). Inhibition of leucocyte migration by a staphylococcal factor. *Journal of Bacteriology,* **98**, 1030–5.

WHANG, H. Y. & NETER, E. (1967). Immunosuppression by endotoxin and its lipoid A component. *Proceedings of the Society for Experimental Biology and Medicine,* **124**, 919–24.

WILSON, G. S. & MILES, A. A. (1964). *Topley and Wilson's Principles of Bacteriology and Immunity,* 5th edition. London: Edward Arnold.

YAMAMOTO, T., NISHIURA, M., HARADA, N. & IMAEDA, T. (1958). Electron microscopy of *Mycobacterium lepraemurium* in ultra thin sections of murine leprosy lesions. *International Journal of Leprosy,* **26**, 111–13.

SPECIFIC AND NON-SPECIFIC EFFECTS OF BACTERIAL INFECTION ON THE HOST

H. B. STONER

Experimental Pathology of Trauma Section, MRC Toxicology Unit, Medical Research Council Laboratories, Woodmansterne Road, Carshalton, Surrey

INTRODUCTION

At first sight the bacterial infections to which animals are susceptible appear to be distinct entities distinguished from one another by the responses they evoke. In fact, the matter is more complicated. Although many organisms produce specific effects, much of what happens in bacterial infection reflects the standard responses of the body to the types of cellular damage produced. The distinction between pathogens is often more closely related to the site of the attack and the limitations on the response imposed by the site, e.g. bone, than to the actual nature of the response. The target tissue may be much more pathognomonic of an infecting organism than the reaction it produces. The effects of pathogenic bacteria could probably be reproduced by chemical and physical injuries if they could be applied to the correct point in the metabolic pathways of the right cells. Bacterial injury (infection) is distinguished from other types by the ability of the infecting agent to deliver the injury to specific and often localised sites in the body.

As with other forms of injury, the host response to bacterial infection can be divided into the local response at the site of injury and the general response in the rest of the body. Both responses have specific and non-specific features which must be appreciated and defined if the disease is to be understood and treated rationally. In this context, the specific features of the response are those related to the primary effect of the bacteria produced, directly or through a toxin, on a defined metabolic process in specific cells. The non-specific features are those which occur secondarily to the primary injury and represent the common response to that particular form of injury such as the inflammatory response to the liberation of endogenous permeability factors and the homeostatic response to hypovolaemia. It is difficult to generalise beyond this point, and the thesis is best developed by the discussion of examples. It will, indeed, be shown that the distinction is an artificial one, for a lesion which gives rise to a common response may be brought about in a very specific way. Conversely, what may appear as a specific response to a bacterial toxin, e.g. paralysis, may be the only possible consequence of the metabolic derangement it causes and would always occur however it was produced. Separation of effects into specific and non-specific may simply reflect the depth of knowledge about an infection. I hope to show that although these terms are used rather loosely, this separation can be useful in dissecting the host response to bacterial infection. It must be remembered that there is still a great deal to be learnt about the mechanisms of these host responses.

PATHOGENIC BACTERIA WITH EXOTOXINS

Many pathogenic bacteria produce characteristic and specific protein exotoxins and the associated illnesses supply a wide spectrum of responses. In some the changes occurring in the host seem completely dominated by the specific action of the exotoxin. In others, although the mode of action of the exotoxin may be specific, the changes produced are those which lead to non-specific responses in the host. In yet others it is a matter of debate which parts of the illness are due to specific exotoxins and which to the non-specifically mediated effects of the pathogen.

Botulism and tetanus

These two conditions are good examples of one end of this spectrum, where the toxins are of overriding importance in the disease. Botulism is an intoxication. The exotoxin is formed during the growth of *Clostridium botulinum* in food before it enters the body. The contaminating toxin is absorbed when the food is eaten and produces its effects by inhibiting the release of acetylcholine at myoneural junctions (van Heyningen & Arseculeratne, 1964). There is no local lesion in the gut and, although its mechanism of action at the myoneural junction is not completely understood, all the primary effects on the host stem from its action at this site. This would be classed as a specific host response in the sense in which this term is being used although the same effects would be seen if acetylcholine release was similarly inhibited by other means.

In tetanus the toxin is formed in the body after infection of wounds with *Clostridium tetani*. The wound may be very small and the infection does not produce any characteristic local response. The toxin is absorbed, passes up the regional motor nerves to the central nervous system where it binds to ganglioside and acts, like strychnine, by suppressing synaptic inhibition (van Heyningen & Arseculeratne, 1964). The mode of action of the toxin is not completely understood but its effect on the central nervous system accounts for the early condition of the patient.

In both botulism and tetanus the interference by a toxin at a specific site in the peripheral or central nervous system is of overwhelming importance in producing the picture of the disease and these effects can be regarded as specific. In both conditions, as the patient progresses towards death further symptoms will appear, but these are clearly secondary to the prime effect of the toxin, e.g. hypoxia from respiratory paralysis. There should be no difficulty in distinguishing between the specific neurological changes produced by these toxins and the terminal sequelae of those actions.

Cholera

Vibrio cholerae does not invade the intestinal mucosa but acts through a potent exotoxin which will probably prove just as specific in its mode of action as those just mentioned. Nevertheless its effect is to produce a non-specific state of hypovolaemia which determines the patient's condition and his treatment. In this respect cholera is at the opposite end of the spectrum to botulism and tetanus. Although antibiotics are used in its treatment, the main event is the loss of fluid and unless this is replaced, the patient will die in a state of hypovolaemic shock showing all the characteristics of that condition irrespective of the immediate cause.

The mode of action of the cholera toxin is not yet clear (Craig, this volume). What is certain is that without causing disruption of the intestinal mucosa (Patnaik & Ghosh, 1966; Chen, Reyes & Fresh, 1971) it leads to the loss of enormous amounts of more-or-less isotonic electrolyte solution (Phillips, 1964) from the body through the anus. There is no loss of protein in the stools. The symptoms and signs of the general response to infection with *V. cholerae* are those of the response to a very pure form of fluid loss. Any difference from the hypovolaemia found after physical injury such as burns and haemorrhage can be attributed to the absence of any loss of erythrocytes or plasma protein. This may account for the success of adequate fluid and electrolyte replacement therapy in cholera. The general appreciation of this form of therapy has reduced the mortality rate to between 1–3 % in those centres where it is practised. The deaths which occur under these conditions are in patients in which treatment has been delayed.

Cholera has been emphasised for two reasons. It is an excellent example of a disease in which the most important clinical features are due to a non-specific, in the sense in which that term is being used, effect of the infection, namely fluid loss. It also shows how loss of fluid from the circulation can be important in bacterial infection. Ignoring the specific ways in which bacteria may bring this about, once it has occurred, its effects and the body's response to it may be classed as non-specific since they are the same as those following fluid loss from other causes. Two other bacterial infections in which fluid loss can play an important part will now be considered.

Anthrax

Anthrax is a complicated condition and death in the generalised form of the disease is believed to be due to the three components of the toxic complex produced by *Bacillus anthracis* (Smith & Stoner, 1967). There are species differences in the responses to the toxic complex, but an important effect is that on the vascular system leading to an increase in permeability and fluid loss. The mechanism of this action is not known. Oedema and haemorrhage in the tissues are important pathological findings in this disease (Gleiser, 1967) and much of what happens in the host can be looked upon as a non-specific response to the fluid loss.

Smith and his co-workers (Smith, 1960) showed that in fatal infection in guinea-pigs there was a progressive fall in the blood volume with haemoconcentration and a decline in the arterial blood pressure. The reported changes in the blood chemistry, such as hyperglycaemia followed by hypoglycaemia, terminal increase in blood lactate and fall in pH (Smith, 1960; Lincoln & Fish, 1970) are what one would expect as part of the standard response to this form of injury. In the untreated guinea-pigs studied by Smith, death occurred in a state of oligaemic shock and this dominated the clinical picture of the disease in those animals. If the infection was halted by streptomycin when the guinea-pigs were in a state of shock but some hours before they were expected to die, death still occurred but it was delayed and the mechanism was different. These treated animals appeared able to compensate for the oligaemia but died of renal failure. This alteration in renal function does not represent an additional action of the anthrax toxic complex. Histological examination of the kidneys revealed tubular damage (Ross, 1955) and changes very similar to those seen in other forms of shock (Stoner, 1955). The situation in these animals is very like that in patients who have suffered a severe crushing injury to a limb and who survive the initial shock phase to die later of renal failure from the kidney damage which occurred during the shock phase.

It is now perhaps possible to see how this separation of host responses into specific and non-specific can be used in the analysis of the disease process. In generalised anthrax there is loss of fluid from the circulation leading to non-specific effects which can cause death in two ways. The disease is due to the toxin and the toxic complex has been shown to cause fluid loss quite apart from its particular action in causing pulmonary oedema in the Fischer 344 rat (Dalldorf *et al.* 1969). Is this the whole explanation? It has been claimed that the toxin complex has a specific effect on the central nervous system and that this plays a part in the pathogenesis of anthrax (Lincoln & Fish, 1970). It should be possible to decide whether actions of the toxin complex additional to that on vascular permeability are required in several ways.

(1) By measuring the reduction in the circulating fluid volume and, by comparison with other conditions, seeing whether it is sufficient to explain the various changes in and death of the animal.

(2) By comparing the changes in central nervous system function in infected or toxin treated animals with those produced by hypovolaemia (Kovách, 1970).

(3) By observing the train of events after effective replacement of the lost fluid. It must, of course, be realised that in the presence of a continuing infection in which the injury is constantly renewed, it is more difficult to replace the fluid than after a physical or toxic injury. The value of attempting to separate the specific and non-specific effects and the responses is also seen when one considers clostridial infections.

Clostridial infections

The clostridial group of bacteria produce a wide range of exotoxins with well characterised properties. It might be thought, therefore, that infection by members of this group would be distinguished by specific host effects produced by these toxins. In fact, this is not the case. The position has been well discussed recently by Bullen (1970) using infection by *Clostridium welchii* type D and type A as illustrations.

Cl. welchii type D produces enterotoxaemia in sheep, a disease characterised by central nervous system changes such as convulsions, coma, etc. The symptoms and signs of the disease can be reproduced by the ε-toxin produced by the organism and adequate levels of circulating ε-antitoxin provide very good protection against the disease. Thus it is thought that this condition is produced by the specific effects of the toxin absorbed from the gut of the infected sheep.

The position in *Cl. welchii* type A infections is entirely different. This is the organism associated with gas gangrene. It is known to produce a number of toxins of which the α-toxin is the most important. However, the role of these toxins, particularly the α-toxin, in the production of this disease is far from clear. This question has been discussed in detail by MacLennan (1962) and Bullen (1970) has re-iterated the difficulties of correlating the virulence of a strain with its toxigenic potency and the limited value of antitoxin therapy in this disease.

Experimental studies on passively immunised guinea-pigs strongly suggested that the exotoxins did not play any essential part in the pathogenesis of this condition (Bullen & Cushnie, 1962, 1963; Bullen, Cushnie & Stoner, 1966; Stoner, Bullen, Cushnie & Batty, 1967). Death could be produced in these animals by injecting the organisms intramuscularly or intraperitoneally in such a way that they survived and multiplied. They then became invasive, caused tissue destruction and loss of fluid from the circulation in amounts which were sufficiently large to be fatal.

Thus we see that the clostridial group of organisms contains both pathogens which produce their effects by the specific action of their exotoxins and others which, while possessing potent exotoxins, can give rise to disease through the production of non-specific tissue damage. Since the way in which this tissue damage is produced is not known it is, of course, begging the question to describe it as non-specific. However, the changes it causes are the same as if it had been produced by non-bacterial agents so that it may be justly termed non-specific in the sense in which that term is being used here. The distinction is important because of its therapeutic implications. In the first case therapy must be immunological but in the second the main attack must be with surgery and fluid replacement and attempts to arrest the growth of the organisms with hyperbaric O_2 and antibiotics.

These examples of pathogens with exotoxins have been chosen because sufficient is known about them to illustrate how in some cases the exotoxin causes the changes in the host associated with the disease while in others it does not and how, in both cases, care must be taken to distinguish between specific pharmacological responses to the exotoxin and responses which are common to the type of injury produced. It would be nice to be able to say that many other bacterial diseases could have been chosen for discussion in this way. This would probably not be true although many other pathogens produce exotoxins. Our aim must be to acquire sufficient data so that the mode of action of any pathogen can be discussed in these terms. In doing so the temptation to say at the end of the long struggle to isolate, purify, identify and characterise an exotoxin, that the exotoxin is the cause of the host response must be resisted. It may be so, but before that conclusion can be reached the biological properties of the toxin must be studied in detail. This means knowing much more than the LD_{50} value. One must know how the body is affected, ideally the mechanism of its action.

Throughout one must distinguish between what can reasonably be termed a primary, specific pharmacological response to the toxin and what are better described as the secondary responses to the primary effect. These distinctions may be difficult to make. Is the hyperglycaemia seen after *Cl. welchii* type D toxin injection due to its direct effect on centres in the nervous system or is it secondary to increased sympathetic-adreno-medullary activity following the convulsions? Inspection of the literature indicates that considerably more sophistication is required in these investigations, particularly in the pathophysiological ones. The presence of abnormal electrocardiographic complexes shortly before death does not necessarily mean that the toxin has a direct action on the heart. Such abnormalities can be recorded in animals dying from many causes not primarily affecting the heart. At the same time one must guard against 'over-sophistication'. From the point of view of patho-genicity, studies on the effects of toxins on cellular organelles, mito-chondria, etc., are only relevant if they can be shown to reach them *in vivo* in effective amounts (Smith, 1969).

Having explored the biological potentialities of the purified toxin the next step is to compare the syndrome produced by the toxin with the host response to the infection, bearing in mind that the single intra-venous injection of a lethal dose of toxin may not imitate its slower liberation from the bacteria too closely. The identification of chemical factors concerned in disease processes simply by comparison between their effects and that of the disease is always unsatisfactory. However, in this case an additional weapon is available since the exotoxins are antigenic and there is the possibility of obtaining a potent antitoxin. If the toxin plays an important part in the pathogenic effects of the bacteria prophylaxis and/or therapy with the antitoxin would be expected to be beneficial.

MEDIATED RESPONSES

Non-specific responses to bacteria and their products may be due to the liberation of endogenous pharmacological agents. These responses are the second part of a two stage reaction. In the first stage the bacteria or its products, including antigen–antibody complexes for those responses which depend on the previous knowledge of the organism by the host, react with certain host cells. These cells then proceed to liberate pharmacological agents such as histamine, serotonin etc. which cause their typical effects. Thus, although these host cells may be stimulated by bacteria in a wide variety of ways, their response does not vary and what is seen is the standard host response to the particular agent which is liberated. The same response is seen when the same agent is liberated by non-bacterial means.

A good example of this is fever which is a common host response to many bacterial infections both Gram-negative and positive. The production of fever by bacteria has now been established as a two stage process. Firstly, the organisms liberate bacterial pyrogens of which the endotoxin lipopolysaccharides are examples. The bacterial pyrogens act on polymorphonuclear leucocytes, monocytes, macrophages, Kupffer cells and possibly others as well, to generate endogenous pyrogens. These in turn act on the thermoregulatory centres in the hypothalamic region of the brain to produce fever. Many features of this intricate system have been discussed in a recent Ciba Foundation symposium (Wolstenholme & Birch, 1971). A non-specific response is not necessarily a simple one.

Bacterial pyrogens are produced by a wide variety of bacteria and their chemical structures are not fully known. The most studied have been the endotoxins of the Gram-negative bacteria which probably consist of complexes of protein, phospholipid and lipopolysaccharide (Work, 1971). Toxicity appears to be related to the lipopolysaccharide component. Although there are differences between the bacterial pyrogens they are sufficiently similar to be able to react with leucocytes and some other cells to lead to the production of endogenous pyrogen. The chemical composition of this is not known but it has a specific action on the thermoregulatory centres and this provides the final common path in the production of fever.

Although these bacterial products are spoken of as pyrogens, it is worth remembering that at ordinary room temperatures, they are not pyrogenic in all species. In the rat (Stoner, 1961; Porter & Kass, 1965; Filkins & Di Luzio, 1968) and the mouse (Prashker & Wardlaw, 1971) they usually depress body temperature. This may depend on the environmental temperature at the time of the experiment and its relation to the thermoneutral zone of the species concerned.

In these mediated responses it is important to take account of species variations in the reactions to these pharmacological agents. This is clearly illustrated by the differences in the form taken by anaphylactic shock in different species. In the guinea-pig it is seen as bronchospasm, whereas in the dog the main symptoms are related to contraction of the hepatic veins, yet the reaction is fundamentally the same. Histamine is liberated and leads to the contraction of smooth muscle, and the result depends on the distribution of smooth muscle in the different species. It is probably correct to look upon all the immune responses of this type in the host as non-specific and mediated. They possess immunological specificity but once the reaction has been set in motion, it runs its normal course irrespective of the causative organism. The Shwartzman reactions, if they can be considered immunological responses, could be included in this group. The main problems concern the nature of the mechanisms which trigger off the reactions and the point at which the final common path begins. These reactions, like some in the following paragraph, are related in varying degree to the 'triggered enzyme reactions' of Macfarlane (1969).

A further important type of mediated response is inflammation. The mechanism of the acute inflammatory reaction still lacks a complete explanation. The current view is that the changes in vascular permeability, the key part of the reaction, are brought about by the action of pharmacological mediators. The response is divided into two stages, early and delayed, and while the changes of the early stage are attributed to the action of compounds such as histamine, serotonin and plasma kinin, the mediators of the delayed response are not yet known. Nevertheless, it is widely believed that the final explanation will be in these terms. The generation of plasma kinin has been particularly intensively studied (Miles, 1969).

Many, probably the majority, of the pathogenic bacteria cause an acute inflammatory response at the site of infection. This response is non-specific in the sense that it does not differ essentially from acute inflammatory responses caused by chemical or physical agents. Nevertheless, there are differences in the details of the responses to different organisms. The early stage of the inflammatory response to bacteria and their products is usually short-lived, but there are considerable differences in the duration of the delayed response (Miles & Wilhelm, 1960). These are unexplained but may reflect differences in the speed with which different pathogens bring about a common cellular change which provokes the response.

The investigation of the delayed response is proving very difficult and it may be that pathogenic bacteria, or more probably their toxins, could serve as useful tools in this work. Most work on inflammation has been done on skin, and Steele & Wilhelm (1970) have wondered whether the usual diphasic nature of the response there could not be explained by the diversion of blood flow to different parts of its complicated micro-circulation. The blood supply to the parietal peritoneum is much simpler yet a diphasic permeability response is also seen there after the intraperitoneal injection of *Cl. welchii* α-toxin (Stoner *et al.* 1967). These experiments had a further interest. As in most inflammatory responses, the permeability defect in the early stage was more-or-less confined to the venules whereas in the delayed response capillaries were affected as well. When the guinea-pigs were passively immunised the intra-peritoneal injection of α-toxin still caused slight venular changes but there was no capillary damage. Similarly the intraperitoneal injection of *Cl. welchii* type A into the peritoneal cavity of passively immunised guinea-pigs only produced visible (carbon-labelling technique) changes in the venules, as did the injection of dead organism although to a lesser degree. It may be that with bacteria and their products one could devise models which would give useful separation of the two stages of the response.

'SEPTIC SHOCK'

'Septic shock' is a term applied to a clinical condition associated with infection, usually with bacterial invasion of the blood stream, and distinguished by fever, circulatory collapse with hypotension, respiratory distress and acute changes in mental state (Shubin & Weil, 1963). This condition, first described by Laennec in 1831 (quoted by Altemeier & Cole, 1956), may appear and develop so rapidly that the differential diagnosis includes acute medical emergencies such as pulmonary embolism (Walters & McGowan, 1963). It must be classed as a non-specific host response since its form and course give no indication of the causative organism. While it may arise in the course of many infections, it seems particularly common after instrumentation of an infected urinary tract in the elderly and is best known in association with Gram-negative bacteria.

Because of this association it is frequently spoken of as endotoxin shock and this is the type which has been most studied. In the case of *Escherichia coli* the condition produced by the live organism has been shown to be the same as that produced by its endotoxin (Guenter, Fiorica & Hinshaw, 1969). This justifies the concentration of research effort on the more easily handled endotoxin. It is perhaps a pity that so much work has been done on the effects of single intravenous injections of large doses rather than on those produced by intravenous infusions which might simulate natural events more closely.

The same condition, or one at present indistinguishable, has been reported in Gram-positive infections and also in leptospiral, rickettsial and viral conditions (Ebert & Abernathy, 1961). Their original list included malaria but I have excluded this since other mechanisms are involved. Although the 'septic shock' produced by *E. coli* and *Staphylococcus aureus* has been described as very similar (Elsberry, Rhoda &

Biesel, 1969; Guenter & Hinshaw, 1970) this may only refer to the final state and the approaches to that final state may vary with different organisms. Further research on the possible routes to the final shock-like state is needed and the identification of the differences between them could be of therapeutic importance. At present the best-known pathway is that in endotoxin shock.

There is now a vast literature on the pathophysiological effects of bacterial lipopolysaccharide endotoxin which produces many other effects besides 'septic shock'. For a particular species endotoxin shock is a well-defined condition, albeit non-specific in the sense that the same state is produced by endotoxin from different sources. At the same time endotoxins also produce specific host responses, particularly immunological ones. The whole question of the role of bacterial endotoxin in disease is a complicated one and, as pointed out by Work (1971) the position today is little different from the one elegantly summarised by Bennett in 1964. However, it seems clear that if large amounts of endotoxin are released into the circulation during an infection a shock-like state will be produced in the host.

The mechanism of endotoxin shock is not yet fully understood. The degree of hypotension is out of proportion to any oligaemia which is usually slight (Gilbert, 1960; Walters & McGowan, 1965). The shock-like state would seem to arise from an effect on the vascular system most probably mediated through endogenous pharmacological agents, noradrenaline, histamine, serotonin etc. (Gilbert, 1960). In view of this it is hardly surprising that there are well-marked species differences in the vascular regions most affected by endotoxin (Gilbert, 1960). In the dog an early fall in arterial pressure and pooling of blood in the splanchnic region are the main effects (Gilbert, 1960; Chien et al. 1966) although the latter has not always been found (Park, Baum & Guntheroth, 1970). The cat also shows an early precipitous fall in blood pressure but this can be ascribed to pulmonary vascular constriction, there being no hepatic vein constriction or splanchnic pooling (Kuida et al. 1961). In the monkey and man the fall in blood pressure is slower and the main changes are found in the lungs (Vaughn, Gunter & Stookey, 1968) as in the sheep (Halmagyi, Starzecki & Horner, 1963). Endotoxin shock is a further example of the complexity of non-specific host responses. It is unfortunate that so much of the physiological study of endotoxin shock has been carried out in the dog where the resemblance to the human condition is least. The closest resemblance is in the simian response.

Endotoxin shock should perhaps have been included among the

mediated responses and that will be its final position in any classification of host responses to bacterial infection. At present it is best retained as part of a wider group labelled 'septic shock' until the confusion of that subject is more resolved. Mediators and mechanisms cannot be suggested in all cases. In general, much further investigation is needed of the primary reactions between microbial products and the cells of the host which trigger off a consequential series of responses.

INFLUENCE OF NORMAL BODY FLORA ON NON-BACTERIAL CONDITIONS

In a discussion of the specific and non-specific effects of bacteria on the host, it may not be out of place to consider how non-bacterial conditions could be influenced by the normal bacterial flora of the body. Several examples of this interaction can be given.

Not all species have sterile tissues. The skeletal muscles of dogs and goats harbour *Cl. welchii* which can be activated by ischaemia or trauma and it has been shown that the action of these bacteria contributes to the effect of these injuries on the host (Aub *et al.* 1944, 1945; Lindsey, Wise, Knecht & Noyes, 1959).

Most interest in this connection concerns the role of the intestinal flora. These organisms may be a source of infection in wounds and burns but they may also affect the host in more subtle ways. For instance, it has been shown that the death of rats in uraemia after bilateral nephrectomy is affected by the presence and nature of the intestinal flora although the mechanism of the effects is obscure (Einheber & Carter, 1966; Carter *et al.* 1966). While the intestinal wall offers a fairly efficient barrier to the endotoxin released in the lumen small amounts are continuously absorbed into the portal blood (Ravin, Rowley, Jenkins & Fine, 1960). It has been suggested that this endotoxin could play a significant part in certain conditions such as the irreversible, terminal stage of haemorrhagic and other forms of shock (Fine *et al.* 1960; Fine, 1961; Fine, Palmerio & Rutenberg, 1968; Palmerio & Fine, 1969). The mechanism proposed may be summarised as follows.

Under normal conditions the endotoxin absorbed from the gut is inactivated by the reticulo-endothelial system and the capacity of that system is sufficient to ensure that the host is unaffected. If the activity of the reticulo-endothelial system is depressed the absorbed endotoxin will survive to act on the host and if that depression is sufficiently severe, death will ensue from the neurotoxic effects of the endotoxin. The main source of the endotoxin is the *E. coli* in the gut.

This theory had a considerable appeal since it could explain several difficult problems in the response to fatal trauma. It has consequently been intensively investigated. Although it is established that trauma (haemorrhage, limb ischaemia, etc.) is associated with depression of the activity of the reticulo-endothelial system the effect of this on the host response is not clear. Many attempts have been made to substantiate this theory, using many different types of injury and a wide range of techniques in both conventional and germ-free animals, without success (e.g. Einheber, 1961; Stoner, 1961; Altura, Thaw & Hershey, 1966; Little & Stoner, unpublished results). Despite the general inability to confirm Fine's theory, it would be unwise to say that the intestinal bacteria play no part in the effects of physical trauma. There is a persistent trickle of evidence (Markley, Smallman, Evans & McDaniel, 1965; Markley, Smallman & Evans, 1967; Einheber, Wren & Dobek, 1970) which suggests that they are involved in some way. At present this must be put among the non-specific effects of bacteria on the host.

CONCLUSIONS

In the present state of knowledge there is still merit in distinguishing between the specific and non-specific effects of pathogenic bacteria on the host. In the final analysis the action of a pathogen will be described in terms of the primary reactions between the organism and its products and the cells of the host and the secondary reactions set in motion by them. At present a full description on these lines can only be given in a few cases. Meanwhile, the study of the host responses to identify the specific and non-specific elements should continue as a preliminary to their complete explanation. As shown above such investigations can help both in understanding the mechanisms of a disease and in its treatment.

REFERENCES

ALTEMEIER, W. A. & COLE, W. (1956). Septic shock. *Annals of Surgery*, **143**, 600.

ALTURA, B. M., THAW, C. & HERSHEY, S. G. (1966). Role of bacterial endotoxins in intestinal ischemic (SMA) shock. *Experientia*, **22**, 786.

AUB, J. C., BRUES, A. M., DUBOS, R., KETY, S. S., NATHANSON, I. T., POPE, A. & ZAMECNIK, P. C. (1944). Bacteria and the 'Toxic Factor' in shock. *War Medicine*, **5**, 71.

AUB, J. C., BRUES, A. M., KETY, S. S., NATHANSON, I. T., NUTT, A. L., POPE, A. & ZAMECNIK, P. C. (1945). The toxic factors in experimental traumatic shock. IV. The effects of the intravenous injection of the effusion from ischemic muscle. *Journal of Clinical Investigation*, **24**, 845.

BENNETT, I. L. (1964). Approaches to the mechanism of endotoxin action. In *Bacterial Endotoxins*, ed. M. Landy and W. Braun, xiii–xvi. New Jersey: Institute of Microbiology Rutgers University.

BULLEN, J. J. (1970). Role of toxins in host–parasite relationships. In *Microbial Toxins*, ed. S. J. Ajl, S. Kadis and T. C. Montie, vol. 1, 233. New York: Academic Press.

BULLEN, J. J. & CUSHNIE, G. H. (1962). Experimental gas gangrene: The effect of antiserum on the growth of *Clostridium welchii* type A. *Journal of Pathology and Bacteriology*, **84**, 177.

BULLEN, J. J. & CUSHNIE, G. H. (1963). The failure of antitoxin to protect guinea-pigs against intraperitoneal injection with *Clostridium welchii* type A. *Journal of Pathology and Bacteriology*, **86**, 345.

BULLEN, J. J., CUSHNIE, G. H. & STONER, H. B. (1966). Oxygen uptake by *Clostridium welchii* Type A: its possible role in experimental infections in passively immunised animals. *British Journal of Experimental Pathology*, **47**, 488.

CARTER, D., EINHEBER, A., BAUER, H., ROSEN, H. & BURNS, W. F. (1966). The role of the microbial flora in uremia. II. Uremic colitis, cardiovascular lesions and biochemical observations. *Journal of Experimental Medicine*, **123**, 251.

CHEN, H. C., REYES, V. & FRESH, J. W. (1971). An electron microscope study of the small intestine in human cholera. *Virchow's Archives B*, **7**, 236.

CHIEN, S., DELLENBACK, R. J., USAMI, S., TREITEL, K., CHANG, C. & GREGERSEN, M. I. (1966). Blood volume and its distribution in endotoxin shock. *American Journal of Physiology*, **210**, 1411.

DALLDORF, F. G., BEALL, F. A., KRIGMAN, M. R., GOYER, R. A. & LIVINGSTON, H. L. (1969). Transcellular permeability and thrombis of capillaries in anthrax toxaemia. *Laboratory Investigation*, **21**, 42.

EBERT, R. V. & ABERNATHY, R. S. (1961). Septic Shock. *Federation Proceedings*, **20**, Supplement 9, 179.

EINHEBER, A. (1961). Discussion. *Federation Proceedings*, **20**, Supplement 9, 170.

EINHEBER, A. & CARTER, D. (1966). The role of the microbial flora in uremia. I. Survival times of germ-free, limited flora and conventionalized rats after bilateral nephrectomy and fasting. *Journal of Experimental Medicine*, **123**, 239.

EINHEBER, A., WREN, R. E. & DOBEK, A. S. (1970). Mortality, morphologic changes and saline therapy after scald injury of germ-free mice and pseudomonas-free conventionalized mice with or without *Proteus mirabilis*. An inquiry into a possible non-infective role of the microbial flora. *Journal of Trauma*, **10**, 135.

ELSBERRY, D. D., RHODA, D. A. & BEISEL, W. R. (1969). Hemodynamics of staphylococcal B enterotoxemia and other types of shock in monkeys. *Journal of Applied Physiology*, **27**, 164.

FILKINS, J. P. & DI LUZIO, N. R. (1968). Endotoxin-induced hypothermia and tolerance in the rat. *Proceedings of the Society of Experimental Biology and Medicine, N.Y.*, **129**, 724.

FINE, J. (1961). Endotoxins in traumatic shock. *Federation Proceedings*, **20**, Supplement 9, 166.

FINE, J., FRANK, E. D., RAVIN, H. A., RUTENBERG, S. H. & SCHWEINBURG, F. B. (1960). The bacterial factor in traumatic shock. In *The Biochemical Response to Injury*, ed. H. B. Stoner and C. J. Threlfall, 377. Oxford: Blackwell.

FINE, J., PALMERIO, C. & RUTENBURG, S. (1968). New developments in therapy of refractory traumatic shock. *Archives of Surgery*, **96**, 163.

GILBERT, R. P. (1960). Mechanisms of the hemodynamic effects of endotoxin. *Physiological Reviews*, **40**, 245.

GLEISER, C. A. (1967). Pathology of anthrax infection in animal hosts. *Federation Proceedings*, **26**, 1518.

GUENTER, C. A., FIORICA, V. & HINSHAW, L. B. (1969). Cardiorespiratory and metabolic responses to live *E. coli* and endotoxin in the monkey. *Journal of Applied Physiology*, **26**, 780.

GUENTER, C. A. & HINSHAW, L. B. (1970). Comparison of septic shock due to gram-negative and gram-positive organisms. *Proceedings of the Society of Experimental Biology and Medicine, N.Y.*, **134**, 780.

HALMAGYI, D. F. J., STARZECKI, B. & HORNER, G. J. (1963). Mechanism and pharmacology of endotoxin shock in sheep. *Journal of Applied Physiology*, **18**, 544.

KUIDA, H., GILBERT, R. P., HINSHAW, L. B., BRUNSON, J. G. & VISSCHER, M. B. (1961). Species differences in effect of gram-negative endotoxin on circulation. *American Journal of Physiology*, **200**, 1197.

KOVÁCH, A. G. B. (1970). The function of the central nervous system after haemorrhage. *Journal of Clinical Pathology*, **23**, Supplement (Royal College of Pathology) 4, 202.

LINCOLN, R. E. & FISH, D. C. (1970). Anthrax toxin. In *Microbial Toxins*, ed. T. C. Montie, S. Kadis and S. J. Ajl, vol. 4, 362. New York: Academic Press.

LINDSEY, I., WISE, H. M., KNECHT, A. T. & NOYES, H. L. (1959). The role of clostridia in mortality following an experimental wound in the goat. *Surgery*, **45**, 602.

MACFARLANE, R. G. (1969). A discussion on triggered enzyme systems in blood plasma. *Proceedings of the Royal Society of London, Series B*, **173**, 259.

MACLENNAN, J. D. (1962). The histotoxic clostridial infections of man. *Bacteriological Reviews*, **26**, 177.

MARKLEY, K., SMALLMAN, E. & EVANS, G. (1967). Mortality due to endotoxin in germ free and conventional mice after tourniquet trauma. *American Journal of Physiology*, **212**, 541.

MARKLEY, K., SMALLMAN, E., EVANS, G. & MCDANIEL, E. (1965). Mortality of germ-free and conventional mice after thermal trauma. *American Journal of Physiology*, **209**, 365.

MILES, A. A. (1969). A history and review of the kinin system. *Proceedings of the Royal Society of London. Series B*, **173**, 341.

MILES, A. A. & WILHELM, D. L. (1960). The activation of endogenous substances inducing pathological increases of capillary permeability. In *The Biochemical Response to Injury*, ed. H. B. Stoner and C. J. Threlfall, 51. Oxford: Blackwell.

PALMERIO, C. & FINE, J. (1969). The nature of resistance to shock. *Archives of Surgery*, **98**, 679.

PARK, M. K., BAUM, D. & GUNTHEROTH, W. G. (1970). Lack of pooling in portal vein in endotoxin shock in the dog. *Journal of Applied Physiology*, **28**, 156.

PATNAIK, B. K. & GHOSH, H. K. (1966). Histopathological studies on experimental cholera. *British Journal of Experimental Pathology*, **47**, 210.

PHILLIPS, R. A. (1964). Water and electrolyte losses in cholera. *Federation Proceedings*, **23**, 705.

PORTER, P. J. & KASS, E. H. (1965). Role of the posterior hypothalamus in mediating the lethal action of bacterial endotoxin in the rat. *Journal of Immunology*, **94**, 641.

PRASHKER, D. & WARDLAW, A. C. (1971). Temperature responses of mice to *Escherichia coli* endotoxin. *British Journal of Experimental Pathology*, **52**, 36.

RAVIN, H. A., ROWLEY, D., JENKINS, C. & FINE, J. (1960). On the absorption of bacterial endotoxin from the gastro-intestinal tract of the normal and shocked animal. *Journal of Experimental Medicine*, **117**, 783.

ROSS, J. M. (1955). On the histopathology of experimental anthrax in the guinea-pig. *British Journal of Experimental Pathology*, **36**, 336.

SHUBIN, H. & WEIL, M. H. (1963). Bacterial Shock. *Journal of American Medical Association*, **185**, 850.

SMITH, H. (1960). The biochemical response to bacterial injury. In *The Biochemical Response to Injury*, ed. H. B. Stoner and C. J. Threlfall, 341. Oxford: Blackwell.

SMITH, H. (1969). Toxic activities of microbes. *British Medical Bulletin*, **25**, 288.

SMITH, H. & STONER, H. B. (1967). Anthrax toxic complex. *Federation Proceedings*, **26**, 1554.

STEELE, R. H. & WILHELM, D. L. (1970). The inflammatory reaction in chemical injury. III. Leucocytosis and other histological changes induced by superficial injury. *British Journal of Experimental Pathology*, **51**, 265.

STONER, H. B. (1955). Studies on the Mechanism of Shock. The effect of limb ischaemia on phosphomonoesterase distribution. *British Journal of Experimental Pathology*, **36**, 306.

STONER, H. B. (1961). Critical analysis of traumatic shock models. *Federation Proceedings*, **20**, Supplement 9, 38.

STONER, H. B., BULLEN, J. J., CUSHNIE, G. H. & BATTY, I. (1967). Fatal intra-peritoneal infection with *Clostridium welchii* type A in passively immunized guinea-pigs. The effect on vascular permeability. *British Journal of Experimental Pathology*, **48**, 309.

VAN HEYNINGEN, W. E. & ARSECULERATNE, S. N. (1964). Exotoxins. *Annual Review of Microbiology*, **18**, 195.

WALTERS, G. & MCGOWAN, G. K. (1963). Pulmonary embolism or bacterial shock? *The Lancet*, **ii**, 17.

WALTERS, G. & MCGOWAN, G. K. (1965). Significance of oligaemia in hypotensive surgical patients. *The Lancet*, **i**, 1236.

VAUGHN, D. L., GUNTER, C. A. & STOOKEY, J. L. (1968). Endotoxin shock in primates. *Surgery, Gynecology & Obstetrics*, **126**, 1309.

WOLSTENHOLME, G. E. W. & BIRCH, J. (1971). *Pyrogens and Fever*. London: Churchill-Livingstone.

WORK, E. (1971). Production, Chemistry and Properties of Bacterial Pyrogens and Endotoxins. In *Pyrogens and Fever*, ed. G. E. W. Wolstenholme and J. Birch, 23. London: Churchill-Livingstone.

THE ENTEROTOXIC ENTEROPATHIES

J. P. CRAIG

Downstate Medical Centre, State University of New York, Brooklyn, New York

INTRODUCTION

My task in this symposium is to discuss the role of toxins in microbial pathogenicity in animals. Instead of attempting to cover such a vast field or even to touch lightly upon each of the major groups of bacterial toxins, I have decided to discuss the role of toxins by selecting a relatively small group of toxin-mediated diseases and presenting them as an example of one way in which microbial toxins may be key factors in evoking disease in animal hosts.

I have chosen to call the group I have in mind 'the enterotoxic enteropathies'. Perhaps this term sounds pedantic and unnecessarily alliterative, but it may be useful in conveying a concept. I have defined the enterotoxic enteropathies as those acute diarrhoeal diseases of man or animals in which the fluid losses into the gut are caused by an exotoxin elaborated by the causative microbe. They are characterised by multiplication of the organism in the lumen of the bowel with little or no invasion of the mucosa and no spread to other tissues. The exotoxin is elaborated locally in the gut lumen and in natural disease acts only locally on gut epithelium. Since it is an exotoxin which acts upon the 'enteron', it is properly called an enterotoxin. Enterotoxins characteristically induce a temporary, reversible functional defect in epithelial cells involving water and electrolyte transport without causing any structural damage.

I have not included staphylococcal enterotoxin in this grouping, although the term 'enterotoxin' has probably been applied to this toxin longer than it has to the group I am about to discuss. It was called an enterotoxin because the obvious clinical manifestations of staphylococcal food poisoning are nausea and vomiting, and because the toxin, which is often preformed, is absorbed through the gastro-intestinal mucosa. It now seems clear, however, that the effects on gastro-intestinal function are indirect, and that the principal and primary targets are the central and autonomic nervous systems. The term neurotoxin would be as appropriate here as it is for botulinum toxin (Lamanna & Carr, 1967).

5

In view of the work on enterotoxins which has taken place during the past decade, I would hope that we could henceforth restrict this term to those exotoxins which exert a local effect on gut epithelial function; that is to say, those toxins which have the 'enteron', embryologically speaking, as their chief target.

At this point in history, cholera is by far the most thoroughly studied of the enterotoxic enteropathies. After summarising the facts and pointing out the numerous unanswered questions which have been raised by recent work on cholera, I shall present some of the other acute diarrhoeal disorders which have recently been shown to be associated with enterotoxin production.

CHOLERA ENTEROTOXIN

The effects of cholera enterotoxin on gut tissue

The seventh pandemic of Asiatic cholera, which began in 1961 and is still with us, has provided the major impetus for the study of this disease, and by so doing, has set in motion a spate of investigations into the basic mechanisms of pathogenesis of acute diarrhoeal diseases of many causes throughout the world. Since it now appears that the common factor shared by many of these diseases is that they are mediated by an enterotoxin elaborated by a microbial agent in the lumen of the bowel, it may be said that cholera was the first of the confirmed enterotoxic enteropathies, and has become the model for this particular kind of mechanism of pathogenicity.

The need for a more rational treatment of cholera cases first led to basic studies of the pathophysiological alterations seen in the disease. Simultaneously, efforts were made to determine the exact role of the cholera vibrio itself in pathogenesis. Through these studies it has become clear that all the clinical manifestations of cholera can be attributed to an enterotoxin elaborated in the gut. As it became fully accepted that cholera was indeed a toxinosis, new questions have been raised concerning immune mechanisms in toxin diseases in general, and especially in those diseases mediated by enterotoxins which appear to remain confined to the gut.

On the practical side, there is an obvious need for better immuno-prophylactic agents, which has forced us to take a new look at the relative importance of anti-bacterial and antitoxic immunity, and of serum versus secretory antibody.

It is ironic that the classic toxinoses, tetanus, botulism and diphtheria, were already recognised as such during the latter years of the last century, whereas cholera, which is in every way a purer and less adulterated toxinosis, has only been recognised as such during the past decade. It is true that Koch considered cholera to be a toxin disease, but his followers in the ensuing decades persisted in their search for endo-toxins or other cell-body substances, and seemed loath to consider a true exotoxin as a likely cause of the major clinical manifestations of disease. What is more astounding is that the recognition of the classic bacterial exotoxins was followed by a concerted search for microbial poisons as an explanation for all of the manifestations of infectious disease, and yet somehow cholera toxin eluded the searchers. Hindsight is always easy, but I think it is fair to say that cholera toxin was not recognised simply because *relevant* animal models were not used. It was good fortune that the classic exotoxins killed experimental animals in rather distinctive ways when they were injected in tiny doses by any parenteral route. In natural tetanus and diphtheria the toxin is similarly absorbed into the circulation from local sites of synthesis and is transported to critical target cells by blood or lymph. Not so with cholera and the other enterotoxins. The enterotoxin-producers multiply only in the gut lumen, and all the alterations we see in natural disease are caused by local effects of the toxin on the intestinal mucosa. Parenteral injection of enterotoxins does not cause diarrhoea. Therefore, it was not until investigators focused on the gut, and put the organisms or their products into the gut lumen that progress in our understanding of enterotoxins began.

The first and most useful step in this direction was the demonstration that a ligated segment of small intestine in the living rabbit would fill with fluid, following intraluminal injection of certain organisms and their products (De & Chatterje, 1953). Using cell-free filtrates of cholera cultures in this model, De and his associates demonstrated that cholera vibrios elaborate *in vitro* and *in vivo* a heat-labile, non-diffusible, trypsin-resistant, antigenic protein which they named, quite appropri-ately, an enterotoxin, since its only measurable effect was to cause accumulation of fluid in the gut lumen (De, 1959; De, Ghose & Chandra, 1962). This model has been extensively used for the study of cholera and other enterotoxins and for their pathophysiological effects on gut mucosa (Burrows, 1968; Leitch & Burrows, 1968), and as a basis for the study of antitoxic immunity (Kasai & Burrows, 1966).

It was also found that both live cholera vibrios and enterotoxins caused fluid accumulation and diarrhoea in suckling rabbits when introduced into the gut lumen by gastric tube (Dutta & Habbu, 1955). This model has been widely used, especially in monitoring studies on

5-2

the purification and properties of the enterotoxin (Finkelstein, Norris & Dutta, 1964; Finkelstein & LoSpalluto, 1970).

An advance of great importance was the development of the dog model for cholera (Sack, Carpenter & Pierce, 1966). Dogs proved to be susceptible to both infection and to the intraluminal administration of enterotoxin in the gut, and they were large enough for physiological studies to be carried out on plasma and intestinal fluids similar to those in man. The response to infection was essentially identical to that of man (Sack & Carpenter, 1969a, b).

With the development of a series of animal models for the study of the effects of vibrio products, it became possible to compare the responses in these animals to the changes seen in naturally occurring cholera in man, and in this way test the validity of the models. It was clear that although human volunteer studies would probably eventually be carried out, there would always be large areas of physiological inquiry into mechanisms of enterotoxic action which would forever be relegated to animal experimentation. The validity of animal models was therefore critical.

Numerous clinical studies have clearly shown that all the disease manifestations of cholera are the direct result of massive losses of water and electrolytes from the gut, leading to reduction of plasma volume, hypovolemic shock, and often death. The so-called rice-water stool is an isotonic fluid which is very low in protein and essentially identical in composition with juices of the normal small intestine. It contains about twice as much bicarbonate as normal plasma, and this loss of alkali leads to metabolic acidosis. The loss of fluid by way of the gut may exceed one litre per hour. Without replacement, this rate of loss rapidly leads to death, but intravenous infusion of adequate fluid and electrolytes almost universally leads to rapid and complete disappearance of all the signs and symptoms of disease except for the diarrhoea itself. Even without specific antibiotic therapy, the diarrhoea spontaneously ceases within 2–5 days (Carpenter et al. 1966; Greenough & Carpenter, 1969; Greenough, Carpenter, Bayless & Hendrix, 1970; Banwell et al. 1970; Carpenter, 1971). The rapid relief from all systemic effects of disease following fluid replacement demonstrates clearly that in natural disease the effect of the enterotoxin is purely a local one restricted to the gut mucosa, and that shock and other systemic manifestations are all secondary to loss of fluid and electrolyte.

Another important fact relating to the pathogenesis of cholera is that the vibrios are confined to the intestinal lumen and are never found in tissues. Moreover, a number of observers have noted that neither cholera infection nor enterotoxin causes significant histological change in the gut mucosa even during active purging. The lesion appears to be completely functional. The presence of intact intestinal mucosa in fatal cholera was reported by Cohnheim (1890) and Goodpasture (1923) and their findings were confirmed in the present pandemic by biopsy of human cases (Gangarosa et al. 1960) and electron microscope studies of canine cholera (Elliot, Carpenter, Sack & Yardley, 1970). The minimal alterations which could be seen were all attributed to functional changes rather than to structural changes due to the enterotoxin. In one recent study, however, oedema of the lamina propria was said to be accompanied by rarefaction of endothelial cells lining the vessels of the villi, suggesting increased vascular permeability as the source of water (Chen, Reyes & Fresh, 1971). Regardless of what these or future studies of ultrastructure may reveal, there can be no doubt that mucosal denudation does not occur, and there is no inflammatory response either in infection or following the application of enterotoxin to the mucosa (Norris & Majno, 1968).

Studies on the nature of the enterotoxin itself have, of course, depended upon animal models for bioassay. The toxin has now been purified by serial column chromatography (Finkelstein & LoSpalluto, 1970; Coleman et al. 1968), precipitation with dextran followed by gel filtration (Richardson, Evans & Feeley, 1970) and by absorption on, and elution from, aluminium hydroxide (Spyrides & Feeley, 1970). The most highly purified preparations are homogeneous by ultracentrifugation, disc electrophoresis and immuno-electrophoresis, and the molecular weight appears to be 90000 (R. A. Finkelstein, personal communication). In addition to the biologically active enterotoxin, liquid cultures contain a toxoid which is antigenic but non-toxic, with a molecular weight of about 60000 (Finkelstein & LoSpalluto, 1970). Since patients recovering from cholera possess circulating antitoxin but do not show the signs of intoxication which have been observed in animals given enterotoxin parenterally (Pierce et al. 1971b, c), it is interesting to speculate whether natural toxoid is selectively absorbed to a greater degree than is toxin during the course of the disease.

Now that purified enterotoxins have become available, it is clear that in their effects on gut function, crude and purified toxins behave identically, indicating that all enterotoxic activity in crude preparations has been due to the effects of the enterotoxin molecule and thus further validating much of the earlier work which preceded purification.

Using both patients with clinical cholera and animal models treated with enterotoxin, a number of studies have now been done in an attempt to define more precisely the pathophysiological alterations caused by cholera enterotoxin. The use of permanent Thiry–Vella loops of isolated segments of small bowel in dogs has made it possible to delineate the site and time course of the fluid response. Although enterotoxin caused net fluid movement from plasma to gut lumen in all segments of small bowel, it was maximal in the duodenum and less marked in jejunum and ileum. This difference was due to increased absorptive capacity of the lower portions rather than to differences in secretion in response to toxin. Movement from plasma to gut probably begins almost immediately after exposure to enterotoxin but fluid output does not reach a maximum until 4–5 h after toxin administration (Carpenter, Sack, Feeley & Greenberg, 1968; Greenough & Carpenter, 1969; McGonagle *et al.* 1969). Neither glucose absorption nor glucose-enhanced sodium absorption was impaired, suggesting that enterotoxin acts primarily by augmenting a secretory process rather than impairing absorption (Carpenter *et al.* 1968; Banwell *et al.* 1970).

Other experiments in dogs indicate that it is highly unlikely that alterations in hydrostatic pressure in gut vasculature play any role in fluid loss. Reduction of mesenteric arterial pressure to 30 % of usual values caused no reduction in fluid output in response to enterotoxin (Carpenter, Greenough & Sach, 1969*b*).

It has been possible to localise the effect of cholera enterotoxin more precisely by studying its effect in isolated ileal mucosa in perfusion chambers. When enterotoxin was placed on the luminal side of such chambers it caused a net secretion of chloride across the membrane, but, as in the intact dogs, glucose-enhanced sodium absorption was not impaired (Field, Fromm, Wallace & Greenough, 1969). The 'sodium pump', by which sodium is transported from lumen to plasma, was not inhibited (Grady *et al.* 1967). These findings all suggested that the major defect produced by enterotoxin was an enhancement of secretion of essentially normal succus entericus caused by a functional lesion in the gut epithelial cell itself.

Since the effect of enterotoxin on ion movement across epithelial membranes was similar to that caused by adenosine-3′-5′-cyclic monophosphate (cAMP), it has been suggested that the effect of cholera enterotoxin may be mediated by cAMP in the intestinal epithelial cell (Field *et al.* 1969; Field, 1971). A great deal of evidence has recently been gathered to support this concept. It was known that prostaglandins raise cAMP levels in some tissues by direct stimulation of adenyl cyclase. Studies in isolated small intestinal loops of dogs showed that when prostaglandin $F_2\alpha$ was given by the mesenteric artery, fluid and electrolyte output was similar to that seen following intraluminal administration of cholera enterotoxin (Pierce, Carpenter, Elliott & Greenough, 1971a). Ethacrynic acid, an adenyl cyclase inhibitor, caused a reduction in volume of fluid produced by the small bowel in response to cholera enterotoxin (Carpenter, Curlin & Greenough, 1969a).

These findings suggested that cholera enterotoxin might exert its effect by activating adenyl cyclase present in the intestinal epithelial cell membrane, thereby raising intracellular cAMP levels which would in turn affect electrolyte transport. In the case of the gut lining, this would involve active movement of chloride into the lumen with the water and other electrolytes which would necessarily follow. Recent investigations have indeed shown that cholera enterotoxin is capable of causing an increase in adenyl cyclase levels (Sharp & Hynie, 1971; Kimberg et al. 1971) and of cAMP levels in intestinal mucosal cells of laboratory animals (Schafer, Lust, Sircar & Goldberg, 1970). Jejunal mucosal tissue of dogs given intraluminal enterotoxin had increased adenyl cyclase activity which rose and fell concurrently with the increase and decrease in net water and sodium movement (R. L. Guerrant, personal communication). Moreover, biopsy specimens of jejunal mucosa obtained from patients with naturally acquired cholera had elevated levels of adenyl cyclase during the acute stage of the disease which returned to normal during convalescence (Chen, Rhode & Sharp, 1971).

It has recently been shown that ganglioside is capable of inactivating cholera enterotoxin. Gut epithelial scrapings and brain suspensions were shown to inactivate toxin, while homogenates of liver, kidney, lung and muscle failed to do so. By fractionating the active tissues it was determined that the inactivating substance was ganglioside (van Heyningen, Carpenter, Pierce & Greenough, 1971). This is of special interest because tetanus toxin is fixed by ganglioside, but its toxicity is not abolished during fixation (van Heyningen & Mellanby, 1971). Since there is good evidence that cholera toxin activates adenyl cyclase in cell membrane, the nature of its affinity for ganglioside, and the role of cell-membrane ganglioside in the mechanism of action of the enterotoxin will be important questions for future study.

There is some evidence that enterotoxin may act by inducing some sort of protein synthesis in intestinal cells. Pretreatment of rabbits with intravenous cycloheximide, a potent inhibitor of protein synthesis, decreased fluid output in response to cholera toxin, but it was incapable of altering the effect when it was given after enterotoxin (Serebro, Iber, Yardley & Hendrix, 1969; Harper, Yardley & Hendrix, 1970).

The effects of cholera enterotoxin on non-intestinal tissues

Crude cholera culture filtrates and rice-water stools from cholera patients cause a marked increase in the permeability of the small blood vessels of the skin following intracutaneous injection in rabbits and guinea pigs. Increased permeability can be demonstrated by intravenous injection of a blue dye as an indicator. The lesion is accompanied by moderate erythema and marked, sharply circumscribed, firm swelling. Skin lesions do not appear for several hours and reach a maximum only after 18–24 h. There is no necrosis. The factor responsible for these skin lesions has been called 'vascular permeability factor' or PF. The diameters of the blue lesions are proportional to PF concentration; thus a measure of toxin potency can be made by comparing test materials with a reference preparation (Craig, 1965, 1966).

Purified PF preparations are active in skin in amounts of less than a nanogram. Recent work suggests strongly that PF and enterotoxin are identical (Finkelstein & LoSpalluto, 1970; Richardson, Evans & Feeley, 1970; Craig, 1971), although there remains a possibility that the two biological activities can be separately inactivated (Grady & Chang, 1970). Immunological evidence indicates that PF and enterotoxin are antigenically identical (Mosley, Aziz & Ahmed, 1970), and indeed at the present time neutralisation tests of PF in rabbit skin are widely used for the titration of cholera antitoxin.

Because of the simplicity and economy of the PF assay, it has been possible to carry out more quantitative studies on cholera enterotoxin production by the skin-test method than by assaying enterotoxic activity in the animal gut. Maximum PF yields have been achieved in casamino acids-glucose-yeast extract medium in shaken flasks at 28–30°. PF in cell-free filtrate appears within 4 h and reaches a maximum at about 24 h, very soon after the end of the log phase of growth. With some strains PF is maintained at high levels for 4 days; in others it is destroyed by unidentified factors in the culture (Kusama & Craig, 1970).

PF appears to fix to vascular walls, presumably to endothelial cells, very rapidly after injection, and to initiate an irreversible process which eventually leads to increased permeability many hours later. Antitoxin administered intravenously one hour after intracutaneous PF injection is only about 40–50 % as effective in neutralising the permeability effect, as it would be if given just prior to PF injection. Similar examples of partial irreversibility have been noted with other bacterial exotoxins in skin (Craig, 1970).

The mechanism by which PF exerts its effect on endothelial cells is not understood. As in the case of the enterotoxic activity, the PF effect is inactivated by ganglioside, suggesting that the initial binding site on endothelial cells may be a cell membrane ganglioside (van Heyningen *et al.* 1971). No data are available concerning the possible role of the adenyl cyclase-cAMP system in vascular permeability. Assays of adenyl cyclase activity in endothelial cells will pose technical problems not encountered with epithelial scrapings from mucosal surfaces.

Because of the evidence that cholera toxin mimics cAMP in gut epithelium, it was of interest to determine its effect on non-intestinal cells known to possess systems mediated by cAMP. The isolated fat cell from rat epididymis has proven to be highly responsive. Cholera toxin, like cAMP, activates an intracellular lipase causing lipolysis with release of glycerol and fatty acids. The activity is abolished by cholera antitoxin and all evidence again indicates that this lipase-stimulating factor is identical with enterotoxin and PF. The effect is delayed; the maximum rate of lipolysis is not attained for 3 to 4 hours after exposure to toxin (Vaughan, Pierce & Greenough, 1970; Greenough, Pierce & Vaughan, 1970). Fat cells exposed to cholera toxin contain increased levels of adenyl cyclase activity, suggesting that, as in gut tissue, the toxin exerts its effect by activating this enzyme (Curlin & Chen, 1971).

Inwood & Tyrell (1970) have reported the isolation from crude cholera filtrates of a heat-labile toxic polypeptide which prevented HeLa cells from spreading on plastic surfaces, and caused increased vascular permeability. Insufficient material has been isolated to test it for enterotoxic activity. It will be interesting to find out whether this material represents the active portion of a larger PF molecule. In our hands, PF activity has never shown diffusion through cellophane.

Most early work on cholera toxin had indicated that it exerted no deleterious effect following parenteral injection in animals. Early on in this discussion I remarked that one of the reasons why the recognition of an exotoxin in cholera was so late in coming was because the toxin was ineffective when given by parenteral routes which had proven useful for the study of tetanus and diphtheria toxins. We have had to revise our notions somewhat about this since recently finding that cholera entero-toxin kills mice when given in adequate doses intravenously. The key words are 'adequate doses'. As it turns out one LD_{50} for most mouse strains is equivalent to about a quarter to a half ml of the most potent crude filtrate, or about 10 μg of purified toxin. Compared with tetanus toxin (about 40000 LD_{50}/μg) the potency is very low and it is easy to see why the lethal effects of this toxin were missed for so long. Amounts equivalent to 50 LD_{50} doses by the intravenous route fail to kill mice when given intraperitoneally.

After 2–3 LD_{50} doses mice show a marked fall in weight during the first 24 h accompanied by a decreased haematocrit value suggesting haemodilution; they appear healthy. Between 24 and 48 h, haematocrit values return toward normal, weight continues to fall and animals appear generally sick, without localising signs and without diarrhoea; death occurs on the third to fourth day. The fact that this toxin must be administered intravenously to have a lethal effect suggests that endothelial surfaces are a critical binding point. Ganglioside inactivates the lethal activity.

Cholera enterotoxin may be lethal for dogs if given in doses of 100 μg or more intramuscularly. As in mice, there is no diarrhoea. Sublethal doses of enterotoxin caused hyperglycaemia, hyponatraemia and increased levels of alkaline phosphatase. Formalinised cholera toxin and heated toxin produced none of these effects (Pierce *et al.* 1971*b*, *c*). As with all other effects produced by cholera toxin, those evoked by parenteral injection were delayed and prolonged. For example, hyperglycaemia did not appear until one or two hours after toxin administration and lasted more than five days in fasting dogs.

The recognition of these widespread effects on many tissues following parenteral injection of enterotoxin in animals has once again raised the question of whether or not some cholera enterotoxin is absorbed from the gut in natural cholera. Vaughan Williams & Dohadwalla (1969*a*, *b*) have presented data indicating that in rabbits, enterotoxin may be absorbed from one segment of gut and transported in the blood to cause diarrhoea in a distant segment. Observations in dogs receiving intravenous toxin indicate, however, that even doses producing marked systemic metabolic effects failed to cause diarrhoea. In human cholera, the metabolic effects noted in dogs which received parenteral enterotoxin are not seen. It is therefore highly unlikely that there would be sufficient toxin absorbed from the gut to produce distant effects on intestinal secretion.

VIBRIO PARAHAEMOLYTICUS ENTEROTOXIN

The use of the term 'enterotoxin' in this subtitle may be a bit presumptious and misleading because a bona fide enterotoxin elaborated by *V. parahaemolyticus* has not yet been described in the scientific literature. The epidemiological, clinical and bacteriological facts at hand suggest so strongly, however, that the enteropathogenicity of this organism will someday be attributable to such a toxic factor, that a brief discussion in the context of the enterotoxic enteropathies seemed to be appropriate.

V. parahaemolyticus is a marine, halophilic vibrio which causes the majority of the aetiologically identifiable cases of food poisoning in the summer months in Japan (Zen-Yoji *et al.* 1970). It is widespread in eastern and especially coastal Asia, where it appears as a common contaminant of sea foods. It is coming to be recognised as a causative agent of outbreaks of acute gastroenteritis with increasing frequency (Neogy, 1970). Its detection in food and faecal samples depends upon the use of bacteriological media which are simple, but have nonetheless not been in routine use in clinical laboratories in Europe and the New World. It will not be surprising if this group of organisms (the species has been divided into a number of serotypes) becomes recognised in the near future as one of the commonest and most cosmopolitan agents of acute diarrhoeal disease and food poisoning.

The spectrum of disease manifestations may range, as in cholera, from mild diarrhoea to a severe dehydrating disease leading to shock and death. Indeed, when this disease occurs during cholera outbreaks the two may be clinically indistinguishable in the individual patient. The proportion of severe cases is, however, greater in cholera.

In fatal cases the lack of structural damage is striking. There may be mild oedema and hyperaemia of the small intestinal wall but, again as in cholera, the mucosa is intact and shows no histological change. The one finding which differs from cholera has been the presence of fatty infiltration and cloudy swelling in the liver (H. Zen-Yoji, personal communication).

These observations all suggest that the major manifestations of *V. parahaemolyticus* diarrhoea are caused by an enterotoxin elaborated in the small intestine as in cholera; but which perhaps differs from cholera enterotoxin in that it may cause structural damage in the liver when absorbed in large quantity in a few highly susceptible individuals.

Enteropathogenicity of living organisms in the ligated rabbit ileum has already been established, and the response in this animal model system is similar to that seen with cholera vibrios. Moreover, human volunteers developed typical clinical manifestations of nausea, vomiting, diarrhoea and abdominal pain after ingestion of whole cultures of strains isolated from cases of food poisoning, while volunteers who ingested marine isolates of the same species which had not been associated with food poisoning failed to develop signs of disease. Enteropathogenicity was associated with the production of a heat-stable haemolysin (Sakazaki *et al.* 1968).

The nature of the relationship between enteropathogenicity and haemolysin production has not been determined. It would seem unlikely that enterotoxic and haemolytic activity should reside in the same molecule. It should not be too long, however, before the appropriate experiments will be conducted to demonstrate an enterotoxic factor in cell-free filtrates and then to isolate it and determine its other biological activities and its physicochemical characteristics.

ENTEROPATHOGENICITY OF NON-CHOLERA VIBRIOS

Vibrios which share certain cultural and biochemical properties with *Vibrio cholerae*, but which are not agglutinated by sera prepared against this species as it is now defined, have been called non-agglutinable or non-cholera vibrios. The former term is poorly chosen because these vibrios are quite 'agglutinable' in their own homologous antiserum. Since the somatic antigens which they possess are indeed different from those found in vibrios associated with epidemic or pandemic cholera, both of the classical and el Tor biotype, the term non-cholera vibrio seems appropriate.

These non-cholera vibrios have been associated repeatedly with sporadic cases and small outbreaks of acute enteric disease, particularly in areas of cholera endemicity (Lindenbaum *et al.* 1965; McIntyre, Feeley & Greenough, 1966; Chatterjee, Gorbach & Neogy, 1970) but also in areas remote from known foci of Asiatic cholera (Aldova, Laznickova, Stepankova & Leitava, 1968). Organisms of this group have not, however, displayed any tendency toward epidemic or pandemic spread as *V. cholerae* has done on numerous occasions.

The clinical manifestations of disease associated with this group of organisms have ranged from mild diarrhoea to severe cholera-like disease with watery diarrhoea, dehydration and shock; but in general the severity has been much less than that of cholera. It has not been proved, strictly speaking, that these agents are causally related to the diseases with which they have been associated; the evidence is presumptive, based mainly on epidemiological grounds and association.

Laboratory work on possible mechanisms of enteropathogenicity has begun only very recently. Utilising the various animal models established during the past decade by cholera investigators, it has now been shown that at least two strains of non-cholera vibrios isolated from cases of acute diarrhoea are capable of producing an enterotoxin. Cell-free culture filtrates of these vibrios caused fluid accumulation in ligated rabbit ileum and caused increased vascular permeability following intracutaneous injection in guinea-pigs. Of particular interest is the fact that there appears to be a close antigenic relationship, as demonstrated by cross-neutralisation tests, between cholera enterotoxin and the enterotoxin elaborated by these non-cholera vibrios (Zinnaka & Carpenter, 1971; M. Ohashi, personal communication).

Classifications of pathogenic micro-organisms have often been based upon biochemical and serological tests which were simple but seldom related to the organism's capacity to evoke disease. Perhaps in vibrio taxonomy more attention will be given in the future to factors such as enterotoxin production, rather than to somatic antigen constitution.

Acute diarrhoeal disease associated with a number of vibrio strains possessing wide biochemical and antigenic heterogeneity has been observed on two continents. Since case fatality has been low, and these organisms are not yet familiar to most bacteriologists outside Asia, it is highly probable that many outbreaks of diarrhoeal disease and food poisoning due to non-cholera vibrios have gone unrecognised. Present evidence suggests that the disorders they produce will prove to be enterotoxin-mediated. The nature of their relationship to cholera will bear watching.

ESCHERICHIA COLI ENTEROTOXINS

The role of enterotoxins in the pathogenesis of *Escherichia coli* diarrhoeas was first demonstrated by veterinary workers using strains isolated from young pigs and calves suffering from an acute diarrhoeal disease sometimes referred to as 'colibacillosis', (Smith & Halls, 1967*a*, *b*; Kohler, 1968). They found that both live bacterial suspensions and cell-free filtrates of *E. coli* cultures of strains isolated from pigs and calves with diarrhoea produced fluid accumulation when injected into ligated small intestinal segments of the same animals, (Smith & Halls, 1967*b*) and diarrhoea when administered intragastrically to young pigs (Kohler, 1968). In the latter case, onset of diarrhoea appeared in $1\frac{1}{2}$–3 h, lasted only 3–10 h, and the diarrhoea fluid was pale, yellow and watery. It is to be emphasised, as in all typical enterotoxic enteropathies, that no lesions could be detected histologically in gut tissues and the disorder appeared to be a strictly functional one involving alteration in ion transport across the gut mucosa. This is true of animals suffering from colibacillosis infection as well as those experimentally intoxicated by instillation of enterotoxin into the gut lumen. The local nature of the derangement was again demonstrated, as in cholera, by the lack of response to intravenous injection of enterotoxin (Kohler, 1968).

Some of the properties of *E. coli* enterotoxins are summarised in

Table 1. *Some properties of* Escherichia coli *enterotoxins*

Reference	Strains	Test Animal	Resistance to heat	Diffusible through cellulose	Onset of enterotoxic effect	Duration of effect	Antigenicity
Smith & Halls (1967*a, b*)	Pig, calf	Pig	100+,* 121−	No	—	—	—
Kohler (1968)	Pig, calf	Pig	100±, 121−	Partially	1½–3 h†	3–10 h	—
Gyles & Barnum (1969)	Pig	Pig	60−	No	—	—	Antitoxin produced in rabbits; reciprocal cross-reactivity with cholera enterotoxin
Sack *et al.* (1971)	Human	Rabbit	80±, 100−	No	—	—	Antitoxin produced in rabbits; no cross-reactivity with cholera enterotoxin
Wallace and colleagues (personal communication)	Human	Dog and rabbit	100+ —	No No	<1 h	<1 h	Antigenicity not examined; not neutralised by cholera antitoxin

* Resistant (+) or inactivated (−) at the temperature shown.
† Onset of diarrhoea following oral administration.

Table 1. Both of the above groups of investigators found that the enterotoxin resisted boiling (100°) for 30 min but was destroyed by autoclaving (121°) for 1 h. Smith & Halls (1967b) found that the toxin did not diffuse through cellulose membranes, while Kohler (1968) detected enterotoxic activity in diffusates. Both groups also demonstrated that typical cell wall lipopolysaccharides (endotoxins) played no role in diarrhoea production by showing that endotoxins prepared from cell bodies of enterotoxin-producing strains failed to cause fluid accumulation in ligated intestinal segments. Moreover, endotoxin is known to cause diarrhoea and vomiting in young pigs when administered intravenously, and Kohler (1968) showed that alcohol extracts capable of eliciting diarrhoea when administered intragastrically caused no disease when given intravenously.

There is now evidence that both heat-stable and heat-labile enterotoxins are elaborated by diarrhoeagenic strains of E. coli. A heat-labile toxin capable of causing fluid accumulation in ligated segments of the small intestine of pigs has been found in both whole cell lysates and cell-free filtrates of cultures of organisms isolated from enteritis in piglets (Gyles & Barnum, 1969). Of some twenty strains examined there was good association between enterotoxigenicity and capacity to evoke fluid accumulation in ligated gut segments following injection of whole living cultures. Preparations of endotoxin, crude capsular polysaccharide and surface protein antigen (K88) from strains isolated both from normal animals and piglets with diarrhoea all failed to evoke fluid accumulation in gut segments, once again consistent with the notion that somatic components of enteropathogenic organisms play little or no role in diarrhoea production. The enterotoxic cell-free filtrates were prepared by cultivation in a peptone dialysate medium followed by centrifugation, filtration, dialysis and concentration by evaporation. Unlike the enterotoxins described by Smith & Halls (1967b) and Kohler (1968) the whole cell lysate toxin prepared by Gyles & Barnum (1969) was inactivated by heating at 60° for 15 min. It was resistant to trypsin and precipitated by both 50 % and 100 % saturation with ammonium sulphate.

Immunogenicity of the enterotoxic moiety has been clearly demonstrated by neutralisation tests in pig intestinal segments using rabbit antisera against various E. coli products. Rabbits immunised with live suspensions of several enterotoxin-producing strains as well as whole cell lysates and cell-free filtrates, all made serum antitoxin which neutralised the toxin from whole cell lysate; enterotoxin-producing organisms which had been heat-killed or acetone-dried, or live suspensions of non-enterotoxigenic strains failed to evoke antitoxin production. Rabbits immunised with an enterotoxin-containing lysate of *V. cholerae* possessed circulating antitoxin which neutralised the *E. coli* enterotoxin in pig intestinal segments (Gyles & Barnum, 1969). This is of particular interest since enterotoxin production may be governed by a plasmid which is transmissible by conjugation from toxigenic to non-toxigenic strains (Smith, 1969),

It has been widely recognised for some time that severe diarrhoeal disease in man, especially in babies, may also be associated with colonisation of the small bowel with *E. coli*, often of a single serotype. Living cultures of many of these strains have been shown to cause fluid accumulation in ligated segments of rabbit small bowel (De, Bhattacharya & Sarkar, 1956; Taylor, Wilkins & Payne, 1961; Ogawa, Nakamura & Sakazaki, 1968), but the mechanism of pathogenicity was not clarified and cell-free toxic materials were not identified.

Recently Sack and Gorbach and their coworkers (Sack *et al.* 1971; Gorbach *et al.* 1971) have demonstrated that *E. coli* strains isolated in Calcutta from patients with an acute cholera-like disease elaborated a heat-labile enterotoxin similar in nature to the enterotoxin described by Gyles & Barnum (1969) from porcine strains of *E. coli*. With 18 h cultures of the human strains grown in a casamino acids-sucrose-salts medium, dialysates of cell-free filtrates produced fluid accumulation in ligated rabbit ileum with a dose-response curve very similar to that seen with cholera enterotoxin. Toxic activity was partially destroyed by heating at 80° for 30 min and totally destroyed by boiling for 2 min. All activity was precipitated by 40 % saturation with ammonium sulphate (Sack *et al.* 1971).

Enterotoxic activity has been demonstrated for a number of *E. coli* strains isolated from cases of infantile diarrhoea in Chicago by use of the suckling rabbit model – earlier revived for cholera studies by Dutta & Habbu (1955). Cell-free filtrates as well as living cultures of diarrhoea-producing strains caused fluid accumulation, but there was no evidence of invasion and the intestinal epithelium appeared intact by both light and electron microscopy. Responses tended to be less marked with toxins produced by strains isolated from American infants than from adults in Calcutta, indicating that *E. coli* enterotoxins may not all be identical, or that organisms vary quantitatively in their output of toxin (S. L. Gorbach, personal communication). The ligated ileum of the mouse has also been used to assay enteropathogenicity and enterotoxigenicity of *E. coli* strains of both human and porcine origin (R. A. Finkelstein, personal communication).

E. coli enterotoxin appears to evoke a more rapid and transient fluid output into bowel lumen than does cholera toxin when both are compared in chronic Thiry–Vella jejunal loops in dogs. Also unlike cholera toxin, ganglioside was not able to neutralise the secretory effect of *E. coli* enterotoxin. (C. K. Wallace and N. F. Pierce, personal communication). These properties suggest that cholera and *E. coli* toxins may have different binding sites in gut epithelium. On the other hand, a number of other recent observations suggest that both these enterotoxins may affect the same transport process in the gut. In isolated rabbit ileal mucosa, both toxins caused an inhibition of net sodium absorption and a reversal of net chloride movement resulting in net chloride secretion (Al-Awqati, Wallace & Greenough, 1971). Neither toxin interfered, however, with absorption of glucose or glucose-enhanced sodium absorption in the dog jejunum (C. K. Wallace and N. F. Pierce, personal communication). Both toxins caused increased adenyl cyclase activity in gut mucosal tissue (G. T. Curlin, personal communication) and lipolysis in rat epidydimal fat cells (W. B. Greenough, personal communication). It therefore appears likely that the fundamental molecular derangement elicited by *E. coli* and cholera enterotoxins may be the same, but that the initial binding sites and hence certain intermediary steps may differ.

Thus far, immunological studies on human strains have failed to demonstrate any cross-neutralisation between *E. coli* and cholera enterotoxins. Heterologous sera have consistently failed to neutralise either the enterotoxin in ligated rabbit ileum or dog jejunum (Sack *et al.* 1971). Because of earlier reports of cross-reactivity (see above), it will be important to continue to search for antigenic differences among *E. coli* enterotoxins. The picture may not be as simple as in the case of cholera, in which only one antigenic type of enterotoxin appears to exist.

CLOSTRIDIUM PERFRINGENS ENTEROTOXIN

It has been recognised for nearly three decades that *Clostridium perfringens*, type A could cause outbreaks of food poisoning in man (McClung, 1945; Hobbs *et al.* 1953). It is interesting that although the term 'food poisoning' was generally used to describe the clinical manifestations of diarrhoea and abdominal cramps with consumption of contaminated food, no toxin or 'poison' capable of reproducing these manifestations was identified until very recently. As in the case of cholera, a systematic search for enterotoxins had to await the development of satisfactory animal models for the demonstration of the effects of infection with living organisms. Early work suggested that in man only heat-resistant strains of *Cl. perfringens* were associated with food poisoning, but it is now clear that outbreaks may be associated with both heat-resistant and heat-sensitive strains (Nakamura & Schulze, 1970). Hauschild, Niilo & Dorward (1967) found that strains isolated from cases of human food poisoning produced diarrhoea in lambs following oral and intra-intestinal inoculation and caused fluid accumulation when injected into ligated lamb intestine; but no enterotoxic activity could be demonstrated by either technique using cell-free culture supernatant fluids. Pathological examination of the lamb intestines showed no gross lesions or invasion of the mucosa by bacteria suggesting that an undetected toxin was responsible (Hauschild, Niilo & Dorward, 1968).

Duncan and Strong and their associates, in a series of systematic studies, showed clearly that the ligated segment of rabbit small intestine could be satisfactorily used as an experimental model for the study of *Cl. perfringens* food poisoning. They first showed that the majority of strains isolated from food poisoning outbreaks caused fluid accumulation when whole cultures were inoculated into ligated segments of rabbit gut, and that strains derived from other sources usually did not. Fluid accumulated steadily over a 24 h period following inoculation. Diarrhoea could be produced in the rabbit by intraluminal injection of live organisms into the ileum but not by oral challenge, and there was good correlation between the ability of strains to produce fluid accumulation in ligated segments and overt diarrhoea after inoculation into the normal ileum. Further study indicated that strains which had produced fluid accumulation in the ligated ileal segments of rabbits, likewise produced diarrhoea and vomiting in monkeys, and diarrhoea typical of *Cl. perfringens* food poisoning in human volunteers. Rabbit-negative strains failed to produce disease in man or monkeys. There was no correlation between enteropathogenicity and heat-resistance (Duncan, Sugiyama & Strong, 1968; Duncan & Strong, 1969a; Duncan & Strong, 1971; Strong, Duncan & Perna, 1971).

In the course of these studies it was found that cell extracts and cell-free culture filtrates of enteropathogenic strains of *Cl. perfringens* contained a factor which was capable of reproducing the effects of whole live cultures in rabbit; namely fluid accumulation in the ligated ileum and overt diarrhoea when injected into normal ileum. The enterotoxic factor was present in cultures only when the organisms were grown under conditions which caused sporulation, suggesting that production of the toxin may be associated with sporulation. Cell-free filtrates active in rabbits were also shown to cause diarrhoea in human volunteers, while filtrates from negative strains failed to cause disease (Duncan & Strong, 1969*b*; Strong, Duncan & Perna, 1971). In addition, enterotoxin-containing filtrates evoked erythema when injected intracutaneously into guinea-pigs (Hauschild, 1970), and cell extracts caused a wide variety of manifestations leading to death when injected intravenously in lambs, rabbits and guinea-pigs (Niilo, 1971).

Table 2. *Properties of some bacterial enterotoxins*

Property	Vibrio cholerae	Escherichia coli	Clostridium perfringens	Shigella dysenteriae
Animals in which *enterotoxic* effect has been shown	Dog, man rat, monkey rabbit, pig mouse	Pig, rabbit, calf, dog mouse	Rabbit, man monkey	Rabbit
Lowest temperature of inactivation	56°	80–100°	60°	90° (Keusch, pers. comm.)
Diffusible through cellulose membranes	No	No	No	No
Molecular weight	90000 (Finkelstein, pers. comm.)			82000 (van Heyningen, 1971)
Antitoxigenicity	Yes	Yes	Yes	Yes
'Toxoidable' with formalin	Yes	?	?	Yes
Inactivation by ganglioside	Yes	No	?	
Adenyl cyclase activation in intestinal tissue	Yes	Yes		
Other biological activities associated with enterotoxin				
Skin reactivity (intradermal)	Vascular permeability factor	?	Erythema	?
Lethality (intravenous)	Mouse, dog	?	Mouse	Rabbit, mouse
Lipase-stimulation in rat (lipocyte)	Yes	Yes	?	?

The biological characteristics of partially purified *Cl. perfringens* enterotoxin have been examined systematically by Stark & Duncan (1971). Some of the important properties are summarised and compared with other enterotoxins in Table 2. The toxin is a heat-labile, non-diffusible protein, which can induce antitoxin and is immunologically distinct from any previously recognised toxic antigen elaborated by any of the types of *Cl. perfringens*. Immunodiffusion studies have thus far revealed only one antigenic type of enterotoxin. In addition to its enterotoxic activity, the toxin is lethal for mice and elicits erythema 18–24 h after intracutaneous injection in guinea-pigs.

All three biological activities were eluted simultaneously from Sephadex G-200 columns and behaved identically in disc electrophoresis (Stark & Duncan, 1971). A heat-labile, non-diffusible enteropathogenic factor active in both ligated intestinal segments and normal small intestine of lambs has also been extracted from sporulating cells (Hauschild, Niilo & Dorward, 1970a, b). It seems likely that this material is identical with the enterotoxin described by Stark & Duncan (1971).

The discovery of an immunologically distinct enterotoxin in cultures of food poisoning strains of *Cl. perfringens* has clearly settled the controversy concerning the role of phospholipase-C (α-toxin or lecithinase) in enteropathogenesis. Nygren (1962) had postulated that phosphoryl choline, an end product of the action of phospholipase-C on lecithin, caused a decrease in intestinal passage time resulting in diarrhoea. Since most strains of *Cl. perfringens* type A produce phospholipase-C it was difficult to rule this out. Subsequent work, however, showed that phosphoryl choline had no effect on intestinal passage time in mice or monkeys (Weiss, Strong & Groom, 1966), and purified lecithinase failed to produce fluid accumulation in ligated rabbit ileum (Duncan, Sugiyama & Strong, 1968). Moreover, passive immunisation of lambs with antiserum to phospholipase-C failed to prevent fluid accumulation in response to injection of live organisms, and a strain of *Cl. perfringens* which produced only traces of phospholipase-C *in vitro* was fully capable of causing diarrhoea in man (Hauschild, Niilo & Dorward, 1968). It seems likely that, although most food poisoning strains can make phospholipase-C *in vitro*, it is either not synthesised in the bowel, or is rapidly destroyed in the presence of intestinal enzymes. This seems to be a case in which a poison (which would indeed be very destructive if it were somehow introduced *into* the tissues of the gut, or anywhere else) simply does not reach a physiologically significant target, although the organism obviously has the genetic equipment to synthesise toxin under appropriate conditions.

SHIGELLA ENTEROTOXIN

Recent studies by Keusch, Mata & Grady (1970) have demonstrated that certain strains of *Shigella dysenteriae* elaborate *in vitro* a heat-labile enterotoxin which causes fluid accumulation in ligated rabbit small intestinal segments. Enterotoxic fractions isolated by Sephadex G-150 chromatography were lethal in mice, but caused no increase in skin vascular permeability in rabbits. A summary of some of the properties of shigella enterotoxin is included in Table 2.

In retrospect, it now seems very likely that this enterotoxin is identical to the lethal exotoxin of the Shiga bacillus which has been referred to for some time as the Shiga neurotoxin. The term 'neurotoxin' was applied because rabbits, which are the laboratory animals most susceptible to its action, develop flaccid paralysis of the fore and hind limbs following intravenous injection.

The properties of this exotoxin were thoroughly reviewed recently by van Heyningen (1971) who pointed out that the term 'neurotoxin' was in all probability a misnomer because the neurological manifestations seen in rabbits following intravenous injection were without doubt secondary to vascular damage in the central nervous system. Bridgwater and his coworkers (Bridgwater, Morgan, Rowson & Wright, 1955) and Howard (1955) concluded that the active material was not a neurotoxin but a vascular toxin attacking the endothelium of blood vessels. It is interesting that, as in the case of cholera toxin, Shiga exotoxin was found to be more lethal to mice by the intravenous than by the intraperitoneal route, suggesting that both the Shiga and cholera toxins, which are now known to be enterotoxins when tested in the proper biological system, may also have selective affinity for vascular endothelium. This is particularly intriguing in view of the fact that cholera enterotoxin is known to increase vascular permeability in the skin, whereas no such effect has been demonstrated for shigella enterotoxin (Keusch et al. 1970).

It is indeed curious that recognition of the enterotoxic activity of shigella exotoxin should have come so late considering that the major clinical manifestation of bacillary dysentery is diarrhoea. This illustrates vividly the need to develop relevant animal models in our attempts to understand natural mechanisms of pathogenesis. In order to be relevant, animal models must allow for exposure of the appropriate tissues to the toxin being tested. There is no doubt that characterisation of shigella exotoxin, using lethality as the biological indicator system, has provided useful information. On the other hand, an earlier recognition of the enterotoxic activity of this exotoxin would have led to a clearer understanding of its role in the pathogenesis of shigellosis.

These recent findings suggest a number of similarities between cholera and shigella enterotoxins. Both are heat-labile, antigenic proteins, of similar molecular size; they demonstrate maximum lethality in mice when given by the intravenous route, suggesting decisive affinity for vascular endothelium. Yet two important differences remain: (1) shigella exotoxin is reported to have no vascular permeability activity in skin (Keusch et al. 1970) and (2) shigella toxin causes ulceration of the intestinal mucosa and acute inflammation in the lamina propria (G. T. Keusch, personal communication).

It is well known that shigella organisms penetrate and multiply in intestinal epithelial cells in the course of natural or laboratory infections, whereas in cholera, the organisms remain confined to the intestinal lumen. It may be that in the natural disease, shigellae must be in an intracellular environment in order to multiply or at least elaborate sufficient enterotoxin to be effective as a diarrhoeagenic agent. If this proves to be the case, shigellosis could not be considered, strictly speaking, a true enterotoxic enteropathy, but would represent a portion of the enteric spectrum in which enterotoxin plays a major but not exclusive role in pathogenesis. On the other hand, we know that enterotoxin is produced *in vitro*. Further investigations may show that individuals with shigellosis may themselves display a spectrum ranging from pure intoxications resembling cholera, to infections characterised by intracellular multiplication of bacilli in which the effect of enterotoxin is superimposed upon other less readily definable effects associated with entry and multiplication of bacilli inside epithelial cells.

Although the term 'invasion' is often applied to such events, it seems inappropriate to attribute invasiveness to a non-motile organism which has no known means of actively penetrating a cell membrane. Although epithelial cells are not usually thought to be aggressive engulfers of particulate matter, perhaps the active role in this instance is, after all, played by the host cell which pulls the microbe into its interior. If this should be the case, the property of a microbe which would determine whether it ultimately enjoys an intracellular existence might better be called 'seductiveness' rather than 'invasiveness'. In view of man's tendency to choose his terms from a purely anthropocentric point of view, however, it is unlikely that we shall be discussing 'seductive' bacteria in the foreseeable future.

REFERENCES

AL-AWQATI, A., WALLACE, C. K. & GREENOUGH, W. B., III. (1971). Stimulation of intestinal secretion *in vitro* by culture filtrates of *E. coli*. *The Journal of Infectious Diseases*, (In press).

ALDOVA, E., LAZNICKOVA, K., STEPANKOVA, E. & LEITAVA, J. (1968). Isolation of non-agglutinable vibrios from an enteritis outbreak in Czeckoslovakia. *The Journal of Infectious Diseases*, **118**, 25.

BANWELL, J. B., PIERCE, N. F., MITRA, R. C., BROGJA, K. L., CARANASOS, G. J., KEIMOWITZ, R. I., FEDSON, D. S., THOMAS, J., GORBACH, S. L., SACK, R. B. & MONDAL, A. (1970). Intestinal fluid and electrolyte transport in human cholera. *Journal of Clinical Investigation*, **49**, 183.

BRIDGWATER, F. A. J., MORGAN, R. S., ROWSON, K. E. K. & WRIGHT, G. P. (1955). The neurotoxin of *Shigella shigae*. Morphological and functional lesions produced in the central nervous system of rabbits. *British Journal of Experimental Pathology*, **36**, 447.

BURROWS, W. (1968). Cholera toxins. *Annual Review of Microbiology*, **22**, 245.

CARPENTER, C. C. J. (1971). Cholera entertoxin – recent investigations yield insights into transport processes. *American Journal of Medicine*, **50**, 1.

CARPENTER, C. C. J., CURLIN, G. T. & GREENOUGH, W. B. III. (1969a). Response of canine Thiry–Vella jejunal loops to cholera exotoxin and its modification by ethacrynic acid. *The Journal of Infectious Diseases*, **120**, 332.

CARPENTER, C. C. J., GREENOUGH, W. B. III & SACK, R. B. (1969b). The relationship of superior mesenteric artery blood flow to gut electrolyte loss in experimental cholera. *Journal of Infectious Diseases*, **119**, 182.

CARPENTER, C. C. J., MITRA, P. P., SACK, R. B., HINMAN, E. J., SAHA, T. K., DANS, P. E., WELLS, S. A. & CHAUDHURI, R. N. (1966). Clinical studies in asiatic cholera. III. Physiologic studies during treatment of the acute cholera patient: comparison of lactate and bicarbonate in correction of acidosis: effects of potassium depletion. *Bulletin of Johns Hopkins Hospital*, **117**, 197.

CARPENTER, C. C. J., SACK, R. B., FEELEY, J. C. & STEENBERG, R. W. (1968). Site and characteristics of electrolyte loss and effect of intraluminal glucose in experimental canine cholera. *Journal of Clinical Investigation*, **47**, 1210.

CHATTERJEE, B. D., GORBACH, S. L. & NEOGY, K. N. (1970). Characteristics of non-cholera vibrios isolated from patients with diarrhea. *Journal of Medical Microbiology*, **3**, 677.

CHEN, H. -C., REYES, V. & FRESH, J. W. (1971). An electron microscopic study of the small intestine in human cholera. *Virchow's Archiv. Abteilung B, Zellpathologie*, **7**, 236.

CHEN, L. C., ROHDE, J. E. & SHARPE, G. W. G. (1971). Intestinal adenyl-cyclase activity inhuman cholera. *The Lancet*, **i**, 939.

COHNHEIM, J. (1890). Lectures on General Pathology (translated by A. B. McKee), New Sydenham Society, 953.

COLEMAN, W. H., KAUR, J., IWERT, M. E., KASAI, G. J. & BURROWS, W. (1968). Cholera toxins: purification and preliminary characterization of ileal loop reactive type 2 toxin. *Journal of Bacteriology*, **96**, 1137.

CRAIG, J. P. (1965). A permeability factor (toxin) found in cholera stools and culture filtrates and its neutralization by convalescent cholera sera. *Nature, London*, **207**, 614.

CRAIG, J. P. (1966). Preparation of the vascular permeability factor of *Vibrio cholerae*. *Journal of Bacteriology*, **92**, 793.

CRAIG, J. P. (1970). Some observations on the neutralization of cholera vascular permeability factor *in vivo*. *Journal of Infectious Diseases*, **121**, (Supplement), 100.

CRAIG, J. P. (1971). Cholera Toxins. In *Microbial Toxins*, vol. 2A, ed. S. Kadis, T. C. Montie and S. J. Ajl, 189. London and New York: Academic Press.

CURLIN, G. T. & CHEN, L. C. (1971). Cholera toxin stimulation of rat lipocyte adenyl cyclase activity. *Clinical Research*, **19**, 456.

DE, S. N. (1959). Enterotoxicity of bacteria-free culture-filtrate of *Vibrio cholerae*. *Nature, London*, **183**, 1533.

DE, S. N., BHATTACHARYA, K. & SARKAR, J. (1956). A study of the pathogenicity of strains of *Bacterium coli* from acute and chronic enteritis. *Journal of Pathology and Bacteriology*, **71**, 201.

DE, S. N. & CHATTERJE, D. N. (1953). An experimental study of the mechanism of action of *Vibrio cholerae* on the intestinal mucous membrane. *Journal of Pathology and Bacteriology*, 66, 559.

DE, S. N., GHOSE, M. L. & CHANDRA, A. (1962). Further observations on cholera enterotoxin. *Transactions of the Royal Society of Tropical Medicine and Hygiene*, 56, 241.

DUNCAN, C. L. & STRONG, D. H. (1969a). Experimental production of diarrhea in rabbits with *Cl. perfringens*. *Canadian Journal of Medicine*, 15, 765.

DUNCAN, C. L. & STRONG, D. H. (1969b). Ileal loop fluid accumulation and production of diarrhea in rabbits by cell-free products of *Cl. perfringens*. *Journal of Bacteriology*, 100, 86.

DUNCAN, C. L. & STRONG, D. H. (1971). *Clostridium perfringens* Type A food poisoning. I. Response of the rabbit ileum as an indication of enteropathogenicity of strains of *Clostridium perfringens* in monkeys. *Infection and Immunity*, 3, 167.

DUNCAN, C. L., SUGIYAMA, H. & STRONG, D. H. (1968). Rabbit ileal loop response to strains of *Cl. perfringens*. *Journal of Bacteriology*, 95, 1560.

DUTTA, N. K. & HABBU, M. K. (1955). Experimental cholera in infant rabbits: a method for chemotherapeutic investigation. *British Journal of Pharmacology*, 10, 153.

ELLIOTT, H. D., CARPENTER, C. C. J., SACK, R. B. & YARDLEY, J. H. (1970). Small bowel morphology in experimental canine cholera. A light and electron microscopic study. *Laboratory Investigation*, 22, 112.

FIELD, M. (1971). Intestinal secretion: effect of cyclic AMP – its role in cholera. *New England Journal of Medicine*, 284, 1137.

FIELD, M., FROMM, D., WALLACE, C. K. & GREENOUGH, W. B. III. (1969). Stimulation of active chloride secretion in small intestine by cholera exotoxin. *Journal of Clinical Investigation*, 48, 24A.

FINKELSTEIN, R. A. & LOSPALLUTO. (1970). Production of highly purified choleragen and choleragenoid. *Journal of Infectious Diseases*, 121 (Supplement), 63.

FINKELSTEIN, R. A., NORRIS, H. T. & DUTTA, N. K. (1964). Pathogenesis of experimental cholera in infant rabbits. I. Observations on the intraintestinal infection and experimental cholera produced with cell-free products. *Journal of Infectious Diseases*, 114, 203.

GANGAROSA, E. F., BEISEL, W. R., BENYAJATI, C., SPRINZ, H. & PIYARATIN, P. (1960). The nature of the gastrointestinal lesion in Asiatic cholera and its relation to pathogenesis: a biopsy study. *American Journal of Tropical Medicine*, 9, 125.

GOODPASTURE, E. W. (1923). Histopathology of the intestine in cholera. *The Philippine Journal of Science*, 22, 413.

GORBACH, S. L., BANWELL, B. D., CHATTERJEE, B. C., JACOBS, B. & SACK, R. B. (1971). Acute undifferentiated human diarrhea in the tropics. 1. Alterations in intestinal microflora. *Journal of Clinical Investigation*, (In press).

GRADY, G. F. & CHANG, M. C. (1970). Cholera enterotoxin free from permeability factor? *Journal of Infectious Diseases*, 121 (Supplement), 92.

GRADY, G. F., MADOFF, M. A., DUHAMEL, R. C., MOORE, E. W. & CHALMERS, T. C. (1967). Sodium transport by human ileum *in vitro* and its response to cholera enterotoxin. *Gastroenterology*, 53, 737.

GREENOUGH, W. B. III & CARPENTER, C. C. J. (1969). Fluid loss in cholera. A current perspective. *Texas Reports on Biology and Medicine*, 27, 203.

GREENOUGH, W. B. III, CARPENTER, C. C. J., BAYLESS, T. M. & HENDRIX, T. (1970). The role of cholera exotoxin in the study of intestinal water and electrolyte transport. In *Progress in Gastroenterology*, vol. 2, ed. G. B. J. Glass, 236. New York: Grune & Stratton.

GREENOUGH, W. B. III, PIERCE, N. F. & VAUGHAN, M. (1970). Titration of cholera enterotoxin and antitoxin in isolated fat cells. *Journal of Infectious Diseases*, **121**, (Supplement), 111.

GYLES, C. L. & BARNUM, D. A. (1969). A heat-labile enterotoxin from strains of *E. coli* enteropathogenic for pigs. *Journal of Infectious Diseases*, **120**, 419.

HARPER, D. T. JR, YARDLEY, J. H. & HENDRIX, T. R. (1970). Reversal of cholera exotoxin induced jejunal secretion by cycloheximide. *Johns Hopkins Medical Journal*, **126**, 258.

HAUSCHILD, A. H. W. (1970). Erythemal activity of the cellular enteropathogenic factor of *Clostridium perfringens* Type A. *Canadian Journal of Microbiology*, **16**, 651.

HAUSCHILD, A. H. W., NIILO, L. & DORWARD, W. J. (1967). Experimental enteritis with food poisoning and classical strains of *Clostridium perfringens* Type A in lambs. *Journal of Infectious Diseases*, **117**, 379.

HAUSCHILD, A. H., NIILO, L. & DORWARD, W. J. (1968). *Clostridium perfringens* Type A infection of ligated intestinal loops in lambs. *Applied Microbiology*, **16**, 1235.

HAUSCHILD, A. H., NIILO, L. & DORWARD, W. J. (1970a). Enteropathogenic factors of food-poisoning *Clostridium perfringens* Type A. *Canadian Journal of Microbiology*, **16**, 331.

HAUSCHILD, A. H., NIILO, L. & DORWARD, W. J. (1970b). Response of ligated intestinal loops in lambs to an enteropathogenic factor of *Clostridium perfringens* Type A. *Canadian Journal of Microbiology*, **16**, 339.

HOBBS, B. C., SMITH, M. E., OAKLEY, C. L., WARRACK, G. H. & CRUIKSHANK, J. C. (1953). *Clostridium welchii* food poisoning. *Journal of Hygiene, Cambridge*, **51**, 75.

HOWARD, J. G. (1955). Observations on the intoxication produced in mice and rabbits by the neurotoxin of *Shigella shigae*. *British Journal of Experimental Pathology*, **35**, 439.

INWOOD, J. & TYRRELL, D. A. J. (1970). A cytotoxic factor in cholera toxin. *British Journal of Experimental Pathology*, **51**, 597.

KASAI, G. J. & BURROWS, W. (1966). The titration of cholera toxin and antitoxin in the rabbit ileal loop. *Journal of Infectious Diseases*, **116**, 606.

KEUSCH, G. T., MATA, L. J. & GRADY, G. F. (1970). Shigella enterotoxin: isolation and characterization. *Clinical Research*, **18**, 442.

KIMBERG, D. V., FIELD, M., JOHNSON, J., HENDERSON, A. & GERSHON, E. (1971). Stimulation of intestinal mucosal adenyl cyclase by cholera enterotoxin and prostaglandins. *Journal of Clinical Investigation*, **50**, 1218.

KOHLER, E. M. (1968). Enterotoxic activity of filtrates of *Escherichia coli* in young pigs. *American Journal of Veterinary Research*, **29**, 2263.

KUSAMA, H. & CRAIG, J. P. (1970). Production of biologically active substances by two strains of *Vibrio cholerae*. *Infection and Immunity*, **1**, 80.

LAMANNA, C. & CARR, C. J. (1967). The botulinal, tetanal and enterostaphylococcal toxins: a review. *Clinical Pharmacology and Therapeutics*, **8**, 286.

LEITCH, G. J. & BURROWS, W. (1968). Experimental cholera in the rabbit ligated intestine: ion and water accumulation in the duodenum, ileum and colon. *Journal of Infectious Diseases*, **118**, 349.

LINDENBAUM, J., GREENOUGH, W. B. III, BENENSON, A. S., OSEASOHN, R., RIZVI, S. & SAAD, A. (1965). Non-vibrio cholera. *The Lancet*, i, 1081.

McCLUNG, L. S. (1945). Human food poisoning due to growth of *Clostridium perfringens* (*Cl. welchii*) in freshly cooked chicken: preliminary note. *Journal of Bacteriology*, **50**, 229.

McGonagle, T. J., Bayless, T., Hendrix, T., Iber, F. & Serebro, H. (1969). Time of onset of action of cholera toxin in the rabbit and dog. *Gastroenterology*, **57**, 5.

McIntyre, O. R., Feeley, J. C., Greenough, W. B. III. (1966). Diarrhea caused by non-cholera vibrios. *American Journal of Tropical Medicine and Hygiene*, **14**, 412.

Mosley, W. H., Aziz, K. M. S. & Ahmed, A. (1970). Vascular permeability factor and ileal loop toxin of *Vibrio cholerae*. *Journal of Infectious Diseases*, **121**, 243.

Nakamura, M. & Schulze, J. A. (1970). *Clostridium perfringens* food poisoning. *Annual Review of Microbiology*, **24**, 359.

Neogy, K. N. (1970). Vibrio parahaemolyticus in Calcutta. *Journal of the Indian Medical Association*, **55**, 322.

Niilo, L. (1971). Mechanism of action of the enteropathogenic factor of *Clostridium perfringens* Type A. *Infection and Immunity*, **3**, 100.

Norris, H. T. & Majno, G. (1968). On the role of the ileal epithelium in the pathogenesis of experimental cholera. *American Journal of Pathology*, **53**, 263.

Nygren, B. (1962). Phospholipase C-producing bacteria and food poisoning. *Acta Pathologica et Microbiologica Scandinavica*, **160** (Supplement), 1.

Ogawa, K., Nakamura, A. & Sakazaki, R. (1968). Pathogenic properties of 'enteropathogenic' *Escherichia coli* from diarrheal children and adults. *Japanese Journal of Medical Science and Biology*, **21**, 333.

Pierce, N. F., Carpenter, C. C. J., Elliot, H. L. & Greenough, W. B. III. (1971*a*). Effects of prostaglandins, theophylline and cholera exotoxin upon transmucosal water and electrolyte movement in the canine jejunum. *Gastroenterology*, **60**, 22.

Pierce, N. F., Graybill, J. R., Kaplan, M. M. & Bouwman, D. L. (1971*b*). Systemic effects of parenteral cholera enterotoxin in dogs. *Journal of Laboratory and Clinical Medicine* (in press).

Pierce, N. F., Greenough, W. B. III & Carpenter, C. C. J. (1971*c*). *Vibrio cholerae* enterotoxin and its mode of action. *Bacteriological Reviews*, **35**, 1.

Richardson, S. H., Evans, D. G. & Feeley, J. C. (1970). Biochemistry of *Vibrio cholerae* virulence. I. Purification and biochemical properties of PF/cholera enterotoxin. *Infection and Immunity*, **1**, 546.

Sack, R. B. & Carpenter, C. C. J. (1969*a*). Experimental canine cholera. I. Development of the model. *Journal of Infectious Diseases*, **119**, 138.

Sack, R. B. & Carpenter, C. C. J. (1969*b*). Experimental canine cholera. II. Production by cell-free culture filtrates of *Vibrio cholerae*. *Journal of Infectious Diseases*, **119**, 150.

Sack, R. B., Carpenter, C. C. J. & Pierce, N. F. (1966). Experimental cholera. A canine model. *The Lancet*, **ii**, 206.

Sack, R. B., Gorbach, S. L., Banwell, J. G., Jacobs, B., Chatterjee, B. D. & Mitra, R. C. (1971). Enterotoxigenic *Escherichia coli* isolated from patients with severe cholera-like disease. *The Journal of Infectious Diseases*, **23**, 378.

Sakazaki, R., Tamura, K., Kato, T., Obara, Y., Yamai, S. & Hobo, K. (1968). Studies on the enteropathogenic, facultatively halophilic bacterium. *Vibrio parahaemolyticus* III. Entero-pathogenicity. *Japanese Journal of Medical Sciences and Biology*, **21**, 325.

Schafer, D. E., Lust, W. D., Sircar, B. & Goldberg, N. D. (1970). Elevated concentration of adenosine 3'5' cyclic monophosphate in intestinal mucosa after treatment with cholera toxin. *Proceedings of the National Academy of Sciences*, **67**, 851.

Serebro, H. A., Iber, F. L., Yardley, J. H. & Hendrix, T. R. (1969). Inhibition of cholera toxin action in the rabbit by cycloheximide. *Gastroenterology*, **56**, 506.

SHARP, G. W. G. & HYNIE, S. (1971). Stimulation of intestinal adenyl cyclase by cholera toxin. *Nature, London*, **229**, 266.

SMITH, H. W. (1969). The enteropathogenicity of *E. coli*. *Journal of General Microbiology*, **59**, x.

SMITH, H. W. & HALLS, S. (1967a). Observations by the ligated intestinal segment and oral inoculation methods on *Escherichia coli* infections in pigs, calves, lambs and rabbits. *Journal of Pathology and Bacteriology*, **93**, 499.

SMITH, H. W. & HALLS, S. (1967b). Studies on *Escherichia coli* enterotoxin. *Journal of Pathology and Bacteriology*, **93**, 531.

SPYRIDES, G. J. & FEELEY, J. C. (1970). Concentration and purification of cholera exotoxin by absorption on aluminum compound gels. *Journal of Infectious Diseases*, **121** (Supplement), 96.

STARK, R. L. & DUNCAN, C. L. (1971). Biological characteristics of *Clostridium perfringens* Type A enterotoxin. *Infection and Immunity*, **4**, 89.

STRONG, D. H., DUNCAN, C. L. & PERNA, G. (1971). *Clostridium perfringens* Type A food poisoning. II. Response of the rabbit ileum as an indication of enteropathogenicity of strains of *Clostridium perfringens* in human beings. *Infection and Immunity*, **3**, 171.

TAYLOR, J., WILKINS, M. P. & PAYNE, J. M. (1961). Relation of rabbit gut reaction to enteropathogenic *Escherichia coli*. *British Journal of Experimental Pathology*, **42**, 43.

VAN HEYNINGEN, W. E. (1971). The exotoxin of *Shigella dysenteriae*. In *Microbial Toxins*, vol. II A, ed. S. Kadis, T. C. Montie and S. J. Ajl, 255. New York and London: Academic Press.

VAN HEYNINGEN, W. E. & MELLANBY, J. (1971). Tetanus toxin. In *Microbial Toxins*, vol. II A, ed. S. Kadis, T. C. Montie and S. J. Ajl, 69. New York and London: Academic Press.

VAN HEYNINGEN, W. E., CARPENTER, C. C. J., PIERCE, N. F. & GREENOUGH, W. B. III. (1971). Inactivation of cholera toxin by ganglioside. *Journal of Infectious Diseases*, **124**, 415.

VAUGHAN, M., PIERCE, N. F. & GREENOUGH, W. B. III. (1970). Stimulation of glycerol production in fat cells by cholera toxin. *Nature, London*, **226**, 658.

VAUGHAN WILLIAMS, E. M. & DOHADWALLA, A. N. (1969a). Diarrhoea and intestinal fluid accumulation in uninfected rabbits cross-perfused with blood from donor rabbits intra-intestinally infected with cholera. *Nature, London*, **222**, 586.

VAUGHAN WILLIAMS, E. M. & DOHADWALLA, A. N. (1969b). The appearance of a choleragenic agent in the blood of infant rabbits infected intestinally with *Vibrio cholerae*, demonstrated by cross-circulation. *The Journal of Infectious Diseases*, **120**, 658.

WEISS, K. F., STRONG, D. H. & GROOM, R. A. (1966). Mice and monkeys as assay animals for *Clostridium perfringens* food poisoning. *Applied Microbiology*, **14**, 479.

ZEN-YOJI, H., SAKAI, S., KUDOH, Y., ITOH, T. & TERAYAMA, T. (1970). Antigenic schema and epidemiology of *Vibrio parahaemolyticus*. *Health and Laboratory Science*, **7**, 100.

ZINNAKA, Y. & CARPENTER, C. C. J. (1971). Permeability factor produced by so-called NAG vibrio. *Johns Hopkins Medical Journal*, (In press).

HOST DAMAGE RESULTING FROM HYPERSENSITIVITY TO BACTERIA

W. E. PARISH

Lister Institute of Preventive Medicine, Elstree, Herts

INTRODUCTION

Definitions of allergy and immunity

Immunological terminology has evolved from early descriptions, and the meaning of some terms has extended beyond that originally intended until they are understandable by the context in which they are used rather than by definition. The terms in this chapter are used as follows.

Allergy designates a specific altered reactivity of the tissues to substances compared with the result of the first exposure to the same substances, or the reactivity of the tissues of other individuals of the same species not previously exposed to the substances. Re-exposure to an antigen is not necessary if it persists in the blood or tissues while antibody is formed and reacts with it, as occurs in serum sickness or some bacterial infections. In this instance allergic change follows the first, but prolonged, exposure to the antigen. Allergy also applies to decreased reactivity which may for example occur in the reduced delayed sensitivity responses in syphilis or leprosy. I usually restrict the interpretation of altered reactivity to mean clinically observable allergic signs. People frequently form antibodies to antigens in their environment, and occasionally even to their own tissues, without showing clinical signs of altered reactivity in the presence of the antigen. Such people are not clinically allergic, though they may react differently from people without antibodies to a substance by eliminating it more quickly from the tissues or blood. In order to avoid misunderstanding arising from the use of the term allergic, where possible in this chapter, the various allergic reactions are referred to by name, e.g. anaphylaxis.

Sensitisation and sensitivity refer to formation of antibody or activation of lymphocytes. It is a state of antigen awareness and priming of the tissues. An allergic reaction may follow re-exposure of the sensitised tissues to antigen if the tissues are sufficiently sensitised. This term covers all types of allergic sensitivity; not just that of anaphylaxis. Hypersensitivity is an increased or heightened sensitivity to antigen above that normally encountered and refers to all types of allergic response, e.g. Arthus hypersensitivity, or delayed hypersensitivity.

The term 'immune response' originated when immunologists were concerned mainly with immunity to infection. The term is now used more flexibly to designate any response involving antibody formation or specific activity of lymphocytes; and it is used as an alternative for delayed sensitivity which is also known as 'cell-mediated immune response'. In this context, immune response may refer to the allergic damage in sensitised individuals exposed to a chemical which, in the amounts used, is harmless to normal, unsensitised individuals. Immunity to infection is rarely absolute, and the terms resistance or acquired resistance are more appropriate though less popular.

Resistance to infection is an allergic state in the strict sense of the term allergic, in that there is an altered reactivity to the infecting organism. However, antibodies can neutralise toxins and opsonise or lyse bacteria in the presence of complement without overt tissue damage and therefore without clinical allergy. In older reports, as in those of studies on lobar pneumonia, it was claimed that allergy differed from immunity because animals could be desensitised without impairing their resistance to infection. In fact, these authors had only suppressed the anaphylactic sensitivity that accompanied the opsonins or bactericidal antibodies induced by vaccination. If the animals had been completely desensitised, they would have lost all resistance. In other words, the authors had suppressed the observable tissue damage due to anaphylaxis and the animals became clinically non-allergic.

In delayed sensitivity responses to infection, the altered reactivity of the tissues (allergic response) is an integral part of the host resistance. Suppression of the delayed sensitivity by large doses of antigen, anti-lymphocyte globulin, or drugs, leaves the animal susceptible to infection. Delayed sensitivity responses resulting in clinical allergy are those accompanied by clinical signs or manifested by skin tests.

It is not possible to define allergy strictly and be understood by clinicians thinking in terms of potential disease as well as by experimental immunologists thinking in terms of altered reactivity even at the level of single cells.

Host damage and bacterial hypersensitivity

Every individual is continually exposed to antigens from bacteria that colonise his tissues or are encountered in the environment. Many environmental bacteria induce only low titres of antibody or delayed sensitivity responses, though later exposure to large numbers may result in a rapid increase of these responses. Some of these bacteria are harmless having no ability to infect, others appear harmless because their virulence is so weak that non-specific defences or only weak specific responses are necessary to prevent infection. However, tissue damage may follow exposure to the harmless bacteria if the host becomes allergic to their antigens.

In certain circumstances allergic responses to harmless bacteria can result in damage to the host, as can be illustrated by infections with the non-pathogenic *Bacillus subtilis*. Although in young animals reared in a sterile environment *B. subtilis* can be a lethal pathogen, in most experimental animals infections with this organism are resisted rapidly though there are only low titres of antibody to *B. subtilis* and little lymphocyte activity *in vitro* to its antigens. If such animals are frequently exposed to large numbers of *B. subtilis* in organic dusts, or are injected with the living organisms or their extracts, subsequent exposure to the organisms or their extracts induces focal lesions attributable to precipitating antibody and delayed sensitivity. Moreover inhalation or intravenous injections of *B. subtilis* protein antigens in sensitised guinea-pigs induces fatal anaphylaxis. It should be emphasised that *B. subtilis* does not survive in the sensitised animals whose serum and macrophages have enhanced bactericidal properties. Following liberation of *B. subtilis* antigens, the tissue damage results from anaphylaxis, Arthus inducing antigen–antibody complexes, and delayed or cell-mediated sensitivity. Thus a normally harmless organism harms the host which has become allergic to its antigens.

Host damage following allergy to the non-pathogenic *B. subtilis* is an extreme example. Host damage due to allergy to pathogenic bacteria however may always occur. The pathogenic bacteria invade the tissues releasing cellular or extracellular antigenic substances, or without invasion, release similar antigens when present in the pharynx or intestine. The cell-mediated and humoral responses of the host resist the infection and neutralise the toxins but every 'primed' lymphocyte, by activating macrophages or releasing lymphocyte tissue substances, may damage the host tissue and every precipitating antibody may form a harmful antigen–antibody complex.

Certain bacteria, e.g. *Streptococcus pyogenes* or *Escherichia coli*, either through long association with, and adaptation to their hosts, or by chance, have antigens with determinants similar to those of the tissues of the host. This may result in partial tolerance of the organism by the host, but when the host forms antibodies to these organisms, the same antibodies may also combine with the similar antigens in the host's tissues, e.g. in the endocardium, glomerular basement membrane or the intestinal mucosa. Such antibodies may induce cytotoxic changes in those organs resulting in auto-allergic disease.

The tissue changes occurring in each of the four types of allergic reaction induced by bacterial antigens, namely anaphylaxis, Arthus response, cytotoxic reaction and delayed sensitivity are examined in turn to illustrate the various forms of host damage that may be due to hypersensitivity to bacteria. Finally, the difficulty of assessing the importance of damage due to hypersensitivity and to auto-allergic phenomena in bacterial disease, is discussed.

ANAPHYLAXIS ASSOCIATED WITH BACTERIAL INFECTION

Anaphylaxis

The main features of anaphylaxis are the passive sensitisation by antibody of mast cells, basophils and possibly other cells. Exogenous antigen reacts with antibody + cell resulting in the release of histamine, slow-reacting substance and other pharmacological mediators. These mediators induce the smooth muscle contraction and increased vascular permeability with oedema and shock characteristic of the various anaphylactic disorders. The disorders, e.g. asthma, are sudden in onset, of short duration and, unless recurrent, resolve without persistent tissue change. In man, most anaphylactic tissue-passive sensitising antibody is the reagin (Ig E). However some human Ig G or Ig G-associated antibodies can passively sensitise homologous tissue for a short time, mediating a weal and flare response in skin (Parish, 1970a). Anaphylactic Ig G-type antibodies were found in the sera of some patients recovering from streptococcal pneumonia, tonsilitis associated with vasculitis, and after vaccination with tetanus toxoid. They probably do not prepare tissues for the classical reagin-type mediated diseases, such as asthma, hay fever and urticaria, but by inducing increased vascular permeability they may facilitate penetration of vessel walls by tissue-damaging antigen–antibody complexes.

Bacterial antigens and anaphylaxis

Many bacterial antigens are anaphylactogenic in suitably sensitised animals when injected, inhaled or ingested in sufficient amount. However, unlike exposure to fungal antigens in dusts, sudden exposure to bacterial antigens rarely occurs apart from accidents during vaccination, and several reasons can be suggested to support a contention that bacteria (and viruses) do not induce anaphylactic disease though they induce anaphylactic sensitivity. Bacteria and viruses, which normally multiply on or in the host's tissues, release small amounts of antigens which increase with the number of the organisms or synthesis of their extracellular products. Such prolonged antigenic exposure is unlikely to induce anaphylaxis: the first release of small amounts of antigen is more liable to desensitise the host. Moreover, a normal host, not anaphylactically sensitive, is unlikely to become sensitised by the first exposure to small amounts of antigen, and then succumb to anaphylaxis as large amounts are released, unless the exposure is interrupted and the subsequent release sudden; an unusual event during infection.

Anaphylactic sensitisation to bacteria has nevertheless been claimed to cause chronic eczema and intrinsic asthma, mainly because many patients give anaphylactic skin responses to bacterial antigens. The evidence does not favour this claim. Injections of preparations of auto-genous or other bacteria sometimes improve the condition, but the improvement may result from a non-specific stimulation of the reticulo-endothelial system rather than from desensitisation. Antibiotics do not effect a cure and removal of infected foci, e.g. teeth or tonsils, is rarely helpful. The immediate or delayed skin responses to bacterial antigens, on which so much emphasis has been placed, are no more frequent in these patients than in normal persons. Moreover it is improbable that fluctuations in the growth of the resident flora would induce anaphylaxis, for reasons already given.

Anaphylaxis in lobar pneumonia

Lobar pneumonia is exceptional in being the one disease in which anaphylactic sensitisation to bacteria significantly contributes to the tissue damage. The disease is sudden in onset, with a copious oedematous exudation followed by deposition of fibrin and infiltration by leucocytes, which leads to consolidation. Loeschke (1931) suggested that susceptible adults are anaphylactically sensitive to pneumococci as a result of previous, non-infective exposure. Inhalation of pneumococci into the sensitised alveoli induces an anaphylactic oedematous exudation followed by leucocyte infiltration and fibrin formation. The organisms spread with the oedema fluid through the alveolar pores until checked by the interlobular septa. Thus, the characteristic lobar distribution of the pneumococci depends upon the large volume of oedema fluid appearing before the infiltration of sufficient leucocytes to inhibit multiplication and spread of the organisms.

Skin tests with pneumococcal antigens support the contention that many normal persons are anaphylactically sensitive to one or more

6

types of this organism (Finland & Sutliff, 1931; Alston & Lowdon, 1933). During pneumonia the anaphylactic skin response is reduced or absent, but occurs again for at least a short time during convalescence: in fatal cases there is no such skin response (Finland & Sutliff, 1931; Francis, 1933). The depression of the anaphylactic response during acute pneumonia probably results from desensitisation as the multiplying organisms release large amounts of antigen, theoretically, it may also result from a general toxic depression of the vascular response.

Persons recovered from lobar pneumonia or vaccinated with pneumococci have protective antibodies measurable by agglutination, capsular swelling and other tests. These protective antibodies are distinct from those mediating anaphylactic cutaneous sensitivity (Finland & Brown, 1938). This distinction between anaphylactic sensitivity and protective antibodies is also evident experimentally. Anaphylactically sensitised immunized rabbits, and desensitised immunised rabbits were resistant to infection, irrespective of allergic sensitisation, though the sensitised rabbits had a more severe inflammatory response than that occurring in the desensitised animals (Rich, 1933; Rich, Jennings & Downing, 1933).

It has been reported many times that challenge by inhalation of sterile pneumococcal antigens in anaphylactically sensitive animals induces severe oedema, sometimes resulting in changes resembling lobar pneumonia in man, which does not occur after the first antigenic exposure of normal unsensitised animals. Furthermore, in rabbits the severity of the lesion can be related to the magnitude of the pre-challenge skin response or to the presence of serum antibody (Sharp & Blake, 1930; Julianelle & Rhoads, 1932), though the tissue-passive sensitising properties of the antibodies were not determined and their significance unknown at that time. Almost identical lobar lesions were induced in guinea-pigs demonstrated to be anaphylactically sensitive to pneumococcal antigens by skin tests before the challenge inhalation (Lindau, 1933). It is interesting to speculate whether the anaphylactic response to inhalation in sensitised guinea-pigs varies with the nature of the antigen because in similar experiments in guinea-pigs sensitised to milk proteins, the animals died rapidly with bronchial constriction (Parish, Barrett & Coombs, 1960); and with horse dander the animals died or had a transient 'asthmatic' episode (Ratner, Jackson & Gruehl, 1927).

Though tests with sterile antigens show that anaphylactic sensitivity predisposes a host to oedematous pneumonic lesions, the same cannot be said of experimental infection. Normal unsensitised rabbits when infected with living pneumococci have changes resembling lobar pneumonia with the characteristic serofibrinous exudation (Gaskell, 1925; Stuppy & Falk, 1931). Similarly normal dogs, which tend to be resistant to pneumococci, have serofibrinous oedema of the alveoli within 1 h of infection (Robertson, Coggeshall & Terrell, 1933). It is improbable that animals develop anaphylactic sensitivity within hours of first contact with the antigen, and unlikely that all normal animals are equally sensitised to organisms that happen to share some antigens with pneumococci. Pulmonary serofibrinous exudation, which is mainly responsible for the rapid spread of the organism, is the usual response to pneumococcal infection even in unsensitised individuals. If the individual is anaphylactically sensitive, the oedema following infection is much greater and the spread of the organism thereby enhanced. Moreover anaphylactic sensitivity leading to oedema may predispose the individual to infection with small numbers of pneumococci, or with pneumococci of low virulence which would be quickly eliminated in a non-sensitive individual. There is experimental support for the contention that allergy predisposes the lungs to more severe inflammation during infection, because rabbits sensitised with killed pneumococci had more extensive lesions after inhaling living pneumococci than the unsensitised controls (Stuppy, Cannon & Falk, 1931), as did dogs recovering from a second, third, or more numerous exposure to infection than had dogs recovering from the first infection (Coggeshall & Robertson, 1935). It would be interesting to examine the course of infection in the absence of allergy by repeating these experiments in animals made tolerant to pneumococci, and in animals in which the anaphylactic response is inhibited by drugs.

Until such experiments are made, it is not possible to conclude that anaphylaxis is an essential feature of lobar pneumonia, but there is little doubt that its occurrence increases the inflammatory changes and the severity of the disease, and probably also undermines resistance at the start of the infection.

Anaphylaxis in piglet bowel oedema

Bowel oedema and haemorrhagic gastro-enteritis of piglets are claimed to be anaphylactic diseases in which certain strains of *Escherichia coli* multiply so rapidly in the intestine that their antigens react anaphylactically with the sensitised tissues (Buxton & Thomlinson, 1961; Thomlinson & Buxton, 1963). Among the post-mortem changes of bowel oedema are mucosal haemorrhages, venular thrombosis and infiltration by lymphocytes, plasma cells and eosinophils. Similar changes occur during anaphylaxis in pigs challenged intravenously with proteins (Thomlinson

& Buxton, 1963). This similarity however is insufficient to support a contention that bowel oedema is mainly an anaphylactic disease. Pigs readily respond to harmful substances and protein deficiencies with oedema, and eosinophilia of the bowel mucosa and lymph nodes is frequently present in apparently normal pigs, even at birth.

A serious gap in the evidence for anaphylaxis is that no antibodies conferring anaphylactic sensitivity on the homologous tissue have been found in the serum in the disease. The detection of antibodies to *E. coli* in the sera of diseased pigs by antiglobulin haemagglutination tests and by passive cutaneous anaphylaxis in guinea-pigs (Buxton & Thomlinson, 1961) is only evidence that the organisms had induced formation of precipitating-type antibodies. Transfer of cutaneous anaphylactic sensitivity to guinea-pigs with pig sera is no evidence that the same antibodies will anaphylactically sensitise the homologous pig tissue. Antibodies of man and rabbit that sensitise guinea-pigs do not sensitise the homologous tissue, and it is very unlikely that those of the pig will do so because transfer of anaphylactic sensitivity is restricted to closely related species. Moreover, the phenomenon described as reversed passive anaphylaxis, when *E. coli* lipopolysaccharide was injected intravenously into pigs, followed 30 minutes later by pig anti-*E. coli* antibody resulting in oedema and haemorrhages, (Thomlinson & Buxton, 1963) probably resulted from one or a combination of three non-anaphylactic responses, namely cytotoxicity due to acquired antigen, Arthus-like response due to antigen–antibody complexes, and anaphylactoid shock due to anaphylatoxin.

The evidence supporting the claim for anaphylaxis in piglet bowel oedema is worthy of further discussion, because it illustrates the difficulties in distinguishing not only between changes due to infection and to allergy, but also between different allergic responses which clinically show similar signs. Bacterial lipopolysaccharide is an unfortunate choice of antigen for a study of anaphylaxis, as it reacts in other forms of tissue response superficially similar to anaphylaxis. It combines quickly and firmly with cells *in vitro* and *in vivo* in such a way that antibody to it damages the cells. Experiments on such cytotoxicity are described later. This cytotoxicity, resulting from antibody reacting with cell-acquired antigen, differs from reversed passive anaphylaxis, in which the antigen, in, though not necessarily adsorbed to the tissues, reacts with a later injection of anaphylactic antibody. The cytotoxic response is not inhibited by antidotes to histamine and the other pharmacological mediators of anaphylaxis, it is usually complement-dependent which anaphylaxis is not, and the signs in an animal are sometimes much slower in appearing, 1 to 24 h, depending upon the system, whereas anaphylactic signs appear in minutes.

Among the tissue changes described by Thomlinson & Buxton (1963) after anaphylaxis in actively sensitised pigs and in 'reversed passive anaphylaxis' were venular thrombosis and haemorrhagic intestinal Peyer's patches. These changes more resemble those of the early Arthus-type response than of anaphylaxis. Actively sensitised pigs will have precipitating antibodies which form complexes with antigen as well as antibodies anaphylactically sensitising tissue, and in the 're-versed' test the antibodies introduced intravenously were very probably precipitins. Intravascular formation of antigen–antibody complexes which bind complement will, in turn, induce formation of anaphyla-toxin resulting in anaphylactoid (anaphylaxis-like) changes in the absence of passive sensitising antibody. The complexes also induce the Arthus-like changes of venular thrombosis and haemorrhage as described below.

Bacterial lipopolysaccharide can moreover induce anaphylactoid and Arthus-like changes in the absence of specific antibody, by fixing complement and forming an anaphylatoxin. This form of 'endotoxin shock' results in oedema, thrombosis, and haemorrhages particularly in lymphoid nodules such as the Peyer's patches. It is noteworthy that one of the control pigs of Thomlinson & Buxton (1963) injected with lipopolysaccharide alone had mild anaphylactic (or anaphylactoid) signs with typical oedematous lesions when killed.

These observations do not refute the contention that bowel oedema is an anaphylactic disease, or even that anaphylaxis occurs and may contribute in small part to the infective changes, but they show that much clearer evidence is required to substantiate the claim. It is essential to identify a porcine antibody that anaphylactically sensitises homo-logous tissues. Antibody of this type to *E. coli* antigens must be shown to occur in bowel oedema, or at least, the tissues of affected pigs when tested with a bland *E. coli* antigen must show an anaphylactic response. Then, passive transfer of this antibody, free of precipitating antibody, to normal pigs should make them susceptible to the oedematous changes when exposed to *E. coli* antigen or organisms. Such tests, though still not giving conclusive evidence, could go far to demonstrate the role of anaphylaxis in piglet bowel oedema.

It is, however, unlikely that this is an anaphylactic disease. Pigs, like other animals, appear susceptible to only a few strains of *E. coli*, though exposed to many others which can also multiply in the intestine. Cultures or cell-free culture filtrates of the pig-pathogenic *E. coli*, induce severe inflammatory changes with dilatation of ligated segments of the small intestine of normal pigs (Smith & Halls, 1967a, b). It is very unlikely that this response reflects anaphylactic sensitisation in all normal pigs. Pathogenicity appears to depend upon formation of an enterotoxin by these strains to which pigs are particularly susceptible.

ARTHUS-LIKE RESPONSES AND ANTIGEN–ANTIBODY COMPLEXES

The classical Arthus response follows repeated injections of antigen into an animal which becomes sensitised, until the later injections progressively induce erythema, haemorrhages and necrosis. The response results from the local formation of aggregates or complexes of precipitating antibody with antigen and complement in the lumen of blood and lymphatic vessels. These complexes, especially the fixed complement, attract neutrophils which ingest them and at the same time release lysosomal enzymes which degrade the vascular internal elastic lamina, and vascular or glomerular basement membranes. Thus neutrophils are an essential feature of the acute stage of Arthus reactions. This acute stage is followed by a more persistent accumulation of mononuclear cells including numerous plasma cells.

Arthus-type reactions also follow formation and dissemination of antigen–antibody complexes in the blood, as occurs in experimental serum sickness, resulting in endocarditis, vasculitis and glomerulonephritis. To be harmful, the complexes must form in antigen excess of antibody equivalence, fix complement, be above a minimum size (sedimenting at about 19 S in the ultracentrifuge) and penetrate beneath the endothelium. Endothelial penetration by complexes is facilitated by increased vascular permeability due to concomitant anaphylaxis or complement-mediated release of vaso-active amines from platelets (Cochrane & Henson, 1970). Neutrophil infiltration, tissue damage and subsequent accumulation of mononuclear cells occurs as in the classical Arthus response.

Vasculitis associated with complexes of bacterial antigen with antibody
The histological similarity between several human spontaneous vascular diseases and experimental serum sickness indicate that human vasculitis sometimes results from antigen–antibody complexes, but though gamma globulin and the β1C complement component were sometimes found in the lesions, no antigen had been identified until recently.

In a study of cutaneous vasculitis (Parish & Rhodes, 1967; Parish, 1970b, 1971a) the lesions were examined by immunofluorescence for bacterial antigens which might have been carried there by the blood, or which may have accumulated after local multiplication of the skin flora. *Streptococcus* group A antigen was found most frequently, appearing in the lesions of 12 of 67 patients, followed in order of frequency by *Candida albicans* in 8 of 67; antigens of *Streptococcus* group D, *Staphylococcus aureus* and *Mycobacterium tuberculosis* were found less often. The antigens appeared to be soluble substances, usually spread diffusely within the cytoplasm of mononuclear cells; finely granular antigen was found in only 10 % of the lesions. Both forms were easily distinguished from the occasional finding of larger particulate antigens, probably representing whole organisms. Ig G, Ig M and the complement β1C were frequently found in the same lesions, particularly those of 'allergic cutaneous vasculitis' in which the mononuclear cells containing antigen had superimposed upon them discrete deposits of Ig G, indicating the presence of complexes (Pl. 1, Figs. 1 and 2).

The sera of the patients were examined for antibodies able to form tissue-damaging complexes. Patients whose vasculitis followed streptococcal infection and whose lesions contained streptococcal antigens, had antibodies to streptococcal cellular proteins and polysaccharides which formed complexes *in vitro* with the appropriate antigens. These complexes fixed complement, released acid phosphatase and acid protease from neutrophils, and induced vascular lesions when injected into monkey or guinea-pig skin (Parish & Rhodes, 1967; Parish, 1971b). There was also indirect evidence of the formation of complexes in that several of the sera contained rheumatoid factor, anti-Ig G, and immunoconglutinin which is an auto-antibody to the fixed form of complement. However, antibodies in the sera of persons recovering from streptococcal infection without vasculitis were more effective in forming tissue-damaging complexes than those of persons whose infection was followed by vasculitis (Parish, 1971b). Similarly in tests on persons whose vasculitis followed a staphylococcal infection, antibodies to staphylococci from persons with and without vasculitis were equally effective in fixing complement. It is possible that the lesser activity of the sera of patients with vasculitis is due to removal of antibodies capable of forming harmful complexes from the blood while inducing the lesions. Persons escaping vasculitis may not have been exposed to sufficient antigen to form the harmful complexes.

Further examination of complexes containing some bacterial antigens showed that they penetrated vascular endothelium more easily than complexes containing bland protein antigens which are commonly used in experimental vasculitis. Moreover, some bacterial cellular substances, particularly endotoxins, and several exotoxins from streptococci, staphylococci and *Corynebacterium diphtheriae*, predisposed vessels to penetration by complexes which would otherwise remain intravascular and harmless (Parish, 1970b). In some instances, e.g. after exposure to streptococcal erythrogenic toxin, increased vascular penetration by complexes was followed by vasculitis, in others, as after exposure to diphtheria toxin, it was not, unless very small doses of toxin were used.

Another similarity between the experimental and spontaneous human vasculitis indicates that failure to produce antibody to an exotoxin predisposes the individual to tissue damage by complexes containing other non-toxic antigens released during the infection. Diphtheria toxin predisposes the blood vessels of guinea-pigs to penetration by unrelated antigen–antibody complexes. Previous immunisation of the guinea-pigs with diphtheria toxoid prevents this effect of the toxin. The sera of normal persons and those convalescent from streptococcal infections have been shown to contain precipitins to many extracellular streptococcal substances some of which are enzymes or toxins (Halbert & Keatinge, 1961). Parish (1971c) found that the sera of persons with post-streptococcal vasculitis tended to have less neutralising antibody to some streptococcal toxins and enzymes, e.g. to proteinase, when compared to post-infection sera of persons without vasculitis. Thus it is possible that some persons are susceptible to complex-induced vasculitis because they fail to neutralise toxins or enzymes that predispose vessels to the harmful effects of complexes.

Superficially these investigations give a convincing picture of vasculitis resulting from complexes of antibody with bacterial soluble antigens. Streptococcal antigen is released from the pharynx, and candidal antigen from pharynx or intestine, to combine with haematogenous antibody forming vasculitis-inducing complexes. Occasionally, staphylococcal antigen may be absorbed through the skin to form local complexes, and the production of vasculitis is promoted by some bacterial cellular substances and exotoxins which predispose the vessels to complex-mediated tissue damage.

Unfortunately the evidence that spontaneous vasculitis is ever a complex-mediated disease is far from conclusive. All of us have circulating antigen–antibody complexes, frequently changing in type and concentration. Nonetheless, they are harmlessly removed. Circulating complexes and complement-fixing antibody that occur in the sera during post-infection vasculitis may not cause damage and their presence in lesions be accidental and harmless. Harmless uptake of intravenously injected complexes has been shown to occur in cutaneous lesions of guinea-pigs previously induced by adjuvants or delayed sensitivity. If the granulomatous lesions are less than 48 h old and the concentration of complexes small, the activated macrophages of the lesion ingest the complexes harmlessly, thereby avoiding reaction with the neutrophils. The complexes found in human lesions may have been removed in an equally harmless manner.

It can be further argued that the detection of bacterial antigen and globulin in a lesion is not evidence of combination in antigen–antibody complexes. In attempts to demonstrate that the globulin in the lesions was anti-bacterial antibody, globulins were eluted from skin of patients with or without vasculitis, but the anti-bacterial antibodies in the eluates correspond to serum antibodies and not to the nature of the cutaneous lesions. It was difficult to test the specificity of the immuno-globulins in lesions by immunofluorescence because of non-specific adherence of bacterial antigens, but in one case, antibody to a strepto-coccal extracellular protein was detected (Parish, 1971a).

There is another unresolved problem, that of the persistence in man of spontaneous vascular lesions for weeks or months. Experimental Arthus-type lesions elicited by complexes of antibody with protein antigens resolve soon after the antibody is degraded in 24 to 48 h (Cochrane, Weigle & Dixon, 1959); and even if such lesions could be perpetuated by continual exposure to such complexes, which does not occur (Parish, 1970b), their histology is different from the mainly mononuclear cell changes of the chronic spontaneous lesions. The mononuclear cell changes may result from delayed sensitivity, from antigen–antibody complexes in antigen excess or from deposition of polysaccharides of group A streptococci which induce nodular lesions in the skin of normal unsensitised animals (Schwab, 1964).

The importance of complexes in vasculitis is still unclear. Neverthe-less, despite the incomplete evidence, it seems probable that antigen–antibody complexes induce vasculitis in man. The role of bacterial antigens has been emphasised here, but the findings also apply to the antigens of other micro-organisms (with or without obvious infection) and parasites, and even cryoglobulins, food antigens and drugs. Once the lesions have been initiated they may be perpetuated by non-allergic inflammatory processes. One such process is the decreased fibrinolytic activity of the plasma and, or, of the vessel walls (Cunliffe, 1968;

Isacson, Linell, Möller & Nilsson, 1970). Presumably the persistent fibrin induces chronic lesions.

Complex-induced glomerulonephritis

Glomerulonephritis may be induced experimentally by antigen–antibody complexes, resulting in acute or chronic lesions depending upon the concentration and persistence of the complexes in the circulation. Acute lesions are inducible by two methods, by active immunisation when one large dose of antigen stimulates antibody formation and then combines with the antibody to form harmful complexes causing serum sickness, or by passive transfer of preformed antigen–antibody complexes. Chronic lesions result from the repeated injection of small amounts of antigen. (Unanue & Dixon, 1967). The antigen–antibody–complement complexes, with properties similar to those inducing Arthus reactions, are deposited along the outer surface of the renal capillary basement membranes. They attract neutrophils whose enzymes degrade the membrane and initiate the lesions in the glomerular tuft. Renal lesions may persist after exposure to the antigen has ceased (Dixon, Feldman & Vazquez, 1961; Unanue & Dixon, 1967; Cochrane, 1969). Vaso-active amines from platelets and kinins may also contribute to the renal damage (Cochrane, 1969; Cochrane & Henson, 1970).

Antibody response to bacterial antigens is associated with renal lesions in animals. Repeated sub-cutaneous injections of heat-killed *Proteus mirabilis* was followed by acute or subacute glomerulonephritis in about half the mice tested (Wood & White, 1956). The mice had fairly high levels of agglutinins to the O and H antigens but without any significant difference in those with or without nephritis. *P. mirabilis* antigen was found by immunofluorescence in the affected glomeruli where it persisted at least three weeks after the last injection. As the renal lesions occurred at varying intervals after the start of the injections and when the animals had formed antibody to the antigens, they were almost certainly induced by complexes. In another study, Howes & Pincus (1963) reported that repeated injections of formalin-killed *Escherichia coli* induced high levels of agglutinins in rabbits at 4 to 6 weeks, rheumatoid-like factor at 10 weeks and a mild glomerulitis at 21 weeks.

It is surprising that so few investigations have been made of nephritis associated with complexes containing bacterial antigens. The filtration function of the glomeruli should be particularly susceptible to the harmful effects of deposited complexes containing bacterial antigens, particularly those of commensals, which are frequently released into the blood. However, harmful complexes may seldom be formed and other factors may also be necessary to predispose the tissues to damage. Some unknown factor or 'helper' substance was believed to alter the basement membrane and allow complexes to become bound in rats with intraperitoneal diffusion chambers containing living group A streptococci from a human case of glomerulonephritis (Vosti, Lindberg, Kosek & Raffel, 1970). Proteinuria and renal lesions appeared after the formation of agglutinating antibody and when bound gamma globulin, $\beta 1C$ globulin and streptococcal M protein were first detected in the glomerular basement membranes. Rats inoculated with cultures of a non-nephritic strain of streptococcus had similar levels of antibody appearing at the same time, but no nephritis. Moreover, eluates from the damaged kidneys contained type-specific antibodies, but not those from the undamaged kidneys. Either the M substance from the nephritogenic strain differed from that in the non-nephritogenic strain in quality or quantity, or some other factor influenced the harmful deposition of the complexes in the glomerular basement membrane.

Though the conditions for, and the mediators of, complex-induced glomerulonephritis are still to be established, there is no doubt that complexes cause renal disease in some cases. However, complex-induced nephritis must be distinguished from the cytotoxic nephritis due to the antigenic similarity between the glomerular basement membrane and some strains of streptococci, as described below, and also distinguished from that resulting from bacterial toxins. Streptolysin S, for example, damages kidneys causing proteinuria and haemorrhages (Tan, Hackel & Kaplan, 1961; Tan & Kaplan, 1961), but the direct action of the toxin differs from the damage provoked by complexes in that the tubules are affected rather than the glomeruli, the lesions occur soon after infection instead of weeks later and antibodies are not required.

Arthus involvement in the local or generalised flare phenomenon

A local or generalised erythematous, sometimes papular, skin reaction appears to result from a modified Arthus response in which antigen reacts with antibody formed locally in tissues instead of with circulating precipitins. This phenomenon, which may mediate human local or generalised eczematous reactions to chemicals, fungi or bacteria, illustrates the difficulty in recognising allergy when the response does not conform to the accepted features of a particular allergic response, and in distinguishing between allergy and bacterial toxaemia.

The flare phenomonon is an erythematous reaction occurring at previous sites of delayed sensitivity skin test responses in man or guinea-pig when the antigen is re-introduced. It was thought previously that the lesions resulted from antigen in the blood reacting with specifically primed lymphocytes remaining at the site of a previous skin-test inducing a delayed sensitivity response. It now appears to be an Arthus response in which antigen reacts with local cell-bound antibody.

The flare phenomenon is best studied in animals sensitised by chemicals with single or few antigenic determinants rather than bacteria with many. Guinea-pigs sensitised by repeated topical applications or intracutaneous injections of dinitrochlorobenzene or picryl chloride develop delayed sensitivity to these antigens, determined by skin testing. Re-introduction of the antigen in 2 weeks to 4 months by skin testing, by intraperitoneal or intravenous injection, or by ingestion, induces an erythematous, oedematous change at some earlier sites of sensitisation and at the site of the skin test response, or if the animal is strongly sensitised, a generalised erythematous reaction. The erythema appears 4 to 6 h after injection of antigen and persists for about 48 h, though more persistent features are a slight thickening of the skin and desquamation of epidermal scales for one to two weeks. The early histological changes are congestion, oedema and infiltration by neutrophils and a few eosinophils. Later changes are thickening of the epidermis with slight hyperkeratosis.

The flare response in guinea-pigs is only slightly reduced by antihistamine drugs which differentiates it from anaphylaxis which is modified or ablated by these drugs. It is also unaffected by antilymphocyte sera which inhibit contact sensitisation mediated by delayed hypersensitivity. It is, however, inhibited by antisera and drugs that reduce the number of circulating neutrophils and that also inhibit the Arthus response. This similarity, together with the finding of neutrophils in the lesion and the appearance of the erythema 4 to 6 h after re-introduction of antigen, suggest that the reaction is neutrophil-dependent, which is an essential feature of the Arthus response. In further tests (Parish, unpublished) the reaction was found to occur in guinea-pigs which either had no antibody or far too low a concentration of agglutinating antibody in the serum to mediate an Arthus response. However, labelled antigens added to sections of the skin lesions were fixed to small mononuclear cells and to plasma cells indicating the presence of cell-bound antibody. An interesting finding, to be further examined, was that more cells fixed antigen in samples taken 2 h after challenge, than in samples from the same animal taken just before challenge, possibly indicating fresh antibody synthesis stimulated by the re-introduced antigen.

This type of allergic sensitivity also occurs after fungal and bacterial infections. Neutrophil-dependent flare responses were induced after cutaneous infection with *Trichophyton mentagrophytes* which needed no additional adjuvant, and with *Staphylococcus aureus* which did. Injection of the antigens of these organisms in Freund's adjuvant, induced delayed sensitivity but no susceptibility to the flare response.

Living *Staph. aureus* in water in oil emulsions applied frequently to small areas of abraded skin, or living cultures applied to 1 cm² skin areas, 1 cm away from a skin site injected with Freund's adjuvant, induced delayed sensitisation and susceptibility to the flare response when the animal was injected with phenol-extracted lipopolysaccharide fractions of *Staph. aureus* containing 15 % amino-acid residues. In strongly sensitised animals, the desquamation of epidermal scales at sites remote from previous treatment, e.g. the ears or paws, continued for 10 to 14 days. This response was specific for *Staph. aureus* and could be inhibited by treatment to remove neutrophils before challenge, as in the tests with animals sensitised to chemicals.

We found that a similar clinical response can be induced in guinea-pigs by streptococcal erythrogenic toxin. Injection of this toxin induces generalised erythema without epidermal cell desquamation; but if the animal had recovered from mild delayed-sensitivity skin response about 7 days previously, desquamation followed the erythema at the sites. This response was not specific because sites tested with chemicals, tuberculin, trichophytin, or *Staph. aureus* extracts were all susceptible and the effect was not neutrophil-dependent.

These studies suggest that bacterial, or fungal, infections associated with delayed sensitivity, may predispose the skin to an erythematous rash on further exposure to the antigen. In man it is possible that a similar rash could result in eczema. On present evidence, the rash appears to be due to neutrophils infiltrating areas where antigen reacts with cell-bound antibody. This, however, makes it difficult to identify the cause of the rash. Skin tests will only reveal delayed sensitivity, which occurs in many normal persons to several common bacteria, unless sufficient antigen is injected to induce the rash. There are no serum precipitins which usually indicate Arthus-type sensitivity. Moreover one bacterial toxin can mimic the reaction if there is pre-existing skin damage and perhaps others will be found to do so.

The experimental model cannot be compared too closely to human eczema, because the mode of sensitisation and of challenge, particularly with staphylococcus antigens, is exaggerated beyond that likely to occur spontaneously. Nevertheless a modified form of this phenomenon may occur spontaneously and tests are being devised to detect it.

CYTOTOXIC REACTIONS

In the terminology of pathology, cytotoxic refers to the action of any agent that damages cells. In a classification of allergic reactions the term is restricted to cell damage due to antibody or to sensitised lymphocytes that is specific for antigen in the susceptible cell ('target cell'), or specific for an antigen that has become intimately adsorbed to or incorporated in the structure of the cell. The lysis of group A or B red cells treated with immune-type anti-A or anti-B antibody in the presence of complement is a simple example of a cytotoxic reaction. The lysis of red cells that have been coated with a bacterial polysaccharide by antibody specific for the polysaccharide in the presence of complement is an example of cell damage due to antibody reacting with acquired antigen.

Cytotoxic reactions may follow sensitisation to bacteria that have antigens with determinants similar to those of the host, or may follow sensitisation to bacterial antigens acquired by the host. Where damage follows sensitisation to shared antigens, auto-allergic-like disease results. Auto-allergy, strictly defined, results from sensitisation to the tissues of the individual: there is first autostimulation in which antibody formation or a cell-mediated (delayed sensitivity) response is stimulated by host antigen, and then auto-reaction in which the antibody or primed cells combine specifically with the tissue antigen which may or may not cause a cytotoxic (autoclastic, cytolytic) effect. It is debatable whether a bacterial antigen that is so similar to a tissue antigen that it stimulates formation of antibodies that react with the tissue antigen is strictly an auto-allergic phenomenon. Nevertheless, the result, harmful or not, is indistinguishable from that resulting from autostimulation by a tissue antigen.

Similar antigens in enterobacteria and tissues

Antigens in *Escherichia coli* and *Salmonella typhosa* are also present in mouse L cells. These surface antigens are sufficiently similar that human sera containing bacteriostatic antibodies to the bacteria are toxic to the cells and the antibodies can be removed by cross-adsorption by bacteria or cells (Fedoroff & Webb, 1962; Webb & Fedoroff, 1963). Though these antigens and antibodies probably have no significance in disease in man or mouse, another is thought to be responsible for the suscepti- bility of mice to infection by *Salmonella*. Mice, partially tolerant to *S. typhimurium* due to the similar antigen, may respond slowly and insufficiently when infected (Rowley & Jenkin, 1962). However partial tolerance to one antigen in an organism containing several others foreign to the host, is unlikely to influence host resistance unless the partially tolerated antigen protects the organism from the host's defences and thereby influences its virulence, like the Vi antigen.

A further example of similar antigens in bacteria and tissues is the lipopolysaccharide of *E. coli* O 14, which shares the type-specific O antigen and other heterogenetic antigens with lipopolysaccharide extracts of human colon (Perlman, Hammarström, Lagercrantz & Gustaffson, 1965). The sera of patients with ulcerative colitis usually have high titres of antibody to this antigen complex. It is however unlikely that sensitisation to *E. coli* is responsible for the mucosal lesions because rabbits injected with *E. coli* form antibodies that combine harmlessly with antigens of the rabbit's colon (Asherson & Holborow, 1966).

Endocarditis and streptococcal antigens

It has been postulated that the antigens of some streptococci and the heart are so similar that anti-streptococcal antibodies formed during streptococcal infection induce auto-allergic endocarditis.

Persons dying from acute rheumatic fever often have γ-globulin bound to the cardiac auricular appendages with other deposits in the myofibres, the sarcolemma and the walls of small vessels. The globulin is present in sites of fibrinoid change but not in the Aschoff nodules (Kaplan & Dallenbach, 1961). It was found (Kaplan & Suchy, 1964) that goat antisera to human or rabbit heart homogenates precipitate with cell walls of group A streptococci; and that (Kaplan & Meyerserian, 1962) rabbit antibody to streptococcal cell walls reacts with sections of human heart in the same immunofluorescence pattern as the γ-globulin bound *in vivo* in rheumatic fever. The link connecting these findings was the detection in the sera of 24 % of patients convalescent from rheumatic fever of precipitins reacting with streptococcal and cardiac antigen (Kaplan & Svec, 1964). The antigen in the streptococci cross- reacting with heart was believed to be M protein (Kaplan, 1967, 1969). This formed the basis of an attractive hypothesis that streptococci stimulated formation of antibodies that reacted harmfully with similar cardiac antigens.

Later investigations of the cross-reacting antigens and antibodies showed that they are more numerous and complex than originally thought. There is debate whether or no the cross-reacting antigen lies in the streptococcal cell membrane instead of being a cell-wall M protein (Zabriskie, 1969); and four different streptococcal group A antigens have been found to share determinants with different elements of the myocardia of man, rabbit and guinea-pig. All four antigens were proteins; three were in the cell wall and one corresponded with that of Kaplan (Lyampert, Vvedenskaya & Danilova, 1966). The identity of the cross-reacting antigen becomes more confusing when it is reported, not to be protein, but a group specific streptococcal carbohydrate with the same terminal grouping as that of a heart valve glycoprotein (Goldstein, Halpern & Robert, 1967). Furthermore, either the 'group specificity' of the cross-reacting antigen is to be doubted, or Nakhla & Glynn (1967) found yet another antigen which was common to 9 of 10 types of group A and to one type of group G streptococci. This antigen, which was not M protein but probably in the cell wall as it was absent from an L form, was also present in cardiac and voluntary muscle of man, rabbit and guinea-pig.

The antibodies reacting with streptococci and with cardiac muscle cannot be assumed to cause the heart damage. Rabbits and guinea-pigs with high titres of anti-streptococcal antibody, which reacted with cardiac antigen *in vitro*, appeared to be unharmed (Nakhla & Glynn, 1967). Also rabbits were unharmed when injected with homologous and heterologous heart antigens after forming anti-heart antibodies, which, in five animals tested, precipitated *in vitro* with extracts of the autologous hearts; a few of the anti-heart sera also precipitated with group A streptococcal antigens (Halbert, Holm & Thompson, 1968). It is regrettable that the opportunity was missed for examining the hearts histologically and by immunofluorescence. Gel plate precipitation tests showed that cardiac auto-antibody may be unrelated to the antibodies that cross-react with streptococci and with hearts of other animals.

If streptococci elicit antibodies that cross-react with the tissues of the host there must be some impairment of tolerance. Not all persons exposed to streptococci containing a cross-reacting antigen subsequently have rheumatic fever (Nakhla & Glynn, 1967). Antibody to this cross-reacting antigen also occurs without rheumatic fever so other factors must determine the tissue damage. There is no doubt that some strepto-coccal antigens share determinants with organ specific antigens of animals but the significance of antibodies to these antigens in the mediation of the cardiac damage of rheumatic fever is unknown.

Glomerulonephritis and streptococcal antigens

Glomerulonephritis following streptococcal infection is stated to occur in a manner similar to that occurring in the endocarditis of rheumatic fever. Rheumatic fever however is associated with infection by group A streptococci of any type but nephritis particularly follows infection by type 12 (Dingle, Rammelkamp & Wannamaker, 1953; Wilmers, Cunliffe & Williams, 1954) and occasionally type 5 streptococci.

In experimental models, renal cytotoxic auto-antibodies have been induced in sheep by injection of heterologous kidney antigens resulting in a progressive epithelial cell proliferative glomerulonephritis, and in rats by injection of homologous kidney antigens, resulting in a membrane-destructive glomerulonephritis. In both models the auto-antibodies are believed to be specific for the glomerular basement membranes (Unanue & Dixon, 1967).

The possible occurrence of a similar auto-allergic nephritis in man is supported by the finding of antibodies to human kidney in patients with nephritis (Lange, Gold, Weiner & Simon, 1949; Liu & McCrory, 1958; Kramer, Watt, Howe & Parrish, 1961). It is not known whether these antibodies are elicited by autologous renal or streptococcal antigens, because antigenic similarities have been demonstrated in oligosaccharide chains of human glomerular basement membrane and the cell membrane of type 12 group A streptococci isolated from a patient with nephritis (Lange, 1969).

The renal cytotoxic effect of antiserum to whole streptococcal membrane was examined by Rapaport & Markowitz (1969) who perfused one kidney of dogs with sheep anti-streptococcal membrane serum and the other with normal sheep serum before replacing the kidneys. Subsequently thirteen of sixteen perfused kidneys had functional and histological changes, and in eight of them sheep globulin was found by immunofluorescence as linear staining or granular deposits along the basement membrane. Control kidneys perfused with normal serum were undamaged. The results were interpreted as evidence that antisera to group A type 12 streptococci cause nephritis. These tests are an important contribution to the study of nephritis but they are not conclusive. The antigenic similarity between type 12 streptococcal membranes and glomerular basement membrane accounts for the fixation of the sheep anti-streptococcal serum to the basement membranes of the perfused kidneys but this need not have caused renal damage. It is equally possible that the dogs formed antibody to the sheep globulin fixed in the kidney and this anti-sheep globulin antibody induced the glomerular damage. The glomeruli would be particularly susceptible in this instance because the foreign globulin was fixed to the component most susceptible to damage and only small amounts of avid antibody to the sheep globulin would be necessary.

Serious doubts were expressed by Cruickshank (1959) that auto-antibodies were a cause of nephritis in man. The serum factor reacting with kidney antigens may not be specific antibody because in the few serum samples that he found to contain ostensible antibody to kidney, absorption with homogenates containing renal antigens failed to reduce the antibody titre. Moreover, he was unable to elute auto-antibody from the kidneys of fatal cases of acute glomerulonephritis; an inability confirmed by Freedman & Markowitz (1959).

Therefore, there is no conclusive evidence that group A type 12 streptococci induce renal cytotoxic auto-antibodies because they share antigenic determinants with glomerular basement membranes.

Cytotoxic changes in tissue cells with acquired bacterial antigens when exposed to antibody

Erythrocytes incubated in lipopolysaccharide extracts of several bacteria, or streptococcal culture supernatant fluids, adsorb the antigens on their surface, so that the washed erythrocytes are agglutinated by the corresponding antibodies and are lysed if the antibodies also fix complement. Few attempts have been made to examine this phenomenon in nucleated cells. The cytolytic changes in tissue cells with acquired serum proteins when exposed to antibody (Hamburger, Pious & Mills, 1963; Hamburger & Mills, 1965) could not be reproduced in my laboratory. I have confirmed however the report by Dishon, Finkel, Marcus & Ginsburg (1967) that streptococci release into the culture medium during the logarithmic phase of growth antigenic substances that are readily adsorbed by tissue cells; and have further shown that lipopolysaccharide–protein fractions of some bacteria are also readily acquired by tissue cells predisposing them to damage by the relevant antibodies. In contrast to the lipopolysaccharides, non-toxic bacterial proteins are firmly retained by tissue cells, but antibodies to these antigens, with complement, are harmless or induce only mild transient changes.

Changes in cells acquiring non-toxic bacterial proteins

HeLa cells exposed to nontoxic protein antigens of *Staph. aureus* or *E. coli* adsorbed the antigens in such a way, that on subsequently exposing the cells to rabbit antibody to the acquired bacterial antigen, they were overtly damaged. There was a much greater release of nucleic acid into the medium than in the control cells without antigen but treated with the antibody, or with the antigen but treated with normal rabbit serum. Many cells died as indicated by the dye exclusion technique or by acridine orange staining. Human sera containing complement-fixing Ig M antibodies to the relevant antigens were harmless and those containing Ig G antibodies caused no overt cell damage and only a slight increase in the number of dead cells.

Damage to the HeLa cells, in these tests with rabbit antisera based on those of Hamburger *et al.* (1963), was partly due to non-allergic causes. In order to avoid non-specific effects the study was continued on primary cell cultures of guinea-pig kidney or lung grown in Hank's solution or Eagle's medium containing 20 % or 10 % normal guinea-pig serum. These cells were exposed to the harmless protein antigens for varying periods before being washed to remove excess antigen and tested with complement containing guinea-pig antibody to the acquired antigen. The antibody was added to the same concentration (20 % or 10 %) as the normal serum used in the tissue culture medium. The test cells showed either no morphological changes when examined by phase contrast illumination or a foam-like perinuclear vacuolation and release of occasional homogeneous blebs from the cell membrane, a milder version of the changes in Plate 1, Fig. 4. These changes disappeared in 3 to 6 h; after which the cells appeared normal in every respect. Examination of the cells by several techniques within 3 h of testing with the antiserum often showed no differences from the control cells exposed to antigen only, or to antibody only, but the average results of over 120 tests indicated a slight transient cytotoxic effect. There was a slight increase in cell death as determined by the dye-exclusion test, though this was reversible as some cells taking up dye excreted it later in a bleb. The increase in the amount of nucleic acid released into the medium was slight but there was a greater release of acid phosphatase and adenosine-5'-triphosphate, 200 to 250 of the treated cells were required to start a fresh culture compared to 70 to 100 of the control cells and there was decreased respiration in the test cells for about ten minutes after treatment as measured by an oxygen electrode. The cell damage however was insufficient to release radio-isotope-labelled chromium.

The bacterial antigens acquired by the cells could not readily be traced by immunofluorescence but after three hours exposure they were retained by the cells for at least five days because cells washed twenty times and injected into rabbits elicited antibodies to their acquired antigen. Trypsin treatment of the cells only slightly reduced the susceptibility to antibody of those acquiring bacterial antigens but it greatly reduced the susceptibility of those acquiring albumin or β-lactoglobin antigens. This indicated that some of the antigen persisted in or on the membrane. Complement potentiated the effect of the antibody; in its absence cytotoxic changes still occurred but were milder and more transient. Young cultures of dividing cells were more susceptible than older cultures.

The transient changes induced by anti-bacterial antibodies in cells acquiring bacterial non-toxic proteins almost certainly have no counterpart in human disease. Tests on the sera of patients with generalised eczema, who have antibodies to several bacteria commonly found on skin, show that these antibodies do not differ from those found in normal persons and cause no damage to cultures of human skin or to embryo kidney which have absorbed the relevant antigen (Welbourn, Champion & Parish, unpublished).

Changes in cells acquiring extracellular streptococcal antigens and bacterial lipopolysaccharides

The extracellular products of group A streptococci and lipopolysaccharide fractions of *Staph. aureus* and *E. coli* predispose tissue cells to severe damage by relevant anti-bacterial antibodies. The severity of the damage depends upon the antigen, because according to the origin and age of the culture, different samples vary in their ability to sensitise cells.

Cells exposed to bacterial antigen for 1 h and then to medium containing normal serum either showed no changes or had slight perinuclear vacuolation even when left overnight (Pl. 1, Fig. 3). Cells treated with antibody only or bacterial antigen and complement without antibody showed similar changes or none at all. Cells acquiring bacterial antigen and treated with antibody and complement showed changes in 15 to 30 min, with foam-like perinuclear vacuolation, dispersal of the cytoplasmic organelles almost to the cell membrane, thickening and shrinking of the intercellular bridges and occasional blebs (Pl. 1, Fig. 4). In some systems these changes occurred slowly and progressed no further; some of the cells eventually recovered. In others the changes occurred more rapidly and progressed further. The remaining intercellular bridges were shed, more blebs of homogeneous material were released and the nuclear detail was obscured (Pl. 2, Fig. 1). Eventually the cell membrane ruptured releasing the granular cytoplasmic contents or the cell became grossly swollen (Pl. 2, Fig. 2). In a primary culture of guinea-pig kidney, which is a mixed population of cells, there was much variation in susceptibility. The cells nearest the original seed of the culture were more susceptible than those on the edge of the clone. There were always degenerate cells near the original seed (Pl. 2, Fig. 3), which probably made them more susceptible, but cells on the periphery of the clone did not escape damage even if they remained on the coverslip (Pl. 2, Fig. 4). The first attempts to test this system *in vivo* resulted in damage to skin or peritoneal cells but indicated more potential interfering phenomena than yielded reliable results, for example the need to avoid changes due to Schwartzman- and Arthus-type reactions.

Some bacterial antigens are probably released in sufficient amount during normal infections to be acquired by cells thus predisposing them to cytotoxic damage. The importance of such damage in diseases associated with bacterial infection is still to be determined. Sera from normal or eczematous patients containing antibodies to bacterial lipopolysaccharides have been shown to damage monolayers of human embryo cells acquiring the relevant antigens, but not skin organ explants. Mycoplasmas become firmly and persistently attached to the surface of cells *in vitro* (Thomas, 1969) which might predispose the cells to antibody-induced damage.

DELAYED SENSITIVITY AND BACTERIAL INFECTION

Delayed sensitivity, or cell-mediated immune response, of which the tuberculin skin test response is the best known example, plays a major part in defence against infection by bacteria, many viruses, fungi and protozoa. The mycobacteria readily induce delayed sensitivity and *Brucella abortus*, *S. typhi*, *Pfeifferella mallei* and *Listeria monocytogenes* infections frequently do so. Indeed most bacteria sensitise some persons to a delayed response.

Every delayed sensitivity response to bacterial antigen is to some extent a cytotoxic reaction. Reactions that proceed further than erythema, scanty mononuclear cell infiltration and activation of macrophages are accompanied by cell degeneration or even tissue necrosis. The tubercle is the classic histological change occurring in a delayed sensitivity response following infection. Rich (1951) emphasised repeatedly that the symptoms of tuberculosis represent in large measure the hypersensitive reaction to tuberculoprotein. Resistance acquired by the activation of lymphocytes and macrophages is accompanied by tissue damage.

It is inappropriate to review here the physiology of the delayed sensitivity response but a summary of some of its features is necessary to illustrate the causes of the tissue damage in the diseases mentioned later.

Delayed sensitivity and lymphocyte toxins

The delayed response is an immunologically specific reaction which takes several hours to reach a maximum and occurs in the absence of demonstrable globulin antibody in the serum. It is effected by lymphocytes 'sensitised', 'activated' or 'primed' to react specifically with an antigen or hapten, and by macrophages, which participate indirectly and non-specifically, mainly under the influence of the lymphocytes. Delayed sensitivity is sometimes transferable to normal individuals by cells or fractions thereof, but not by serum. Thus cell-mediated immune response is another convenient term for this reaction.

Specific antigen elicits in a skin test an erythematous and/or indurated lesion, which reaches its maximum intensity in 10 to 72 h. The essential histological features are a diffuse infiltration of mononuclear cells with characteristic perivascular accumulations. Neutrophils also infiltrate the site, especially when high concentrations of antigen are present, but they usually emigrate or degenerate by the time the lesion reaches its maximum intensity. Nevertheless their lysosomal enzymes probably contribute to the tissue changes.

When a normal animal is sensitised passively with labelled cells from an actively sensitised donor few labelled cells are found in the delayed reaction skin test site of the recipient. It has been proposed that the reaction is initiated by the random infiltration of a few specifically sensitised cells which are modified by the antigen at the site to release substances which cause inflammatory changes involving immunologically non-specific cells. The influence of primed lymphocytes on other cells is also exerted generally on the reticulo-endothelial system of the recipient. Lymphocytes from mice highly resistant to *L. monocytogenes*, when transferred to normal mice confer on them resistance to *Listeria* and a proportionate level of delayed sensitivity. The lymphocytes from the resistant animal appear to activate the macrophages of the recipient (Mackaness, 1969; Mackaness & Hill, 1969).

The changes resulting from lymphocyte–antigen interaction are mediated by substances secreted by the lymphocytes. Primed lymphocytes incubated with antigen *in vitro* synthesise substances which have several biological activities, cell-stimulating and attracting, or cytotoxic (Dumonde, 1970). There is a tendency to attribute each biological activity to a separate substance but my preliminary tests indicate that some substances have several activities.

Among the substances synthesised are cytotoxins which undoubtedly contribute to the host damage occurring in delayed sensitivity to bacteria. Lymphocytes from sensitised animals destroy tissue cells *in vitro* when lymphocytes and specific antigen are added together to the culture: lymphocytes from unsensitised normal animals will also do this when they have been stimulated by phytohaemagglutinin or filtrates of *Staph. aureus* cultures (Holm & Perlman, 1967; Ruddle & Waksman, 1967). Contact between the lymphocytes and tissue cells is not necessary because lymphocytes incubated with specific antigen *in vitro*, or stimulated by phytohaemagglutinin synthesise a substance which in a cell-free solution is damaging to tissue cells (Kolb & Granger, 1968; Williams & Granger, 1969). The lymphocyte toxin lyses erythrocytes and damages tissue cells from many species. However, overt destruction of the test cells does not always occur and the damage may be so slight as to require sensitive techniques, e.g. release of radioisotopes from the test cells, to detect it.

Leucocytes are also susceptible to antigen induced damage. The tests of Rich & Lewis (1927), confirmed many times, showed that tuberculin kills spleen cells and blood leucocytes from sensitised guinea-pigs *in vitro* in a concentration harmless to cells from normal animals.

Tuberculosis

The classic lesion of *Mycobacterium tuberculosis* infection is the tubercle. The activation of the lymphocytes and macrophages, and to a lesser extent any infiltration of neutrophils, is a reflection of delayed sensitivity. Though necrosis may not be apparent in small tubercles, the presence of epithelioid cells (macrophages modified by antigen or by toxin) and fibroblasts reflects mild degeneration. Necrosis, leading to caseation in larger tubercles, results mainly from the diminished blood supply from vessels occluded by infiltrating mononuclear cells and by leucocyte thrombi. Lymphocyte toxins may contribute to the necrosis as do toxic substances from dying cells and the bacteria. Liquefaction of the tubercles may occur with multiplication of organisms protected within the tubercle. Subsequent release of the organisms into the sensitised tissues leads to the mechanical damage and physiological dysfunction of organs which are only too well known as forms of host damage due to hypersensitivity.

Leprosy

Leprosy is one example of a disease in which the type and intensity of the allergic response influences the nature of the tissue damage as well as resistance to the infection. Leprosy exists in a wide variety of forms. At one extreme there is a 'tuberculoid' type with granulomatous lesions comprising dense collections of small lymphocytes and histiocytes rarely containing *Mycobacterium leprae*, which occurs in people who give a strong delayed sensitivity reaction to lepromin. At the other extreme, is the 'lepromatous' type with nodular lesions containing many macrophages with numerous organisms, but few small lymphocytes, which occurs in people with weak delayed sensitivity responses to lepromin or even to unrelated contact-sensitising antigens (Turk, 1970). Between the two extremes are patients with intermediate forms characterised as they approximate to the lepromatous type by a decreasing delayed hyposensitivity, first to lepromin, then to other antigens.

People with tuberculoid leprosy are very resistant to the infection, which accounts for the scarcity of the organisms in the lesions; those with lepromatous leprosy are extremely susceptible and there are numerous organisms in the tissues despite the large amounts of precipitating antibody sometimes found, which is presumably opsonic, or bactericidal. These antibodies probably contribute to the lesions in the form of lepromatous leprosy termed erythema nodosum leprosum, in which cutaneous nodules are infiltrated by large numbers of neutrophils in a manner resembling Arthus reactions (Turk, 1970).

Syphilis

Syphilis is a disease with a wide variety of lesions and several forms of allergic response. Its consideration makes a useful conclusion to the chapter because it illustrates the difficulty of attributing tissue damage in a complicated disease to particular types of allergic response.

The usual diagnostic tests for syphilis, flocculation in the Kahn test, or complement fixation in the Wassermann Reaction, detect anti-cardiolipins in the sera. *Treponema pallidum* contains phospholipids with antigenic determinants in common with the cardiolipins in the inner membrane of mitochondria of all living cells. Thus, infection with *T. pallidum* provokes cross-reacting haptenic auto-antibody. Several auto-antibodies are associated with syphilis, to brain, to erythrocytes, to mitochondria, and to a cardiolipin F (Wright & Doniach, 1971). These antibodies are useful diagnostic aids but they have not been proved to be harmful.

Antitreponemal antibodies distinct from the anticardiolipins are also formed during infection. These may contribute to resistance, but though they have not been demonstrated to be harmful, it has been suggested that complexes are formed with soluble antigen which induce the clinical manifestation of secondary syphilis (Levine, Turk, Wright & Grimble, 1969).

Resistance to syphilis is associated with a delayed sensitivity response. Many forms of the lesions are attributable to primed lymphocytes reacting with large amounts of treponema antigen known to be spread through the body in primary and secondary syphilis. However, as in lepromatous leprosy, it is possible that in congenital and secondary syphilis, delayed sensitivity responses may be impaired due to large amounts of antigen reducing lymphocyte receptivity in the early stages of the disease. Lymphocytes from persons with primary or secondary syphilis respond only about half as actively to stimulation by phyto-haemagglutinin, as those from normal persons. Impairment of lymphocyte transformation in this test is believed to reflect reduced ability to respond in delayed sensitivity reactions. Moreover a substance was found in the plasma of patients with secondary syphilis that reduced the ability of normal lymphocytes to transform; this could be a treponemal product or an anti-lymphocyte factor (Levene et al. 1969; Levene, Wright & Turk, 1971).

As lymphocyte activity in secondary syphilis is reduced but not abolished, it is possible that the severe lesions and death may result from a progressive delayed sensitivity response insufficient to curb the infection. Alternatively, complexes, auto-allergy and non-allergic processes may combine to induce further tissue damage when the resistance associated with delayed sensitivity is reduced.

DIFFICULTIES IN DISTINGUISHING BETWEEN ALLERGIC AND NON-ALLERGIC CHANGES INDUCED BY BACTERIA

Some allergic change occurs in nearly all bacterial infections but it is usually difficult to distinguish between the changes induced directly by the bacterial cells and their toxins and the changes resulting from allergic responses to their antigens. Several examples have been discussed in this chapter. The direct effect of bacteria and the influence of allergy varies during the course of the infection. Usually most early changes are directly induced by bacteria; the later changes are predominantly allergic responses which either hasten recovery or perpetuate the inflammation. Examples have been given of diseases in which the activated lymphocytes and macrophages, or complement-fixing bactericidal antibodies and antitoxins, by which the individual resists infection, may also induce cytotoxic changes or harmful antigen–antibody complexes. These allergic changes may evoke others resulting in autoallergy or in non-allergic inflammation manifested for example by decreased fibrinolysis in lesions.

In this account, reasons are given for the belief that anaphylactic sensitivity plays little part in bacterial disease in that it probably does not result in asthma, generalised oedema or urticaria. Anaphylactic sensitivity to bacteria may however occur in tissues resulting in local oedema thereby increasing spread of infection as in lobar pneumonia. But, similar changes may be evoked by anaphylatoxin, as for example that generated by lipopolysaccharides or by activated kinins and platelet substances (Cochrane & Henson, 1970), without mediation of anaphylactic passive sensitising antibody. Moreover, some bacterial substances disrupt mast cells, or release substances from neutrophils that disrupt mast cells, with release of histamine. A person anaphylactically sensitised to a bacterium may be more susceptible at the onset of infection but would become desensitised by rapid multiplication of the organisms and release of their antigens. It is extremely difficult to assess the contribution of anaphylaxis compared with that of anaphylatoxins and the bacterial toxins in inducing local tissue oedema.

The difficulty in determining the initial cause of focal neutrophil vascular lesions is even more complex. Apart from those due to the antigen–antibody–complement complexes described, similar lesions may be induced by infective bacterial emboli in the presence of little or no antibody; by local non-allergic Shwartzman effects associated with endotoxin mediated fibrinogen precipitation; or conceivably by the immunologically non-specific combination of bacterial substance and globulin as shown to occur between protein A of *Staph. aureus* and the non-antigen binding Fc portion of human Ig G resulting in Arthus-like reactions (Gustafson, Sjöquist & Stalenheim, 1967).

Similar difficulties attend the identification of the cause of focal mononuclear cell lesions. They could represent the late stage of Arthus reactions, especially when induced by small amounts of complexes which elicit only sparse neutrophil infiltration (Parish & Rhodes, 1967), or result from complexes formed in the region of antibody equivalence of antigen (Spector & Heesom, 1969). Delayed sensitivity may participate or be wholly responsible for formation of nodular lesions. Nevertheless some streptococcal cell-wall fragments elicit nodular granulomatous lesions after local injection (Schwab, 1964) or after generalised dissemination (Ohanian, Schwab & Cromartie, 1969) which are not due to hypersensitivity.

The importance of non-allergic 'acute phase' proteins in eliciting lesions is still to be determined, and though an allergic response in some form invariably occurs when pathogenic or harmless bacteria enter the tissues, it cannot be concluded that the allergic response harms the tissues just because experimental models elicit similar changes.

REFERENCES

ALSTON, J. M. & LOWDON, A. S. R. (1933). Studies of skin reactions to the specific soluble substances of the pneumococcus types I and II. *British Journal of Experimental Pathology*, 14, 1.

ASHERSON, G. L. & HOLBOROW, E. J. (1966). Auto-antibody production in rabbits. VII. Auto-antibodies to gut produced by the injection of bacteria. *Immunology*, 10, 161.

BUXTON, A. & THOMLINSON, J. R. (1961). The detection of tissue-sensitizing antibodies to *Escherichia coli* in oedema disease, haemorrhagic gastro-enteritis and in normal pigs. *Research in Veterinary Science*, 2, 73

COCHRANE, C. G. (1969). Mediation of immunologic glomerular injury. *Transplantation Proceedings*, 1, 949.

COCHRANE, C. G. & HENSON, P. M. (1970). Experimental immune complex disease. In *Immune complex diseases*, ed. L. Bonomo and J. L. Turk, 11. Milan: Carlo Erba Foundation.

COCHRANE, C. G., WEIGLE, W. O. & DIXON, F. J. (1959). The role of polymorphonuclear leucocytes in the initiation and cessation of the Arthus vasculitis. *Journal of Experimental Medicine*, 110, 481.

COGGESHALL, L. T. & ROBERTSON, O. H. (1935). Study of repeated attacks of experimental pneumococcus lobar pneumonia in dogs. *Journal of Experimental Medicine*, 61, 213.

CRUICKSHANK, B. (1959). Localisation of nephrotoxic antigens. *The Lancet*, i, 50.

CUNLIFFE, W. J. (1968). An association between vasculitis and decreased blood-fibrinolytic activity. *The Lancet*, i, 1226.

DINGLE, J. H., RAMMELKAMP, C. H. & WANNAMAKER, L. W. (1953). Epidemiology of streptococcal infections and their non-suppurative complications. *The Lancet*, i, 736.

DISHON, T., FINKEL, R., MARCUS, Z. & GINSBURG, I. (1967). Cell-sensitizing products of streptococci. *Immunology*, 13, 555.

DIXON, F. J., FELDMAN, J. D. & VAZQUEZ, J. J. (1961). Experimental glomerulonephritis. The pathogenesis of a laboratory model resembling the spectrum of human glomerulonephritis. *Journal of Experimental Medicine*, 113, 899.

DUMONDE, D. C. (1970). Mediators of cellular immunity in man. *Proceedings of the Royal Society of Medicine*, 63, 899.

FEDOROFF, S. & WEBB, S. J. (1962). Natural cytotoxic antibodies in human blood sera which react with mammalian cells and bacteria. *Nature, London*, 193, 80.

FINLAND, M. & BROWN, J. W. (1938). Reactions of human subjects to the injection of purified type specific pneumococcus polysaccharides. *Journal of Clinical Investigation*, 17, 479.

FINLAND, M. & SUTLIFF, W. D. (1931). Specific cutaneous reaction and circulating antibodies in the course of lobar pneumonia. I. Patients receiving no serum therapy. *Journal of Experimental Medicine*, **54**, 637.

FRANCIS, T. (1933). The value of the skin test with type specific capsular polysaccharide in the serum treatment of type 1 pneumococcus pneumonia. *Journal of Experimental Medicine*, **57**, 617.

FREEDMAN, P. & MARKOWITZ, A. S. (1959). Immunological studies in nephritis. *The Lancet*, **ii**, 45.

GASKELL, J. F. (1925). Experimental pneumococcal infections of the lung. *Journal of Pathology and Bacteriology*, **28**, 427.

GOLDSTEIN, I., HALPERN, B. & ROBERT, L. (1967). Immunological relationship between streptococcus A polysaccharide and the structural glycoproteins of heart valve. *Nature, London*, **213**, 44.

GUSTAFSON, G. T., SJÖQUIST, J. & STALENHEIM, G. (1967). Protein A from *Staphylococcus aureus*. II. Arthus-like reaction produced in rabbits by interaction of protein A and human γ-globulin. *Journal of Immunology*, **98**, 1178.

HALBERT, S. P. & KEATINGE, S. L. (1961). The analysis of streptococcal infections. VI. Immunoelectrophoretic observations on extracellular antigens detectable with human antibodies. *Journal of Experimental Medicine*, **113**, 1013.

HALBERT, S. P., HOLM, S. E. & THOMPSON, A. (1968). Cardiac auto-antibodies. I. Immunodiffusion analysis of multiple responses evoked homologously and heterologously. *Journal of Experimental Medicine*, **127**, 613.

HAMBURGER, R. N., PIOUS, D. A. & MILLS, S. E. (1963). Antigenic specificities acquired from the growth medium by cells in tissue culture. *Immunology*, **6**, 439.

HAMBURGER, R. N. & MILLS, S. E. (1965). Passive immune kill of cells in tissue culture. *Immunology*, **8**, 454.

HOLM, G. & PERLMANN, P. (1967). Cytotoxic potential of stimulated human lymphocytes. *Journal of Experimental Medicine*, **125**, 721.

HOWES, E. L., JR. & PINCUS, T. (1963). The development of a combined glomerulonephritis and amyloidosis following the injection of formalin killed *B. coli*. *Federation Proceedings*, **22**, 257.

ISACSON, S., LINELL, F., MÖLLER, F. & NILSSON, I. M. (1970). Coagulation and fibrinolysis in chronic panniculitis. *Acta Dermatovener, Stockholm*, **50**, 213.

JULIANELLE, L. A. & RHOADS, C. P. (1932). Reactions of rabbits to intracutaneous injections of pneumococci and their products. VII. Relation of hypersensitiveness to lesions in lungs of rabbits infected with pneumococci. *Journal of Experimental Medicine*, **55**, 797.

KAPLAN, M. H. (1967). Multiple nature of the cross-reactive relationship between antigens of Group A streptococci and mammalian tissue. In *Cross Reacting Antigens and Neo-antigens*, ed. J. J. Trentin, p. 48. Baltimore: Williams and Wilkins.

KAPLAN, M. H. (1969). Cross-reaction of Group A streptococci and heart tissue: varying serologic specificity of cross-reactive antisera and relation to carrier-hapten specificity. *Transplantation Proceedings*, **1**, 976.

KAPLAN, M. H. & DALLENBACH, F. D. (1961). Immunologic studies of heart tissue. III. Occurrence of bound gamma globulin in auricular appendages from rheumatic hearts, relationship to certain histopathologic features of rheumatic heart disease. *Journal of Experimental Medicine*, **113**, 1.

KAPLAN, M. H. & MEYESERIAN, M. (1962). An immunological cross-reaction between Group A streptococcal cells and human heart tissue. *The Lancet*, **i**, 706.

KAPLAN, M. H. & SUCHY, M. L. (1964). Immunologic relation of streptococcal and tissue antigens. II. Cross-reaction of antisera to mammalian heart tissue with a cell wall constituent of certain strains of Group A streptococci. *Journal of Experimental Medicine*, 119, 643.

KAPLAN, M. H. & SVEC, K. H. (1964). Immunologic relation of streptococcal and tissue antigens. III. Presence in human sera of streptococcal antibody cross-reactive with heart tissue. Association with streptococcal infection, rheumatic fever, and glomerulonephritis. *Journal of Experimental Medicine*, 119, 651.

KOLB, W. P. & GRANGER, G. A. (1968). Lymphocyte *in vitro* cytotoxicity: Characterization of human lymphotoxin. *Proceedings of the National Academy of Sciences, U.S.A.*, 61, 1250.

KRAMER, N. C., WATT, M. F., HOWE, J. H. & PARRISH, A. E. (1961). Circulating antihuman kidney antibodies in human renal disease. *American Journal of Medicine*, 30, 39.

LANGE, C. F. (1969). Chemistry of cross-reactive fragments of streptococcal cell membrane and human glomerular basement membrane. *Transplantation Proceedings*, 1, 959.

LANGE, K., GOLD, M. M. A., WEINER, D. & SIMON, V. (1949). Auto-antibodies in human glomerulonephritis. *Journal of Clinical Investigation*, 28, 50.

LEVENE, G. M., TURK, J. L., WRIGHT, D. J. M. & GRIMBLE, A. G. S. (1969). Reduced lymphocyte transformation due to a plasma factor in patients with active syphilis. *The Lancet*, ii, 246.

LEVENE, G. M., WRIGHT, D. J. M. & TURK, J. L. (1971). Cell-mediated immunity and lymphocyte transformation in syphilis. *Proceedings of the Royal Society of Medicine*, 64, 426.

LINDAU, A. (1933). Studies on varying response to pneumococcal infections, especially lobar pneumonia. *Acta pathologica et microbiologica Scandinavica*, 10, 1.

LIU, C. T. & McCRORY, W. M. (1958). Autoantibodies in human glomerulonephritis and nephrotic syndrome. *Journal of Immunology*, 81, 492.

LOESCHKE, H. (1931). Untersuchungen über die kruppöse Pneumonie. *Beiträge zur pathologischen Anatomie und zue allgemeinen Pathologie*, 86, 201.

LYAMPERT, I. M., VVEDENSKAYA, O. I. & DANILOVA, T. A. (1966). Study on streptococcus Group A antigens common with heart tissue elements. *Immunology*, 11, 313.

MACKANESS, G. B. (1969). The influence of immunologically committed lymphoid cells on macrophage activity *in vivo*. *Journal of Experimental Medicine*, 129, 973.

MACKANESS, G. B. & HILL, W. C. (1969). The effect of anti-lymphocyte globulin on cell-mediated resistance to infection. *Journal of Experimental Medicine*, 129, 993.

NAKHLA, L. S. & GLYNN, L. E. (1967). Studies on the antigen in β-haemolytic streptococci that cross-reacts with an antigen in human myocardium. *Immunology*, 13, 209.

OHANIAN, S. H., SCHWAB, J. H. & CROMARTIE, W. J. (1969). Relation of rheumatic-like cardiac lesions of the mouse to localization of Group A streptococcal cell walls. *Journal of Experimental Medicine*, 129, 37.

PARISH, W. E. (1970a). Short-term anaphylactic IgG antibodies in human sera. *The Lancet*, ii, 591.

PARISH, W. E. (1970b). Complexes of bacterial antigens with IgG or IgM antibodies in cutaneous vasculitis. In *Immune Complex Diseases*, ed. L. Bonomo and J. L. Turk, 98. Milan: Carlo Erba Foundation.

190 W. E. PARISH

PARISH, W. E. (1971a). Studies on vasculitis. I. Immunoglobulins, β1C, C-reactive protein, and bacterial antigens in cutaneous vasculitis lesions. *Clinical Allergy*, **1**, 97.

PARISH, W. E. (1971b). Studies on vasculitis. II. Some properties of complexes formed of antibacterial antibodies from persons with or without cutaneous vasculitis. *Clinical Allergy*, **1**, 111.

PARISH, W. E. (1971c). Studies on vasculitis. III. Decreased formation of antibody to M protein, group A polysaccharide and to some exotoxins, in persons with cutaneous vasculitis after streptococcal infection. *Clinical Allergy*, **1**, 295.

PARISH, W. E., BARRETT, A. M. & COOMBS, R. R. A. (1960). Inhalation of cow's milk by sensitized guinea-pigs in the conscious and anaesthetized state. *Immunology*, **3**, 307.

PARISH, W. E. & RHODES, E. L. (1967). Bacterial antigens and aggregated gamma globulin in the lesions of nodular vasculitis. *British Journal of Dermatology*, **79**, 131.

PERLMAN, P., HAMMARSTRÖM, S. LAGERCRANTZ, R. & GUSTAFFSON, B. E. (1965). Antigen from colon of germfree rats and antibodies in human ulcerative colitis. *Annals of the New York Academy of Sciences*, **124**, 377.

RAPAPORT, F. T. & MARKOWITZ, A. S. (1969). Streptococcal antigens and antibodies in transplantation. *Transplantation Proceedings*, **1**, 638.

RATNER, B., JACKSON, H. C. & GRUEHL, H. L. (1927). Respiratory anaphylaxis; sensitization; shock; bronchial asthma and death induced in guinea-pigs by nasal inhalation of dry horse dander. *American Journal of Diseases of Children*, **34**, 23.

RICH, A. R. (1933). Mechanism responsible for prevention of spread of bacteria in the immune body. *Bulletin of Johns Hopkins Hospital*, **52**, 203.

RICH, A. R. (1951). *The Pathogenesis of Tuberculosis*, 2nd ed. Oxford: Blackwell.

RICH, A. R. & LEWIS, M. R. (1927). Mechanisms of allergy in tuberculosis. *Proceedings of the Society for Experimental Biology and Medicine*, **25**, 596.

RICH, A. R., JENNINGS, F. B. & DOWNING, L. M. (1933). The persistence of immunity after the abolition of allergy by desensitization. *Bulletin of Johns Hopkins Hospital*, **53**, 172.

ROBERTSON, O. H., COGGESHALL, L. T. & TERRELL, E. E. (1933). Experimental pneumococcus lobar pneumonia in the dog. III. Pathogenesis. *Journal of Clinical Investigation*, **12**, 467.

ROWLEY, D. & JENKIN, C. R. (1962). Antigenic cross-reaction between host and parasite as a possible cause of pathogenicity. *Nature, London*, **193**, 151.

RUDDLE, N. H. & WAKSMAN, B. H. (1967). Cytotoxic effect of lymphocyte–antigen interaction in delayed hypersensitivity. *Science, New York*, **157**, 1060.

SCHWAB, J. H. (1964). Analysis of the experimental lesion of connective tissue produced by a complex of C polysaccharide from Group A streptococci. II. Influence of age and hypersensitivity. *Journal of Experimental Medicine*, **119**, 401.

SHARP, E. A. & BLAKE, F. G. (1930). Host factors in the pathogenesis of pneumococcus pneumonia. *Journal of Experimental Medicine*, **52**, 501.

SMITH, H. W. & HALLS, S. (1967a). Observations by the ligated intestinal segment and oral inoculation methods on *Escherichia coli* infections in pigs, calves, lambs and rabbits. *Journal of Pathology and Bacteriology*, **93**, 499.

SMITH, H. W. & HALLS, S. (1967b). Studies on *Escherichia coli* enterotoxin. *Journal of Pathology and Bacteriology*, **93**, 531.

SPECTOR, W. G. & HEESOM, N. (1969). The production of granulomata by antigen-antibody complexes. *Journal of Pathology*, **98**, 31.

STUPPY, G. W., CANNON, P. R. & FALK, I. S. (1931). Nature of immunity in lungs of rabbits following immunization with pneumococci. *Journal of Preventive Medicine*, **5**, 97.

STUPPY, G. W. & FALK, I. S. (1931). Experimental pneumonia in rabbits produced by intrabronchial insufflation of pneumococci. *Journal of Preventive Medicine*, **5**, 89.

TAN, E. M., HACKEL, D. B. & KAPLAN, M. H. (1961). Renal tubular lesions in mice produced by Group A streptococci grown in intraperitoneal diffusion chambers. *Journal of Infectious Diseases*, **108**, 107.

TAN, E. M. & KAPLAN, M. H. (1962). Renal tubular lesions in mice produced by streptococci in intraperitoneal diffusion chambers. Role of streptolysin S. *Journal of Infectious Diseases*, **110**, 55.

THOMAS, L. (1969). Mechanisms of pathogenesis in mycoplasma infection. *The Harvey Lectures*. Academic Press N.Y., **63**, 73.

THOMLINSON, J. R. & BUXTON, A. (1963). Anaphylaxis in pigs and its relationship to the pathogenesis of oedema disease and gastro-enteritis associated with *Escherichia coli*. *Immunology*, **6**, 126.

TURK, J. L. (1970). Reaction states in leprosy. In *Immune Complex Diseases*, ed. L. Bonomo and J. L. Turk, 165. Milan: Carlo Erba Foundation.

UNANUE, E. R. & DIXON, F. J. (1967). Experimental glomerulonephritis: immunological events and pathogenetic mechanisms. *Advances in Immunology*, **6**, 1.

VOSTI, K. L., LINDBERG, L. H., KOSEK, J. C. & RAFFEL, S. (1970). Experimental streptococcal glomerulonephritis: longitudinal study of a laboratory model resembling human acute post-streptococcal glomerulonephritis. *Journal of Infectious Diseases*, **122**, 249.

WEBB, S. J. & FEDOROFF, S. (1963). Natural cytotoxic antibodies in human blood sera which react with mammalian cells and bacteria. II. Effect of heated human serum on micro-organisms. *Canadian Journal of Microbiology*, **9**, 155.

WILMERS, M. J., CUNLIFFE, A. C. & WILLIAMS, R. E. O. (1954). Type-12 streptococci associated with acute haemorrhagic nephritis. *The Lancet*, **ii**, 17.

WILLIAMS, T. W. & GRANGER, G. A. (1969). Lymphocyte *in vitro* cytotoxicity: mechanism of lymphotoxin-induced target cell destruction. *Journal of Immunology*, **102**, 911.

WOOD, C. & WHITE, R. G. (1956). Experimental glomerulo-nephritis produced in mice by subcutaneous injections of heat-killed *Proteus mirabilis*. *British Journal of Experimental Pathology*, **37**, 49.

WRIGHT, D. J. M. & DONIACH, D. (1971). The significance of cardiolipin immunofluorescence (CLF). *Proceedings of the Royal Society of Medicine*, **64**, 419.

ZABRISKIE, J. B. (1969). The relationship of streptococcal cross-reactive antigens to rheumatic fever. *Transplantation Proceedings*, **1**, 968.

EXPLANATION OF PLATES

PLATE 1

Fig. 1. Allergic vasculitis (from Parish, 1971a; Blackwell Scientific Publications, by kind permission). Neutrophils accumulate and degenerate on the endothelium while others penetrate the wall. A group of endothelial cells is on the edge of the vessel (*right*) and a large unidentified cell (*top right*). Haematoxylin and eosin.

Fig. 2. Double staining immunofluorescence test of same lesion as that in Fig. 1 (from Parish, 1971a). Streptococcal antigen stained by rhodamine lies in the large mononuclear cells (*right*) and fluorescein stained IgG in discrete particles lies on or in the cells containing antigen; this indicates the presence of complexes. The cell (*top right*) also contains antigen and IgG.

Fig. 3. Guinea-pig kidney primary culture monolayer exposed to streptococcal extracellular nontoxic antigen for 1 h, and in medium with 10% normal complement containing guinea-pig serum overnight. Slight perinuclear vacuolation: intercellular processes are intact. Phase contrast.

Fig. 4. Culture treated with antigen as in Fig. 3. Then tested with medium containing 10% guinea-pig complement containing antiserum to streptococcal antigens. In 30 min the cells contract, with thickening and disruption of the intercellular processes, and have foam-like perinuclear vacuolation. Phase contrast.

PLATE 2

Fig. 1. Culture treated with antigen and antibody as in Fig. 4. After 1 h the intercellular processes are destroyed, cytoplasmic material is released in blebs, and nuclear detail obscured. Phase contrast.

Fig. 2. Culture treated with antigen and antibody as in Fig. 4. In 3 h the cells are almost completely destroyed. Phase contrast.

Fig. 3. Clone of kidney cells cultured overnight with guinea-pig antistreptococcal serum. (no antigen). The few degenerate cells near the original seed occur in normal cultures. Giemsa.

Fig. 4. Clone of kidney cells cultured overnight with guinea-pig antistreptococcal antibody after treatment with antigen. Most of the cells near the original seed are destroyed. Cells on the periphery are damaged. A few fibroblast-like cells appear unharmed. Giemsa.

PLATE 1

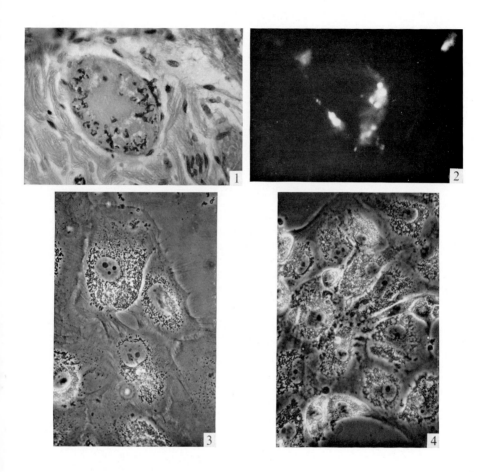

PLATE 2

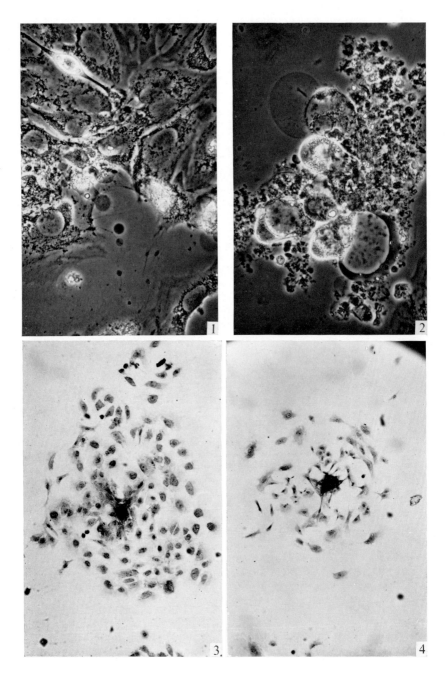

TISSUE AND HOST SPECIFICITY IN BACTERIAL INFECTION

J. H. PEARCE AND D. B. LOWRIE

Department of Microbiology, University of Birmingham, Birmingham, B15 2TT

INTRODUCTION

Our knowledge of the particular interactions which influence the susceptibility to infection of a given tissue within a host or of one host species relative to others has progressed little since the topic was reviewed by Keppie (1964) at an earlier symposium of the Society. We can see in principle how these specificities must be determined but know little of the details – except in a few instances. Yet appreciation of the way in which organisms localise in certain tissues or infect certain hosts could advance our understanding of the general problem of pathogenicity and might provide us with new approaches to problems of therapy.

In 'host specificity' the disease-producing capacity of an organism is restricted to a single host species or to a narrow host range in which the pattern of infection is comparable. In diseases where a tissue specificity is apparent the organism grows mainly, but not necessarily exclusively, in a particular tissue and typical disease is associated with infection of that tissue. Definition of terms here, although arbitrary, is important because it determines the limits of the phenomena for which we have to account. For example, in host specificity we exclude situations where an organism shows a narrow host range but causes markedly different patterns of infection in each host species. And our definition of tissue specificity has excluded those instances where tissue localisation is distinct from the site of pathological change or where localisation occurs without disease production. Nevertheless both these situations are relevant and we shall consider them later. Ultimately, although the two phenomena are quite distinct, host or tissue specificity must be expressed through the ability of the organism to resist host defence mechanisms, to multiply and to damage host tissues. Fundamentally then, we are concerned with factors which restrict or enhance microbial growth and how the balance of these accounts for susceptibility of one host species or one tissue rather than another.

7

One further point needs to be made. In many cases we have to deal with 'natural' infections where a given host species, the infecting organism and the 'infection chain' are the outcome of evolutionary adaptation. Thus the tissue localisation of any particular organism may be influenced by the route of infection – which essentially is predetermined in the natural situation but variable in experimental work. Correspondingly, the host range of a bacterial species in nature may reflect restriction to a particular route of infection. Thus exposure to another route, under laboratory tests, may reveal a different disease pattern within the known host range or give rise to disease in hosts previously thought to be insusceptible. Clearly, for diseases in which tissue or host specificity is maintained irrespective of route of infection the underlying mechanisms of pathogenesis must be powerful in effect.

The major part of our article is devoted to the problem of tissue specificity since, in the bacterial field, comparative data on factors relating to species differences are limited. We begin with a discussion of the underlying mechanisms that determine tissue localisation. To illustrate these we then consider in detail local infection at mucosal surfaces, mixed infection of the foot-rot disease complex in sheep and localisation in the placenta and kidney.

TISSUE SPECIFICITY

Accompanying or perhaps as an outcome of localised growth of a micro-organism one of three possible situations may develop: (1) localised growth and local damage due to toxic microbial products (or possibly to hypersensitivity reactions), (2) local growth without apparent damage, and (3) local growth with damage at some distant site due to released microbial products.

The situation of local growth with local damage is perhaps the most important. It includes the majority of the localising diseases of medical or veterinary importance; for example, localisation of proteus species in the kidney, *Escherichia coli* and *Shigella dysenteriae* in the mucosal surface of the gut, and of brucella species in placental infections of domestic animals. However, local growth of micro-organisms without evident pathological change is not infrequent – witness the mixed populations of the mucous surfaces of the mouth, throat or intestine.

Local growth with damage only at a distance is rare in microbial infection. It implies an extreme specificity of interaction of the released microbial product with cell or tissue. An example is tetanus, where the lesion is in the central nervous system (Wilson & Miles, 1964), and involves damage to anterior horn cells of the spinal cord; however bacterial growth, while localised, is not restricted to any particular tissue. In diphtheria, growth of the organism is largely confined to the tonsillar region of the throat (Wilson & Miles, 1964) with toxin acting primarily on heart muscle, although tissue specificity of action may depend on toxin concentration (Bowman & Bonventre, 1970).

The basis of localisation

Nutrition

The fundamental problem in assessing the relative significance of nutrition or defence mechanisms in localisation is that we know very little about the sequence of events in infection. Localisation might result from rapid proliferation owing to an unusually favourable nutritional environment. Alternatively the prime event could be the operation of factors leading to survival of organisms on lodgement – the size of the microbial population subsequently attained reflecting a delicate balance between division and death rates.

It would be technically very difficult to determine whether tissue localised growth was due to higher growth or lower death rate at the site. Where division rate *in vivo* has been determined, by the ingenious method of measuring the dilution rate of a segregating non-replicating genetic marker (Maw & Meynell, 1968), values were very different from those obtainable under optimal conditions *in vitro*. A doubling time of about 1 h was observed for *Salmonella typhimurium* in the mouse spleen in the first 90 min after inoculation, slowing to 5–10 h over the next 4 days. This type of study cannot readily be applied to all problems of tissue specificity and in our gross ignorance of the physico-chemical environment at the micro-level of host–pathogen interaction we have little to go on in reconstructing conditions for analysis *in vitro*.

Nutrients may frequently be growth-limiting within host tissues. Indeed phenotypic change, which undoubtedly can occur *in vivo* (Smith, 1968; Meynell, 1961), may reflect a shift in growth-limiting nutrient – growth in chemostat culture has shown that bacterial cell composition is altered by change of the nutrient present in limiting concentration (Ellwood & Tempest, 1969).

In several instances a nutritional basis for localisation has been inferred from growth-rate measurement in media derived from different tissue extracts (Keppie, 1964; Smith, 1968; Lowrie & Pearce, 1970). Observation of maximum growth-rate in extracts of the susceptible tissue, which suggests the presence of a growth stimulant, needs to be accompanied by isolation of the factors responsible and measurement of their distribution in other tissues. This approach entails the assumption that the nutrients present in extracts are readily available to the invading organism in the tissue and has the potential disadvantage that the organism is exposed to stimulants or inhibitors which are not in fact available *in vivo*. It might for this reason be more useful to search for tissue stimulation of bacterial growth in organ culture or within cells maintained *in vitro* when the organism is capable of intracellular growth.

Tissue damage and local defence mechanisms

Non-specific anti-bacterial defence is mediated by a variety of mechanisms, including mucus, lysozyme, polyamines and phagocyte migration and not all of these occur at any one site. Thus tissue localisation could entail microbial adaptation to the differences; for example, the destructive activity of alveolar macrophages to *Haemophilus influenzae* in the mouse is much less than that shown by peritoneal macrophages (Degré, 1969).

Toxin-mediated tissue damage can presumably affect growth indirectly through inhibition of host defences or liberation of nutrients. An analogous situation which favours microbial growth is seen in wound tissue or haematoma (Ferguson, 1970).

A quite different mechanism affecting host defences or susceptibility to tissue damage in localisation also deserves consideration, namely, that a particular tissue nutrition might influence the phenotype of the invading organism in respect of its synthesis of aggressins or toxins. Thus, *Brucella abortus*, which localises in the bovine placenta (see later), shows an enhanced capacity to survive intracellular killing by bovine phagocytes if grown in media supplemented with placental fluid (Fitzgeorge & Smith, 1966).

Infections of mucosal surfaces

The large mixed microbial populations present on mucosal surfaces provide examples of tissue-localised growth occurring without disease production. Although mucosal infections are dealt with in greater detail elsewhere in this symposium we consider them here because they illustrate three problems relevant to the general phenomenon of tissue specificity: (1) why does a micro-organism remain localised (in this instance on a mucosal surface), (2) what occurs when an organism, normally localised, gets away and establishes elsewhere, and (3) how does an incoming exogenous pathogen establish itself in the face of a pre-existing microflora? These problems resolve into questions of resistance to defence mechanisms and availability of suitable nutrients. Special interest, particularly from the point of view of localisation, arises from a characteristic feature of micro-organisms on mucosal surfaces: in mixed microbial populations nutritional interdependence and synergy assume major importance. Synergy will be encountered again in the discussion of ovine foot-rot and 'heel abscess' which follows this section.

Defence mechanisms as factors in localisation

Mucus, surface phagocytes and associated anti-microbial products of cellular origin are active against a range of pathogenic micro-organisms (Wilson & Miles, 1964; Vedros, Robinson & Gutekunst, 1966) but little is known of their role in defence against the various mucosal microflora. Presumably the organisms comprising a normal microflora are adapted to survive local surface bactericidins but not to penetrate the defences which confine them to the surface. Some microflora constituents bind closely to underlying epithelial cells and the attachment is cell-specific (Hoffman & Valdina, 1968; Savage, 1970; Gibbons & van Houte, 1971). If attachment confers protection from host defences, for example by preventing ingestion by phagocytes, then survival elsewhere may be limited by inability to bind to other cells. Immunity mechanisms play a part in confining growth of the common mucosal commensal *Candida albicans* to mucosal surfaces since, in deficiency of cell-mediated immunity, the organism destroys large areas of mucosa and produces systemic infection in progressive candidiasis (Chilgren *et al.* 1967; Valdimarsson *et al.* 1970; Schoch, 1971); progression may also result from hormone-induced alterations in host metabolism (Spellacy *et al.* 1968; Catterall, 1971).

Damage to the gut mucosa of normal turkeys caused by intestinal infection with a virulent strain of *Streptococcus faecalis* has been shown to enable non-sporing anaerobes of intestinal origin to spread elsewhere and cause liver granulomas; the introduction of either the anaerobes, re-isolated from the granulomas, or *Strep. faecalis* alone into the intestine of germ-free turkeys did not cause liver lesions; both organisms were required (Moore & Gross, 1968; Moore, Cato & Holdeman, 1969).

Slight physical damage to the integrity of mucosal surfaces which results in dissemination of components of the microflora may be a fairly common occurrence but defence mechanisms in other regions of the host are normally effective in eliminating the organisms. Only in exceptional circumstances are these defences inadequate and the initial localisation is lost. This is seen for example when damaged heart valves favour settlement of α-haemolytic streptococci released into the blood stream during tonsillectomy or tooth extraction.

Nutrient availability as a factor in localisation

On mucosal surfaces microbial nutrients appear to determine localisation in both a positive and a negative sense; essential metabolites must be present at a site of localisation for growth to occur; absence of essential nutrients from other regions might provide the basis of restriction of growth. Many mucosal commensals are nutritionally exacting yet their metabolic requirements clearly must be fulfilled by the local mucosal micro-environment. A degree of dependence upon metabolic products from underlying host cells is implied by the intimate association of specific bacteria in the mouth and alimentary canal with particular epithelial cell types (Savage, 1970; Gibbons & van Houte, 1971). Other evidence suggests that constituents of the microflora are dependent *in vivo* upon the metabolism of accompanying micro-organisms (Socransky *et al.* 1964; Rosebury, 1965; Hanks, 1966; Hardy & Munro, 1966). Thus the fastidious members of a mucosal microflora are less likely to encounter adequate nutrition in other host tissues. This is seen in the example mentioned earlier where growth of a non-sporing anaerobe was restricted to the turkey liver after systemic spread from the intestine (Moore & Gross, 1968).

Bacteroides melaninogenicus provides another example. Originating from mucosal surfaces, it is sometimes found in lesions at a variety of sites in man and animals but always as a mixed infection with other obligate or facultative anaerobes. The metabolic requirements for growth of the separate bacteria are not apparently fulfilled by non-mucosal tissues; only together are the organisms pathogenic (Socransky & Gibbons, 1965). Macdonald, Gibbons & Socransky (1960) showed that for invasion of sub-mucosal tissue in periodontal disease, which was simulated by subcutaneous infection in guinea-pigs, the four organisms, *B. melaninogenicus*, another bacteroides species, a motile Gram-negative rod and a facultatively anaerobic diphtheroid, were all required. *B. melaninogenicus* was regarded as the primary pathogen and the diphtheroid was shown to produce vitamin K, an essential growth factor for the primary pathogen; the contributions of the other two organisms were undefined.

Mucosal localisation of exogenous pathogens

Infection of the mucosal surface of the throat with *Corynebacterium diphtheriae* and of the lower intestine with cholera and dysentery bacilli in man and with *Mycobacterium paratuberculosis* (Johne's bacillus) in cattle provide clear examples of tissue localised infectious disease; little is known of the basis of these important localisations. Successful competition with normal mucosal flora, which has been shown to be important in the establishment of some exogenous pathogenic micro-organisms, e.g. salmonella and shigella bacilli (Meynell, 1963; Hentges, 1970), cariogenic and β-haemolytic streptococci (Krasse *et al.* 1967; Tillotson & Finland, 1969; Johanson *et al.* 1970), may also be required in the initiation of diphtheria. Competition may play a part in the restriction of growth of *C. diphtheriae* within the upper respiratory tract to the throat where intracellular growth may confer some protection from competition as well as from host defence mechanisms.

The predilection of *M. paratuberculosis* for intestinal mucosa in cattle and sheep is shown in young lambs even after experimental intravenous or intratracheal infection (Kluge *et al.* 1968); the determining factor is unknown but clearly it is not the route of exposure. Again competition with normal microflora may be largely avoided by intracellular growth and possibly also by haematogenous arrival at the susceptible mucosal cells. The ability of the bacterium to invade and survive systemic defence mechanisms, at least to a limited extent, emphasises that conditions within the intestinal mucosa are exceptionally favourable for growth; the very large population sizes attained at this site suggest that nutritional factors may be of major importance in the localisation.

Mixed infection in bulbar necrosis and foot-rot of sheep

Study of these inter-related diseases of the sheep's foot, by Roberts, Egerton and their colleagues in Australia (see Roberts, 1969) provides another example of synergistic interaction between organisms in local growth. The hoof and interdigital tissue are exposed to a mixed microbial flora – notably aerobic diphtheroids and anaerobic fusiforms – and under wet conditions both foot-rot and infective bulbar necrosis (IBN) or 'heel abscess' (Roberts, Graham, Egerton & Parsonson, 1968) are prevalent in flocks. Both diseases appear to be extensions of ovine interdigital dermatitis, in which inflammation of the interdigital skin is associated with epidermal invasion of the anaerobe *Fusiformis necrophorus*, (Parsonson, Egerton & Roberts, 1967) and which rapidly resolves in dry conditions.

In IBN, *F. necrophorus* penetrates into the dermis, causing a severe lesion, particularly in the soft bulbar tissue of the digit; this dermal invasion is accompanied by *Corynebacterium pyogenes* in the central necrotic area (Roberts *et al.* 1968).

In foot-rot, infection is limited to the epidermis and *F. necrophorus* invasion is invariably associated with the presence of *Fusiformis nodosus* which may be responsible for the chronic nature of the disease since, in contrast to *F. necrophorus*, it shows a capacity for slow persisting growth in tissues (Egerton, Roberts & Parsonson, 1969). In 'virulent foot-rot' infection spreads from the interdigital epidermis to the epidermal matrix of the hoof (Egerton *et al.* 1969; Roberts & Egerton, 1969) leading to separation of the hoof itself.

The conclusion that interaction between the component organisms determines both diseases was drawn from histological examination of lesions, experimental infection with single and mixed cultures and the demonstration of nutrient or anti-bactericidal factors in culture filtrates. The main points are summarised below.

Infective bulbar necrosis

The typical lesion was reproduced by inoculation with the mixed culture; *F. necrophorus* or *C. pyogenes* alone produced little inflammation. In culture filtrates from *F. necrophorus* an exotoxin was detected which damaged leucocytes and induced inflammation – possibly through release of lysosomal components. The toxin enhanced *C. pyogenes* infection and reduced phagocytosis of organisms at the injection site (Roberts, 1967*a*). *C. pyogenes* released a factor which stimulated the growth of *F. necrophorus* both in tissue and on nutrient agar (Roberts, 1967*b*).

Foot-rot

The disease could not be reproduced by *F. nodosus* alone when care was taken to exclude faecal material – which contained *F. necrophorus*. Growth of the latter was stimulated by a factor in broth filtrates from *F. nodosus* but *F. nodosus* infection was not enhanced by the toxin from *F. necrophorus* (Roberts & Egerton, 1969). A stimulant for *F. nodosus* may have been released by an unidentified motile fusiform seen in natural lesions; the frequent presence also of *C. pyogenes* suggests that it can stimulate *F. necrophorus* in foot-rot (Roberts, 1969).

Restriction of infection

None of the organisms mentioned above is powerfully invasive. However, hydration of the keratinised surface of the interdigital tissue in seasonally wet conditions appears to assist colonisation by aerobic catalase-producing diphtheroids which in turn provide a suitably anaerobic environment for invasion by *F. necrophorus* (Parsonson *et al.* 1967). Deeper penetration of fusiforms would be precluded by greater O_2 tension and *Eh* attainable within the vascular dermis.

That deeper infection does occur in IBN may depend on the development of anaerobic conditions due to reduced blood circulation in the foot. Circulatory stasis was inferred from histological material in which pigment deposits were seen in the hyperaemic region surrounding IBN lesions. Roberts and his colleagues (Roberts *et al.* 1968) suggest that low seasonal temperatures predispose to blood stasis in the extremities through peripheral vascular constriction and that the prevalence of IBN in lambing ewes – especially in the hind-feet – might arise from stasis simply as a result of increased weight or from constriction of the vascular trunk by the developing foetus. The more recent finding that haemoglobin and haematin enhance the growth of *F. necrophorus* on agar suggests that blood pigment may act as a stimulatory factor *in vivo* and assist the deeper penetration of infection (Roberts, 1970).

Infections of ungulate placenta

Animals of the order Artiodactyla (ungulates) seem to be unusually prone to placental infection by a variety of infectious agents. Micro-organisms which grow prolifically in the placenta of cattle or sheep to cause abortion include brucella species (Molello *et al.* 1963*a*, *b*, *c*), *Vibrio fetus* (Jensen, Miller & Molello, 1961; Osburn & Hoskins, 1970), the agents of enzootic abortion in ewes (Novilla & Jensen, 1970) and epizootic bovine abortion (Kwapien *et al.* 1970), *Toxoplasma gondii* (Beverley, Watson & Payne, 1971), *Listeria monocytogenes* (Molello & Jensen, 1964; Smith, Reynolds & Clarke, 1968), *Aspergillus fumigatus* (Cysewski & Pier, 1968; Hillman & McEntee, 1969), *Serratia marcescens* (Smith & Reynolds, 1970) and *Bacillus* spp. (Mason & Munday, 1968). In contrast, micro-organisms infecting the human placenta are seldom found in numbers likely to cause abortion (although mycoplasmas may yet be shown to play a significant role (see Kundsin & Driscoll, 1970; Harwick *et al.* 1970)).

Possible reasons for microbial localisation in ungulate placenta are here discussed in the light of circumstantial evidence that defence mechanisms may be comparatively ineffective and experimental evidence that nutrient supply is unusually favourable at this site.

Defence mechanisms

Placental deficiency in one or more of the known mechanisms of host defence could explain susceptibility to infection with the range of organisms quoted. The ungulate placenta is indeed unusual in presenting a near-complete barrier to the transfer of maternal immunoglobulins to the immunologically deficient foetus. However, the site of the barrier is unknown and initiation of infection, at least with *V. fetus* and brucella species in sheep (Jensen, Miller & Molello, 1961; Molello *et al.* 1963c), appears to occur in the maternal parts of the placenta: the relevance of the immunoglobulin transfer barrier in the susceptibility of ungulate placenta therefore remains uncertain. Placental haematomas probably provide an environment protected from maternal defence mechanisms and their development in the placenta during the later stages of gestation may play a part in the simultaneous increase in susceptibility to infection with, for example, *V. fetus* (Osburn & Hoskins, 1970). Haematomas and anatomical equivalents to the late-gestation ungulate placenta can however be found in other mammalian placentas (Solomon, 1971) and these are perhaps less prone to infection. The mammalian foetus, which can be regarded as an allograft, is able in some as yet undefined way to resist maternal cellular defence mechanisms (Solomon, 1971). Possibly micro-organisms in the ungulate placenta, but not those in the placenta of other animals, can coincidentally benefit from foetus-protecting factors in the vicinity of the foetal-maternal junction; the possibility has not however been investigated. Clearly some direct investigations of the activity of defence mechanisms against micro-organisms within placental tissues are required.

Nutrient supply

An unusual abundance of nutrients might provide an alternative explanation of prolific growth of micro-organisms in ungulate placenta. But there is no reason to suppose that the nutrient level is above that in other species. There is at present no experimental evidence that any single feature, nutritional or host defensive, predisposes the ungulate placenta to infection with *all* of the micro-organisms cited. However, Smith *et al.* (1962) obtained evidence that a high concentration of a single nutrient, erythritol, in the placenta was important for the severe placentitis observed in ungulate brucellosis. Discussion of this and subsequent work on the relevance of erythritol to growth of other micro-organisms in the ungulate placenta forms the remainder of this section.

The role of erythritol in ungulate placentitis

The work of Smith *et al.* (1962) illustrates a valuable approach to the problem of tissue specificity in microbial disease and has been discussed by Keppie (1964). The essential findings were as follows: erythritol, a preferred metabolite and growth stimulant for brucellae *in vitro*, was present in significant amounts only in the ungulate placenta and not in other ungulate tissues or in the placenta of man, rat, rabbit or guinea-pig which are resistant to the localised placental form of brucellosis; administration of erythritol enhanced brucella infection in calves and erythritol analogues decreased the severity of infection in guinea-pigs.

It is reasonable to conclude that brucella strains which utilise erythritol will grow better in the ungulate placenta than those strains which do not, and thus that erythritol utilisation is a virulence attribute for placental infection. This is borne out by the infrequency with which vaccine strains of *Br. abortus*, cause abortion in cattle; little or no growth stimulation of these strains is shown by erythritol (Keppie, Witt & Smith, 1967). However, that erythritol is not the sole factor in localisation of brucella species is indicated by the fact that *Brucella ovis* causes a placental infection in sheep which is similar to that of *Br. abortus* in cattle or sheep (Osburn & Kennedy, 1966) yet it is unable to metabolise erythritol (Meyer, 1969*a*, *b*).

The importance of erythritol in placental localisation may extend to the recently described *Brucella canis* which is responsible for outbreaks of abortion in dogs; the form of the disease and the pathology of the canine placenta are typical of ungulate brucellosis (Carmichael & Kenney, 1970). Although attempts to demonstrate an enhancing effect of erythritol on growth *in vitro* have been unsuccessful (Jones *et al.* 1968) the organism can metabolise erythritol (Meyer, 1969*a*; Renoux & Philippon, 1969) and erythritol is present in canine placenta (D. B. Lowrie and J. F. Kennedy, unpublished).

Erythritol is probably not responsible for placental localisation of *L. monocytogenes* or *V. fetus* since it failed to enhance growth of these organisms in nutrient media (Trivett & Meyer, 1967; Lowrie & Pearce, 1970). Also respiration studies failed to demonstrate metabolism of erythritol by *L. monocytogenes* (Trivett & Meyer, 1967) or *V. fetus* (D. B. Lowrie, unpublished).

Other nutritional factors may however play a part in placental localisation. Nutrient stimulation does appear to be involved in localisation of *V. fetus* in ovine placenta (Lowrie & Pearce, 1970). Growth rate of the organism in saline extracts of the placenta was greater than in extracts of most other ovine tissues and after dialysis of the extracts the diffusate from placental tissue when added to a basal medium enhanced growth rate more than did diffusates from other tissues. The nature and distribution of the diffusible factor(s) responsible for this effect are the subject of further studies.

Kidney infections

Bacterial growth localised in the kidneys is a common cause of renal failure in man and animals, and has been the subject of numerous experimental studies in laboratory animals. In man the most common cause of pyelonephritis is *E. coli* but other bacteria have been implicated including staphylococci, enterococci, pseudomonads and proteus species; also important are *Corynebacterium renale* in cattle and leptospira species in domestic animals, rodents and man.

The pattern of localisation in the kidney of experimental animals is different according to the infecting micro-organism but two stages can be discerned: (1) initial lodgement in glomeruli and capillaries after haematogenous spread or, with ascending infection, lodgement in the ureter, renal pelvis and tubules; (2) growth in medullary tissue. Little is known of the factors important for survival at initial lodgement; microbial preference for growth in the medulla appears to be related either to deficiency of host defences or provision of nutrients.

Initial lodgement

Before proliferation in the kidney can occur a micro-organism has to evade or withstand the host defences encountered in the initial stages of infection. This is reflected in the difficulty of establishing pyelonephritis in normal experimental animals infected intravenously. Large doses are required, presumably to saturate kidney defence mechanisms. Organisms which become lodged in the renal cortex by the filtering action of glomeruli are subsequently eliminated (Gorrill, 1968). This is seen in the kidney of rats and mice infected with enterococci (Guze, Goldner & Kalmanson, 1961), *Pseudomonas aeruginosa*, *Proteus mirabilis* and *E. coli* (Gorrill & De Navasquez, 1964; Paplanus, 1964). *Staphylococcus aureus* is unusual in producing an acute cortical infection; survival in the cortex appears to be related to a trypsin-labile factor possessed by the organism but its mode of action is unknown (Gorrill & McNeil, 1968).

Establishment of kidney infection by the ascending route seems generally to require predisposing factors such as renal massage in experimental animals (Sanford *et al.* 1965) or urinary tract abnormality in man. How these factors operate to increase bacterial survival is unknown although the importance of reflux in conveying infection from bladder to renal pelvis in man seems clearly established (Rosenheim, 1965).

Recent investigations suggest that the possession of K antigen by *E. coli* may be important for the renal establishment of the organism in natural infection in man (Glynn, Brumfitt & Howard, 1971). Different serotypes of *E. coli* appear to become established in infections of the bladder in proportion to their frequency in the faecal flora, but strains rich in K antigen are more likely to succeed in invading the kidney. The reasons are not yet established but organisms which possess K antigen are known to be more resistant to phagocytosis and the bactericidal action of antibody and complement (see also A. A. Glynn, this volume). K antigen may be important for resistance to defence mechanisms either in the ureter, pelvis and tubules during ascending infection or in the cortex if blood-borne infection is involved.

Medullary predilection

In pyelonephritis medullary tissue is usually the preferred site of microbial growth once organisms attain access. It is seen in experimental animals after infection by the intravenous (Guze, Goldner & Kalmanson, 1961), or ascending routes (Jackson, José & Kozij, 1965), or after direct injection into the kidney substance (Freedman & Beeson, 1958).

Defences. The results of several studies suggest that deficiencies in host defence mechanisms within the medulla are important. Rocha & Fekety (1965) showed that the rate of granulocyte mobilisation in the medulla was much slower than in the cortex after thermal injury or direct inoculation with micro-organisms or carbon particles.

Hypertonicity of renal tissue may be a factor in inhibition of local defence mechanisms (Delucchi & Kass, 1965) since, *in vitro*, equivalent solute concentrations inhibit the phagocytic activity of human leucocytes (Chernew & Braude, 1962). The high concentration of ammonia and ammonium ion in the medulla (Lyon & Tuttle, 1965) may also be important since ammonium ion can prevent the action of serum complement (Beeson & Rowley, 1959). While decreased effectiveness of humoral and cellular defences is probably a major factor in the renal localisation of bacteria as different as *C. renale* and *E. coli* the relative importance of one or other mechanism for each kidney pathogen is unknown.

Nutrients. Investigation of the role of nutritional factors in medullary localisation has centred upon the function of urea as a growth stimulant (see Keppie, 1964). Addition of urea enhanced the intracellular growth of urease-active *Pr. mirabilis in vitro* (Braude & Siemienski, 1960) and simultaneous release of anti-complementary ammonia might further assist growth *in vivo*. However, urease activity is not essential for localisation in the medulla to occur. In mice the renal pathology produced by a urease-less strain of *Strep. faecalis* was typical of that produced by urease-active streptococci (Guze, Goldner & Kalmanson, 1961) and urease-less mutants of *C. renale* and *Pr. mirabilis* retained the renal predilection of the parent organisms (Lister, 1957; MacLaren, 1968). However, virulence of the mutants was diminished since growth in the medulla was more limited, with correspondingly less damage, and recovery of animals from infection was rapid. These findings emphasise that the ability to use available nutrients is important for organism proliferation and development of renal infection, even if not essential for the initial lodgement in the medulla. It may be relevant here that the growth rates, *in vitro*, of strains of *Ps. aeruginosa* correlate with their virulence for the kidney (Klyhn & Gorrill, 1967).

The role of nutritional factors other than urea has been little investigated. Glucose enhancement of growth of *Leptospira pomona* in appropriate laboratory media and of *E. coli* in urine *in vitro* (Ellinghausen, 1968; Roberts, Clayton & Bean, 1968) has led to suggestions that the presence of glucose in the renal tubules may assist growth at that site.

Lipids are growth factors for leptospira species *in vitro* and lipid distribution *in vivo* might correlate with the preferential growth and disease caused by leptospira in the kidney. Tissue differences in lipid structure may be the important factor in view of the demonstration that lipids on some cell surfaces were more susceptible to breakdown by leptospiral enzymes than those on other different cells and that such cells were more readily destroyed (Kasǎrov, 1970; Miller, Froehling & White, 1970; Yam, Miller & White, 1970).

HOST (SPECIES) SPECIFICITY

Lack of detailed information restricts us to very general remarks on explanations for species specificity in bacterial infection. Smith (1968) has made the point that species-determined resistance far outweighs any resistance that a susceptible host acquires by immunisation. It may be supposed that any single factor, such as host nutrition or defences, could determine whether microbial growth does or does not occur in the 'un-natural' host. However species specificity of infection more commonly involves – as the search for adequate models of human infection teaches us – the problem of why an organism produces in a laboratory host a disease quite different from that in the natural host.

Species differences are seen in, for example, free amino-acid levels (Hernandez & Coulson, 1971) sensitivity to pharmacological mediators of hypersensitivity (Chase, 1965) and phagocyte bactericidins (Padgett & Hirsch, 1967; Paul et al. 1970) but the relevance of these to specificity of infection is unclear. Earlier work, reviewed by Keppie (1964), illustrates how variation in nutrition, and probably defence mechanisms, may account for differences in host susceptibility; more recent evidence supports this.

The absence of free asparagine in guinea-pig tissues is suggested by Burrows & Gillett (1971) to provide an explanation for the failure of eleven Brazilian strains of *Pasteurella pestis* to kill guinea-pigs. The typical fully-virulent strain, MP6, of *P. pestis* kills both mice and guinea-pigs and the deficiency in the atypical Brazilian strains was seen in all of some 200 isolates. The only difference detectable *in vitro* was the reduced growth-rate of the isolates on a minimal agar and this was restored to normal by supplementation with L-asparagine. The fact that guinea-pigs had an active circulating asparaginase in their bloodstream suggested that the consequent absence of free asparagine would so reduce the growth rate of the atypical strains that developing immunity would prevent infection. Consistent with this, guinea-pigs previously infected with the Brazilian strains proved resistant to infection by the MP6 strain. On blood agar growth, asparagine supplementation of human and mouse blood and heated guinea-pig blood allowed growth rates of the Brazilian strains equal to that of strain MP6. Also injection of asparaginase into mice infected with Brazilian strains substantially delayed the time to death.

The natural resistance of pigeons to infection with type III pneumococci correlates with the very rapid clearance of organisms from the blood stream compared with that in normal mice, which are very susceptible (Bateman & Rowley, 1969). The clearance in pigeons appears to reflect opsonisation by capsular antibody, since clearance rates were reduced after injection of type III capsular polysaccharide. Bateman and Rowley suggest that, compared with mice, pigeons have an inherently greater capacity to produce anti-pneumococcal opsonins in response to antigen and that this determines their resistance to infection.

The work of Rees and his colleagues (see Rees, 1970) on experimental leprosy in the mouse, demonstrates the control which may be exerted by host defences over the clinical form of a disease. Foot-pad inoculation of the mouse with *Mycobacterium leprae* (responsible for human leprosy) leads to slowly developing infection, which after several years resembles the disease state in man which is termed 'border line' (between the tubercular and lepromatous forms); in lepromatous leprosy organisms are present in large numbers within nodular lesions spread throughout the tissues. Lepromatous leprosy could be induced in mice when thymectomy and whole body irradiation were carried out before infection, and reversed to the borderline situation by injection of normal lymphoid cells. The condition of impaired immunological function required to achieve the spreading infection correlates with the finding that 'natural' lepromatous leprosy induced in mice by *M. lepraemurium* and the lepromatous form of disease in man are both associated with depression of cell-mediated immunity.

We conclude this section on host specificity with an account of current work on baboon and guinea-pig susceptibility to chlamydial infections.

Chlamydial infections of the conjunctiva

The genus *Chlamydia* (Page, 1966) represents the agents forming the psittacosis–lymphogranuloma–trachoma group of organisms. They are obligate intracellular parasites which contain both DNA and RNA, have cell walls similar in composition to those of Gram-negative bacteria, reproduce by binary fission but have only limited metabolic activity and are lacking in energy-generating systems; they may be regarded as host-dependent bacteria (Moulder, 1966). Organisms causing trachoma and inclusion conjunctivitis (TRIC agents) show marked host specificity in their infectivity for epithelial cells of the conjunctiva and genital tract. The natural disease is limited to man (Jawetz, 1964) and experimental infection of the conjunctiva has been produced only in other primates (Thygeson *et al.* 1960). Another organism, the agent of guinea-pig inclusion conjunctivitis (gp-ic), causes conjunctivitis only in guinea-pigs (Murray, 1964). The mechanism of these specificities is unknown but work in progress (Moore, Griffiths & Pearce, 1970 and unpublished) suggests that some form of defence mechanism plays a decisive part.

The problem of specificity here is similar to that posed in virus infection in that factors influencing specificity could operate at any stage of the infection cycle – either before uptake, at adsorption or penetration, at some stage during multiplication or at release of infective particles. Analysis requires a culture system in which structural and functional properties of differentiated cells are maintained and this appears to be met by organ culture (Bang & Luttrell, 1961; Hoorn & Tyrrell, 1969).

Organ culture studies

Using the MRC-4f strain, growth of TRIC agent in organ cultures of baboon conjunctiva was demonstrated by egg titration of homogenised tissue fragments taken at intervals after inoculation. Further evidence of multiplication was given by the presence of occasional inclusion bodies in epithelial cells of infected samples examined histologically. The low numbers accorded with the general finding that TRIC strains produce few inclusions in the baboon conjunctiva *in vivo*. On the other hand gp-ic, which causes severe infection in the guinea-pig eye and large numbers of inclusions, also caused many inclusions in organ cultures of guinea-pig conjunctiva. It seemed therefore that organ culture reasonably reproduced the pattern of infection *in vivo* by both agents.

Heterologous infection

Maintenance of host specificity of infection in conjunctival organ culture was examined by infection of baboon tissue with gp-ic and guinea-pig tissue with TRIC agent. It became apparent that specificity was altered; egg titration showed that TRIC agent multiplied in guinea-pig cultures – although inclusions have not so far been detected with certainty. Gp-ic multiplied in baboon cultures with detectable inclusion bodies but these were less frequent than in the homologous system.

In control experiments to check host specificity of infection *in vivo* large inocula of TRIC agent (suspension containing 10^8 50 % egg infective doses (EID_{50}) per ml rubbed into the eye) produce slight inflammation in guinea-pigs; agent was detected 48 h after inoculation but not thereafter. Baboons receiving large inocula of gp-ic (suspension of $10^{9.5}$ EID_{50} per ml) developed slight infection: agent was detected in conjunctival scrapings at 7 and 14 days after inoculation, although inclusion bodies were not found, and typical infection was produced on transfer to guinea-pigs. Nevertheless 30–100fold lower doses of gp-ic failed to produce infection in baboons. This contrasts with the situation in the guinea pig where an inoculum as small as 10–10^2 EID_{50} is sufficient to produce severe conjunctivitis.

Analysis of dose-responses in conjunctival cultures showed that much the same doses of gp-ic were required to achieve detectable response at 48 h (when one full cycle of growth should have occurred) in either guinea-pig or baboon tissues (see Table 1). Preliminary data for TRIC agent also showed comparable dose-response patterns in guinea-pig or baboon cultures. This indicates that conjunctival fragments in culture are more susceptible to *heterologous* infection than the

Table 1. *Growth* of gp-ic in organ cultures of guinea-pig or baboon conjunctiva*

Dose per culture dish	Incubation time (h)				
	2	24	48	72	144
			Guinea-pig tissue		
6·8	—	3·2	—	4·5	5·1
5·8	1·5	—	3·3	—	—
4·8	<1·0	—	2·8	—	—
3·8	ND	—	<1·0	—	—
2·8	ND	—	ND	ND	—
5·8†	1·4	—	ND	—	—
			Baboon tissue		
6·5	3·3	—	2·6	—	2·3
5·5	1·3	—	1·4	—	—
4·5	ND	—	ND	—	—
3·5	ND	—	ND	—	ND
5·5†	1·4	—	ND	—	—

* Values are expressed as $\log_{10}EID_{50}$; after 2h adsorption, tissue fragments (4 per dish) were rinsed 3× and incubation continued in fresh medium; values at each time are per 4 tissue fragments.

† Fragments frozen at $-20°$ and thawed before inoculation. ND = agent not detectable.

intact conjunctiva *in vivo*. Clearly a restrictive mechanism which influences host susceptibility *in vivo* is diminished or absent in organ culture. We cannot exclude the possibility that an alteration in the metabolism of epithelial cells in culture leads to an increase in susceptibility. However, bearing in mind that baboon resistance to infection with gp-ic is overcome by sufficiently large inocula a more likely explanation for the change is perhaps that some kind of barrier or host defence (cellular or humoral) is reduced in organ culture. These possibilities are being investigated.

CONCLUDING REMARKS

The essence of specificity in infection is adaptation of the pathogen to the host, and the means by which this is achieved are evidently as varied as the invading organisms themselves.

If there is a lesson to be learned at this stage it is that the host–pathogen interaction is exceedingly complex, that no single mechanism is likely to decide localisation and that no one experimental approach is going to provide all the answers.

Of the evidence we have presented, much the most convincing has been that bearing on nutrition. Clearly we need more intelligent ways of dissecting out events *in vivo* but the technical problem looks massive; more might usefully be done here with mutants to test out hypotheses of interaction.

Two other points strike us in retrospect. First, that the guidelines for study of synergism between organisms in infection are remarkably well advanced. Second, that the notion of phenotypic shift of organisms in different tissue environments is likely to prove an important concept – although as yet we have no evidence for it – especially now that continuous culture studies have established the inherent plasticity of the bacterial cell.

If the uncommitted reader is persuaded that these problems are important – and assailable – then we may look forward to a productive period.

REFERENCES

BANG, F. B. & LUTTRELL, C. N. (1961). Factors in the pathogenesis of virus diseases. *Advances in Virus Research*, **8**, 199.

BATEMAN, R. G. & ROWLEY, D. (1969). The natural resistance of pigeons to type III pneumococcus. *Australian Journal of Experimental Biology and Medical Science*, **47**, 733.

BEESON, P. B. & ROWLEY, D. (1959). The anticomplementary effect of kidney tissue: its association with ammonia production. *Journal of Experimental Medicine*, **110**, 685.

BEVERLEY, J. K. A., WATSON, W. A. & PAYNE, J. M. (1971). The pathology of the placenta in ovine abortion due to toxoplasmosis. *Veterinary Record*, **88**, 124.

BOWMAN, C. G. & BONVENTRE, P. (1970). Studies on the mode of action of diphtheria toxin. *Journal of Experimental Medicine*, **131**, 659.

BRAUDE, A. I. & SIEMIENSKI, J. (1960). Role of bacterial urease in experimental pyelonephritis. *Journal of Bacteriology*, **80**, 171.

BURROWS, T. W. & GILLETT, W. A. (1971). Host specificity of Brazilian strains of *Pasteurella pestis*. *Nature, London*, **229**, 51.

CARMICHAEL, L. E. & KENNEY, R. M. (1970). Canine brucellosis: the clinical disease, pathogenesis and immune response. *Journal of the American Veterinary Medical Association*, **156**, 1726.

212   J. H. PEARCE AND D. B. LOWRIE

CATTERALL, R. D. (1971). Influence of gestogenic contraceptive pills on vaginal candidiasis. *British Journal of Venereal Diseases*, **47**, 45.

CHASE, M. W. (1965). In *Bacterial and Mycotic Infections of Man*, ed. R. J. Dubos and J. G. Hirsch, 250. Philadelphia: J. B. Lippincott Co.

CHERNEW, I. & BRAUDE, A. I. (1962). Depression of phagocytosis by solutes in concentrations found in the kidney and urine. *Journal of Clinical Investigation*, **41**, 1945.

CHILGREN, R. A., MEUWISSEN, H. J., QUIE, P. G. & HONG, R. (1967). Chronic mucocutaneous candidiasis, deficiency of delayed hypersensitivity and selective local antibody defect. *The Lancet*, **ii**, 688.

CYSEWSKI, S. J. & PIER, A. C. (1968). Mycotic abortion in ewes produced by *Aspergillus fumigatus*. *American Journal of Veterinary Research*, **29**, 1135.

DEGRÉ, M. (1969). Phagocytic and bactericidal activities of peritoneal and alveolar macrophages from mice. *Journal of Medical Microbiology*, **2**, 353.

DELUCCHI, C. & KASS, E. H. (1965). Effect of osmolality on regional bacterial multiplication in the kidney. In *Progress in Pyelonephitis*, ed. E. H. Kass, 280. Philadelphia: F. A. Davis Co.

EGERTON, J. R., ROBERTS, D. S. & PARSONSON, I. M. (1969). The aetiology and pathogenesis of ovine foot-rot. I. A histological study of the bacterial invasion. *Journal of Comparative Pathology*, **79**, 207.

ELLINGHAUSEN, H. C. (1968). Stimulation of leptospiral growth by glucose. *American Journal of Veterinary Research*, **29**, 191.

ELLWOOD, D. C. & TEMPEST, D. C. (1969). Control of teichoic acid and teichuronic acid biosynthesis in chemostat cultures of *Bacillus subtilis* var. *niger*. *Biochemical Journal*, **111**, 1.

FERGUSON, D. J. (1970). Paréian and Listerian slants on infection in wounds. *Perspectives in Biology*, **14**, 63.

FITZGEORGE, R. B. & SMITH, H. (1966). The chemical basis of the virulence of *Brucella abortus*. VIII. The production *in vitro* of organisms with an enhanced capacity to survive intracellularly in bovine phagocytes. *British Journal of Experimental Pathology*, **47**, 558.

FREEDMAN, L. R. & BEESON, P. B. (1958). Experimental pyelonephritis. IV. Observations on infections resulting from direct inoculation of bacteria in different zones of the kidney. *Yale Journal of Biology and Medicine*, **30**, 406.

GIBBONS, R. J. & VAN HOUTE, J. (1971). Selective bacterial adherence to oral epithelial surfaces and its role as an ecological determinant. *Infection and Immunity*, **3**, 567.

GLYNN, A. A., BRUMFITT, W. & HOWARD, C. J. (1971). K-antigens of *Escherichia coli* and renal involvement in urinary-tract infections. *The Lancet*, **i**, 514.

GORRILL, R. H. (1968). Susceptibility of the kidney to experimental infection. In *Urinary Tract Infections*, ed. F. O'Grady and W. Brumfitt, 24. London, New York and Toronto: Oxford University Press.

GORRILL, R. H. & DE NAVASQUEZ, S. J. (1964). Experimental pyelonephritis in the mouse produced by *Escherichia coli*, *Pseudomonas aeruginosa* and *Proteus mirabilis*. *Journal of Pathology and Bacteriology*, **87**, 79.

GORRILL, R. H. & MCNEIL, E. M. (1968). The problem of staphylococcal lodgement in the mouse kidney. *Journal of Pathology and Bacteriology*, **96**, 431.

GUZE, L. B., GOLDNER, B. H. & KALMANSON, G. M. (1961). Pyelonephritis. I. Observations on the course of chronic non-obstructed enterococcal infection in the kidney of the rat. *Yale Journal of Biology and Medicine*, **33**, 372.

HANKS, J. H. (1966). Host-dependent microbes. *Bacteriological Reviews*, **30**, 114.

HARDY, P. H. & MUNRO, C. O. (1966). Nutritional requirements of anaerobic spirochaetes. I. Demonstration of isobutyrate and bicarbonate as growth factors for a strain of *Treponema microdentium*. *Journal of Bacteriology*, **91**, 27.

HARWICK, H. J., PURCELL, R. H., IUPPA, J. B. & FEKETY, F. R. (1970). *Mycoplasma hominis* and abortion. *Journal of Infectious Diseases*, **121**, 260.

HENTGES, D. J. (1970). Enteric pathogen-normal flora interactions. *American Journal of Clinical Nutrition*, **23**, 1451.

HERNANDEZ, T. & COULSON, R. A. (1971). Free amino acids in the carcasses of various animals. *Comparative Biochemistry and Physiology*, **38 B**, 679.

HILLMAN, R. B. & McENTEE, K. (1969). Experimental studies on bovine mycotic placentitis. *Cornell Veterinarian*, **59**, 289.

HOFFMAN, H. & VALDINA, J. (1968). Mechanisms of bacterial attachment to oral epithelial cells. *Acta Cytologica*, **12**, 37.

HOORN, B. & TYRRELL, D. A. J. (1969). Organ cultures in virology. *Progress in Medical Virology*, **11**, 408.

JACKSON, G. G., JOSÉ, A. A. & KOZIJ, V. M. (1965). Retrograde pyelonephritis in the rat and the role of certain cellular and humoral factors in the host defence. In *Progress in Pyelonephritis*, ed. E. H. Kass, 202. Philadelphia: F. A. Davis Co.

JAWETZ, E. (1964). Agents of trachoma and inclusion conjunctivitis. *Annual Reviews of Microbiology*, **18**, 301.

JENSEN, R., MILLER, V. A. & MOLELLO, J. A. (1961). Placental pathology of sheep with vibriosis. *American Journal of Veterinary Research*, **22**, 169.

JOHANSON, W. G., BLACKSTOCK, R., PIERCE, A. K. & SANFORD, J. P. (1970). The role of bacterial antagonism in pneumococcal colonization of the human pharynx. *Journal of Laboratory and Clinical Medicine*, **75**, 946.

JONES, L. M., ZANARDI, M., LEONG, D. & WILSON, J. B. (1968). Taxonomic position in the genus *Brucella* of the causative agent of canine abortion. *Journal of Bacteriology*, **95**, 625.

KASĂROV, L. B. (1970). Degradation of the erythrocyte phospholipids and haemolysis of the erythrocytes of different animal species by leptospirae. *Journal of Medical Microbiology*, **3**, 29.

KEPPIE, J. (1964). Host and tissue specificity. *Symposium of the Society for General Microbiology*, **14**, 44.

KEPPIE, J., WITT, K. & SMITH, H. (1967). Effect of erythritol on growth of S19 and other attenuated strains of *Brucella abortus*. *Research in Veterinary Science*, **8**, 294.

KLUGE, J. P., MERKAL, R. S., MONLUX, W. S., LARSEN, A. B., KOPECKY, K. E., RAMSEY, F. K. & LEHMANN, R. P. (1968). Experimental paratuberculosis in sheep after oral, intratracheal or intravenous inoculation: lesions and demonstrations of etiologic agent. *American Journal of Veterinary Research*, **29**, 953.

KLYHN, K. M. & GORRILL, R. H. (1967). Studies on the virulence of hospital strains of *Pseudomonas aeruginosa*. *Journal of General Microbiology*, **47**, 227.

KRASSE, B., EDWARDSSON, S., SVENSSON, I. & TRELL, L. (1967). Implantation of caries inducing streptococci in the human oral cavity. *Archives of Oral Biology*, **12**, 231.

KUNDSIN, R. B. & DRISCOLL, S. G. (1970). Mycoplasmas and human reproductive failure. *Surgery, Gynecology and Obstetrics*, **131**, 89.

KWAPIEN, R. P., LINCOLN, S. D., REED, D. E., WHITEMAN, C. E. & CHOW, T. L. (1970). Pathologic changes of placentas from heifers with experimentally induced epizootic bovine abortion. *American Journal of Veterinary Research*, **31**, 999.

LISTER, A. J. (1957). *The metabolism of Corynebacterium renale*. Ph.D. thesis: University of Cambridge.

LOWRIE, D. B. & PEARCE, J. H. (1970). The placental localisation of *Vibrio fetus*. *Journal of Medical Microbiology*, 3, 607.

LYON, M. L. & TUTTLE, E. P. (1965). Differential effects of acidifying agents in experimental pyelonephritis. In *Progress in Pyelonephritis*, ed. E. H. Kass, 242, Philadelphia: F. A. Davis Co.

MACDONALD, J. B., GIBBONS, R. J. & SOCRANSKY, S. S. (1960). Bacterial mechanisms in periodontal disease. *Annals of the New York Academy of Sciences*, 85, 467.

MACLAREN, D. M. (1968). The significance of urease in proteus pyelonephritis: a bacteriological study. *Journal of Pathology and Bacteriology*, 96, 45.

MASON, R. W. & MUNDAY, B. L. (1968). Abortion in sheep and cattle associated with *Bacillus* spp. *Australian Veterinary Journal*, 44, 297.

MAW, J. & MEYNELL, G. G. (1968). The true division and death rates of *Salmonella typhimurium* in the mouse spleen determined with superinfecting phage P22. *British Journal of Experimental Pathology*, 40, 597.

MEYER, M. E. (1969a). Brucella organisms isolated from dogs: comparison of characteristics of members of the genus *Brucella*. *American Journal of Veterinary Research*, 30, 1751.

MEYER, M. E. (1969b). Phenotypic comparison of *Brucella ovis* to the DNA-homologous brucella species. *American Journal of Veterinary Research*, 30, 1757.

MEYNELL, G. G. (1961). Phenotypic variation and bacterial infection. *Symposium of the Society for General Microbiology*, 11, 174.

MEYNELL, G. G. (1963). Antibacterial mechanisms of the mouse gut. II. The role of E_h and volatile fatty acids in the normal gut. *British Journal of Experimental Pathology*, 44, 209.

MILLER, N. G., FROEHLING, R. C. & WHITE, R. J. (1970). Activity of leptospires and their products on L-cell monolayers. *American Journal of Veterinary Research*, 31, 371.

MOLELLO, J. A., FLINT, J. C., COLLIER, J. R. & JENSEN, R. (1963a). Placental pathology. II. Placental lesions of sheep experimentally infected with *Brucella melitensis*. *American Journal of Veterinary Research*, 24, 905.

MOLELLO, J. A. & JENSEN, R. (1964). Placental pathology. IV. Placental lesions of sheep experimentally infected with *Listeria monocytogenes*. *American Journal of Veterinary Research*, 25, 441.

MOLELLO, J. A., JENSEN, R., COLLIER, J. R. & FLINT, J. C. (1963b). Placental pathology. III. Placental lesions of sheep experimentally infected with *Brucella abortus*. *American Journal of Veterinary Research*, 24, 915.

MOLELLO, J. A., JENSEN, R., FLINT, J. C. & COLLIER, J. R. (1963c). Placental pathology. I. Placental lesions of sheep experimentally infected with *Brucella ovis*. *American Journal of Veterinary Research*, 24, 897.

MOORE, J. E., GRIFFITHS, M. S. & PEARCE, J. H. (1970). Host specificity in conjunctival infections caused by *Chlamydia*. *Journal of Medical Microbiology*, 3, x.

MOORE, W. E. C., CATO, E. P. & HOLDEMAN, L. V. (1969). Anaerobic bacteria of the gastro-intestinal flora and their occurrence in clinical infections. *Journal of Infectious Diseases*, 119, 641.

MOORE, W. E. C. & GROSS, W. B. (1968). Liver granulomas of turkeys – causative agents and mechanism of infection. *Avian Diseases*, 12, 417.

MOULDER, J. W. (1966). The relation of the psittacosis group (chlamydiae) to bacteria and viruses. *Annual Reviews of Microbiology*, 20, 107.

MURRAY, E. S. (1964). Guinea-pig inclusion conjunctivitis virus. I. Isolation and identification as a member of the psittacosis – lymphogranuloma – trachoma group. *Journal of Infectious Diseases*, 114, 1.

NOVILLA, M. N. & JENSEN, R. (1970). Placental pathology of experimentally induced enzootic abortion in ewes. *American Journal of Veterinary Research*, **31**, 1983.

OSBURN, B. I. & HOSKINS, R. K. (1970). Experimentally induced *Vibrio fetus* var. *intestinalis* infection in pregnant cows. *American Journal of Veterinary Research*, **31**, 1733.

OSBURN, B. I. & KENNEDY, P. C. (1966). Pathologic and immunologic responses of the fetal lamb to *Brucella ovis*. *Pathologia Veterinaria*, **3**, 110.

PADGETT, G. A. & HIRSCH, J. G. (1967). Lysozyme: Its absence in tears and leukocytes of cattle. *Australian Journal of Experimental Biology and Medical Science*, **45**, 569.

PAGE, L. A. (1966). Revision of the family *Chlamydiaceae* Rake (Rickettsiales); unification of the psittacosis – lymphogranuloma venereum – trachoma group of organisms in the genus *Chlamydia* Jones, Rake and Stearns, 1945. *International Journal of Systematic Bacteriology*, **16**, 223.

PAPLANUS, S. H. (1964). Bacterial localisation in the kidney. *Yale Journal of Biology and Medicine*, **37**, 145.

PARSONSON, I. M., EGERTON, J. R. & ROBERTS, D. S. (1967). Ovine interdigital dermatitis. *Journal of Comparative Pathology*, **77**, 309.

PAUL, B. B., STRAUSS, R. R., JACOBS, A. A. & SBARRA, A. J. (1970). Function of H_2O_2, myeloperoxidase, and hexose monophosphate shunt enzymes in phagosytizing cells from different species. *Infection and Immunity*, **1**, 338.

REES, R. J. W. (1970). Immunological aspects of experimental leprosy in the mouse. *Proceedings of the Royal Society of Medicine*, **63**, 1060.

RENOUX, G. & PHILIPPON, A. (1969). Position taxonomique dans le genre *Brucella* de bactéries isolées de brebis et de vaches. *Annales de l'Institut Pasteur, Paris*, **117**, 524.

ROBERTS, A. P., CLAYTON, S. G. & BEAN, H. S. (1968). Urine factors affecting bacterial growth. In *Urinary Tract Infections*, ed. F. O'Grady and W. Brumfitt. London, New York and Toronto: Oxford University Press.

ROBERTS, D. S. (1967a). The pathogenic synergy of *Fusiformis necrophorus* and *Corynebacterium pyogenes*. I. Influence of the leucocidal exotoxin of *F. necrophorus*. *British Journal of Experimental Pathology*, **48**, 665.

ROBERTS, D. S. (1967b). The pathogenic synergy of *Fusiformis necrophorus* and *Corynebacterium pyogenes*. II. The response of *F. necrophorus* to a filterable product of *C. pyogenes*. *British Journal of Experimental Pathology*, **48**, 674.

ROBERTS, D. S. (1969). Editorial: Synergic mechanisms in certain mixed infections. *Journal of Infectious Diseases*, **120**, 720.

ROBERTS, D. S. (1970). Toxic, allergenic and immunogenic factors of *Fusiformis necrophorus*. *Journal of Comparative Pathology*, **80**, 247.

ROBERTS, D. S. & EGERTON, J. R. (1969). The aetiology and pathogenesis of ovine foot-rot. II. The pathogenic association of *Fusiformis nodusus* and *F. necrophorus*. *Journal of Comparative Pathology*, **79**, 217.

ROBERTS, D. S., GRAHAM, N. P. H., EGERTON, J. R. & PARSONSON, I. M. (1968). Infective bulbar necrosis (Heel abscess) of sheep, a mixed infection with *Fusiformis necrophorus* and *Corynebacterium pyogenes*. *Journal of Comparative Pathology*, **78**, 1.

ROCHA, H. & FEKETY, F. R. (1965). Delayed granulocyte mobilisation in the renal medulla. In *Progress in Pyelonephritis*, ed. E. H. Kass, 211. Philadelphia: F. A. Davis Co.

ROSEBURY, T. (1965). Bacteria indigenous to man. In *Bacterial and Mycotic Infections of Man*, ed. R. J. Dubos and J. G. Hirsch, 4th edition, 326. Philadelphia: J. B. Lippincott Co.

ROSENHEIM, M. L. (1965). Vesico-ureteric reflux. In *Progress in Pyelonephritis*, ed. E. H. Kass, 599. Philadelphia: F. A. Davis Co.

SANFORD, J. P., HUNTER, B. W., AKINS, L. L. & BARNETT, J. A. (1965). Immunity and obstructive uropathy as determinants in the pathogenesis of experimental pyelonephritis with observations on the distribution of antibody in hydronephrotic kidneys. In *Progress in Pyelonephritis*, ed. E. H. Kass, 255. Philadelphia: F. A. Davis Co.

SAVAGE, D. C. (1970). Associations of indigenous micro-organisms with gastrointestinal mucosal epithelia. *American Journal of Clinical Nutrition*, **23**, 1495.

SCHOCH, E. P. (1971). Thymic conversion of *Candida albicans* from commensalism to pathogenism. *Archives of Dermatology*, **103**, 311.

SMITH, H. (1968). Biochemical challenge of microbial pathogenicity. *Bacteriological Reviews*, **32**, 164.

SMITH, H., WILLIAMS, A. E., PEARCE, J. H., KEPPIE, J., HARRIS-SMITH, P. W., FITZGEORGE, R. B. & WITT, K. (1962). Foetal erythritol: A cause of the localisation of *Brucella abortus* in bovine contagious abortion. *Nature, London*, **193**, 47.

SMITH, R. E. & REYNOLDS, I. M. (1970). *Serratia marcescens* associated with bovine abortion. *Journal of the American Veterinary Medical Association*, **157**, 1200.

SMITH, R. E., REYNOLDS, I. M. & CLARK, G. W. (1968). Experimental ovine listeriosis. I. Inoculation of pregnant ewes. *Cornell Veterinarian*, **58**, 169.

SOCRANSKY, S. S. & GIBBONS, R. J. (1965). Required role of *Bacteroides melaninogenicus* in mixed anaerobic infections. *Journal of Infectious Diseases*, **115**, 247.

SOCRANSKY, S. S., LOESCHE, W. J., HUBERSAK, C. & MACDONALD, J. B. (1964). Dependency of *Treponema microdentium* on other oral micro-organisms for isobutyrate, polyamines and a controlled oxidation–reduction potential. *Journal of Bacteriology*, **88**, 200.

SOLOMON, J. B. (1971). *Foetal and Neonatal Immunology*. Amsterdam and London: North-Holland.

SPELLACY, W. N., CARLSON, K. L., BIRK, S. A. & SCHADE, S. L. (1968). Glucose and insulin alterations after one year of combination-type oral contraceptive treatment. *Metabolism*, **17**, 496.

THYGESON, P., DAWSON, C., HANNA, L., JAWETZ, E. & OKUMOTO, M. (1960). Observations on experimental trachoma in monkeys produced by strains of virus propagated in yolk sac. *American Journal of Ophthalmology*, **50**, 907.

TILLOTSON, J. R. & FINLAND, M. (1969). Bacterial colonization and clinical superinfection of the respiratory tract complicating antibiotic treatment of pneumonia. *Journal of Infectious Diseases*, **119**, 597.

TRIVETT, T. L. & MEYER, E. A. (1967). The effect of erythritol on the *in vitro* growth and respiration of *Listeria monocytogenes*. *Journal of Bacteriology*, **93**, 1197.

VALDIMARSSON, H., HOLT, L., RICHES, H. R. C. & HOBBS, J. R. (1970). Lymphocyte abnormality in chronic mucocutaneous candidiasis. *The Lancet*, **i**, 1259.

VEDROS, N. A., ROBINSON, P. J. & GUTEKUNST, R. A. (1966). Isolation and characterisation of a bacterial inhibitor from human throat washings. *Proceedings of the Society for Experimental Biology and Medicine*, **122**, 249.

WILSON, G. S. & MILES, A. A. (1964). *Topley and Wilson's Principles of Bacteriology and Immunity*, 5th edition. London: Edward Arnold.

YAM, P. A., MILLER, N. G. & WHITE, R. J. (1970). A leptospiral factor producing a cytopathic effect on L-cells. *Journal of Infectious Diseases*, **122**, 310.

PATHOGENICITY OF MYCOPLASMAS IN ANIMALS

P. WHITTLESTONE

Department of Animal Pathology, University of Cambridge

INTRODUCTION

Mycoplasmas are the smallest organisms known to be capable of autonomous growth and multiplication. The diameter of the smallest viable particles is probably about 0·25 μm. They are procaryotic organisms bounded by a unit membrane but have no cell wall. Their general structure and behaviour suggests that they may have been derived from bacteria which have lost their cell walls, like the stable L-phase variants of bacteria; this view of their origin, however, is unsupported by genetic evidence.

Because their minimal cell size approaches that of the theoretical minimal cell and because their biochemical activity is limited, mycoplasmas are of great interest to cellular and molecular biologists for the study of basic biological processes.

Mycoplasmas were first discovered to be the cause of disease in animals before the beginning of this century, but only within the last ten years has it been appreciated that man may also suffer from mycoplasma-induced disease. This realisation, together with the appreciation that tissue cultures are frequently contaminated with mycoplasmas, has brought about a great increase in mycoplasma research and advance of knowledge has been rapid.

Mycoplasma-like bodies have now been firmly associated with several plant diseases and their insect vectors and at least one is a proven causal agent.

Mycoplasmas often require complex growth substances elaborated by other living organisms and at present there is no such thing as an ideal culture medium for the range of mycoplasmas already known; no doubt there are many species, including pathogens, that will be discovered as appropriate media become available. The development of improved cultural methods is central to the study of the pathogenesis of the full range of these fastidious organisms.

DETECTION AND MEASUREMENT OF THE VIRULENCE OF DIFFERENT MYCOPLASMA ISOLATES

Virulence differences as detected in laboratory animals

Nearly all pathogenic mycoplasmas are either species specific or have a limited host range. If the mycoplasmas which are natural pathogens of the small rodents are excluded, only a few mycoplasmas have been shown to induce recognisable diseases or lesions in laboratory animals. Thus, for the most part virulence differences between various strains of mycoplasmas have not been studied using methods *in vivo* comparable to those used for bacteria.

Many mycoplasmas causing disease in man and domestic animals are pathogens of the respiratory tract, and thus the respiratory route is chosen for the production of lesions in experimental animals. However, laboratory rodents, notably mice and rats, are commonly infected, particularly in the respiratory tract, by their indigenous mycoplasmas. In addition, latent virus infections of the respiratory tract of laboratory mammals limits the usefulness of conventional animals. Ideally, therefore, mycoplasma-free, pathogen-free or gnotobiotic laboratory animals should be used for virulence studies; the difficulty in obtaining suitable high standard animals in most laboratories has limited studies *in vivo* of mycoplasmas causing respiratory disease. A few species, however, have been successfully studied in laboratory animals.

Mycoplasma pneumoniae (Eaton's agent), the causal agent of primary atypical pneumonia in man, induces bronchopneumonia in hamsters and cotton rats following intranasal inoculation (Eaton, Meiklejohn & van Herick, 1944). The ID_{50} of the pathogenic strains in the hamster is only ten colony forming units (c.f.u.) and most infected hamsters, in which multiplication of mycoplasmas occur in the lungs, develop microscopic evidence of bronchopneumonia (Dajani, Clyde & Denny, 1965). The strains of *M. pneumoniae* may thus be classified as virulent or attenuated in terms of their ability to produce pneumonia in the Syrian hamster (Dajani *et al.* 1965; Fernald, 1969; Lipman & Clyde, 1969; Collier, Clyde & Denny, 1969). A strain of *M. pneumoniae* is graded as 'virulent' if 10^5 c.f.u. produce pneumonia in 80–100 % of hamsters killed 14 days after intranasal inoculation; the strains causing no lung lesions are considered to be 'avirulent' or 'attenuated'; strains which induce pneumonia rates between these two grades are considered to be partially attenuated. The laboratory strain, 'Mac' is thus classified as attenuated, whereas strains recently isolated from the respiratory tract of man are consistently virulent.

There has been only limited success in studying the pathogenicity in laboratory animals of *Mycoplasma mycoides* var. *mycoides* (referred to subsequently as *M. mycoides*), the cause of the important disease, contagious bovine pleuropneumonia. Tang, Wei, McWhirter & Edgar (1935) could not detect any reaction to the agent in white mice, hamsters, albino rats, guinea-pigs, rabbits and cats after inoculation by a variety of routes. The intravenous agar-embolus method devised by Tuttle (1933) for the production of experimental pulmonary abscesses in the dog has been used successfully with *M. mycoides* in cattle (Daubney, 1935; Turner, Campbell & Dick, 1935; Mettam & Ford, 1939) and in experimental animals.

The greatest success has been achieved by using subcutaneous inoculation into mice of agar plugs containing virulent cultures (Hyslop, 1955, 1958; Gerlach & Heikkila, 1956; Lindley, 1959; Ross, 1962a, b). The initial subcutaneous lesions disappeared by the 3rd day but were followed a few days later by a secondary swelling which finally regressed between the 18th and 35th days (Hyslop, 1958). Histologically, such lesions have a central caseo-necrotic area containing *M. mycoides*, surrounded by a zone of monocytes and polymorphonuclear neutrophils with an outer zone of fibroblasts.

Using the same method, Hyslop (1958) had similar success in rabbits, guinea-pigs, hamsters and rats, the latter two species being the more susceptible, and Lindley (1959) compared in mice the virulence of three strains of *M. mycoides* of different virulences for cattle. Although parallel inoculations in cattle were not made, Lindley provided evidence that in cattle strain B1 was fully virulent, strain A55 had intermediate virulence and strain KH_3J used as a vaccine in Zebu cattle was avirulent. In mice there was no apparent difference between the number of organisms of the B1 and A55 strains required to set up a local lesion. The ID_{50} for the KH_3J strain was significantly greater but this attenuated strain could be recovered in greater numbers from the mice, in the absence of a palpable lesion, for up to 10 days. The difference in the response of mice to the avirulent and virulent strains indicates that further work along these lines would be justified. It is, therefore, surprising that the method has not been taken up by other workers, for it is cheap and easy to perform, the progression of the reaction can be assessed simply in a living animal and the results are usually uncomplicated by other infections (Hyslop, 1958). A different and fruitful approach has been made in mice by Smith (1967c, 1968, 1969, 1971): cultures of 7 fresh isolates inoculated intraperitoneally induced a prolonged mycoplasmaemia, although the mice showed no observable illness. Repeated subculture attenuated strains for mice. Smith (1968) suggested that there may be a parallel between virulence for mice and cattle. If this is so, this technique could standardise the virulence of the strains; until now there has been difficulty in selecting strains with the correct degree of attenuation for use as field vaccines.

Infections of *M. mycoides* have been induced in rabbits (Sheriff, 1952; Hyslop, 1958) by the intravenous injection of a mixture of the organism and agar. Pulmonary lesions, usually containing the mycoplasma, resulted. Several strains were passaged serially by this method, but the virulence of different isolates was not compared, nor have other workers used the technique. In limited experiments Smith (1965) produced fibrinous pleurisy, pericarditis and lung consolidation in some of the rabbits he inoculated with a mixture of mucin and *M. mycoides* into the thoracic cavity. Subcutaneous inoculation of rabbits with material from suspected cases of contagious bovine pleuropneumonia was used by Gambles (1956) to diagnose the disease, judgement being made on the complement-fixing-antibody response about one week later. Most rabbits developed subcutaneous abscesses and died.

The subcutaneous route was found by Lloyd (1970) to give consistent results in rabbits for comparing the pathogenicity of strains of *M. mycoides*. The effect in rabbits of the three strains, Gladysdale, V5 and KH$_3$J was parallel to their effect in cattle (Brown, 1964; Lindley & Abdulla, 1967; Lloyd, 1970). The Gladysdale strain produced the largest and KH$_3$J the smallest lesions. The representative strain PG1 (formerly NCTC No. 3278) however, which produced small or no lesions in cattle (Lloyd, 1970), produced large lesions in rabbits, comparable with those seen with the Gladysdale strain

Thus a number of techniques *in vivo* are known which might well be suitable for the study of the virulence of different strains of *M. mycoides*. More direct comparisons must be made of the virulence of different strains for cattle and experimental animals before definite conclusions can be drawn. Work along these lines is in progress at the East African Veterinary Research Organisation, Muguga, Kenya (Windsor, 1971).

The caprine pleuropneumonia agent, *Mycoplasma mycoides* var. *capri*, has also been induced to infect rabbits and mice by Smith (1965, 1967a, 1969). When the organism is mixed with gastric mucin (Olitski, 1948) or BVF-OS medium (Turner, Campbell & Dick, 1935) and inoculated intrathoracically into the rabbit, severe fibrinous pleurisy follows (Smith, 1965). In mice infection occurs provided gastric mucin is included with the intraperitoneal inoculum (Smith, 1967a); the effects are dose-dependent and include prolonged mycoplasmaemia, weight loss, obvious illness and occasional deaths.

One laboratory strain of *Mycoplasma agalactiae*, the cause of contagious agalactia of sheep and goats, produced a mycoplasmaemia in mice following intraperitoneal inoculation (Smith, 1967*b*). The organism was almost non-pathogenic unless suspended in mucin. It is not yet known whether these techniques might reveal virulence differences between different strains of *M. mycoides* var. *capri* and *M. agalactiae*.

The virulence of *Mycoplasma neurolyticum* can be detected by inoculation of large doses intravenously into the natural hosts, namely rats and mice. The different pathogenicities exhibited by different strains and the toxin produced by this organism are discussed later in this chapter.

Virulence differences between strains as detected in tissue culture systems

The action of *M. pneumoniae*, and other mycoplasmas of human origin, on ciliary activity has been studied in both hamster tracheal organ cultures (Collier & Clyde, 1969; Collier *et al.* 1969, 1971) and in cultures of human respiratory epithelium (Collier & Clyde, 1971). *M. pneumoniae* produced disturbance of ciliary motion and a distinct cytopathology not seen with the other mycoplasmas tested – namely *Mycoplasma hominis*, *Mycoplasma salivarium*, *Mycoplasma pharyngis* (*Mycoplasma orale*) and *Mycoplasma fermentans*. Collier *et al.* (1969) compared two strains of *M. pneumoniae* – the M129 and the Mac strain which had been classified respectively as virulent and attenuated on their ability to produce pneumonia in the Syrian hamster (Fernald, 1969). The attenuated (Mac) strain did not cause interference with ciliary motility until day 6 when growth had already produced marked acidity, whereas the virulent (M129) strain, although of comparable titre, stopped ciliary activity by day 4. Butler (1969*a*, *b*) used human embryo trachea cultures to study several mycoplasma species: the growth of *M. salivarium*, *M. orale* type 1 and *M. hominis* did not alter ciliary activity or induce structural changes but *M. pneumoniae* damaged the cultures. *M. mycoides* var. *capri* also caused cytopathic effects in the cultures, the first effect being cessation of ciliary activity, followed by rounding up and stripping of the ciliated epithelium. Butler suggested that the tracheal culture system could be used not only for the study of respiratory mycoplasma pathogenesis but also for mixed mycoplasma and viral infections and their control by chemotherapy. Cherry & Taylor-Robinson (1970*a*) used the simpler method of Harnett & Hooper (1968) for producing larger quantities of chick embryo tracheal organ cultures in tissue culture tubes. Of the avian strains studied in this system, seven strains of *Mycoplasma gallisepticum*, at least three strains of *Mycoplasma gallinarum* and the WR1 strain grew but only *M. gallisepticum* stopped ciliary movement. This result therefore parallels the pathogenicity of these strains in the natural host. *M. mycoides* var. *capri* and the related porcine isolate, B3, also stopped ciliary movement. One of five strains of *M. pneumoniae* grew and a *M. hominis* strain probably grew but neither caused inhibition of ciliary movement.

The effect of *M. hominis* on organ cultures of human fallopian tubes has been studied by Taylor-Robinson (1971). In some instances a chronic infection was established for a month but ciliary activity remained as active as in the uninfected control cultures. It is not known whether this indicates a lack of pathogenicity of *M. hominis* for the human fallopian tube.

Further experiments along these lines would probably throw light on the virulence differences between strains isolated from the respiratory and genital tracts of man and animals. For example, a study of the various mycoplasma strains isolated from the genital tract of cattle, with particular reference to the role of *Mycoplasma agalactiae* var. *bovis* (syn. *Mycoplasma bovimastitidis*, bovine group 5) in the production of salpingitis, might be profitable.

VIRULENT AND AVIRULENT STRAINS OF MYCOPLASMAS: THEIR STABILITY AND ATTRIBUTES CONNECTED WITH VIRULENCE

Mycoplasma pneumoniae

Couch, Cate & Chanock (1964), noted that there was an apparent decrease in virulence of *M. pneumoniae* when it was passed on cell-free agar medium. The FH strain was probably less virulent at the 55th passage than at the 9th as judged by the febrile response in volunteers. Strain 898 which had been passed only twice in tissue culture was inoculated at a similar dose level by the same route; it was considerably more pathogenic, producing in addition to the febrile response, respiratory signs, pneumonia and bullous myringitis in a proportion of cases.

The attenuation of *M. pneumoniae* was studied further by Smith, Chanock, Friedewald & Alford (1967). Two strains (FH and PI-1428) were passed in broth or on agar medium and then 10^5–10^7 organisms inoculated nasopharyngeally into volunteers. At low passage levels, both strains produced 100 % infection, as judged by antibody rise, but at the 27th passage the PI-1428 strain failed to infect a third of the men, and at the 55th passage the FH strain failed to infect two thirds of the volunteers.

The illness occurring in the men receiving the high passage level of the PI-1428 strain was less severe than in those receiving low passage culture. With the FH strain the three men infected with the 55th passage developed mild illness. Thus prolonged passage selected variants with both decreased infectivity and virulence.

The production of pneumonia in the hamster by *M. pneumoniae* was studied by Dajani *et al.* (1965), using two strains: the Mac strain which had been passaged 78 times in chick embryos, and a recent isolate, Bru. With both strains further passaging in broth reduced the incidence of pneumonia. Lipman & Clyde (1969) also working with *M. pneumoniae* in the hamster not only converted a virulent strain (M129) into an avirulent strain by 169 serial passages in broth but also showed that the reverse phenomenon occurred. An attenuated strain (Mac), previously passed 78 times in chick embryos and 144 times on or in lifeless media, was restored to full virulence by 8 successive intranasal passages in hamsters.

The virulent and avirulent *M. pneumoniae* strains of Dajani *et al.* (1965) did not differ in their basic biological properties. However, antigenic deficits developed in broth-cultivated mycoplasma strains, compared with the same strains propagated in a tissue culture system. (Dajani *et al.* 1965).

In attempts to recognise pathogenicity markers, Lipman & Clyde (1969) compared peroxide formation and cytadsorption in two pairs of virulent and avirulent or attenuated strains of *M. pneumoniae*. They found no greater haemolytic activity with the virulent strains. The avirulent strain of M129 showed no haemadsorption but the attenuated Mac strain and both virulent strains showed the same marked degree of haemadsorption. Later Lipman, Clyde & Denny (1969) found that the pairs of virulent and avirulent strains varied in their growth, glycolysis, protein electrophoretic patterns, peroxide formation, morphology and cytadsorption. Variations of the last two characteristics were found to be correlated with total loss of virulence. When broth cultures were examined by light microscopy the virulent and attenuated strains both formed smooth, refractile, spherical clusters up to 400 μm in diameter, whereas the two completely avirulent strains derived from M129 showed clusters of organisms that were smaller, amorphous, granular and non-refractile. These strain differences, which could not be seen by electron microscopy, were highly stable and may represent mutations.

Again the loss of haemadsorption by colonies was associated with total loss of virulence suggesting that cytadsorption is a prerequisite of virulence. However, as the strain Mac-K which cytadsorbs is not fully virulent, other factors, possibly involving a toxin must also be involved in the full manifestation of virulence.

Steinberg, Horswood & Chanock (1969) have exposed *M. pneumoniae* to the potent mutagen *N*-methyl-*N'*-nitro-*N*-nitrosoguanidine. They selected mutants which grew at the permissive temperature of 32° but not at the restrictive temperature of 38°, with the aim of obtaining strains for a vaccine that would grow in the upper respiratory tract and stimulate resistance without growing in the lower respiratory tract. Such mutants may prove useful in the analysis of the attributes of *M. pneumoniae*, including those essential for virulence (Steinberg *et al.* 1969).

Mycoplasma mycoides

There are avirulent and virulent strains of *M. mycoides* whose pathogenicity for the natural host is fairly constant. For example the KH_3J strain (Anon, 1960), originally obtained from the Sudan and passed many times in broth, rarely if ever causes lesions in cattle (Hudson, 1965; Lloyd & Trethewie, 1970). Even after subcutaneous inoculation of 50 ml of fresh broth culture into Zebu cattle, no general or local reaction is observed (Lindley, 1959). The representative strain (PG1) is virtually non-pathogenic in cattle. The avianised strain (T1) used in East African cattle by Hyslop (1955) is practically avirulent but survived 150 serial egg passages without loss of immunising powers. In contrast, the Gladysdale strain almost always causes a severe reaction (Turner, 1961; Davies & Hudson, 1968). The H1 strain was maintained for at least 50 subcutaneous passages in mice without losing its virulence for cattle (Hyslop, 1958). Various other virulent strains exist; most have been passaged only a few times since their isolation from cattle. The type of cattle used to test the virulence of different isolates will influence the results; for Purchase (1939) noted that in Africa, different breeds of cattle in different environments differed in their reactions to living vaccine.

The attributes *in vitro* of virulent and avirulent strains of *M. mycoides* have not been compared except in relation to the galactan of this organism (see later).

Mycoplasma neurolyticum

The virulence of different isolates of *M. neurolyticum* varies but tends to be stable, although some workers have found loss of virulence with prolonged passage. The strains Sabin type A (Sabin, 1938*a*) and KSA (Lemcke, 1961) were virulent to mice when inoculated intravenously

(Tully & Ruchmann, 1964); each strain has been passed by Tully (1964) more than 100 times in broth without any detectable changes in neurotoxicity. Thomas (1970a), however, found that the KSA strain received from Tully had lost its virulence for mice and rats and he could not restore it by passage in these animals. Although this avirulent KSA strain was indistinguishable by the growth-inhibition and gel-precipitation reactions from a consistently toxic strain (TA), antibody against KSA did not neutralise TA exotoxin (Thomas, 1970a). Other strains of *M. neurolyticum* have always had low virulence. For example, the L5 (PG28) isolate of Findlay, Klieneberger, MacCullum & Mackenzie (1938) produced rolling disease only when cultures were mixed with agar or with lymphocytic choriomeningitis virus-infected mouse brain and inoculated intra-cerebrally (Findlay *et al.* 1938). After several years of sub-culture this strain completely lost its virulence (Edward, 1954); at this stage it produced granular growth in semi-solid medium whereas originally the growth had been smoother (Klieneberger, 1940).

ENTRY OF MYCOPLASMAS INTO THE HOST: THEIR SURVIVAL ON MUCOUS MEMBRANES AND PENETRATION INTO THE TISSUES

Entry

Most important mycoplasma infections of man and animals involve the respiratory system, and, to a lesser extent, the urogenital system; joints, mammary glands and other sites are also affected (Fallon & Whittlestone, 1969). Mycoplasmas generally enter the body via the respiratory system, the mouth, the genital system, the teat canal or the conjunctiva. Entry via the respiratory tract not only occurs for the mycoplasmas causing respiratory disease but also probably for some organisms causing disease elsewhere in the body. For example, *Mycoplasma arthritidis*, the cause of murine mycoplasmal arthritis, is commonly present in the nasopharynx of healthy rats and *M. neurolyticum* may be harboured on the nasal mucosa of normal mice (Ward & Cole, 1970); these organisms may well penetrate from the respiratory epithelium. The inhalation and retention of particles in the respiratory system of man has been discussed by Druett (1967) in relation to bacterial and viral infection. Although little precise information relevant to mycoplasmas is available, the general principles described by Druett probably apply to these micro-organisms.

8

Survival

An important feature of respiratory tract infections is the prolonged survival of mycoplasmas usually on the epithelium of the trachea or bronchiole. Some respiratory mycoplasmas also survive for long periods in other sites e.g. *M. pulmonis* in the middle ear of rats and *M. mycoides* in sequestrated lesions in the lung. Little appears to be known about how mycoplasmas survive so effectively on the respiratory mucous membrane. The failures, so far, in demonstrating local antibody are mentioned later. Why phagocytic cells do not eliminate infection does not appear to have been studied. Perhaps they are unable to reach or take in mycoplasmas closely attached to the cell membrane between the cilia. (see later).

Production of disease locally

There has been great interest in the mechanisms and factors which influence the establishment of contagious bovine pleuropneumonia because of the difficulty in establishing a disease resembling the natural one by exposure of animals to a spray of virulent mycoplasmas. In the natural cases there are a few large consolidated areas in the lungs, whereas when infection is established by artificially produced aerosol droplets, the lesions are often small, discrete and multiple. Even the introduction of a liquid culture down an intrabronchial tube is not a reliable way of producing bovine pleuro-pneumonia (Lloyd & Trethewie, 1970).

A serological response (complement fixation titre exceeding 1:20) can, however, be reliably induced by an aerosol of either infected thoracic exudate or a second passage broth culture from this exudate of the same hypervirulent strain (HI) grown to the log phase (Hyslop, 1963). Lesions closely resembling the natural disease have been induced in animals which received, after intrabronchial intubation, 15 ml of a 10^{-1} suspension of infected lung followed by 100 ml broth (Anon, 1960). Such infected animals readily transmit the disease naturally to in-contact cattle (Hudson & Turner, 1963).

Factors that might influence the experimental establishment of a pneumonia comparable with that seen naturally are discussed by Lloyd & Trethewie (1970). They suggest that both droplet size and other factors may be involved. Large droplets, which were not in the artificial aerosols, would be expected to produce fewer and larger lesions. The other factors could be those substances present in lung-tissue suspension and natural aerosols, but not present in cultures of organisms. Thus the complex mixture of inflammatory exudates, products of lung degeneration and substances resulting from the multiplication of the organism in lung, rather than in the artificial medium, could be involved.

Penetration into tissues

The mycoplasmas causing respiratory disease are present in large numbers on the surface of the respiratory epithelium; they reach the local lymph nodes but the mechanism does not appear to have been studied. *M. mycoides* and *M. mycoides* var. *capri* are unusual amongst the mycoplasmas causing respiratory disease in that they also spread to the pleural cavity. Lloyd & Trethewie (1970) suggest that *M. mycoides* deposited in the alveoli or terminal bronchioles is phagocytosed by alveolar macrophages which then move up the respiratory tract, penetrate the bronchial wall and enter the lymphatics. The regional lymph node thus contains infected cells which stimulate the development of a germinal centre. The resultant impairment of lymph flow not only gives rise to the pulmonary oedema but also allows organisms to be carried to the pleura by the outward flow of lymph.

The mechanisms of penetration of mycoplasmas which induce arthritis in animals are very poorly understood. It is not known, for example, how *M. arthritidis*, commonly present in the nasopharynx of rats, reaches the joints. Ward & Cole (1970) suggest that some special host–parasite interaction must exist before the natural disease is initiated. Under experimental conditions arthritic disease was more prevalent in rats kept under insanitary conditions associated with high humidity and when the cages were inclined to produce injury. Also, some strains of *M. arthritidis* seemed to be more inclined to produce arthritis than others. If the pathogenesis of rat arthritis could be unravelled, it might help to elucidate the possible role of mycoplasmas in arthritis in man.

Penetration of *M. neurolyticum* from the respiratory tract (if that is the portal of entry) to other sites must occur on occasion. The organism was first detected by passaging other agents intracerebrally and has also been isolated from the blood and liver of asymptomatic mice (Tully & Rask-Nielsen, 1967).

During the last few years, attempts to associate mycoplasmas, in addition to *M. pneumoniae* and *M. hominis*, with disease in man have resulted in the appearance of many different mycoplasma species in cultures prepared from human tissues (Morton, 1970). Some may be laboratory contaminants since they are types not usually associated with man. There have been frequent isolations from deeper tissues, e.g. leukaemic bone marrows, of mycoplasma species usually present in the

oropharynx of man. Thus even apathogenic mycoplasmas may penetrate the tissues, become widely disseminated in the body and survive or
multiply, particularly when host defences are impaired as in leukaemia.

GROWTH FACTORS *IN VITRO* AND POSSIBLE
RELATIONSHIP TO BEHAVIOUR *IN VIVO*

The genome of mycoplasmas is smaller than that of bacteria. For
example, the genome size of the saprophytic mycoplasma, *Acholeplasma
laidlawii* type A is 880×10^6 daltons, and those of two pathogenic
species *M. gallisepticum* and *M. arthritidis* are 1200×10^6 daltons and
444×10^6 daltons respectively (Morowitz, Bode & Kirk, 1967), whereas
the genome of *E. coli* is 2500×10^6 daltons (Razin, 1969, from data of
Cairns, 1963). The more limited genetic information in mycoplasmas
correlates with their exacting nutritional requirements. There have been
great difficulties in the development of defined media and the marked
sensitivity of mycoplasmas to lysis by many factors has increased the
difficulty. Razin (1969) suggests that progress in producing defined
media may have been slow because many of the biosynthetic pathways
can be elucidated in undefined media by the use of labelled precursors.
Hence there are numerous undefined media for cultivating mycoplasmas
(Fallon & Whittlestone, 1969), but these empirical media have contributed little to our knowledge of growth factors.

Completely defined media have recently been worked out for the goat
mycoplasma strain Y and for *A. laidlawii* type B (Rodwell, 1970). Four
partially defined media for *A. laidlawii* type A, Goat Y, Avian J and
M. arthritidis (Campo) have also been prepared (Rodwell, 1970). The
organism best understood, *A. laidlawii*, is generally considered to be a
saprophyte, so the information may be of limited value to the study
of mycoplasma pathogenicity. The available information on the growth
factors required by mycoplasmas and their metabolic pathways (Razin,
1969), has not yet been related to the behaviour of any of the pathogenic
mycoplasmas *in vivo*. Further work could be directed towards relating
the marked tissue and host specificities shown by mycoplasmas to the
possible differential distribution of nutrient factors or inhibitors of
metabolic pathways in the animal body.

HUMORAL AND CELLULAR REACTIONS BY THE HOST
TO MYCOPLASMAS

Antibodies

Circulating antibodies detectable by a wide variety of serological tests develop during the course of mycoplasma infections of the respiratory tract and it has been suggested that antibodies inhibiting growth or metabolism might provide an index of immunity. However with contagious bovine pleuropneumonia, Lloyd (1967) showed that transfer of serum from immune animals failed to confer immunity and Davies & Hudson (1968) demonstrated that growth-inhibiting antibody is not connected with immunity. Similarly, in enzootic pneumonia of pigs caused by *Mycoplasma suipneumoniae* (*Mycoplasma hyopneumoniae*), Goodwin, Hodgson, Whittlestone & Woodhams (1969a, b) found no correlation between protective immunity and serum titres obtained by means of the metabolism-inhibition, passive-haemagglutination and complement-fixation tests. In studies of immunity to *M. pneumoniae* in hamsters (Fernald, 1969) and to *M. suipneumoniae* in the natural host (Goodwin *et al.* 1969b), parenteral vaccination gave rise to high serum antibody titres but was not protective.

In man, there appears to be a direct correlation between the presence of metabolism-inhibiting antibody and resistance to infection with *M. pneumoniae* (Chanock, Steinberg & Purcell, 1970). However there are anomalies. Following the parenteral injection of inactivated *M. pneumoniae* vaccine into men, there was no correlation between the antibody response and the ability to isolate *M. pneumoniae* from throat swabs after challenge (Smith *et al.* 1967). After natural infection, circulating antibody levels may correlate with the levels of resistance without this antibody itself being protective. If this were so, the evidence from all species would indicate that resistance to the mycoplasma-induced pneumonias is not due to circulating antibody but to some local factor in the respiratory tract. In virus diseases causing superficial infection of the respiratory tract, the main determinant of resistance is the 11S Ig A antibody secreted locally in the respiratory tract (Chanock *et al.* 1970). At present in mycoplasma diseases there is little evidence to indicate whether local antibody (or cellular immunity) is involved. Davies (1969) working with *M. mycoides* examined nasal mucus from both vaccinated and infected cattle but was unable to find either complement-fixing or metabolism-inhibiting antibodies in samples. Following *M. pneumoniae* infection, attempts to detect antibody in respiratory tract secretions have been unsuccessful both in volunteers

(Smith *et al.* 1967) and in hamsters with straightforward infection (Fernald, 1969).

The immune mechanism in mycoplasma-induced arthritis in rats appears to differ from that in the respiratory diseases caused by myco-plasmas. Cole, Cahill, Wiley & Ward (1969) found that immunity could be transferred by convalescent serum, which probably correlates with the spread of the organism via the blood stream to the joints. The circulating protective antibody was not significantly affected by adsorption with antigen – a procedure that greatly reduced the comple-ment-fixation titre.

Cellular host defences

There are a few studies *in vitro* on the phagocytosis and degradation of mycoplasmas by various cell types. Zucker-Franklin, Davidson & Thomas (1966*a*) investigated with the electron microscope the reaction of human leucocytes with *M. pneumoniae*, *M. neurolyticum* and *M. gallisepticum*; all three species were phagocytosed by both neutrophil and eosinophil polymorphonuclear cells within a few minutes of incubation. The neutrophils underwent degranulation and both types of cells showed vacuoles containing mycoplasmas; after 1 h the organ-isms could barely be recognised – presumably due to their degradation by lysosomal enzymes. Neutrophil polymorphonuclear cells were labelled after incubation with radio-isotope-labelled *M. mycoides* (Anon, 1967).

The interaction *in vitro* between mouse macrophages and *Mycoplasma pulmonis* has been studied with the electron microscope and by phase-contrast microscopy (Jones & Hirsch, 1971). The mycoplasmas attached to the cell surface and grew into a lawn covering most of the cell surface. The addition of low concentrations of anti-mycoplasma anti-body stimulated massive phagocytosis of the surface micro-organisms. The intracellular organisms continued to incorporate tritiated thymidine into DNA for a few hours, so the antibody effect appears to be opsonic rather than lethal. By electron microscopy the intracellular mycoplasmas already appeared partially degraded within 2 h, and by 24 h digestion of the mycoplasmas was nearly complete.

M. mycoides survived in the presence of bovine macrophages (Lloyd & Trethewie, 1970) but it is not clear whether the organisms were on the cells or inside them.

Blood monocytes from man were phagocytic for *M. pneumoniae*, *M. neurolyticum* and *M. gallisepticum* (Zucker-Franklin, Davidson, Gesner & Thomas, 1965; Zucker-Franklin, Davidson & Thomas, 1966*b*) the organisms becoming unrecognisable when the phagocytic

vacuole had coalesced with the cell's lysosomes. Surprisingly, mycoplasmas were also found in phagocytic vacuoles within peripheral blood
cells which looked like lymphocytes on routine blood smears. Antibody
synthesis was not detected in these cells within 3 h, nor were there
signs of the mycoplasmas being altered morphologically, which may be
related to the paucity of lysosomes in lymphocytes.

The available evidence indicates that observations *in vitro* demonstrating phagocytosis of mycoplasmas by neutrophil polymorphonuclear
cells parallels the situation *in vivo*. In Giemsa-stained preparations of
infected tissues from some mycoplasmal diseases organisms can be seen
in the cytoplasmic area of polymorphonuclear neutrophils. For example,
in the first week of enzootic pneumonia of the pig neutrophil polymorphonuclear cells contain *M. suipneumoniae* in various stages of
disintegration (Whittlestone, 1958 and unpublished observations)
suggesting that the organisms are probably within the cytoplasm rather
than extracellular. In *M. pulmonis* infection of the lungs of gnotobiotic
mice studied by electron microscopy (Organick, Siegesmund & Lutsky,
1966) phagocytosis of organisms by polymorphonuclear leucocytes was
seen one week after infection.

Virtually nothing is known about whether mycoplasmas produce
substances which inhibit host defences. The possibility that the enzyme
L-arginine iminohydrolase from the arginine-metabolising mycoplasmas
might interfere with antibody production by lymphocytes is discussed
later.

MYCOPLASMAL PRODUCTS AND PATHOGENICITY

The toxin of Mycoplasma neurolyticum

The only mycoplasma known to produce a soluble toxin is *M. neurolyticum*. Two strains of this organism (Sabin type A and L5) were
isolated from rolling disease of mice by Sabin (1938*a*, *b*) and by
Klieneberger (Findlay *et al.* 1938). Inoculation of cultures intracerebrally
into mice induced a characteristic turning or rolling on the long axis
of the body which is associated with almost complete necrosis and lysis
of the posterior pole of the cerebellum (Sabin, 1941). Sabin's isolate
produced an exotoxin in glucose serum broth (Sabin, 1938*b*) which on
intravenous inoculation into mice produced the same nervous signs and
cerebellar lesion as did live organisms. The latent period following
inoculation and the duration of clinical signs depended on the dose of
toxin given; it varied from 5 min to 4 h and with moderate doses rolling
may continue for several hours before death, whereas with large doses

of toxin rolling occurs for only a few minutes before the mice develop convulsions and die. The symptoms and course are similar in the rat but the rolling is less conspicuous and chronic convulsions may occur repeatedly for up to 24 h before death. (Thomas, 1970a).

Some properties of the neurotoxin were delineated early by Sabin (1938a, b, 1941). Filtrates of young cultures contained toxin but 48 h cultures, in which the pH had dropped from 7·8 to 6·0, were neither variable nor contained toxin. Mice surviving challenge were immune to further challenge, although Sabin was unable to detect the presence of neutralising antibodies. Little interest was taken in the toxin for more than 20 years until its production and properties were investigated systematically (Tully, 1964; Tully & Ruchman, 1964; Thomas & Bitensky, 1966b; Thomas, Aleu, Bitensky, Davidson & Gesner, 1966; Aleu & Thomas, 1966; Thomas, 1967, 1969; Tully, 1969; Thomas, 1970a; Kaklamanis & Thomas, 1970). The toxin is a labile protein, being destroyed at 45° in 15 min and deteriorating rapidly at room temperature; but it is stable when frozen or lyophilised. Its molecular weight probably exceeds 200000. Only mice and rats are susceptible to the toxin; other laboratory animals have received up to fifty times the minimal lethal dose for mice without effect.

Doses of 10^{10} or more washed cells of M. neurolyticum given intravenously produce the same signs and lesions as the toxin and again the latent period varies from 5 mins to 2 h depending on the dose. Evidence provided by Thomas and his colleagues shows that the living mycoplasmas must synthesise the toxin rapidly after intravenous inoculation, but that the multiplication of the organism is not essential.

The toxin appears to mediate its effect in the vascular bed of the brain, for its activity is much less when inoculated intracerebrally than when given intravenously and no effect is seen following subcutaneous or intraperitoneal inoculation of large doses. The permeability of the blood–brain barrier increases just before the onset of rolling disease; the uptake of radioactive glycine by the brain is increased and trypan blue inoculated intraperitoneally into animals with rolling disease rapidly stains their brains. The toxin may act at the site of attachment of astrocytes to the walls of the brain capillaries, the resultant damage to these membranes allowing fluid to flow into the astrocytic processes. Thus numerous vesicles develop both in the intercommunicating astrocytic processes and the astrocyte body, causing great enlargement with compression of neighbouring structures. Both neuronal damage and demyelination is seen. The interaction between the toxin and its receptors occurs rapidly. Mice can only be protected against the toxin if antibody is given less than 3 min after the toxin. The cerebral permeability changes are detectable some 45 min later, immediately before rolling develops.

A sedimentable component of brain tissue causes irreversible inactivation of the toxin. A purified brain ganglioside has a similar activity and as the active material is heat- and trypsin-stable but destroyed by periodate, the brain receptor for the toxin might be N-acetyl neuraminic acid (sialic acid). Kaklamanis & Thomas (1970) speculate that the toxin may be an enzyme which acts on this constituent at the membrane junction between astrocytic processes and the walls of capillaries.

The toxic effect of Mycoplasma gallisepticum

The occurrence of encephalitis in turkeys affected with chronic respiratory disease was noted by Jungherr (1949) who described lesions in the cerebral arteries of affected birds. The relationship of the organism, now known as *M. gallisepticum* to the disease was established by Adler, Yamamoto & Bankowski (1954). One strain of the organism was shown by Cordy & Adler (1957) to produce nervous signs and death in young turkey poults inoculated intravenously. The main lesions in the brain were confined to, or associated with, the blood vessels. The production of cerebral polyarteritis with Adler's S6 strain of *M. gallisepticum* was studied by Thomas, Davidson & McCluskey (1966) and discussed by Thomas (1967, 1969, 1970a, b) and Kaklamanis & Thomas (1970).

There are many points of similarity between the neurotoxicity of *M. neurolyticum* in mice and *M. gallisepticum* in turkeys. Intravenous inoculation of 10^{11} live organisms of the S6 strain of *M. gallisepticum* into turkey poults induces within 2 hours ataxia, torticollis and paralysis of the extremities followed by death. Some birds exhibited rolling similar to that caused by *M. neurolyticum* in mice. Antibody to S6 is protective only when inoculated within 5 minutes after the organisms. Pre-treatment of birds with tetracycline gives complete protection and organisms killed by heating or freezing and thawing are no longer toxic. Other species of laboratory mammals and birds are not susceptible to the toxicity of S6. The brain lesions in turkeys are associated with fluid engorgement of the astrocytes and their processes.

With *M. gallisepticum* no exotoxin has been demonstrated in cultures. When smaller doses of organisms are injected there is not such a clear distinction between the immediate toxic effects and those resulting from multiplication of the organism. As the dose of *M. gallisepticum* is reduced the latent period increases up to about a day with 10^{10} organisms and a week or more with 10^7 organisms. Turkeys surviving more than 24 h show polyarteritis nodosa, with necrotising and inflammatory involvement of virtually all arteries in the central nervous system, a similar lesion to that seen in chronic infections of the paranasal sinuses with the S6 strain of *M. gallisepticum.*

The neurotoxic effect may be due to an affinity between the organism itself or a component of the organism and the *N*-acetyl neuraminic acid receptors of the central nervous system. The mechanism of the production of the polyarteritis is not known but as mycoplasmas cannot be detected by immunofluorescence in association with affected arteries the action of an extracellular toxin is postulated.

The toxic effect of Mycoplasma pulmonis and Mycoplasma arthritidis

The toxicity of *M. pulmonis* and *M. arthritidis* in rats and mice has been studied by Kaklamanis & Thomas (1970). The general picture is the same as for the toxicity of *M. gallisepticum* except that following inoculation, after an interval of 1–18 h, animals become lethargic, comatose and die, but do not show specific neurological manifestations. Again, only the intravenous route is effective and only the natural hosts mice and rats are susceptible. Prior inoculation with tetracycline provides partial protection and killed organisms are non-toxic. Animals surviving sublethal doses of *M. pulmonis* show no subsequent evidence of disease. Sublethal doses of *M. arthritidis* cause two effects probably not related to toxin; namely polyarthritis within 1–2 days of inoculation and death of foetuses in pregnant mice. Such foetuses contain viable *M. arthritidis* and are either resorbed or delivered stillborn.

Products of Mycoplasma mycoides

A possible exotoxin

The early work of Nocard *et al.* (1898) suggested that *M. mycoides* might produce a toxin: for cultures of the organism in semi-permeable collodion sacs inserted into the peritoneal cavity of rabbits caused the rabbits to lose weight. Lloyd (1966), in similar experiments, found that the loss of weight appeared to be associated with the laparotomy but that necrosis occurred around the chambers containing the culture but not medium. Later Lloyd showed (Lloyd & Trethewie, 1970) that the contents of the chambers produced necrosis of subcutaneous tissues in the rabbit and that the reaction was not inhibited by anti-mycoplasma drugs. He also implanted chambers filled with *M. mycoides* culture intramuscularly into both susceptible cattle and cattle resistant to subcutaneous inoculation with *M. mycoides*. In both cases the infected chambers, but not control chambers became surrounded by connective tissue infiltrated with mononuclear cells, the appearance being similar to that seen in lung sequestra in the natural disease. It is not yet known whether both virulent and avirulent strains of *M. mycoides* produce the 'toxic' factor, nor have the toxic effects of filtrates of cultures *in vivo* been tested.

A possible endotoxin

Villemot, Provost & Queval (1962) claim to have extracted an endotoxin from *M. mycoides* and implied a similarity to the endotoxins of bacteria. Their extract was, however, not purified and its toxic component was not identified. Without additional evidence it is difficult to assess the claim. Razin (1969) points out that it is hard to conceive how mycoplasmas could possess endotoxins like the cell-wall-associated toxins of the Gram-negative bacteria.

The galactan

Kurotchkin (1937) showed that *M. mycoides* elaborated large amounts of a specific carbohydrate substance which could be found in culture supernatants and in the blood of animals in severe cases of the disease. This polysaccharide is a galactan with the unusual $6\text{-}O\text{-}\beta\text{-}\text{D}$-galacto-furanose repeating unit (Plackett & Buttery, 1958; Buttery & Plackett, 1960; Plackett, Buttery & Cottew, 1963). The galactan can be obtained from washed organisms and Gourlay & Thrower (1968) have shown that the visible threads in cultures of *M. mycoides* consist of mycoplasma cells embedded in a homogeneous matrix, probably composed of galactan. So far no other mycoplasmas are known to possess a capsule made of galactan.

In cattle the galactan may have an effect similar to that of the capsular polysaccharides that promote pneumococcal infections, in that it binds some of the antibodies appearing in the blood stream (Razin, 1969). Galactans prepared from the V5 vaccine strain of *M. mycoides* and inoculated intravenously promote a persistent mycoplasmaemia following subcutaneous inoculation with either the attenuated V5 strain or the avirulent KH_3J strain, (Hudson, Buttery & Cottew, 1967). Inoculation of strain V5 usually causes arthritis in young calves only, but when preceded by the galactan, arthritis and synovitis developed in yearling cattle (Hudson *et al.* 1967). Virulent (T_3) and avirulent (KH_3J) strains of *M. mycoides* contain serologically common galactan-containing lipopolysaccharides (Gourlay, 1965). However, there could be a difference in the amount of capsular material present in the different strains since the centrifuge deposit from the virulent Gladyside strain is about 5 times the volume and about 13 times the dry weight of that from the avirulent (KH₃J) strain for a similar concentration of particles (Gourlay & Thrower, 1968).

A serological connection between the galactan of *M. mycoides* and a pneumogalactan isolated from bovine lung, has been demonstrated by Shifrine & Gourlay (1965) and Gourlay & Shifrine (1966). The possibility that this cross reactivity could play a part in the pathogenesis of contagious bovine pleuropneumonia, is mentioned later.

Hydrogen peroxide and other haemolysins

The haemolysin produced by *M. pneumoniae* (Somerson, Taylor-Robinson & Chanock, 1963) was shown to be hydrogen peroxide by Somerson, Walls & Chanock (1965) and Cohen & Somerson (1967). It is now known to be produced by a wide range of mycoplasma species e.g. *M. gallisepticum* (Thomas & Bitensky, 1966a) *M. pulmonis*, *M. mycoides* var. *capri, M. neurolyticum, M. gallinarum, M. arthritidis, M. felis, M. bovigenitalium, A. laidlawii* type B, and some untyped species (Cole, Ward & Martin, 1968). The latter authors, employing an agar containing benzidine with sheep blood, correlated peroxide production and haemolysis on blood agar and showed that the major haemolysin produced by the mycoplasmas was hydrogen peroxide.

Peroxide production alone is not a determinant of virulence because its production is not limited to the pathogenic mycoplasmas but there are quantitative differences between species. Thus, *M. pneumoniae*, compared with the other mycoplasmas of human origin (Sobeslavsky & Chanock, 1968), produced a wider and a more rapid and complete haemolytic zone. To prevent haemolysis, a higher concentration of catalase was required for *M. pneumoniae* than for the other mycoplasmas (except *M. fermentans*). Lipman & Clyde (1969) compared the peroxide formation of pairs of virulent and avirulent strains of *M. pneumoniae*. Although there were differences between heterologous strains they could find no greater activity in the virulent member of each pair. So although the theory of Sobeslavsky, Prescott & Chanock (1968), that pathogenesis follows a sequence of organism adsorption followed by peroxide damage, is plausible, this work of Lipman & Clyde (1969) indicates additional mediators of virulence of *M. pneumoniae*.

The liberation of peroxide was shown by Cherry & Taylor-Robinson (1970b) to be an important factor in the pathogenesis of *M. mycoides* var. *capri* in chick tracheal organ cultures. The addition of glucose to the growth medium increases the peroxide production of *M. pneumoniae*, *M. gallisepticum* (Cohen & Somerson, 1969) and *M. pulmonis* (Brennan & Feinstein, 1969), yet has a sparing effect on loss of ciliary activity caused by *M. mycoides* var. *capri* (Cherry & Taylor-Robinson, 1970*b*). These authors suggest that the latter mycoplasma may have its own peroxidase which is also stimulated by glucose.

The effectiveness of secreted hydrogen peroxide in damaging host cells is probably influenced by a variety of factors: for example, *M. pneumoniae* is closely attached to the cell membrane (Collier & Clyde, 1971) by neuraminic acid receptors (Sobeslavsky *et al.* 1968) so that there is an unusual opportunity for the peroxide secreted by the organisms to attack the tissue cell membrane before being destroyed by catalase or peroxidase present in extracellular body fluids. The surface of the bronchiolar epithelium where *M. pneumoniae* lives would be expected, with its aerobic conditions, to be favourable for the elaboration of hydrogen peroxide. This would also apply to pathogenic respiratory mycoplasmas found in many species of domestic and laboratory animals.

Brennan & Feinstein (1969) showed that hydrogen peroxide contributes to the virulence of *M. pulmonis* in mice. A strain of mice essentially devoid of catalase had more pneumonia 3 days after infection than did wild-type mice. At 5 days, however, fewer acatalactic mice had pneumonia and *M. pulmonis* could not be recovered. This suggests that in the absence of tissue catalase, hydrogen peroxide accumulates and is lethal to *M. pulmonis*. The relationship of hydrogen peroxide to possible auto-immune phenomena seen in *M. pneumoniae* infection is discussed later.

Evidence that haemolysins other than peroxide are possessed by some mycoplasmas was provided by Cole *et al.* (1968). Catalase even in high concentrations did not completely suppress the haemolytic activity of *M. mycoides* var. *capri*. The double zone of haemolysis caused by *M. gallinarum* on guinea-pig blood medium and the anaerobic haemolysis of chicken blood by this organism also indicate additional haemolysins. Further work is needed to elucidate the nature of these haemolysins.

Enzymes

Mycoplasma contamination has been a frequent and major problem encountered in cell culture and virus research. Some of these mycoplasmas were shown by Barile & Schimke (1963) to convert arginine to ornithine by a pathway involving the enzyme L-arginine iminohydrolase (arginine deiminase). The depletion of the essential metabolite arginine interferes with cell growth (Kenny & Pollack, 1963). The lysis of cell cultures by some mycoplasmas was due to the liberation into the medium of a factor (Kraemer, 1964). The effect could be prevented by adding enough L-arginine to the medium. The arginine-metabolising mycoplasmas, *M. hominis* type 1, *M. arthritidis* and *M. orale* inhibit the usual stimulation of mitosis in lymphocyte cultures by phytohaemagglutinin (Copperman & Morton, 1966; Barile & Leventhal, 1968). The inhibitor from *M. arthritidis* suppresses the production of antibody by lymph node explants (Thomas, 1969) but does not prevent antigen binding to the cell. All the mycoplasmas producing such an inhibitor were arginine dependent for growth and the addition of arginine to the lymphocyte cultures restored their response to both phytohaemagglutinin and antigen. The effect would therefore seem to be due to the enzyme L-arginine iminohydrolase.

Thus the possibility arises that this enzyme from mycoplasmas might inhibit antibody production *in vivo* and thereby contribute to the pathogenesis of disease. Preliminary experiments to test this hypothesis have been made by Thomas (1969) but results are not yet conclusive.

The enzyme nucleoside phosphorylase present in mycoplasmas splits thymidine, which is essential for cellular multiplication (Razin, 1969). Russell (1966) suggested that the chromosomal abnormalities occurring in mycoplasma-infected tissue cultures may arise indirectly as a result of the changes he observed in the nucleic-acid metabolism of infected cells. The possible relation of nucleoside phosphorylase to disease caused by mycoplasmas has not yet been studied.

OTHER MECHANISMS HAVING A BEARING ON PATHOGENICITY

Synergism between mycoplasmas and other agents and factors

There are many examples in diseases of animals where a synergistic relationship exists between a mycoplasma and another infectious agent or factor, so that an inapparent or mild disease is transformed into a recognisable syndrome. Klieneberger-Nobel (1954) even expressed the opinion that unless a second factor is present, such as a breakdown in the defence mechanisms of the host, an unspecific adjuvant or another infective agent, disease will not develop with mycoplasma infection. This view arose from the difficulty of establishing mycoplasma diseases with cultures of organisms. For example attempts to induce bronchiectasis in rats with the L3 organism (*M. pulmonis*) were not successful until bronchial ligation was practised (Klieneberger-Nobel & Cheng, 1955).

Already mentioned are: the problems in early experiments to produce lesions resembling those occurring naturally with cultures of *M. mycoides* and the lack of virulence for mice of the L5 strain of *M. neurolyticum* unless inoculated intracerebrally mixed with agar or with mouse brain containing the virus of lymphocytic choriomeningitis virus. More recent work with *M. neurolyticum* (Kaklamanis & Thomas, 1970) has shown that after intracerebral, intravenous and subcutaneous inoculation of doses smaller than the fatal toxic dose, mice do not contain viable organisms in the blood or tissues for more than two or three days. On the other hand, chronic infection, did become established in mice used for passaging a myeloma tumour (Kaklamanis & Thomas, 1970). The tumour was the only site infected with the mycoplasma, and mice with the tumour had a high mortality with the usual signs of rolling disease.

Other examples of synergism in experimental systems are the activation of mycoplasma infection in mice by ectromelia virus (Mooser, 1949), and the enhancing effect of murine hepatitis virus on cerebral infection with a mycoplasma from infectious nasal catarrh (Nelson, 1957).

In natural disease there is evidence that the severity of mycoplasma infections is increased by concurrent infection with other agents. For example, in respiratory diseases of chickens the relatively mild infection induced by *M. gallisepticum* is considerably enhanced by concurrent infection with *E. coli*, Newcastle disease virus or infectious bronchitis virus (Fabricant, 1969). Although these synergistic effects are often explained in general terms, there appears to be little precise information on the actual mechanisms involved.

Relation between mycoplasmas and host cells

Association with cells

The close association between mycoplasmas and the host-cell surface could be of considerable pathogenic significance. The possible effects of liberated hydrogen peroxide and the resultant cell damage are discussed elsewhere. Additional mechanisms by which injury could occur to ciliated respiratory epithelial cells are suggested by the work of Collier *et al.* (1971): the masses of organisms could interfere mechanically with the normal beating of the cilia or they could result in a disturbance of the surface charges needed for the synchronisation of ciliary motion. There could also be nutritional deprivation of the host cells, either because of competition for metabolites or because of physiological injury to the cell membrane. In their study of hamster tracheal organ culture, Collier *et al.* (1971) noted that *M. pneumoniae* infection resulted in loss of cilia and the disappearance of the terminal bars, which normally close the intercellular space at the epithelial border; thus disruption of the tissue architecture of the epithelial surface occurs and mycoplasmas may penetrate the intercellular spaces.

The presence of *M. suipneumoniae* organisms on the bronchiolar epithelium of the pig is also associated with a loss of cilia and a reduction in adherence between the individual epithelial cells (Whittlestone, unpublished observations).

An ultrastructural study of *M. pneumoniae* in human respiratory epithelium cultures by Collier & Clyde (1971), showed filamentous organisms between the cilia. These mycoplasmas showed a differential terminal structure in very close proximity to the host cell membrane which in some instances appeared indistinct at the point of contact, suggesting local injury.

Despite its limited host range in infection, *M. pneumoniae* will attach to a range of cell types, in which the chemical groupings involved in adherence have been studied, particularly by haemadsorption. The receptor site on the tissue cell involves N-acetyl neuraminic acid (Sobeslavsky *et al.* 1968; Manchee & Taylor-Robinson, 1969). The same receptor is responsible for the adsorption of *M. gallisepticum* (Gesner & Thomas, 1965; Manchee & Taylor-Robinson, 1969) but in the case of *M. hominis* and *M. salivarium* the cell receptors are not influenced by purified neuraminidase and are associated with protein (Manchee & Taylor-Robinson, 1969). On the mycoplasma surface, the component which is responsible for attachment can be removed by trypsin so is evidently a protein (Thomas, 1969). Such treatment does not affect viability of the organisms, and when they are reincubated in a glucose medium the component is reconstituted.

Gesner & Thomas (1965) have suggested that there could be a relation between the binding sites and pathogenesis of disease. They postulate that preferred N-acetyl neuraminic acid binding sites may occur in certain tissues and may thus play a part in determining the susceptible host or the localisation of disease in vulnerable hosts. Further, if mouse fibroblasts are incubated with *M. gallisepticum*, they behave as magnets for normal lymphocytes which became attached within 2 h (Thomas, 1969). The number of mycoplasma cells required was comparable with the mycoplasma populations occurring in natural infections. Thomas suggests that lymphocyte aggregation at the side of natural mycoplasma infection might occur through a similar mechanism.

Entry of mycoplasmas into cells

Mycoplasmas usually appear to remain on the surface of cells considered to be non-phagocytic. The phagocytosis of mycoplasmas has been discussed elsewhere in this chapter, as has the entry of mycoplasmas into cells which looked like lymphocytes in routine blood smears (Zucker-Franklin *et al.* 1966a, b). As the organisms were within membrane-bounded vacuoles it seems likely that they had been phagocytosed.

Whether mycoplasmas can actively penetrate other cells is not clear on the present evidence. Observations made with the conventional microscope indicating that mycoplasmas are within cells (e.g. Shepard, 1958; Hayflick & Stinebring, 1960) cannot be taken as conclusive evidence because the organisms could well be in deep indentations within the plasma membrane. Recently, it was found (Taylor-Robinson, 1971) that T-mycoplasmas entered cells in tissue culture but the detailed evidence is not yet available.

Induction of chromosomal abnormalities

There has been speculation that mycoplasmas might be involved in the production of neoplasia, since the discovery of these organisms in various neoplastic conditions of man, particularly leukaemia. Some were types already known to infect man, whereas others were types usually associated with other species of animals. In some instances the neoplastic tissues themselves have probably harboured the mycoplasma; in other cases the chance of laboratory contamination cannot be ignored (Morton, 1970). There is no positive evidence that mycoplasmas are involved aetiologically in neoplasia. However, the possibility remains open, particularly as hamster cells have been transformed by mycoplasmas (Macpherson & Russel, 1966) and chromosomal aberrations develop in human cell cultures infected with mycoplasmas (Fogh & Fogh, 1965, 1967; Paton, Jacobs & Perkins, 1965; Allison & Paton, 1966). The hypothesis that *M. hominis* and *M. fermentans*, mycoplasmas commonly found in the genital tract of women, might be involved in the aetiology of Mongolism (Down's syndrome) has been advanced by Allison & Paton (1966), who outline additional work that might be done.

The demonstration that the mycoplasma, *A. laidlawii* can be infected with a virus (Gourlay, 1970), raises the possibility that the observed chromosome abnormalities induced by other mycoplasmas might be the result of virus infection of the mycoplasma. This suggestion is pure conjecture, particularly as the one mycoplasma virus known is a rod-shaped, DNA virus (Gourlay, Bruce & Garwes, 1971) and therefore does not resemble any of the known oncogenic viruses.

Anti-tissue antibodies in mycoplasma diseases

During the course of *M. pneumoniae* infection in man, complement-fixing antibodies reactive with human lung antigens develop (Thomas, Curnen, Mirich, Ziegler & Horsfall, 1943); a similar situation appears to exist in mycoplasma pneumonia of the pig, for positive complement-fixing reactions to pig lung extract were invariably found in pig sera showing high complement-fixing titres to *M. suipneumoniae* (Roberts & Little, 1970). The tissue antibodies may develop as a consequence of the lung cells being altered antigenically during mycoplasmal infection. The close association between lung cells and *M. pneumoniae* (Collier & Clyde, 1971), might provide the opportunity for cell damage to occur either from the peroxide liberated by the organism or during antigen–antibody–complement reactions at the cell surface.

Also, the cold agglutinins which often develop during *M. pneumoniae* infection could be the result of an immunological response to erythrocytes antigenically altered (Smith, McGinniss & Schmidt, 1967) by the persistent low hydrogen peroxide concentration (Cohen & Somerson, 1969). Additional evidence that erythrocyte damage occurs *in vivo* was provided by Feizi (1967) who showed that *M. pneumoniae* infections were associated with a positive Coomb's test with red cells and an increase in the reticulocyte counts.

Biological mimicry

It was noted by Cole *et al.* (1969) and by Cole, Golightly-Rowland, Ward & Wiley (1970) that with the mycoplasmas of murine origin, *M. arthritidis*, *M. pulmonis* and *M. neurolyticum* there was a direct correlation between the ability of the mycoplasma to induce disease and the lack of metabolism-inhibiting (MI) antibody formation by its natural host. Thus *M. neurolyticum* and *M. pulmonis* failed to produce MI titres in mice but induced high MI titres in guinea-pigs; *M. arthritidis* induced no or low MI titres in rats and mice but high titres in guinea-pigs. They suggested that the concept of biological mimicry, i.e. the sharing of immunological similarities between organism and host, might explain these observations. The inability of the natural hosts to produce MI antibodies might enable the mycoplasmas to become established in their hosts.

The concept has recently been developed by the work of Cahill, Cole,

Wiley & Ward (1971), who demonstrated antigenic relationships between rat tissues and *M. arthritidis*. They also found that rat lymphocytes inoculated into 6-week-old mice induced complement-fixing antibodies to *M. arthritidis*. Mice made tolerant to rat lymphocytes by inoculation at birth were subsequently much more susceptible than normal mice to polyarthritis induced by injection of *M. arthritidis*. Thus a partial explanation of why this organism infects rats is apparent. Other evidence which could be interpreted in the same light includes the demonstration by Taylor-Robinson & Berry (1969) that MI-antibody titres to *M. gallisepticum* were lower in the natural hosts (chicken and turkey) than in rabbits; and the finding that the crude galactan antigen of *M. mycoides* is related immunologically to pneumogalactan (Shifrine & Gourlay, 1965; Gourlay & Shifrine, 1966) and may thus result in a partial immune tolerance of the host to the organism.

REFERENCES

ADLER, H. E., YAMAMOTO, R. & BANKOWSKI, R. A. (1954). A preliminary report of efficiency of various mediums for isolation of pleuropneumonia-like organisms from exudates of birds with chronic respiratory disease. *American Journal of Veterinary Research*, **15**, 463–5.

ALEU, F. & THOMAS, L. (1966). Studies of PPLO infection III. Electron microscopic study of brain lesions caused by *Mycoplasma neurolyticum* toxin. *Journal of Experimental Medicine*, **124**, 1083–7.

ALLISON, A. C. & PATON, G. R. (1966). Chromosomal abnormalities in human diploid cells infected with mycoplasma and their possible relevance to the aetiology of Down's syndrome (Mongolism). *The Lancet*, **ii**, 1229–30.

ANON (1960). *Report of the First Meeting of the Joint FAO/OIE/CCTA Expert Panel on Contagious Bovine Pleuro-pneumonia, Melbourne*, 1–37.

ANON (1967). 'Phagocytosis'. In *Report of the Third Meeting of the FAO/OIE/OAU Expert Panel on Contagious Bovine Pleuropneumonia, Khartoum*, 8.

BARILE, M. F. & LEVENTHAL, B. G. (1968). Possible mechanism for mycoplasma inhibition of lymphocyte transformation induced by phytohaemagglutinin. *Nature, London*, **219**, 751–2.

BARILE, M. F. & SCHIMKE, R. T. (1963). A rapid chemical method for detecting PPLO contamination of tissue cell cultures. *Proceedings of the Society for Experimental Biology and Medicine, N.Y.* **114**, 676–9.

BRENNAN, P. C. & FEINSTEIN, R. N. (1969). Relationship of hydrogen peroxide production by *Mycoplasma pulmonis* to virulence for catalase-deficient mice. *Journal of Bacteriology*, **98**, 1036–40.

BROWN, R. D. (1964). Endobronchial inoculation of cattle with various strains of *Mycoplasma mycoides* and the effects of stress. *Research in Veterinary Science*, **5**, 393–404.

BUTLER, M. (1969a). Pathogenesis of respiratory mycoplasma studied in human embryo trachea cultures. *Journal of General Microbiology*, **593**, vii.

BUTLER, M. (1969b). Isolation and growth of mycoplasma in human embryo trachea cultures. *Nature, London*, **224**, 605–6.

BUTTERY, S. H. & Plackett, P. (1960). A specific polysaccharide from *Mycoplasma mycoides*. *Journal of General Microbiology*, **23**, 357–68.

CAHILL, J. F., COLE, B. C., WILEY, B. B. & WARD, J. R. (1971). Role of biological mimicry in the pathogenesis of rat arthritis induced by *Mycoplasma arthritidis*. *Infection and Immunity*, **3**, 24–35.

CAIRNS, J. (1963). The bacterial chromosome and its manner of replication as seen by autoradiography. *Journal of Molecular Biology*, **6**, 208–13.

CHANOCK, R. M., STEINBERG, P. & PURCELL, R. H. (1970). Mycoplasmas in human respiratory tract disease. In *The role of Mycoplasmas and L forms of Bacteria*, ed. J. T. Sharp, 110–36. Springfield, Illinois: Charles C. Thomas.

CHERRY, J. D. & TAYLOR-ROBINSON, D. (1970*a*). Large quantity production of chicken embryo tracheal organ cultures and use in virus and mycoplasma studies. *Applied Microbiology*, **19**, 658–62.

CHERRY, J. D. & TAYLOR-ROBINSON, D. (1970*b*). Growth and pathogenesis of *Mycoplasma mycoides* var. *capri* in chicken embryo tracheal organ cultures. *Infection and Immunity*, **2**, 431–8.

COHEN, G. & SOMERSON, N. L. (1967). *Mycoplasma pneumoniae:* hydrogen peroxide secretion and its possible role in virulence. *Annals of the New York Academy of Sciences*, **143**, 85–7.

COHEN, G. & SOMERSON, N. L. (1969). Glucose-dependent secretion and destruction of hydrogen peroxide by *Mycoplasma pneumoniae*. *Journal of Bacteriology*, **98**, 547–51.

COLE, B. C., CAHILL, J. F., WILEY, B. B. & WARD, J. R. (1969). Immunological responses of the rat to *Mycoplasma arthritidis*. *Journal of Bacteriology*, **98**, 930–7.

COLE, B. C., GOLIGHTLY-ROWLAND, L., WARD, J. R. & WILEY, B. B. (1970). Immunological response of rodents to murine mycoplasmas. *Infection and Immunity*, **2**, 419–25.

COLE, B. C., WARD, J. R. & MARTIN, C. H. (1968). Hemolysin and peroxide activity of *Mycoplasma* species. *Journal of Bacteriology*, **95**, 2022–30.

COLLIER, A. M. & CLYDE, W. A., JR. (1969). *Mycoplasma pneumoniae* disease pathogenesis studied in tracheal organ culture. *Federation Proceedings*, **28**, 616.

COLLIER, A. M. & CLYDE, W. A., JR. (1971). Relationships between *Mycoplasma pneumoniae* and human respiratory epithelium. *Infection and Immunity*, **3**, 694–701.

COLLIER, A. M., CLYDE, W. A., JR. & DENNY, F. W. (1969). Biologic effects of *Mycoplasma pneumoniae* and other mycoplasmas from man on hamster tracheal organ culture. *Proceedings of the Society for Experimental Biology and Medicine, N. Y.*, **132**, 1153–8.

COLLIER, A. M., CLYDE, W. A., JR. & DENNY, F. (1971). *Mycoplasma pneumoniae* in hamster tracheal organ culture: immunofluorescent and electron microscopic studies. *Proceedings of the Society for Experimental Biology and Medicine, N. Y.*, **136**, 569–573.

COPPERMAN, R. & MORTON, H. E. (1966). Reversible inhibition of mitosis in lymphocyte cultures of non-viable mycoplasma. *Proceedings of the Society for Experimental Biology and Medicine, N. Y.*, **123**, 790–5.

CORDY, D. R. & ADLER, H. E. (1957). The pathogenesis of the encephalitis in turkey poults produced by a neurotropic pleuropneumonia-like organism. *Avian Diseases*, **1**, 235–45.

COUCH, R. B., CATE, T. R. & CHANOCK, R. M. (1964). Infection with artifically propagated Eaton agent (*Mycoplasma pneumoniae*). Implications for development of attenuated vaccine for cold agglutinin-positive pneumonia. *Journal of the American Medical Association*, **187**, 442–7.

DAJANI, A. S., CLYDE, W. A., JR. & DENNY, F. W. (1965). Experimental infection with *Mycoplasma pneumoniae* (Eaton's agent). *Journal of Experimental Medicine*, **121**, 1071–86.

DAUBNEY, R. (1935). Contagious bovine pleuro-pneumonia. Note on experimental reproduction and infection by contact. *Journal of Comparative Pathology and Therapeutics*, **48**, 83–96.

DAVIES, G. (1969). The examination of nasal mucus for antibodies to *Mycoplasma mycoides*. *Veterinary Record*, **84**, 417–8.

DAVIES, G. & HUDSON, J. R. (1968). The relationship between growth inhibition and immunity in contagious bovine pleuropneumonia. *Veterinary Record*, **83**, 256–8.

DRUETT, H. A. (1967). The inhalation and retention of particles in the human respiratory system. In *Airborne Microbes*, ed. P. H. Gregory and J. L. Monteith, 165–202. London: Cambridge University Press.

EATON, M. D., MEIKLEJOHN, G. & VAN HERICK, W. (1944). Studies on the etiology of primary atypical pneumonia. A filterable agent transmissible to cotton rats, hamsters, and chick embryos. *Journal of Experimental Medicine*, **79**, 649–68.

EDWARD, D. G. ff. (1954). The pleuropneumonia group of organisms: A review together with some new observations. *Journal of General Microbiology*, **10**, 27–64.

FABRICANT, J. (1969). Avian mycoplasmas. In *The Mycoplasmatales and the L-phase of Bacteria*, ed. L. Hayflick, 621–41. Amsterdam: North Holland Publishing Company.

FALLON, R. J. & WHITTLESTONE, P. (1969). Isolation, cultivation and maintenance of mycoplasmas. In *Methods in Microbiology*, ed. J. R. Norris and D. W. Ribbons, vol. 3B, 211–67. London and New York: Academic Press.

FERNALD, G. W. (1969). Immunologic aspects of experimental *Mycoplasma pneumoniae* infection. *Journal of Infectious Diseases*, **119**, 255–66.

FIEZI, T. (1967). Cold agglutinins, the direct Coomb's test and serum immuno-globins in *Mycoplasma pneumoniae* infection. *Annals of the New York Academy of Sciences*, **143**, 801–12.

FINDLAY, G. M., KLIENEBERGER, E., MacCALLUM, F. O. & MACKENZIE, R. D. (1938). Rolling disease. New syndrome in mice associated with a pleuro-pneumonia-like organism. *The Lancet*, **235**, 1511–3.

FOGH, J. & FOGH, H. (1965). Chromosome changes in PPLO-infected FL human amnion cells. *Proceedings of the Society of Experimental Biology and Medicine, N. Y.*, **119**, 233–8.

FOGH, J. & FOGH, H. (1967). Irreversibility of major chromosome changes in a mycoplasma-modified line of FL human amnion cells. *Proceedings of the Society for Experimental Biology and Medicine, N. Y.*, **26**, 67–74.

GAMBLES, R. M. (1956). Studies on contagious bovine pleuropneumonia. With special reference to the complement-fixation test. *British Veterinary Journal*, **112**, 34–40, 78–86, 120–7 and 162–9.

GERLACH, F. & HEIKKILA, I. (1956). Passage du virus de la peripneumonie conta-gieuse des bovins, chez la souris blanche. *Bulletin de L'Office International des Epizooties*, **46**, 404–19.

GESNER, B. & THOMAS, L. (1965). Sialic acid binding sites: role in hemagglutination by *Mycoplasma gallisepticum*. *Science*, **151**, 590–1.

GOODWIN, R. F. W., HODGSON, R. G., WHITTLESTONE, P. & WOODHAMS, R. 1969a). Immunity in experimentally induced enzootic pneumonia of pigs. *Journal of Hygiene*, **67**, 193–208.

246 P. WHITTLESTONE

GOODWIN, R. F. W., HODGSON, R. G., WHITTLESTONE, P. & WOODHAMS, R. (1969b). Some experiments relating to artificial immunity in enzootic pneumonia of pigs. *Journal of Hygiene*, **67**, 465–76.

GOURLAY, R. N. (1965). Antigenicity of *Mycoplasma mycoides* II. Further studies on the precipitating antigens in the body fluids from cases of contagious bovine pleuropneumonia. *Research in Veterinary Science*, **6**, 1–8.

GOURLAY, R. N. (1970). Isolation of a virus infecting a strain of *Mycoplasma laidlawii*. *Nature, London*, **225**, 1165.

GOURLAY, R. N., BRUCE, J. & GARWES, D. J. (1971). Characterization of myco-plasmatales virus *laidlawii I*. *Nature New Biology*, **229**, 118–19.

GOURLAY, R. N. & SHIFRINE, M. (1966). Antigenic cross-reactions between the galactan from *Mycoplasma mycoides* and polysaccharides from other sources. *Journal of Comparative Pathology*, **76**, 417–25.

GOURLAY, R. N. & THROWER, K. J. (1968). Morphology of *Mycoplasma mycoides* thread-phase growth. *Journal of General Microbiology*, **54**, 155–9.

HARNETT, G. B. & HOOPER, W. L. (1968). Test-tube organ cultures of ciliated epithelium for the isolation of respiratory viruses. *The Lancet*, **i**, 339–40.

HAYFLICK, L. & STINEBRING, W. R. (1960). Intracellular growth of pleuropneu-monia-like organisms (PPLO) in tissue culture and *in ovo*. *Annals of the New York Academy of Sciences*, **79**, 433–49.

HUDSON, J. R. (1965). Contagious bovine pleuropneumonia: The immunizing value of the attenuated strain KH₃J. *Australian Veterinary Journal*, **41**, 43–9.

HUDSON, J. R., BUTTERY, S. & COTTEW, G. S. (1967). Investigations into the influence of the galactan of *Mycoplasma mycoides* on experimental infection with that organism. *Journal of Pathology and Bacteriology*, **94**, 257–73.

HUDSON, J. R. & TURNER, A. W. (1963). Contagious bovine pleuropneumonia: a comparison of the efficacy of two types of vaccine. *Australian Veterinary Journal*, **39**, 373–85.

HYSLOP, N. ST G. (1955). Review of progress on contagious bovine pleuropneu-monia in Kenya. *Bulletin of Epizootic Diseases in Africa*, **3**, 266–70.

HYSLOP, N. ST G. (1958). The adaptation of *Asterococcus mycoides* to rodents. *Journal of Pathology and Bacteriology*, **75**, 189–99.

HYSLOP, N. ST G. (1963). Experimental infection with *Mycoplasma mycoides*. *Journal of Comparative Pathology*, **73**, 265–76.

JONES, T. C. & HIRSCH, J. G. (1971). The interaction in vitro of *Mycoplasma pulmonis* with mouse peritoneal macrophages and L-cells. *Journal of Experimental Medicine*, **133**, 231–59.

JUNGHERR, E. (1949). The pathology of experimental sinusitis of turkeys. *American Journal of Veterinary Research*, **10**, 372–83.

KAKLAMANIS, E. & THOMAS, L. (1970). The toxins of Mycoplasma. In *Microbial Toxins*, ed. T. C. Montie, S. Kadis and S. J. Ajl, vol. 3, 493–505. New York and London: Academic Press.

KENNY, G. E. & POLLOCK, M. E. (1963). Mammalian cell cultures contaminated with pleuropneumonia-like organisms. *Journal of Infectious Diseases*, **112**, 7–16.

KLIENEBERGER, E. (1940). The pleuropneumonia-like organisms: further compara-tive studies and a descriptive account of recently discovered types. *Journal of Hygiene*, **40**, 204–22.

KLIENEBERGER-NOBEL, E. (1954). Micro-organisms of the pleuropneumonia group. *Biological Reviews of the Cambridge Philosophical Society*, **29**, 154–84.

KLEINBERGER-NOBEL, E. & CHENG, K. K. (1955). On the association of the pleuropneumonia-like L3 organism with experimentally produced bronchiectasis in rats. *Journal of Pathology and Bacteriology*, **70**, 245–6.

KRAEMER, P. M. (1964). Interaction of mycoplasma (PPLO) and murine lymphoma cell cultures: prevention of lysis by arginine. *Proceedings of the Society for Experimental Biology and Medicine, N. Y.*, **115**, 206–12.

KUROTCHKIN, T. J. (1937). Specific carbohydrate from *Asterococcus mycoides* for serologic tests of bovine pleuropneumonia. *Proceedings of the Society for Experimental Biology and Medicine, N. Y.*, **37**, 21–2.

LEMCKE, R. M. (1961). Association of PPLO infection and antibody response in rats and mice. *Journal of Hygiene*, **59**, 401–12.

LINDLEY, E. P. (1959). An investigation into the virulence for mice of certain strains of *Asterococcus mycoides*. *Bulletin of Epizootic Diseases in Africa*, **7**, 235–42.

LINDLEY, E. P. & ABDULLA, A. E. D. (1967). The development and titration in cattle of a dried contagious bovine pleuropneumonia vaccine. *Sudan Journal of Veterinary Science and Animal Husbandry*, **8**, 40–6.

LIPMAN, R. P. & CLYDE, W. A., JR. (1969). The interrelationship of virulence, cytadsorption, and peroxide formation in *Mycoplasma pneumoniae*. *Proceedings of the Society for Experimental Biology and Medicine, N. Y.*, **131**, 1163–7.

LIPMAN, R. P., CLYDE, W. A., JR. & DENNY, F. W. (1969). Characteristics of virulent, attenuated and avirulent *Mycoplasma pneumoniae* strains. *Journal of Bacteriology*, **100**, 1037–43.

LLOYD, L. C. (1966). Tissue necrosis produced by *Mycoplasma mycoides* in intraperitoneal diffusion chambers. *Journal of Pathology and Bacteriology*, **92**, 225–9.

LLOYD, L. C. (1967). An attempt to transfer immunity to *Mycoplasma mycoides* infection with serum. *Bulletin of the Epizootic Diseases in Africa*, **15**, 11–17.

LLOYD, L. C. (1970). Subcutaneous lesions induced in rabbits by *Mycoplasma mycoides*. *Journal of Comparative Pathology*, **80**, 195–209.

LLOYD, L. C. & TRETHEWIE, E. R. (1970). Contagious bovine pleuropneumonia. In *The Role of Mycoplasmas and L-forms of Bacteria in Disease*, ed. J. T. Sharp, 172–97. Springfield, Illinois: Charles C. Thomas.

MACPHERSON, I. & RUSSELL, W. (1966). Transformations in hamster cells mediated by mycoplasmas. *Nature, London*, **210**, 1343–5.

MANCHEE, R. J. & TAYLOR-ROBINSON, D. (1969). Studies on the nature of receptors involved in attachment of tissue culture cells to mycoplasmas. *British Journal of Experimental Pathology*, **50**, 66–75.

METTAM, R. W. M. & FORD, J. (1939). Experiments on the transmission of bovine contagious pleuro-pneumonia with a report on a new method of testing immunity following vaccination. *Journal of Comparative Pathology and Therapeutics*, **52**, 15–28.

MOOSER, H. (1949). Die Mobilisation von Musculomyces durch das Virus der Ektromelie. *Experientia*, **5**, 364.

MORTON, H. E. (1970). Mycoplasmas from man with undetermined specific relationships to their human host. In *The Role of Mycoplasmas and L-forms of Bacteria in Disease*, ed. J. T. Sharp, 147–71. Springfield, Illinois: Charles C. Thomas.

MOROWITZ, H. J., BODE, H. R. & KIRK, R. G. (1967). The nucleic acids of mycoplasma. *Annals of the New York Academy of Sciences*, **143**, 110–14.

NELSON, J. B. (1957). The enhancing effect of murine hepatitis on the cerebral activity of pleuropneumonia-like organisms in mice. *Journal of Experimental Medicine*, **106**, 179–89.

NOCARD, ROUX, BORREL, SALIMBENI & DUJARDIN-BEAUMETZ. (1898). Le microbe de la peripneumonie. *Annales de L'Institut Pasteur*, **12**, 240–62.

OLITZKI, L. (1948). Mucin as a resistance-lowering substance. *Bacteriological Reviews*, **12**, 149–72.

ORGANICK, A. B., SIEGESMUND, K. A. & LUTSKY, I. I. (1966). Pneumonia due to mycoplasma in gnotobiotic mice. II. Localization of *Mycoplasma pulmonis* in the lungs of infected gnotobiotic mice by electron microscopy. *Journal of Bacteriology*, **92**, 1164–76.

PATON, G. R., JACOBS, J. P. & PERKINS, F. T. (1965). Chromosome changes in human diploid-cell cultures infected with *Mycoplasma*. *Nature, London*, **207**, 43–5.

PLACKETT, P. & BUTTERY, S. H. (1958). A galactan from *Mycoplasma mycoides*. *Nature, London*, **182**, 1236–7.

PLACKETT, P., BUTTERY, S. H. & COTTEW, G. S. (1963). Carbohydrates of some mycoplasma strains. In *Recent Progress in Microbiology*, Symposia held at the VIIIth International Congress for Microbiology, Montreal 1962, ed. N. E. Gibbons, 535–47. Toronto: University of Toronto Press.

PURCHASE, H. S. (1939). Vaccination against contagious bovine pleuropneumonia with 'culture-virus'. *Veterinary Record*, **51**, 31–47, 67–75.

RAZIN, S. (1969). Structure and function in mycoplasma. *Annual Review of Microbiology*, **23**, 317–56.

ROBERTS, D. H. & LITTLE, T. W. A. (1970). An auto-immune response associated with porcine enzootic pneumonia. *Veterinary Record*, **86**, 328.

RODWELL, A. (1970). Nutrition and metabolism of the mycoplasmas. In *The Mycoplasmatales and the L-Phase of Bacteria*, ed. L. Hayflick, 413–49. Amsterdam: North Holland Publishing Company.

ROSS, J. G. (1962*a*). *Mycoplasma mycoides* var. *mycoides* in mice. Course of the lesion, and the primary and secondary serological response in an inbred line of mice. *Journal of Comparative Pathology*, **72**, 1–10.

ROSS, J. G. (1962*b*). *Mycoplasma mycoides* var. *mycoides* in mice. A comparison of lesion and serological response in five inbred lines. *Journal of Comparative Pathology*, **72**, 332–6.

RUSSELL, W. C. (1966). Alterations in the nucleic acid metabolism of tissue culture cells infected by mycoplasmas. *Nature, London*, **212**, 1537–40.

SABIN, A. B. (1938*a*). Isolation of a filtrable, transmissible agent with 'neurolytic' properties from toxoplasma-infected tissues. *Science*, **88**, 189–91.

SABIN, A. B. (1938*b*). Identification of the filtrable, transmissible neurolytic agent isolated from toxoplasma-infected tissue as a new pleuropneumonia-like microbe. *Science*, **88**, 575–6.

SABIN, A. B. (1941). The filterable microörganisms of the pleuropneumonia group. *Bacteriological Reviews*, **5**, 1–67.

SHEPARD, M. C. (1958). Growth and development of T strain pleuropneumonia-like organisms in human epidermoid carcinoma cells (HeLa). *Journal of Bacteriology*, **75**, 351–5.

SHERIFF, D. (1952). *Annual report of the Department of Veterinary Services, Kenya*, cited by Hyslop, 1958.

SHIFRINE, M. & GOURLAY, R. N. (1965). Serological relationship between galactans from normal bovine lung and from *Mycoplasma mycoides*. *Nature, London*, **208**, 498–9.

SMITH, G. R. (1965). Infection of small laboratory animals with *Mycoplasma mycoides* var. *capri* and *Mycoplasma mycoides* var. *mycoides*. *Veterinary Record*, **77**, 1527–8.

SMITH, G. R. (1967*a*). Experimental infection of mice with *Mycoplasma mycoides* var. *capri*. *Journal of Comparative Pathology*, **77**, 21–7.

SMITH, G. R. (1967b). Experimental infection of mice with *Mycoplasma agalactiae*. *Journal of Comparative Pathology*, **77**, 199–202.

SMITH, G. R. (1967c). *Mycoplasma mycoides* var. *mycoides*: production of bacteriaemia and demonstration of passive immunity in mice. *Journal of Comparative Pathology*, **77**, 203–9.

SMITH, G. R. (1968). Factors affecting bacteriaemia in mice inoculated with *Mycoplasma mycoides* var. *mycoides*. *Journal of Comparative Pathology*, **78**, 267–74.

SMITH, G. R. (1969). A study of *Mycoplasma mycoides* var. *mycoides* and *Mycoplasma mycoides* var. *capri* by cross-protection tests in mice. *Journal of Comparative Pathology*, **79**, 261–5.

SMITH, G. R. (1971). Experimental mycoplasma infections in unnatural laboratory animal hosts. *Proceedings of the Royal Society of Medicine*, **64**, 36.

SMITH, C. B., CHANOCK, R. M., FRIEDEWALD, W. T. & ALFORD, R. H. (1967). *Mycoplasma pneumoniae* infections in volunteers. *Annals of the New York Academy of Sciences*, **143**, 471–83.

SMITH, C. B., McGINNISS, M. H. & SCHMIDT, P. J. (1967). Changes in erythrocyte I agglutinogen and anti-I agglutinins during *Mycoplasma pneumoniae* infection in man. *Journal of Immunology*, **99**, 333–9.

SOBESLAVSKY, O. & CHANOCK, R. M. (1968). Peroxide formation by Mycoplasmas which infect man. *Proceedings of the Society for Experimental Biology and Medicine, N. Y.*, **129**, 531–5.

SOBESLAVSKY, O., PRESCOTT, B. & CHANOCK, R. M. (1968). Adsorption of *Mycoplasma pneumoniae* to neuraminic acid receptors of various cells and possible role in virulence. *Journal of Bacteriology*, **96**, 695–705.

SOMERSON, N. L., TAYLOR-ROBINSON, D. & CHANOCK, R. M. (1963). Hemolysin production as an aid in the identification and quantitation of Eaton agent (*Mycoplasma pneumoniae*). *American Journal of Hygiene*, **77**, 122–8.

SOMERSON, N. L., WALLS, B. E. & CHANOCK, R. M. (1965). Hemolysin of *Mycoplasma pneumoniae*: tentative identification as a peroxide. *Science*, **150**, 226–8.

STEINBERG, P., HORSWOOD, R. L. & CHANOCK, R. M. (1969). Temperature-sensitive mutants of *Mycoplasma pneumoniae* I. *In vitro* biologic properties. *Journal of Infectious Diseases*, **120**, 217–24.

TANG, F. F., WEI, H., McWHIRTER, D. L. & EDGAR, J. (1935). An investigation of the causal agent of bovine pleuropneumonia. *Journal of Pathology and Bacteriology*, **40**, 391–406.

TAYLOR-ROBINSON, D. (1971). Mycoplasmas and the evidence for their pathogenicity in man. *Proceedings of the Royal Society of Medicine*, **64**, 31–3.

TAYLOR-ROBINSON, D. & BERRY, D. M. (1969). The evaluation of the metabolic-inhibition technique for the study of *Mycoplasma gallisepticum*. *Journal of General Microbiology*, **55**, 127–37.

THOMAS, L. (1967). The neurotoxins of *M. neurolyticum* and *M. gallisepticum*. *Annals of the New York Academy of Sciences*, **143**, 218–24.

THOMAS, L. (1969). Mechanisms of pathogenesis in mycoplasma infection. In *Harvey Lectures*, Series 63, 73–98. London and New York: Academic Press.

THOMAS, L. (1970a). The toxic properties of *M. neurolyticum* and *M. gallisepticum*. In *The Role of Mycoplasmas and L-forms of Bacteria in Disease*, ed. J. T. Sharp, 104–8. Springfield, Illinois: Charles C. Thomas.

THOMAS, L. (1970b), Mycoplasmas as infectious agents. *Annual Reviews of Medicine*, **21**, 179–86.

THOMAS, L., ALEU, F., BITENSKY, M. W., DAVIDSON, M. & GESNER, B. (1966). Studies of PPLO infection. II The neurotoxin of *Mycoplasma neurolyticum*. *Journal of Experimental Medicine*, **124**, 1067–82.

THOMAS, L. & BITENSKY, M. W. (1966a). Methaemoglobin formation by *Mycoplasma gallisepticum*: the role of hydrogen peroxide. *Nature, London,* **210,** 963–4.

THOMAS, L. & BITENSKY, M. W. (1966b). Studies of PPLO infection. IV The neurotoxicity of intact mycoplasmas, and their production of toxin *in vivo* and *in vitro. Journal of Experimental Medicine,* **124,** 1089–98.

THOMAS, L., CURNEN, E. C., MIRICK, G. S., ZIEGLER, J. E. JR. & HORSFALL, F. L. JR. (1943). Complement fixation with dissimilar antigens in primary atypical pneumonia. *Proceedings of the Society for Experimental Biology and Medicine, N. Y.,* **52,** 121–5.

THOMAS, L., DAVIDSON, M. & McCLUSKEY, R. T. (1966). Studies of PPLO infection. I The production of cerebral polyarteritis by *Mycoplasma gallisepticum* in turkeys; the neurotoxic property of the mycoplasma. *Journal of Experimental Medicine,* **123,** 897–912.

TULLY, J. G. (1964). Production and biological characteristics of an extracellular neurotoxin from *Mycoplasma neurolyticum. Journal of Bacteriology,* **88,** 381–8.

TULLY, J. G. (1969). Murine mycoplasmas. In *The Mycoplasmatales and the L-phase of Bacteria,* ed. L. Hayflick, 571–605. Amsterdam: North-Holland Publishing Company.

TULLY, J. G. & RASK-NIELSEN, R. (1967). Mycoplasma in leukemic and non-leukemic mice. *Annals of the New York Academy of Sciences,* **143,** 345–52.

TULLY, J. G. & RUCHMAN, I. (1964). Recovery, identification and neurotoxicity of Sabin's type A and C mouse mycoplasma (PPLO) from lyophilized cultures. *Proceedings of the Society for Experimental Biology and Medicine, N. Y.,* **115,** 554–8.

TURNER, A. W. (1961). Preventive tail-tip inoculation of calves against bovine contagious pleuropneumonia. III Immunity in relationship to age at inoculation. *Australian Veterinary Journal,* **37,** 259–64.

TURNER, A. W., CAMPBELL, A. D. & DICK, A. T. (1935). Recent work on pleuropneumonia contagiosa boum in North Queensland. *Australian Veterinary Journal,* **11,** 63–71.

TUTTLE, W. M. (1933). A new method for production of experimental abscesses of the lung in dogs. *Proceedings of the Society for Experimental Biology and Medicine, N. Y.,* **30,** 462–3.

VILLEMOT, J. M., PROVOST, A. & QUEVAL, R. (1962). Endotoxin of *Mycoplasma mycoides, Nature, London,* **193,** 906–7.

WARD, J. R. & COLE, B. C. (1970). Mycoplasmal infections of laboratory animals. In *The role of Mycoplasmas and L-forms of Bacteria in Disease,* ed. J. T. Sharpe, 212–39. Springfield, Illinois: Charles C. Thomas.

WHITTLESTONE, P. (1958). *Enzootic pneumonia of pigs and related conditions.* Dissertation for Ph.D. degree, University of Cambridge.

WINDSOR, R. S. (1971). Contagious bovine pleuropneumonia. *State Veterinary Journal,* **26,** 103–10.

ZUCKER-FRANKLIN, D., DAVIDSON, M., GESNER, B. & THOMAS, L. (1965). Phagocytosis of PPLO by mononuclear cells. *Journal of Clinical Investigation,* **44,** 1115–16.

ZUCKER-FRANKLIN, D., DAVIDSON, M. & THOMAS, L. (1966a). The interaction of mycoplasmas with mammalian cells. I. HeLa cells, neutrophils, and eosinophils. *Journal of Experimental Medicine,* **124,** 521–32.

ZUCKER-FRANKLIN, D., DAVIDSON, M. & THOMAS, L. (1966b). The interaction of mycoplasmas with mammalian cells. II Monocytes and lymphocytes. *Journal of Experimental Medicine,* **124,** 533–42.

THE PATHOGENICITY OF FUNGI

P. K. C. AUSTWICK

Nuffield Institute of Comparative Medicine, The Zoological Society of London, Regent's Park, London, NW 1

INTRODUCTION

In defining pathogenicity as 'the ability of a micro-organism to cause disease' the mycologist is forced to take a much wider view of animal disease than any of his microbiological counterparts, for the fungi range in their pathogenic activities from primary tissue invasion in contagious infections to the nutritional degradation of foodstuffs resulting in deficiency disease, adding for good measure the high toxicity and unparalleled carcinogenicity of their metabolites. In consequence the criteria of bacterial pathogenicity defined by Robert Koch and later applied in virology have proved inadequate and one must turn to plant pathology for guidance. This was the one branch of biological science concerned with disease which used the ideas of microbial infections built up in the first half of the last century by the careful observations of such masters as Bassi (1835), Berkeley (1845) and Gruby (1841), who, working on muscardine disease of silkworms, on potato blight and on human favus respectively, laid the foundation for our knowledge of fungal pathogenicity in plants and animals.

Any consideration of the role of fungi as animal pathogens inevitably invites comparison with the performance of those bacteria and viruses which similarly cause disease, and this is perhaps unfortunate because of the vast differences in parasitic and saprophytic behaviour within the three groups. In addition one must also emphasise the fundamental difference between disease resistance in plants and that in animals, for although both have well-defined defence mechanisms to deal with pathogens, this defence is achieved by rather different methods although using the same underlying principles. These are, first the restriction of entry and later the rendering of the fungal progress through the tissues as slow as possible, finally, hopefully disposing of the invader, be it organism or metabolite (Wheeler, 1968).

Plant pathology developed a broader outlook on the occurrence of pathogens in nature. Important here was the acceptance of saprophytic activity as an essential part of many disease cycles in which provision is made for a vital increase in the volume of fungal thallus which in turn produces the propagules. Such a concept was necessarily lacking in contagious bacterial and viral diseases and at first in many of the fungal infections of man and animals, and it has been a hard-won struggle to be able to state it as one of the fundamental principles of mycotic disease. Too often even today a fungus growing from an infected tissue is dismissed contemptuously as a 'mould contaminant'.

There are then two aspects of fungal pathogenicity, the involvement with the animal and the existence away from it. One starts with the entry of the organism or its metabolites into the animal body, and the other with the fungus finding a favourable saprophytic substrate for growth; an attempt has been made to distinguish between the sites of these two aspects of fungal activity by using the terms 'internal' and 'external' environment in relation to the animal body (Austwick, 1963). In doing this it should be understood that fungal pathogenicity in animals is almost always a side-line to the normal saprophytic activity of the fungus, arising by chance but probably in a few cases by the evolution of specialised systems of parasitism, e.g. the invasion of hair by the ringworm fungi and perhaps the endosporulation of *Coccidioides immitis*. Ainsworth (1952) has suggested that many mycoses are 'incidental' diseases for they only play such a role in the life cycles of normally saprophytic pathogens.

Fungi cause three main types of animal and human disease:

Mycoses; diseases resulting from the invasion of living tissues by the fungus;

Allergies; diseases resulting from the development of hypersensitivity to fungal antigens;

Toxicoses; consisting of the *mycotoxicoses* resulting from the ingestion of toxic fungal metabolites formed in food and the *mycetisms* resulting from the ingestion of toxic fungal fruiting bodies.

There are also three indirect ways in which fungi can be involved in pathogenic processes in animals, viz. by inducing the formation of toxic substances (including oestrogens) by a host plant which is then eaten by an animal; by so degrading a foodstuff that macro-nutrient deficiencies appear in animals feeding on it; and by forming substances such as enzymes, which although not acting directly on metabolic processes themselves affect the availability of growth substances to the animal, e.g. vitamins, which may lead to micro-nutrient deficiency.

Fungal pathogenicity thus covers a much wider range of activity than that seen in other pathogenic micro-organisms, and its height of complexity may be realised when it is possible for all three types of disease to be present in one individual animal, each caused by the same species of fungus playing three different roles at one and the same time. After all, infection and allergy together due to *Aspergillus fumigatus* are not uncommon in man.

Further complications arise in the existence of two types of fungal thallus morphology *in vivo*. Firstly apical hyphal growth leads to spherical or disc-like colonies whose spread is by peripheral extension and by fragmentation, a process compensating for its slowness of dissemination by the provision of relatively large inocula for its extension, and so of more certainty of success. Secondly, the self-replication of single cells either as yeast cells or endosporulating spherules can more readily play a leading role in the dissemination of infection in the body, but for this the fungus is generally dependent upon phagocytosis. It may be said that, in the first instance, the fungus is essentially active, making its own way through the tissues and, in the second, passive and dependent on the blood stream or blood cells for transport. The latter type of growth is also directly comparable to that in many bacterial and viral infections, being frequently accompanied by acute pyrexial symptoms, whilst the former is generally associated with a chronic inflammatory process extending sometimes over many years.

Failure to understand this fundamental difference in the behaviour *in vivo* of fungi has led to much confusion in the classification of fungal disease, and it still remains a major problem when trying to generalise on the pathogenesis of the mycoses. The ability of fungi to adopt a wide range of vegetative morphology according to their environment is being increasingly observed and a number of filamentous saprophytic fungi are now known to be capable of budding as yeast cells under defined experimental conditions, e.g. *Mucor rouxii* and *Aspergillus parasiticus* (Bartnicki-Garcia & Nickerson, 1962; Detroy & Ciegler, 1971). The possession of hyphae or of a unicellular state therefore bears no relationship to the classification or to the potential pathogenicity of the species.

Because of the ubiquitous nature of many pathogenic fungi, e.g. *Aspergillus fumigatus*, and the local abundance of others, e.g. *Histoplasma capsulatum*, the chances of exposure to an infecting propagule are very much higher than the reported occurrence of the diseases would indicate. Both a high degree of resistance in the animal population or a low degree of pathogenicity in the fungi could have this effect, which is generally summarised under the heading of host susceptibility. A great variety of factors determine this, age probably being the main one in animals and in the primary mycoses as a whole, and in man the state of health being the most significant factor in view of the increasing numbers of cases of iatrogenic mycoses (where fungal infection is imposed on a host already made susceptible by an underlying metabolic, neoplastic or infectious disease). There is also some evidence that host genetics may play a role.

A last factor to be considered in fungal pathogenicity is virulence, a much-vaunted, but nonetheless vital part of other microbial disease systems which, however, seems to play a lower role in fungal infections. The properties of enhancement and of attenuation of viral and bacterial pathogens so frequently exploited in the prophylaxis of their respective diseases do not seem to be so amenable in the fungi, despite the fact that a solid immunity may be conferred on recovery in most of the mycoses. However, if one can truly suggest that variation in toxin production is comparable or even related to variation in virulence then it immediately assumes a far greater significance, and it remains to be seen how this feature can be fitted into the general concept of microbial pathogenicity.

In this account, fungal pathogenicity is looked at from the viewpoint of the fungal propagule or metabolite and its subsequent fate, but with so many different types of potential fungal pathogen it has not been possible to cite all the interesting variants in behaviours which are known. The chief aim is to provide a background for comparing pathogenicity in other organisms.

THE INTERNAL AND EXTERNAL ENVIRONMENTS

The events in the pathogenesis of fungal disease occur strictly in the internal environment, but they connect at their commencement and termination to other events which occur in the external environment. The form and amount of the fungus or its products available to the animal are determined outside the body and similarly the form and amount of the fungus which returns to the external environment are determined by the outcome of the pathogenic activity. Fig. 1 expresses some of this activity and provides for an understanding of the part played by the external environment in the disease. As each event is discussed reference will be made to the various known factors at work in determining the direction that the cycle is taking in a particular disease.

The main distinction made is in the separation of the contagious mycoses from the others because their main cycle is that of direct transmission from affected to unaffected animal by fomites. Ten years ago this might have been all one could say about these diseases, but since then all but a few species of dermatophytes have been isolated from soil or rotting material of animal origin (e.g. hair and wool) and even *Candida albicans* has been isolated from soil and grass leaves. In this group of diseases it would seem that the external environment plays a much greater role in both furnishing infection from soil inhabitants e.g. the *Microsporum gypseum* agg., and in providing shelter and succour to fungal cells sloughing from lesions. This second part of the external cycle in the contagious mycoses therefore has much more significance than the same part of the cycle in the non-contagious diseases, because in these the return of the fungus to the soil is highly speculative and probably plays no role in natural infections.

The first requirement for a fungus to fulfil a pathogenic role is access to the animal body either at its skin surface or on the mucosa at one of its orifices. This is represented in the focal point of Fig. 1 by the propagules arranged strategically and in correct size and number, i.e. the right inoculum potential, to proceed to the next stage of deposition on or in the operative site. Following this, events vary with the pathogenic activity of the fungus. Antigens diffuse from the allergenic particles into the underlying tissue and elicit an immediate or delayed response (Type I or III); the toxin is absorbed by the gut wall; and the infecting propagule germinates. Thereafter the allergen may no longer act and is carried away by the respiratory cilia or removed from its site of deposition by phagocytosis whilst the toxin is translocated to its site of action e.g. the liver.

From here onwards the fungal allergens and toxins are thus excluded, and these remarks become restricted to the development of infection. This proceeds by the penetration of the host tissues and cells by the fungus to the consolidation of the infection by vegetative growth and then to either the death of the host or to the sequestration of the pathogen and its eventual absorption. In ringworm the affected tissue is simply pushed away from the surface by the 'epidermal effluvial current', scab, hair, fungus and all.

ACCESS AND DEPOSITION

The means by which a pathogenic fungus reaches its host is not the concern of this symposium but it does have some bearing on pathogenicity in that the medium in which the propagule is borne may determine its point of arrival. Air, food, soil, plants, other animals and water, may all be involved in this transport. The quality, i.e. the species of fungi, and the quantity, i.e. the amount of inoculum, are the foremost factors in the mycoses and allergies, paralleled in the toxicoses by the amount of the toxic substance available. In determining the site of deposition of a propagule, other factors, such as its size and wettability, may influence its ability to survive long enough to have a pathogenic effect.

The unbroken skin probably presents as inhospitable a surface to the aspiring dermatophyte as does that of the moon despite its pores, follicle mouths, ridges and scales, and unless some rapid means of deep attachment takes place, e.g. a vigorous rubbing in, the relentless sloughing of the skin carries the propagule away. (Perhaps certain aquatic fungi with motile spores can invade the intact skin – although *Saprolegnia* spp. require prior damage – but in the Actinomycetales, *Dermatophilus congolensis*, the causal organism of streptotrichosis, has zoospores which are certainly able to penetrate moist undamaged skin to set up infection.)

The abraded skin provides a very different welcome, with ample opportunity for arthrospores (2–8 μm diam.) to lodge beneath epidermal scales, in follicle necks or beneath congealed blood and serum on superficial wounds. It has been calculated that a hair base on a calf infected with *Trichophyton verrucosum* can carry at least 30000 arthrospores per mm and if placed on the skin of an unaffected animal requires only the slightest rub of a finger-tip to initiate infection. The evidence for contact allergy by fungi is not great but it seems that in certain circumstances it can develop, and contact dermatitis from fungal toxins applied to the skin either deliberately or accidentally is known, e.g. with verrucarin from *Myrothecium verrucaria* (Brian, Hemming & Jefferys, 1948).

Infection of the subcutaneous tissues requires deeper wounding, as for example by the plant thorns shown by Basset, Camain, Baylet & Lambert (1965) to carry *Leptosphaeria senegalensis* into the foot to initiate a mycetoma and these structures are also assumed to carry conidia of *Sporothrix schenckii* into wounds by Mackinnon (1949). The yeast phase of the latter fungus has also been transmitted by rat bite. The remaining skin structures and orifices also receive fungal spores but few reports exist of any pathogenic effect.

Inhalation is probably the greatest factor in enabling fungal disease to become established. Histoplasmosis, coccidioidomycosis, adiaspiromycosis, aspergillosis and cryptococcosis, and North and South American blastomycosis are all considered to be primary pulmonary mycoses whilst allergic asthma, hayfever and farmers' lung also follow this event. All manner of fungal propagules enter the nostrils to be filtered out of the respired air by the series of natural cascade impactors, sedimentation tanks, electrostatic precipitators and microscale spore-papers (working on the principle of Brownian movement) which ensure that nothing over 10 μm penetrates the nose, but that almost everything under 2 μm reaches the alveoli and is retained. The situation was

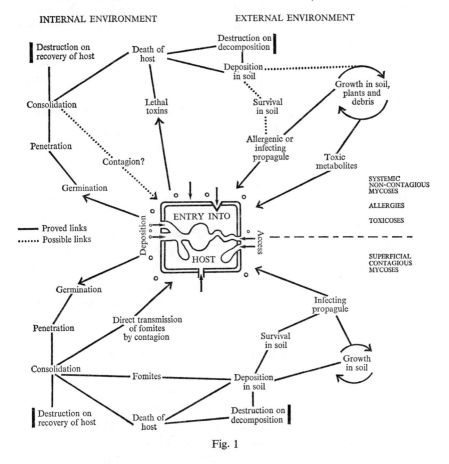

Fig. 1

reviewed by Austwick (1966). Briefly, large particles impact in the upper respiratory tract and can be retained on the mucosa long enough for their antigens to diffuse out and react with the reagenic antibodies to mediate a Type I immediate reaction in sensitised animals, with the typical symptoms of rhinitis. Removal by ciliary action is rapid and the offending particles are expectorated or swallowed.

As the most peripheral parts of the pulmonary tree are beyond the reach of the bronchiolar cilia, the minute (0·5–2 μm diam.) spores of the thermophilic actinomycetes which cause farmers' lung deposited here together with certain fungal spores, e.g. those of *Aspergillus clavatus* (which are capable of causing the same syndrome (Riddle *et al.* 1968)), can only be removed by phagocytosis. Before or during this process, these spores also stimulate an antibody response but this time it is a delayed Arthus-type reaction (Type III) with its attendant capillary dilation and precipitin reaction. Continual exposure from the use of mouldy hay may eventually lead to fibrotic lung lesions. As yet no case of pulmonary toxicosis from inhaled spores has come to light, naturally or experimentally, but in view of the carcinogenic properties of certain heavily-sporing fungi and especially actinomycetes it might be worth looking into this aspect of lung disease.

Infections caused by *Histoplasma capsulatum*, *Coccidioides immitis*, *Blastomyces dermatitidis*, *Aspergillus fumigatus* and *Emmonsia crescens* are the main respiratory diseases of animals of fungal origin. The airborne spores of these fungi are all in the size range 2–4 μm and they easily reach the respiratory bronchioles and alveoli where they are probably promptly phagocytosed. Their subsequent fate is discussed later.

The alimentary tract provides another major portal of entry which seemingly is of less significance than the pulmonary route, because by far the majority of fungal spores pass through to the faeces without harm and even have their germination enhanced. The undamaged tract gives little foothold for a fungal propagule and one naturally assumes that those few lesions seen, invariably result from secondary invasion of the damaged mucosa or follow stasis of the gut content, when spores in the food germinate and their hyphae invade the adjacent tissue. Such may be the case in phycomycotic ulcers produced by *Mucor*, *Rhizopus* and *Absidia* spp. and occasionally *Aspergillus fumigatus*. The one mystery of the intake of fungal propagules is in the aetiology of bovine mycotic abortion where, in the face of negative results from experimental pulmonary and vaginal routes of infection, the indications are that placental infection is most likely derived from ingested spores.

Deposition of spores in other sites, *e.g.* the vagina and prepuce, seems of little consequence and they are removed with the surplus mucus from such organs. Penetration of spores up the urethra must be relatively rare and fungal infections of the bladder in animals have not been reliably reported. This is rather surprising because in a similar physical situation, the problem of fungal propagules being deposited on bovine udders and somehow ascending the teat canals to reach the acini is one of great importance. The main organisms responsible for the consequent mastitis are yeasts which may be derived from many sources in the environment of the cow, especially faeces, grain and grass silage and brewers' grains, but *Allescheria boydii* (stat. impf. *Monosporium apiospermum*), *Absidia ramosa* and *Aspergillus fumigatus* can set up chronic infections which can only have been derived from ascending spores carried perhaps on the nozzles of anti-bacterial antibiotic applicators.

SURVIVAL AND GERMINATION

An essential part of infection by a micro-organism is survival at the site of deposition for sufficiently long to allow germination and penetration of the tissues. In the fungi we know virtually nothing of the factors determining this but natural and experimental cases show that the period may be as little as two hours, the time taken for *Candida albicans* to produce pseudo-germ tubes in serum (Mackenzie, 1964).

In view of the frequent adaptation of fungal propagules to desiccation and to high and low temperatures in their natural environment, survival on the surfaces of an animal body is hardly as critical a factor in pathogenicity as it is for bacteria. The recovery of intravenously inoculated organisms after relatively long periods of dormancy in organs has been noted and one assumes that naturally deposited spores can be similarly long-lived. Germination of the propagules is a matter of finding the favourable conditions for the species including any necessary stimulation, but the niceties of germination in fungi reviewed by Sussmann (1966) have never been pursued in detail for animal pathogens. Similarly, the nutrients available at the germination site, e.g. the respiratory mucosa or the phagocyte cytoplasm, have never been assessed, but it is assumed that each species differs from the next in its requirements. These will range from adequate moisture alone to prior heat treatment to 50–60°, and from glucose alone to an absolute vitamin B_1 requirement (Cochrane, 1966). Oxygen is said to be always required, often in quite small quantities so this is hardly likely to be a limiting factor where living animals are concerned, whilst carbon dioxide may also be necessary. Primary distinction between the hyphal and yeast-phases of tissue invaders is also apparent at this stage; the difference is between what de Bary described in 1887 as 'tube germinators' in hyphal forms and 'sprout germinators' in yeast forms. Although little information may be available on the germination of the spores of those dimorphic species with yeast or endosporulating tissue forms, many of the factors governing the corresponding yeast–hypha transformation in infected organs have been studied in detail and also found to differ for each species. Another aspect of spore germination in the animal body is the unknown significance of the lag phase with its swelling and wall splitting, and also of the circumstances initiating the invasive phase.

In the germination of the spores of pathogenic fungi *in vivo* we can therefore only speculate. Do dermatophyte arthrospores germinate because they have been pushed just within the lip of the hair follicle and are thus in a moist atmosphere? Do the long-chain fatty acids, once supposed to provide much of the resistance to ringworm in man, act at this time and site, or just where and when do they act? Similarly, do the inhibitors to germination present in serum find their way into the mucosa? And are there intracellular inhibitors in the phagocytes themselves?

PENETRATION

The events following germination are largely governed by the species of pathogen, for at first the reaction of the body is similar to that to any other foreign body. Hence two quite different patterns of pathogenesis evolve, one is a hyphal invasion of tissue and the other an increase in the pathogen by self-replication of its unicellular form which either buds or forms endospores. Within a very short time the susceptibility of the host has been tested and on this all subsequent events depend. Penetration then takes the two paths which are described in more detail below, which although mechanically different meet with the same types of immunological, physical and biochemical obstacles.

Hyphal growth

Study of the pathogenesis of mycoses has been dominated by the dimorphic fungi (reviewed by Romano, 1966) and those pathogens relying on hyphal growth have been eclipsed. This is a pity, because the morphological changes of the hyphal tips express a very fine sensitivity to the environment, as discussed by Robertson (1965), hence these changes could provide much information on the conditions in the interior of the lesions which is not at present available. One clear-cut advantage that the hypha has over the yeast cell is the purely mechanical thrust observed when more resistant muscle layers are depressed before penetration by the actively growing hyphal tip. Such penetration, as well as involving mechanical forces, must also be facilitated by enzyme action.

In the dermatophytes there seems little morphological change in the hyphal penetration to the centre of the hair and downwards, even at Adamson's fringe, and none in the epidermal scales. Where pathogenic potential seems to operate is in the degree of reaction produced in the underlying dermis. Some ringworm infections, especially in the normal host, elicit a very small response, others set up irritation which may proceed to micro-abscess formation (*Trichophyton verrucosum*) and then to full-scale follicular granulomata in which the arthrospores are produced in large numbers to give the honeycomb appearance of favus (*T. schoenleinii* in man and *T. quinckeanum* in the mouse). It seems unlikely that these special pathological differences relate to either the intrinsic range of reaction of the host or to the growth potential of the fungus. One feels that it is much more likely that this is a relationship to the type and amount of antigen produced and to the rapidity and severity of the antibody reactions elicited. These lesions are the more remarkable when the general inability of dermatophyte hyphae to penetrate the 'barrier layer' in keratinised tissues is considered. Faviform lesions are perhaps at the end of dermatophyte pathogenicity which grades into that of the geophilic species found only in the soil. The problem of keratinolysis has not yet been solved in hair and skin penetration; in fact English (1963, 1965) has shown that much of the initial growth of hyphae in hair is associated with the mechanical separation of the cells of the hair and probably with the utilisation of the free intercellular proteins. The well-marked difference of behaviour by dermatophytes on hairs *in vivo* and *in vitro* is as yet unexplained, but perhaps attributable to the moisture gradient in the growing hair.

In the viscera the hyphal tip emerging from the propagule is not too large for a phagocyte to ingest and it may be assumed that this event normally ensues. If the food reserves of the propagule outweigh the antagonistic potential of the phagocyte, then sooner or later the hyphal growth will take the initiative by carrying the cell off on its tip, sometimes penetrating through its body. Other cells come to assist and form syncytial giant cells investing the tips of the branches, but generally the impetus of hyphal growth, perhaps by now in its exponential phase, takes over. The morphology of the hyphae of *Aspergillus fumigatus* in acute lesions shows that a fungus does not always have its own way, for the initial swollen cell stages followed by the actinomycetoid branching systems indicate that the antagonistic influences have at least a tempering effect. Later still the changes in the resistance met may be reflected by the appearance of zonation in the colony and by a succession of related morphological forms (Austwick, 1965). The most impressive of these changes is the 'breakthrough' in which unmodified vegetative hyphae start to grow immediately preceding the death of the animal, when all forms of resistance have broken down and the fungus treats the tissues as if they were a culture medium.

Similar morphological modification has been observed in phycomycotic infections but are less understood, and it is possible that the marked hyphal swellings to form structures up to 100 μm in diameter may be demonstrations of response to the host reaction, as perhaps also are the internal wall thickenings often observed in these fungi. One really marked feature of penetration by phycomycete hyphae into tissue is their propensity to enter blood vessels and grow in the lumina, often causing thrombi to form. The blood clearly offers less mechanical resistance to hyphal extension and provides an ample supply of nutrients and oxygen. Dissemination by this means seems surprisingly rare except for the form of bovine mycotic abortion seen in New Zealand and caused by *Mortierella wolfii*. It is thought here that hyphal fragments from the placenta enter the effluent blood at parturition and are carried to the lungs of the dam where they lodge and grow. A rapidly progressive pneumonia is produced which causes a 20 % mortality in the cows concerned and probably a transient pulmonary infection in the remainder. Not only does haematogenous spread occur in the dam but a surviving calf apparently acquires infection from the same source via the umbilical circulation at the time of birth, and normally dies of generalised *Mortierella* infection within three weeks (Cordes & Shortridge, 1968).

The pathogenicity of the twenty or so species of fungi which cause bovine mycotic abortion remains very much a mystery, for although some are well known such as *Aspergillus fumigatus*, others, e.g. *Microascus cirrhosus*, *Humicola lanuginosa* and *Mucor racemosus* hardly rank as pathogens elsewhere. Infection is normally restricted to the placenta but in this organ there is extensive development, possibly indicating a specific tissue preference as described by Smith (1968) in the case of *Brucella abortus* and erythritol, or perhaps simply the absence of a growth inhibitor operating in the other organs of the body. There is no doubt that in the early stages of the infection an acute granulocytic response occurs but there is no evidence that this is at all effective in preventing further hyphal growth. Mycetoma is another disease of multiple causation, but in this the hyphal growth becomes greatly modified with terminal swellings and the dissemination of the infection takes place by the fragmentation of the colonies or granules.

Spherules and yeast cells

The differences between pathogenic fungi which grow as hyphae in tissue and those which are unicellular, do not become apparent until germination, because before this time the nature of the propagules, be they conidia, thallospores, yeast cells or hyphal fragments, is unrelated morphologically to their later development. At germination the problems of penetrating tissue presented to a unicellular organism are infinitely greater than those for a hypha where the advancing tip can call on energy reserves far back in the mycelium, for the volume of metabolising cells in a hyphal colony must always be greater than a single yeast cell. Rapid budding however, ensures that the yeast cell is not alone for long and the synergistic effect of many similar cells dividing together is well known in culture but rarely explained.

A single cell formed on germination has a need first for protection – and what better milieu to provide this in the animal body than phagocytes? It is by no means certain that unicellular organisms are always initially phagocytosed, but this event, followed by a few generations of self-replicated organisms, would at least ensure the passive migration of cells away from the point of arrival. Dissemination throughout the body can also occur by this method, but from the natural cases recorded it seems that localisation is the rule, often with the involvement of the regional lymph node. According to Howard (1965) the generation time for phagocytosed yeast-phase cells of *Histoplasma capsulatum* is approximately 10 hours and, as the ingesting cell probably only lasts for a maximum of three days before disintegrating, the release of the yeast cells must escalate rapidly.

Involvement of the reticulo-endothelial system in histoplasmosis means that extracellular yeast cells are rarely found but with *Cryptococcus neoformans* the cells soon become independent and grow at the expense of the body tissues which surround them. A distinctive feature of this organism is the relative lack of any immunological response in well-advanced brain lesions and the possession of the thick mucopolysaccharide capsule around each fungal cell has been held responsible for this. In other mycoses the pathogenesis of the lesion containing infected phagocytes follows a chronic granulating course with penetration and consolidation by the fungus being countered by the inflammatory response and encapsulation of the lesion. Yeast-phase multiplication in the bovine placenta clearly follows a different course for no granulomata are formed.

One of the more remarkable fungal pathogens is *Emmonsia crescens* which grows in the soil and forms small conidia about 2 μm in diameter, which are inhaled by small burrowing mammals. It is not clear whether phagocytosis takes place but it must be assumed that this is how the spore gets into its final position in the interstitial tissue. Here it enlarges and becomes surrounded by a fibrous capsule. The increase in size may proceed up to 500 times the diameter of the original inhaled spore but eventually breakdown occurs and the spore and its content is disrupted by phagocytes. When removed from the lung and cultured, the adiaspores germinate only below 27° and with a profusion of germ tubes, but it is doubtful if this latter stage – which completes the life cycle – ever occurs after death of the host.

CONSOLIDATION AND DESTRUCTION

The period following successful penetration and establishment of infection is one of the steady growth of the pathogen in which it consolidates its position in the lesions, and, in the contagious mycoses, produces fomites for direct transmission. To the animal this is the time when the immune response goes beyond the initial foreign body reaction to the more sophisticated specific antibody response. It is, then, the outcome of the meeting of these two opposing forces that decides the fate of both animal and fungus. If the immune mechanisms are developed fast enough and in sufficient potential, then destruction of the fungus and the survival of the host are assured; if the defence is not mustered adequately then the host dies or in the non-fatal cases, the lesion lingers. The dividing point of this dichotomy is the most important stage in the pathogenesis of the lesion and yet little is known of the pathological, immunological and morphological changes which occur at this time. One well known related phenomenon is seen in patients in which dissemination of *Coccidioides immitis* has occurred, where the decline of titre in coccidioidin skin tests is accompanied by the maintenance of precipitin titre; this is the reverse of the situation in recovery. In ringworm infections this division occurs when the disease persists beyond the time when natural recovery should have occurred especially in young animals. Support that this might be due to some immunological defect is given by the recent finding of Grappel, Fethiere & Blank (1971) that gamma-globulins from rabbits immunised with autoclaved mycelium of *Trichophyton mentagrophytes* are strongly inhibitory to the growth of this species in culture. It thus seems possible that such substances diffusing into the hair bulbs could be responsible for the degenerative changes observed in the intrapilary hyphae in healing ringworm lesions (P. K. C. Austwick, personal observation). The final disintegration of fungal cells in encapsulated lesions can occasionally be seen in aspergillosis, but as yet there has been no reported study of the environmental conditions in the lesions at the time of the death of the fungal cells.

VIRULENCE

Although some discussion on virulence has been included in this account, there are many other aspects of this important attribute of pathogenic organisms which need clarification in the fungi. First impressions have been that there is far less variation in virulence of fungal strains than there is with bacteria, but gradually evidence is coming forward of quite wide ranges of virulence. Thus Hasenclever & Emmons (1963) found a 1000-fold difference in the ability of soil isolates of *Cryptococcus neoformans* to infect mice. Avirulent or low virulence isolates of *Coccidioides immitis*, *Histoplasma capsulatum* and *Candida albicans* have now all been reported but it seems that *Aspergillus fumigatus* remains generally pathogenic from whatever source it is derived. From our knowledge of the genetics of fungi, one would expect wide natural variations, but as yet the study of the relationship of the genotype to virulence is in its infancy in mycopathology. The advances in the field of cereal rust genetics in relation to virulence indicate the way in which such studies could progress.

The basis of fungal virulence is still too uncertain for generalisations but at least one can speculate on the different intra- and extracellular mechanisms which may be responsible. Toxin secretion by the fungus must play some role, although in species with known toxicogenic and invasive propensities, e.g. *Aspergillus flavus*, there is no evidence that the most active of fungal toxins, the aflatoxins, have any role in the rare infection by this species. Louria, Brayton & Finkel (1963) have demonstrated that the more highly virulent the strain of *Candida albicans* (as tested by intravenous inoculation of mice) the more toxic is the supernatant from cultures. Certain fungi, of which *Mortierella wolfii* is one, seem to release their toxic and antigenic intracellular products very readily in culture through the death of the mycelium and there is every indication from infected cattle that a similar phenomenon occurs in the uterine and pulmonary lesions which this fungus causes (J. L. Longbottom and P. K. C. Austwick, personal observations). Much confusion has arisen in the understanding of the toxigenic action of infecting and non-infecting pathogenic fungi partly because of our lack of knowledge and partly because of the use of the restricted concepts of bacterial toxicity – the latter, based largely on the clostridial toxins.

CONCLUSIONS

A very broad and sequential view of fungal pathogenicity has been taken in this account because only this approach can conveniently cover the very wide range of activity in which pathogenic fungi are concerned. In so doing it has not been possible to make more than passing reference to several aspects of pathogenicity of increasing importance. Among these the biochemistry of fungal aggression and of countering host resistance is beginning to be explored, whilst the relationship of virulence requires much more attention, especially in the light of the advances made in the field of extra-chromosomal inheritance in plant pathogenic fungi. Electron-microscope study of fungal invasion of animal tissue has already revealed many interesting facets of the relationship between host and parasite, but general agreement on interpretation of the results and the use of these to initiate studies at the molecular level of pathogenesis have still to be developed. The approach from the viewpoint of the fungal pathogen which is used here also provides for comparison with the performance of other micro-organisms, and in particular emphasises the importance of the two environments to the pathogenic fungus.

Apart from their natural habitat, their morphology and their chemical composition, fungal pathogens may be distinguished from bacteria and viruses in other less tangible ways, such as the prospects of eradication of their causal diseases. Certainly success in the control of contagious mycoses such as favus by hygiene seems to indicate that a policy of reducing the amount of available infectious material, (i.e. the fungal population) may produce a point when the organism no longer survives. Perhaps for *Trichophyton schoenleinii*, the general spread of cleanliness has contributed to its demise in the United Kingdom! But quite how any such eradication could be implemented in soil-borne, non-contagious mycoses one cannot imagine, for already experiments to render soil habitats unfavourable to *Histoplasma capsulatum* have failed to maintain their early promise.

The wide range of pathogenic activity in fungi also causes difficulties in finding a necessary unifying factor. Possibly a useful one would be the defence mechanisms of the body assembled against the fungal cell and its toxins, for most of the immune responses of the animal body have been observed following contact with fungi. The cellular defences include a granulocytic response in early infection, changing gradually to a lymphocytic one with the encapsulation of the lesion. Fungi also classically induce a skin sensitisation giving a delayed tuberculin type of reaction (Type IV allergy) to intradermal injection of antigen. Humoral defences include the development of precipitating, agglutinating and complement-fixing antibody systems. In atopic subjects (and we assume such occur in animals as well as man) allergenic fungi mostly induce a typical Type I response (immediate wheal) on prick test with fungal extracts, whilst in others an Arthus reaction is given (Type III). These protective reactions are probably highly successful in preventing the establishment of a fungus in an organ but some, especially the Arthus reaction in allergic alveolitis (farmers' lung), lead to excessive tissue damage.

One of the most neglected factors in fungal disease is the quantative relation of the pathogen to the host, i.e. the dose-response, and it has not been included in this account because so little is known. It is clearly a crucial factor in toxicoses and allergies and is thought to be so in at least inhalation mycoses derived from airborne spores. However, consistent guides even to the intravenous LD_{50} levels of dosage have not been built up by experimental study, let alone related to the situation in the field. The mycotoxicoses lend themselves to estimates of intake of toxin in this way and some work has been carried out on propagule intake in respiratory infection, but in other mycoses and in allergy, effective dose levels are quite unknown.

In all aspects of mycopathology the role of the intrinsic attribute of 'fungal pathogenicity' in determining the pathogenesis of the disease has hardly yet been assessed. The way ahead appears to be by looking to developments in plant pathology where fungi have always held their rightful place as pathogens and their pathogenicity is well understood.

I wish to thank Miss L. Sparkes for preparing the figure.

REFERENCES

AINSWORTH, G. C. (1952). *Medical Mycology: An Introduction to Its Problems*, 105, London: Pitman.

AUSTWICK, P. K. C. (1963). Ecology of *Aspergillus fumigatus* and the pathogenic phycomycetes. *Recent Progress in Microbiology*, 8, 644.

AUSTWICK, P. K. C. (1965). Pathogenicity. In *The Genus Aspergillus*, ed. K. B. Raper and D. I. Fennell, 82. Baltimore: Williams & Wilkins.

AUSTWICK, P. K. C. (1966). The role of spores in the allergies and mycoses of man and animals. In *The Fungus Spore*, ed. M. F. Madelin, 321. London: Butterworths.

BARTNICKI-GARCIA, S. & NICKERSON, W. J. (1962). Isolation, composition and structure of filamentous and yeast-like forms of *Mucor rouxii*. *Biochimica et Biophysica Acta*, **58**, 102.

BASSET, A., CAMAIN, R., BAYLET, R. & LAMBERT, D. (1965). Role des épines de Mimosacées dans l'inoculation des mycetomes. *Bulletin Sociétié de la Pathologie Exotique*, **58**, 22.

BASSI, A. (1835). Cited from Ainsworth, G. C. (1956). Agostino Bassi, 1773–1856. *Nature, London*, **177**, 255.

BERKELEY, M. J. (1845). Observations, botanical and physiological on the potato murrain. *Journal of the Horticultural Society*, **1**, 9.

BRIAN, P. W., HEMMING, H. G. & JEFFERYS, E. G. (1948). Production of antibiotics by species of *Myrothecium*. *Mycologia*, **40**, 363.

COCHRANE, V. W. (1966). Respiration and spore germination. In *The Fungus Spore*, ed. M. F. Madelin, 201. London: Butterworths.

CORDES, D. O. & SHORTRIDGE, E. H. (1968). Systemic phycomycosis and aspergillosis of cattle. *New Zealand Veterinary Journal*, **16**, 65.

DETROY, R. W. & CIEGLER, A. (1971). Induction of yeast-like development in *Aspergillus parasiticus*. *Journal of General Microbiology*, **65**, 259.

ENGLISH, M. P. (1963). The saprophytic growth of keratinophilic fungi on keratin. *Sabouraudia*, **2**, 115.

ENGLISH, M. P. (1965). The saprophytic growth of non-keratinophilic fungi on keratinized substrata and a comparison with keratinophilic fungi. *Transactions of the British Mycological Society*, **48**, 219.

GRAPPEL, S. F., FETHIERE, A. & BLANK, F. (1971). Effect of antibodies on growth and structure of *Trichophyton mentagrophytes*. *Sabouraudia*, **9**, 50.

GRUBY, D. (1841). Sur les mycodermes qui constituent le teigne faveuse. *Comptes rendus des séances de l'Académie des Sciences*, **13**, 309.

HASENCLEVER, J. F. & EMMONS, C. W. (1963). The prevalence and mouse virulence of *Cryptococcus neoformans* from urban areas. *American Journal of Hygiene*, **78**, 227.

HOWARD, D. H., (1965). Intracellular growth of *Histoplasma capsulatum*. *Journal of Bacteriology*, **89**, 518.

LOURIA, D. B., BRAYTON, R. G. & FINKEL, G. (1963). Studies on the pathogenesis of experimental *Candida albicans* infections in mice. *Sabouraudia*, **2**, 271.

MACKENZIE, D. W. R. (1964). Morphogenesis of *Candida albicans in vivo*. *Sabouraudia*, **3**, 225.

MACKINNON, J. E. (1949). The dependence on the weather of the incidence of sporotrichosis. *Mycopathologia & Mycologia Applicata*, **4**, 367.

RIDDLE, H. F. V., CHANNELL, S., BLYTH, W., WEIR, D. M., LLOYD, M., AMOS. W. M. G. & GRANT, I. W. B. (1968). Allergic alveolitis in a maltworker, *Thorax*, **23**, 271.

ROBERTSON, N. F. (1965). The mechanism of cellular extension and branching. In *The Fungi*, ed. G. C. Ainsworth and A. S. Sussman, vol. 1, 613. London: Academic Press.

ROMANO, A. H. (1966). Dimorphism. In *The Fungi*, ed. G. C. Ainsworth and A. S. Sussman, vol. 2, 181. London: Academic Press.

SMITH, H. (1968). Biochemical challenge of microbial pathogenicity. *Bacteriological Reviews*, **32**, 164.

SUSSMAN, A. S. (1966). Types of dormancy as represented by conidia and ascospores of *Neurospora*. In *The Fungus Spore*, ed. M. F. Madelin, 235. London: Butterworths.

WHEELER, B. E. J. (1968). Fungal parasites of plants. In *The Fungi*, ed. G. C. Ainsworth and A. S. Sussman, vol. 3, 179. London: Academic Press.

PROTOZOAL PATHOGENICITY

B. A. NEWTON

Medical Research Council Biochemical Parasitology Unit, The Molteno Institute, University of Cambridge, Cambridge

INTRODUCTION

The number of protozoan species is probably more than 50000 (Weisz, 1954) and the majority of these lead a parasitic existence. Considering these numbers alone, man might be considered fortunate since only about twenty species are known to be human pathogens. However, this relatively small number of pathogenic protozoa cause diseases which have an almost worldwide distribution and, at any one time, probably afflict about a quarter of mankind. These diseases (Table 1) range from infections of the gastro-intestinal and urino-genital tracts to diseases of the central and peripheral nervous systems, muscle and the reticulo-endothelial system: they include, amongst others, malaria, trypanosomiasis, leishmaniasis and amoebiasis. There are over a hundred million cases of malaria a year, about a million of which are fatal. There is no cure for Chagas' disease, and seven million individuals are thought to be infected. Sleeping sickness in man is now controlled to a considerable extent but trypanosomiasis in domestic animals is responsible for human malnutrition, and all the diseases it encourages, over an area of four and a half million square miles of Africa south of the Sahara.

These examples emphasise the importance of protozoal diseases today; they also serve to illustrate the magnitude of the exercise facing the writer! To try to review present knowledge of protozoal pathogenicity in 9000 words is a daunting task: the subject could provide material for a whole symposium. I have decided therefore to restrict this contribution to a discussion of three general questions. How do protozoa gain access to their mammalian hosts? What factors affect the establishment of these organisms in their hosts? What do we know about the mechanisms of their pathogenicity? Even with this limitation the discussion will be far from comprehensive. I shall select aspects of recent work which seem to provide important leads for future research and take the Society's 5th Symposium (Howie & O'Hea, 1955) as a base line by referring readers to the four authoritative reviews on protozoal pathogenicity that it contains for a survey of literature up to 1955.

Table 1. *Protozoal diseases of man*

Disease	Causative organism	Principle mechanism of transmission
Amoebic dysentery	*Entamoeba histolytica*	Ingestion of cysts
Amoebic meningoencephalitis	*Hartmanella & Naegleria* spp.	Penetration of mucous membranes
Giardiasis	*Giardia lamblia*	Ingestion of cysts
Balantidiasis	*Balantidium coli*	Ingestion of cysts
Toxoplasmosis	*Toxoplasma gondii*	Ingestion of cysts. Diaplacentar infection of foetus
Malaria	*Plasmodium falciparum* *vivax* *malariae* *ovale*	Female anopheline mosquitos
Leishmaniasis		
Cutaneous	*Leishmania tropica*	*Phlebotomus* spp.
Muco-cutaneous	*braziliensis*	(sandflies) and Direct
Visceral	*donovani*	contact
Trypanosomiasis:		
African (sleeping sickness)	*Trypanosoma gambiense*	*Glossinidae* (tsetse)
S. American (Chagas' disease)	*rhodesiense*	—
	cruzi	*Reduviidae*
Trichomoniasis	*Trichomonas vaginalis*	Venereal

MECHANISMS OF TRANSMISSION

The mechanisms of transmission of protozoa are as varied as the diseases they produce.

Ingestion of cysts

Protozoa which are transmitted by the ingestion of cysts include *Entamoeba histolytica*, *Giardia lamblia* and *Balantidium coli*, all of which infect the gastro-intestinal tract. These organisms have relatively simple life cycles involving a trophozoite or proliferative stage and cysts which survive passage to the external environment in faeces and may subsequently contaminate food and water supplies. *Toxoplasma gondii*, now recognised as a coccidian parasite closely related to the genus *Isospora* (Hutchison, Dunachie & Work, 1970), forms oocysts which can be transmitted by contaminative methods (Jacobs, Remington & Melton, 1960; Jacobs, 1967; Hutchison & Work, 1969). This organism also forms tissue cysts which are resistant to digestion and, since they are frequently found in skeletal muscle of sheep and swine, it is quite possible that ingestion of raw meat is a source of human infection. *Toxoplasma* cysts have been observed in lung alveoli, suggesting that inhalation as well as ingestion by mouth may also be of importance in their transmission. Amoeba of the genus *Hartmanella* provide another example of protozoa which may be transmitted by ingestion of cysts. Until recently this genus was thought to be composed of exclusively free-living organisms, however, they can be isolated from throat swabs, and about 2 % of a group of people surveyed in North America were found to be infected (Wang & Feldman, 1967). Most of the infections were found to be in infants at the crawling stage, suggesting that ingestion of dust or soil containing cysts or trophozoites of the amoebae may be a likely means of infection.

The factors controlling encystment and excystment and the biochemistry of the processes involved are little understood. Excystment of *E. histolytica* cysts appears to take place in the small intestine and is influenced by the velocity of intestinal transit, the quantity of food ingested and by host digestive enzymes. Achlorhydric dogs are reported not to acquire infection and animals with accelerated intestinal transit eliminate cysts before they excyst (Hegner, Johnson & Stabler, 1932; Swartzwelder, 1939). Inadequate culture techniques have hindered biochemical studies of encystment and excystment (McConnachie, 1969) but recent work (Griffiths, 1970) has shown that the hartmanellid amoebae provide a valuable experimental system for the investigation of this problem.

Penetration of mucous membranes

The amoebal throat infections just mentioned were reported to be symptomless but there is now good evidence (Culbertson, 1970, 1971) that normally free-living amoebae of the genera *Hartmanella* and *Naegleria* can be pathogenic to man. The role of *Naegleria* spp. in acute human amoebic meningoencephalitis is well-established and infection can result from swimming or water-skiing in rivers and fresh water lakes. It is generally believed that amoebae invade the nasal mucosa. This process may be aided by the high water pressures which are exerted on the mucosa during participation in water sports, but laboratory experiments leave no doubt that these amoebae can invade the nasal mucosa of animals unaided (Culbertson, 1961; Singh & Das, 1970). The amoebae can be found subsequently in the olfactory nerves and the olfactory portion of the brain, where they proliferate rapidly and produce massive brain damage. Tissue culture experiments have established a correlation between tissue-destroying ability of amoebae and their animal virulence but the factors responsible for virulence remain unknown. Singh & Das (1970) found that only three out of nine *Hartmanella* spp. examined were pathogenic to mice when administered intranasally: it would be interesting to know whether intra-cerebral injection would give the same result.

Trypanosoma cruzi is also able to penetrate intact mucous membranes and commonly penetrates the ocular conjunctiva.

Penetration through skin

Infection of man by malaria parasites, trypanosomes and leishmania occurs principally through the skin but there is no evidence that any of these protozoa can actively penetrate undamaged skin; they are either injected by an insect vector during the feeding process or they enter through a skin abrasion.

Sporozoites of *Plasmodium* spp. are pumped into the bloodstream in a jet of saliva when an infected female anopheline mosquito takes a blood meal. Male mosquitos do not transmit malaria; they can be infected in the laboratory by feeding them on defibrinated blood containing gametocytes and sporogonic development will occur (Shute, 1964) but the structure of the male mouth parts does not permit probing and penetration of skin. African trypanosomes and *Leishmania* spp. are also injected into the mammalian host in a stream of saliva when their insect vectors, *Glossina* spp. and *Phlebotomus* spp. respectively, take a blood meal. An elegant and detailed study by Gordon, Crewe & Willett (1956) showed that when *Glossina morsitans* feeds, some saliva containing infective metacyclic forms of trypanosomes may enter the bloodstream directly but much more is deposited in the tissue spaces as the fly probes for a suitable capillary from which to draw blood. Thus initial growth of the parasites occurs mainly in the tissues where a local chancre develops; later they spread through connective tissues, the lymphatic system, the bloodstream and ultimately the central nervous system. *Leishmania*, following injection through skin by the proboscis of sand flies, are ingested by macrophages and multiply intracellularly as the non-motile amastigote forms.

In contrast to the African trypanosomes *T. cruzi* develops to the metacyclic stage in the hind gut of blood-sucking triatomid bugs. These bugs defaecate whilst feeding and parasites may be rubbed into the puncture wound or transferred to the ocular conjunctiva and other mucous membranes when the area of the bite is scratched. Having penetrated the vertebrate host in this manner *T. cruzi* develops intracellularly in lymphoid–macrophage cells and muscle cells, particularly heart muscle.

Venereal transmission

Trichomonas vaginalis was long considered a harmless commensal of the human vagina but trichomonas vaginitis is now generally accepted as a true venereal disease. *T. vaginalis* occurs in the male urethra and prostate as well as the female vagina. Infection rates as high as 40 % have been recorded in unselected women. Sexual intercourse is the major means of transmission; trichomonads do not form cysts and can survive for only very short periods outside the human body (Jirovec & Petru, 1968).

The only other known protozoal venereal disease occurs in equines and is a form of trypanosomiasis ('dourine') caused by *Trypanosoma equiperdum*.

Diaplacental infection of the foetus

Infection of a foetus with *Toxoplasma gondii* by parasites passing through the placenta is well known to occur if the pregnant female has an acute infection (Van der Waaij, 1964). *T. cruzi* is also known to be transmitted by this route (Gavaller, 1951). In neither case is the mechanism of placental penetration known but recent work on the mechanism of cell penetration by *Toxoplasma* suggests that lysosomal enzymes may be involved (Norrby & Lycke, 1967; Norrby, Lindholm & Lycke, 1968; Lycke, Norrby & Remington, 1968).

Other mechanisms

The majority of protozoal infections are undoubtedly transmitted by one or other of the mechanisms discussed but other routes of infection are of considerable importance in certain diseases and in certain areas of the world. Cutaneous leishmaniasis can be transmitted by direct contact. Transmission of *T. cruzi* and *Plasmodia* spp. from mother to child during breast feeding has been reported. In areas of South America where Chagas' disease is endemic it is believed that blood transfusion may play a significant part in the spread of *T. cruzi* (Köberle, 1968).

SOME FACTORS AFFECTING THE ESTABLISHMENT OF PARASITIC PROTOZOA IN THEIR HOSTS

Changes in infectivity of protozoa during their life cycles

The digenetic protozoal pathogens have complex life cycles involving a number of morphologically distinct forms which may differ markedly in structure, metabolism and infectivity. The development of any one of these forms is often limited to a specific location and cell type in the host. For example, malaria is commonly regarded as a disease of erythrocytes; the parasites divide within these cells and ultimately cause their lysis. However, sporozoites which are injected into the bloodstream by a mosquito are totally unable to infect red cells; in man they migrate to the liver and undergo one or more cycles of division in the parenchyma cells (James & Tate, 1937) before becoming adapted for life in the bloodstream. We have virtually no knowledge of the biochemical changes which take place during a developmental cycle such as this, neither do we understand the mechanisms which control these changes and impose an absolute tissue specificity at different stages of the life cycle. Clearly such knowledge is vital to a complete understanding of the pathogenicity of these organisms and a great deal of work is now being done in this field.

The system about which we know most at present is the life cycle of *brucei* group trypanosomes. This group includes *Trypanosoma brucei*, which infects game animals and cattle but will not infect man; *T. rhodesiense* and *T. gambiense* which cause, respectively, acute sleeping sickness in East Africa and a more chronic form of the disease in Central and West Africa. These organisms are morphologically indistinguishable and the forms which cause human sleeping sickness are generally regarded as genetic variants or sub-species of *T. brucei* (Hoare, 1966). Studies on the biochemistry and ultrastructure of these organisms is beginning to give some indication of the factors which influence their infectivity. Fig. 1 summarises the life cycle of *T. brucei* and illustrates some of the morphological and biochemical changes now known to be associated with it. The mammalian host is infected by 'metacyclic' trypanosomes which develop in the salivary glands of the tsetse. Tsetse are believed to be infected by ingesting stumpy trypanosomes from the bloodstream of the mammalian host (Robertson, 1912; Reichenow, 1921; Wijers & Willett, 1960). If trypanosomes of a 'pleomorphic' strain (i.e. a strain in which intermediate and stumpy forms appear in the bloodstream in the later stages of an infection) are removed from the mammalian bloodstream and incubated in a suitable blood-containing medium (Taylor & Baker, 1968) at 37° they do not continue to divide; however, culture of these organisms *in vitro* may be established by lowering the incubation temperature to 25°. The so-called 'culture' forms which develop resemble, morphologically, the stage found in the midgut of the tsetse, and these forms have generally been found non-infective to mammals. If a pleomorphic strain of *T. brucei* is maintained in laboratory rodents by syringe passage, rather than by cyclic transmission through the insect vector and mammalian host alternately, it ultimately becomes 'monomorphic' (i.e. the forms which develop in the bloodstream are 'slender' throughout the infection). When a strain becomes monomorphic it loses its infectivity for the insect vector. Thus loss of ability to infect both mammalian and insect hosts can be induced by certain cultural conditions and we are now

IN MAMMALS

BLOODSTREAM FORMS: Metabolise glucose mainly
to pyruvate; terminal respiration
by glycerophosphate oxidase system

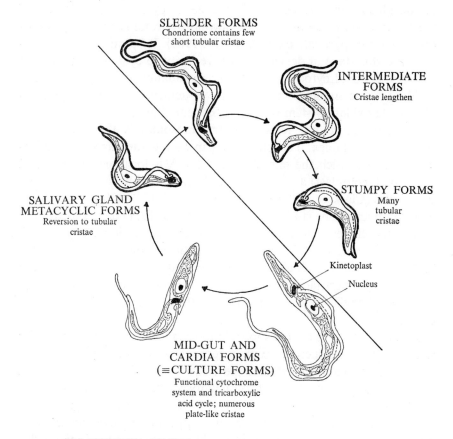

SLENDER FORMS
Chondriome contains few
short tubular cristae

INTERMEDIATE
FORMS
Cristae lengthen

SALIVARY GLAND
METACYCLIC FORMS
Reversion to tubular
cristae

STUMPY FORMS
Many
tubular
cristae

Kinetoplast

Nucleus

MID-GUT AND
CARDIA FORMS
(≡CULTURE FORMS)
Functional cytochrome
system and tricarboxylic
acid cycle; numerous
plate-like cristae

IN TSETSE FLIES

Fig. 1. Diagram of developmental stages of *Trypanosoma brucei* to show changes in the surface coat, state of development of the mitochondrion and position of the nucleus and kinetoplast (based on Vickerman, 1969, 1971). The slender bloodstream form lacks a functional cytochrome system and tricarboxylic acid cycle (TCAC). Some TCAC enzymes have been detected in stumpy forms but cytochromes are still missing. Terminal respiration in bloodstream forms involves a glycerophosphate oxidase system. Forms which occur in the tsetse midgut and in culture have an active cytochrome system and TCAC; they also have a well-developed mitochondrial network containing many plate-like cristae. The forms which occur in the salivary gland have a less extensive mitochondrial structure. Forms which have a surface coat (Plate 1) are indicated by a thick outline. The transformation from slender to stumpy forms occurs within each relapse population in pleomorphic strains of *T. brucei*. When trypanosomes enter the tsetse all variants lose their antigenic identity and revert to a basic antigenic type when they re-acquire a surface coat and become metacyclic forms in the salivary glands.

able to study the changes in cell physiology which are associated with these losses of infectivity. Results to date (Fulton, 1969; Vickerman, 1971) indicate that the metabolism of 'slender' bloodstream forms is characterised by a rapid but incomplete oxidation of glucose, over 80 % of the glucose carbon being recoverable as pyruvate. The failure of these forms to metabolise glucose beyond pyruvate is primarily due to the absence of pyruvate oxidase (Ryley, 1962). Respiration of trypanosomes at this stage of development is resistant to cyanide and the cells lack a functional cytochrome system; reoxidation of reduced nicotinamide adenine dinucleotide (NADH) is achieved by an L-α-glycerophosphate oxidase system (Grant & Sargent, 1960).

A unique structural feature of trypanosomes is the kinetoplast–chondriome complex. This structure can be regarded as a giant mitochondrion in which the mitochondrial DNA is localised in one small area (the kinetoplast) and can be readily visualised by suitable staining techniques (Clark & Wallace, 1960; Steinert, 1960; Mühlpfordt, 1964). Electron microscope studies (Vickerman, 1965, 1971) have shown that, in the bloodstream forms of *T. brucei*, the chondriome is a simple tubular structure containing few cristae. In contrast to these findings 'culture' forms have a highly developed chondriome network containing many plate-like cristae. They also have a functional cytochrome system and tricarboxylic acid cycle and are thus able to oxidise glucose completely.

These results raise the question: do the changes in respiratory activity and chondriome structure occur in response to the change of environment when trypanosomes enter the tsetse gut? Evidence from studies of ultrastructure and respiratory activity suggests that this is not so (Vickerman, 1965; Ryley, 1966; Flynn & Bowman, 1970): at least partial activation of the chondriome appears to have occurred in intermediate and stumpy forms in the bloodstream. What initiates this change remains unknown; a number of workers have speculated that the switch to a more efficient respiratory metabolism in stumpy forms pre-adapts them to life in the tsetse gut, where, compared to the bloodstream, glucose and oxygen will be in relatively short supply. Whether this is so remains to be proved but certainly slender forms of monomorphic strains rarely, if ever, infect tsetse. Also trypanosomes which have lost their kinetoplast DNA (dyskinetoplastic forms), either by spontaneous mutation or as a result of treatment with certain drugs (Werbitzki, 1910; Van Assel & Steinert, 1968; Newton, 1970) do not undergo the respiratory switch, cannot be cultured *in vitro* and will not infect tsetse. In the light of these results it seems reasonable to deduce that a functional mitochondrial or chondriome system is a prerequisite for infection of *Glossina* spp. by *brucei* group trypanosomes.

Electron microscopy has revealed another structural difference between bloodstream and culture forms of *brucei* group trypanosomes which is thought to be of importance in determining whether these organisms can infect the mammalian host. In all bloodstream forms a layer of moderately electron-dense material (12–15 nm thick) exists outside the plasma membrane (Plate 1) (Godfrey & Taylor, 1969; Vickerman, 1969). Organisms growing in culture, or developing in the tsetse midgut, are non-infective to mammals and lack this coat, but it is again present at the infective metacyclic stage of development. Vickerman (1969) suggests that this coat, which contains the surface agglutinogens, is an adaptation to life in the bloodstream. As yet little is known about the composition of the surface coat; it can be removed by treatment with pronase (Vickerman, 1969) but is unaffected by snake venom phospholipase A (Godfrey & Taylor, 1969). Recent cytochemical studies (Wright & Hales, 1970) suggest that, structurally, the coat may be more complex than originally thought; they claim to have detected a prominent inner, and a less prominent outer layer of carbohydrate within the cell coat. Whatever the detailed structure of the coat proves to be, it seems, on the basis of present knowledge, that synthesis of coat material is switched on and off at different times in the developmental cycle, and that its presence is associated with acquisition of infectivity for mammals by trypanosomes.

Loss and reacquisition of virulence

It is well known that many protozoal pathogens lose virulence when they are cultured *in vitro* and that virulence may sometimes be restored by animal passage. The biochemical mechanisms involved in these changes are still unknown but recent work is beginning to uncover some clues.

Neal (1958) has described the loss of virulence of *E. histolytica* during culture *in vitro* and its reacquisition after liver passage in hamsters. Axenically grown strains of *E. histolytica* have recently been used to study factors influencing virulence (Wittner, Rosenbaum & Einstein, 1970). Inoculation of such organisms into the liver of hamsters or the caecum of germfree guinea-pigs rarely caused any pathological effect: if the amoebae were incubated in the presence of various strains of bacteria for 6–12 h before animal inoculation the infections proved fatal. When freshly isolated strains of *E. histolytica* were established in axenic culture virulence was found to decline gradually over a period of 3–5 months. Incubation of avirulent axenic cultures with various preparations of killed bacteria or supernatant fluids from bacterial cultures failed to restore virulence as did growth of axenic amoebae and bacteria in parabiotic chambers separated by membranes of various pore sizes. Thus increased virulence seems to result only from direct contact between amoebae and live bacteria. The authors suggest that an episome-like factor may be involved (quoted by Neal, 1971).

Electron microscopy is beginning to reveal some details of the ultra-structure of *E. histolytica* (Eaton, Meerovitch & Costerton, 1969, 1970; El-Hashmit & Pittam, 1970; Lowe & Maegraith, 1970) and some differences in the surface of amoebae grown in culture and *in vivo* have been reported. (El-Hashmit & Pittam, 1970; Rondanelli, Carosi, Gerna & De Carneri, 1968) but the significance of these findings is still obscure.

Honigberg and his associates (Stabler, Honigberg & King, 1964; Honigberg, Becker, Livingston & McLure, 1964; Honigberg, 1967) have made a detailed study of factors affecting virulence of *Trichonomas gallinae* (a parasite of birds which infects the mouth, pharynx and oesophagus) and have concluded that virulence is only maintained in the presence of living host cells. Cultivation of a virulent strain in serum containing nutrient medium for 4–5 months resulted in loss of virulence but cultivation of the same strain in the presence of chicken liver cells for over a year caused no such loss. Virulence could be restored by several serial passages through pigeons or by exposing avirulent organisms to a cell-free homogenate prepared from the virulent strain. The virulence-restoring activity of this homogenate was destroyed by treatment with ribonuclease and deoxyribonuclease and restored by native high molecular weight RNA and DNA prepared from a virulent strain (Honigberg, 1970). It seems likely that work in progress on this system may soon lead to an understanding of the chemical and genetic basis of pathogenicity in this organism.

Intercurrent infections

While there is still relatively little information about the effect of intercurrent infections on the course of protozoal diseases, there is, I believe, reason to think that they may be extremely important.

It is well known that agents which damage or functionally exhaust the reticulo-endothelial system affect the course of malaria and other parasitic infections. There is evidence that Marek's disease (a lympho-proliferative disease of chickens caused by a herpes virus) suppresses immune responses in chickens and enhances coccidial infections (Biggs, Long, Kenzy & Rootes, 1968). The reverse situation is thought to apply to the suggested relationship between Burkitt's lymphoma and malaria (Burkitt, 1969); here, the immunosuppressive activity of plasmodia is thought to cause an abnormal response to an oncogenic virus. A similar relationship has been demonstrated between *Plasmodium berghei yoelii* and a murine oncogenic virus (Salaman, Wedderburn & Bruce-Chwatt, 1969). In the light of these observations it is tempting to suggest that *Toxoplasma* infections may also be influenced by intercurrent infections. *Toxoplasma gondii* infections after birth are generally so mild as to pass unnoticed. Serological evidence suggests that as many as one third of the population may have been infected but only a few hundred clinical cases occur annually. However, on occasions, and for reasons unknown, a more general infection may result in which the brain, liver and lungs become involved and death may result from the damage caused to these organs. Could this sudden 'flare-up' of infection be triggered by an as yet unidentified viral infection?

The development of some intestinal protozoal infections is certainly influenced by the presence of other micro-organisms. It has been reported that *E. histolytica* cannot be propagated in germfree animals (Phillips & Bartgis, 1954) but the role played by the intestinal bacterial flora in supporting amoebal growth has not yet been clearly defined. Bacteria may provide essential nutrients. Until recently it was believed that bacteria were mandatory for the growth of *E. histolytica in vitro* but Diamond (1968) has now maintained cultures under axenic conditions. However, this advance does not rule out the possibility that bacteria provide essential nutrients *in vivo*. Another possibility is that bacteria create the right physico-chemical conditions for amoebal growth in the gut. *In vitro E. histolytica* will only grow under conditions of low oxygen tension (-350 to -450 mV) and bacteria may be responsible for lowering the oxidation–reduction potential to this level in the colon (Phillips *et al.* 1955). It has also been suggested that the presence of certain bacteria in the gut may determine whether *E. histolytica* exists as a harmless commensal or invades the intestinal wall. Hoare (1950) has concluded that in up to 90 % of persons harbouring this parasite it is present as a commensal, but, under certain conditions, the amoebae may quite suddenly reveal their virulence; trophozoites will then penetrate the mucosa, invade the submucosa, and form a characteristic flask-shaped ulcer. If one accepts Hoare's view the dramatic change from avirulence to virulence could be explained by the presence of another agent, possibly a pathogenic bacterium, which can cause primary damage to the intestinal wall and so provide a site through which the amoebae can gain access to the submucosa. While this explanation remains a possibility it seems that in the majority of cases of acute intestinal amoebiasis the intestines are free of pathogenic bacteria (Okamoto, 1954). Another possibility is that dietary factors may affect invasibility. Certain diets favour the mucosal invasion of amoebae in the guinea-pig (Taylor, Greenberg & Josephson, 1952); the addition of cholesterol to the diet was found to increase the frequency and size of ulcerous lesions in the intestine (Biagi-F, Robledo, Servin & Martuscelli-Q, 1962). A more detailed discussion of work on this problem will be found in the reviews of Biagi-F & Beltran-H (1969) and Neal (1971).

A similar picture is presented by *Balantidium coli* which is found in the large intestine of man and animals in many parts of the world. Experiments on pigs have shown that the ciliate is non-invasive unless the intestinal mucosa is damaged by another infection such as *Salmonella* (Baker, 1969).

Resistance of the host

This topic can be considered under two heads: acquired immunity and innate resistance. The former I consider to be outside the scope of the present article; it has been the subject of two symposia in recent years and the reader is referred to the published accounts for information on this topic (Garnham, Pierce & Roitt, 1963; Taylor, 1968). Some of the most interesting and widely studied examples of innate resistance concern susceptibility to malaria infection and current views about these are worth considering in some detail.

There are several genetic traits, known to affect the biochemistry of erythrocytes, which confer a biological advantage on persons exposed to malaria infections, the most well known being the sickle cell trait. Carriers of this trait are not generally resistant to malaria but are protected against severe infections of *Plasmodium falciparum* malaria. They are as susceptible to infection as people with normal haemoglobin but show lower parasite counts and lower mortality. Thus persons who are heterozygous for the sickle cell gene have a selective advantage in regions where malaria is hyperendemic; Allison (1954) has suggested that this fact may explain why the sickle cell gene remains in these areas, in spite of the elimination of genes in patients dying of sickle cell anaemia. We cannot yet explain how the abnormal haemoglobin-*S* molecule, which contains valine in place of certain glutamic acid residues, can inhibit the growth of *P. falciparum* trophozoites but two hypotheses have been proposed. The change in amino-acid composition of haemoglobin results in a decreased solubility and an increase in the viscosity of solutions. Heterozygous individuals have both types of haemoglobin, but a quarter to a half of the total is haemoglobin-*S* and this is sufficient to cause a considerable increase in viscosity of the erythrocyte contents. Haemoglobin is the primary source of amino acids for malaria parasites during the erythrocytic stage of development, and Moulder (1962) has suggested that an increase in viscosity may interfere with ingestion of this molecule by the parasites. If this hypothesis is correct it might be expected that the sickle cell trait would protect against all species of malaria, but this is not so; protection is only provided against *P. falciparum.* In an attempt to explain this selective protection Garnham (1966) pointed out that erythrocytic stages of *P. falciparum* spend most of their time in the capillaries of internal organs where the blood supply is slowed down. Under these conditions, he suggests, the corpuscles might become sufficiently anoxaemic to sickle and liberate immature schizonts into the plasma where they would be unable to complete their development.

Two other red cell abnormalities are known to retard *P. falciparum* development. Glucose-6-phosphate dehydrogenase (G6PD) deficiency, which may be associated with β-thalassaemia, has been observed in about 16 % of male adults in Uganda. There is now little doubt that this is a favourable selective factor in *P. falciparum*-endemic regions, (Motulski, 1964; Siniscalco *et al.* 1966; Gilles *et al.* 1967). G6PD-deficient red cells have a shorter life and a lower capacity to reduce haemoglobin than normal cells, but how the deficiency affects *P. falciparum* development remains to be determined. In view of these findings it is interesting that red cells of owl monkeys (*Aotus trivirgatus*), which support the growth of *P. falciparum*, have a high G6PD level and, like human and chimpanzee blood, they contain haemoglobin A_2 (Schnell, Siddiqui & Geiman, 1969; Voller, Richards, Hawkey & Ridley, 1969). In contrast the blood of *Macacus* monkeys has a lower level of G6PD, no haemoglobin A_2 and will not support the growth of *P. falciparum* (Geiman, Siddiqui & Schnell, 1969). However, some caution is called for in interpreting these findings since *Macacus* monkeys are not susceptible to *P. falciparum* in spite of the fact that the level of G6PD in their blood is similar to that of humans.

The third inherited factor now thought to give some protection against *P. falciparum* is a low erythrocyte ATP level (Brewer & Powell, 1965) Brewer (1967) has shown that this is a genetically determined feature and has found that the ATP level is lower in American Negroes than in Caucasians.

Clearly we are only just beginning to realise the importance of the role factors such as these may play in resistance or susceptibility to protozoa diseases.

Effects of hormones

There are numerous, and often conflicting, reports of the effect of hormones on the course of protozoal infections (Culbertson, 1941; Solomon, 1969). Although none of the studies to date has elucidated a biochemical mechanism for the action of host hormones on parasitic protozoa, there can be no doubt about their importance in some infections. For example, conditions in the vagina only become suitable for the development of *Trichonomas vaginalis* at the onset of puberty when the ovaries start to produce hormones (Jirovec & Petru, 1968). In this case it is not known whether the hormones exert a direct effect on *Trichomonas* or cause changes in the bacterial flora of the vagina so producing a more favourable environment.

Many workers have reported differences in the susceptibility of males and females to malaria, trypanosomiasis, leishmaniasis and amoebiasis (Solomon, 1969). Some of the most detailed work has been done with *T. cruzi*. Chagas noted that cardiomyopathies and mega conditions of the alimentary and respiratory tracts occur more often in men than in women and, in keeping with this, Hauschka (1947) and Goble (1951) found male mice to be more susceptible to *T. cruzi* than females. However, it seems that these differences in response are not due entirely to steroids; when heterologous hormones were given to immature or mature animals it was found that they had no effect on the course of *T. cruzi* infections (Goble, 1966), a result which can perhaps be taken to support Draper's (1925) comment that '. . . sex may be something more than the presence of a gonad'!

The influence of host nutrition

Two other factors of undoubted importance which influence the *in vivo* development of parasitic protozoa are the diet and nutritional state of the host. In general, malnutrition leads to a reduced ability to overcome the challenge of infection by micro-organisms, but, as we learn more about the biochemistry of parasitic protozoa, it is becoming clear that obligate parasites are obligate parasites because of biochemical, or biophysical, deficiencies for which the host compensates. These compensatory capabilities of the host may be impaired by dietary deficiencies so that host cells no longer provide an adequate milieu for growth of the parasite. In the case of malaria, it has frequently been observed in India that, during a famine, parasitaemia in the undernourished increases relatively slowly compared with development in the well-fed. Laboratory experiments support this observation: development of *Plasmodium berghei* in mice is suppressed by feeding the animals on a milk diet (Maegraith, Deegan & Sherwood-Jones, 1952). Further work (Hawking, 1953) established that the important factor missing from this diet was *p*-amino benzoic acid (PABA). Mortality from *P. berghei* infections was increased when the diet was supplemented with PABA (Kretchmar, 1965). These findings have now been confirmed for rodent, simian and avian malarias and other vitamin deficiencies, particularly vitamin C, are known to produce similar effects (Jacobs, 1964). Recently Gilbertson, Maegraith & Fletcher (1970) have reported that suppression of a *P. berghei* infection in mice by a milk diet can result in animals developing resistance to a subsequent infection. Whether such a phenomenon can occur in man by suppression of a *P. falciparum* infection during breast feeding is an interesting question which remains to be answered.

PATHOGENESIS

Many of the disturbances which occur in a host during a protozoal infection occur in other acute medical states and are basically non-specific to the parasite causing the infection (cf. Stoner, this volume). Maegraith (1968), discussing the development of malaria in the host, has stressed this point and stated that '. . . the reactions of a given organ to any form of stress are limited by its structure and function and by the reactions of its blood vessels and other auxiliary tissues'. However, the parasite initiates these reactions by starting a chain of events which leads first to local and then to general disturbances; these disturbances may initially be reversible but later lead to death of tissues or even the host. The question we should be concerned with here is: how do protozoa initiate this chain reaction? In every case the answer is short: we do not know. Some possible mechanisms based on recent work will be discussed but in all cases they must still be regarded as hypotheses.

Mechanical effects

Some protozoa are believed to produce pathological effects by sheer weight of their numbers, causing mechanical blockage of some essential physiological process rather by invasion and destruction of host tissues. The flagellate *Giardia lamblia* may act in this way. It is a parasite of the small intestine, especially the duodenum. Its effects on the host are frequently very slight but severe illness may occur and when it does, it is characterised by chronic diarrhoea, steatorrhoea and the malabsorption syndrome (O'Donovan, McGraith & Boland, 1942). The flagellate has a shallow disc on its anterior ventral side which it appears to use as a 'suction disc' to attach itself firmly to the host's intestinal epithelium. In acute cases of giardiasis the inner surface of the upper small intestine may be literally blanketed with vast numbers of trophozoites and it has been suggested that they cause a mechanical blockage of fat absorption, resulting in steatorrhoea and secondary effects caused by a deficiency of fat-soluble vitamins (Peterson, 1957; Yardley, Takano & Hendrix, 1964); however, more recent work suggests that some mucosal damage also occurs (Alp & Hislop, 1969).

Cytolysis and endocytosis

The cytolytic activity of *E. histolytica* was first reported by Councilman & Lafleur (1891). Since that time many workers have attempted, without success, to correlate virulence of isolates of this organism with production of proteinases, hyaluronidase and other enzymes which might be responsible for invasiveness. Lytic factors have been described in cell extracts but there is no evidence for the release of extracellular lytic substances (Jarumilinta & Maegraith, 1969). A new concept of the mechanism of amoebic attack has been proposed recently, based on studies of the behaviour of *E. histolytica* in cell culture systems. Jarumilinta & Kradolfer (1964), examining the leucocytocidal action of *E. histolytica* observed that leucocytes were only killed when they came into direct contact with amoebae and similar observations have now been made in experiments with HeLa and RK_{13} cell cultures (Eaton, Meerovitch & Costerton, 1969). Amoeba appear to kill mammalian cells within a minute or so of contacting them and later they engulfed them. Cytochemical and electron microscope studies (Eaton, Meerovitch & Costerton, 1970) have revealed a system of lysosomes at the surface of *E. histolytica* and each lysosome appears to have a vermiform protrusion which the authors believe is a trigger mechanism (Plate 2). Discussing these observations, Eaton, Meerovitch & Costerton propose that contact between the 'trigger' and another living cell may bring lysosomal enzymes into contact with the surface of this cell, perhaps by the external lysosomal membrane becoming depolarised following contact with a charged cell surface; this membrane may then rupture and expose an inner enzyme-bearing membrane. Normal intracellular pressure could cause the inner membrane to evert and contact the triggering organism. They suggest that in pathogenic strains of *E. histolytica* this organelle has become adapted to function under the conditions found at the surface of the intestinal mucosa thus enabling amoebae to attack and penetrate host tissues. Endocytosis may occur by the formation of a food vacuole in the area of the plasmalemma originally occupied by the lysosome, but this remains to be established. A mechanism such as this could explain the relative absence of leucocytic reaction in amoebic lesions, since any leucocyte contacting an amoeba would be killed immediately; but it is not clear from these studies how such a sensitive 'trigger' might be sheathed when *E. histolytica* is present in the intestinal lumen as a commensal.

Damage to erythrocytes by plasmodia

The developmental cycle of malaria parasites in man involves five distinct stages. Sporozoites, injected by the vector, remain in the blood-stream for a short time only before they invade liver parenchyma cells where they undergo exoerythrocytic schizogony. Merozoites released from liver cells may escape into the plasma and penetrate erythrocytes. Asexual division in the erythrocyte ends with lysis of the host cell and release of more merozoites, a few of which develop into gametocytes. The first question we must ask in considering pathogenesis in malaria is: which of these developmental stages is responsible for the symptoms of the disease? There is no evidence that sporozoites are pathogenic agents. They are structurally quite complex (Garnham, Bird & Baker, 1962) and are clearly highly efficient at locating and penetrating liver paren-chyma cells, but beyond destroying a few of these, they are not known to be responsible for other host reactions. Gametocytes are also regarded as being pathogenically inert. Maegraith (1966) discussing this question concludes that the stage developing in erythrocytes is responsible for the majority of host reactions and pathological effects of the infection. Much recent work has been concentrated on this stage and questions which are being investigated include: How do merozoites penetrate erythrocytes? How are nutrients, including host cell haemoglobin, ingested? By what mechanism are erythrocytes lysed to release mature merozoites? Are only parasitised cells lysed? Can all the varied reactions of the host observed during a malaria infection result from destruction of erythrocytes? We cannot yet give unequivocal answers to any of these questions but considerable advances have been made in the last decade.

Studies on the fine structure of merozoites (Aikawa, 1966; Hepler, Huff & Sprinz, 1966; Rudzinska & Vickerman, 1968) have shown that this stage is highly differentiated and contains a number of structures not seen in other erythrocytic forms. Of particular interest is the finding that one end of merozoites (now designated the anterior end) is conical in shape. The way in which merozoites penetrate erythrocytes was a matter for debate for many years, but the studies of Ladda, Aikawa & Sprinz (1969) seem to have resolved the question for *P. berghei* in rats

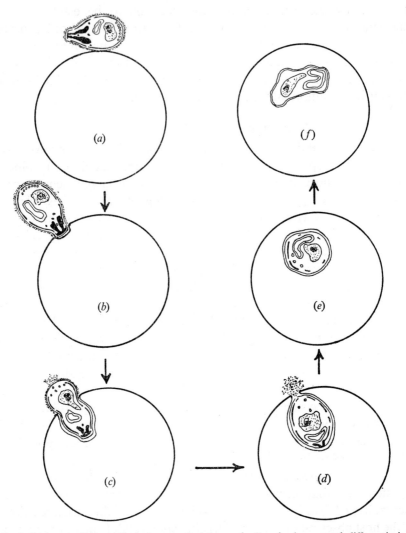

Fig. 2. Sequence of penetration of merozoite into a red cell and subsequent dedifferentiation. The extracellular merozoite (*a*) may come in contact with many red cells but the process of penetration is initiated only after the parasite becomes properly oriented (*b*) with the conoid region against the red cell. From the site of initial parasite contact the red cell membrane progressively expands to form a cavity (*c*, *d*), which eventually separates from the red cell plasmalemma to form a vacuole in which the parasite subsequently is enclosed (*d*, *e*). The site of initial contact remains relatively constricted causing distortion of the merozoite as it enters the red cell cavity. The constriction scrapes off the adherent granular material from the merozoite. Following penetration, the conoid region, paired organelles and dense inner pellicle are rapidly lost (*e*) leading to the highly amoeboid dedifferentiated trophozoite (*f*). (Ladda, Aikawa & Sprinz, 1969).

and *Plasmodium gallinaceum* in chickens. Their results confirm and extend earlier phase contrast studies (Trager, 1956; Huff, Pipkin, Weathersby & Jensen, 1960) and show that the process of penetration is initiated when merozoites contact the erythrocyte membrane with the conoid region. A depression is created which rapidly becomes a cavity as the merozoite advances. The continuity of the host cell membrane is not disrupted and eventually the parasite becomes enclosed in a vacuole within the erythrocyte by the invaginated part of the host cell membrane sealing off (Fig. 2, Plate 3). At this stage the merozoite loses its characteristic thick inner membrane and the conoid region disappears. These findings establish the origin of the envelope observed round intracellular parasites and show that they are not in direct contact with host cell cytoplasm. It seems likely that many nutrients and metabolites will be able to diffuse freely between the erythrocyte cytoplasm and the parasite but there is evidence that degradation of haemoglobin occurs in food vacuoles formed in the cytoplasm of the parasite (Rudzinska & Trager, 1959; Rudzinska, Trager & Bray, 1965).

After a period of growth, division of intracellular parasites occurs to give up to 24 daughter merozoites which are eventually freed by rupture of the erythrocyte membrane. How this occurs has not been established. Laser (1948, 1950) reported that cis-vaccenic acid is produced in increased amounts in monkey erythrocytes parasitised by *Plasmodium knowlesi*; this substance has haemolytic activity and he suggested that its accumulation in infected red cells may be responsible for their lysis. Consistent with this suggestion is the finding that plasma of animals infected with malaria contains increased quantities of $C18:1$ nonesterified fatty acids (Angus, Thurnham, Fletcher & Maegraith, 1967; Ginger, 1967; Angus, Fletcher & Maegraith, 1971). Other changes in the contents of infected monkey erythrocytes have been reported; these include increases in Na^+ (Dunn, 1969) and ATP (Fletcher, Fielding & Maegraith, 1970) and a decrease in reduced glutathione (Fletcher & Maegraith, 1970). These changes may well reflect damage to transport mechanisms and permeability barriers by circulating or intracellular haemolytic sybstances. Fletcher & Maegraith (1970) point out that reduced glutathione is responsible for the maintenance of membrane sulphydryl groups in a reduced form (Weed & Reed, 1966) and suggest that a depletion of this substance may contribute to erythrocyte lysis at schizogony. In support of this Sass (1967) has reported a close correlation between the levels of erythrocyte glutathione and a tendency to haemolysis. However, other factors may also be involved in the destruction of parasitised erythrocytes. The majority of the free fatty acids found in infected erythrocytes are produced by hydrolysis of host cell lipids rather than by *de novo* synthesis (Cenedella, Jarrell & Saxe, 1969). This intracellular breakdown of host cell material may contribute to the increased fragility of parasitised cells.

Many workers have reported that the anaemia which accompanies an infection of malaria is greater than can be accounted for by direct rupture of parasitised red blood cells (Zuckerman, 1964). Phagocytosis of both parasitised and normal erythrocytes occurs and the number of uninfected red blood cells observed within macrophages may out-number infected cells (Zuckerman, 1966). It has been proposed that an auto-immune mechanism is involved (McGhee, 1964; Cox, Schroeder & Ristic, 1965; Zuckerman, 1964, 1966), but the search for anti-erythrocyte antibodies in serum from infected individuals has yielded conflicting results (Kreier, 1969). An alternative explanation is that phagocytosis of uninfected red cells occurs because they become coated with parasite antigen. In support of this it has been found that anaemia can be induced by the injection of soluble antigens into normal animals (McGhee, 1965; Cox, 1966). Adopting a different approach Seed & Kreier (1969) found that serum from rats infected with *P. berghei* agglutinated trypsinised, but not normal erythrocytes; they suggest that the damage caused to erythrocyte membranes during a malaria infection may be qualitatively similar to that caused by trypsin treatment. Further work (Seed & Kreier, 1969) has shown that lipids isolated from trypsin-treated erythrocytes interact with antibodies found in the serum of infected animals. The authors speculate that these antibodies may be produced in response to membrane fragments released when parasitised cells rupture. Normal red cells could then be 'marked' for phagocytosis either by adsorbing lipid material from ruptured cells or by the action of proteolytic enzymes released from such cells; these enzymes might remove protein units from the cell surface so exposing membrane lipids which could then interact with antibody. Which of these various possibilities provides the true explanation for the phagocytosis of uninfected erythrocytes remains to be established.

Release of pharmacologically active substances

Malaria

In addition to fever and anaemia, other host reactions resulting from a malaria infection include general circulatory disturbances (such as increases in Ig M and Ig G), general pathological changes in bone marrow, spleen, kidney and other organs. In the terminal stages of the disease a breakdown of the blood–brain barrier occurs resulting in cerebral malaria. Many of these changes are typical of a generalised anoxia; whether this is due to an inadequate supply of oxygen to the tissues (anoxic anoxia) or to an inability of cells to use the oxygen they receive has been considered in detail by Maegraith and co-workers (Maegraith, Jones & Andrews, 1951; Jones, Maegraith & Sculthorpe, 1951; Maegraith, 1966). They found no evidence that oxygen carriage by, or discharge from, haemoglobin is reduced by an infection of malaria and they conclude that the anaemia which occurs is never great enough to cause the total oxygen carriage to fall below the basic tissue requirements. If this is so inhibition of tissue respiration must arise from other more direct causes; and Maegraith (1966) considers that there is evidence of histotoxic anoxia in late stages of acute plasmodial infections in mammals. Changes in the respiration and oxidative phosphorylation of liver cell mitochondria have been found during malaria infections (Riley & Deegan, 1960; Riley & Maegraith, 1962) and serum from infected animals and man was found to cause similar biochemical lesions in normal liver mitochondria (Riley & Maegraith, 1961). This serum also contained factors which cause vasoconstriction and break-down of the blood–brain barrier; the production of these factors appears to be initiated by the erythrocytic cycle of development of *Plasmodium*, but their identity has not yet been established. The non-esterified fatty acids already discussed have been implicated (Maegraith, 1966; Angus, Fletcher & Maegraith, 1971) and there is evidence that kinins may have important effects on the circulation in acute malaria (Tella & Maegraith, 1966). An increased concentration of kininogenase has been found in the blood of Rhesus monkeys infected with *P. knowlesi*. This proteolytic enzyme is known to split off small peptides, kinins, from plasma kininogen (Rocha e Silva, Beraldo & Rosenfeld, 1949); these increase endothelial permeability and act as vasodilators. Both these activities may contribute to the impedance of blood flow and stasis which occurs in small blood vessels of the brain and certain other tissues in an acute malaria infection. Kininogenase fractions prepared from serum of infected monkeys have recently been reported to cause a breakdown of the blood–brain barrier when injected intracerebrally into guinea-pigs (Onabanjo & Maegraith, 1970a, b). All these and other findings indicate that pharmacologically active substances do appear in the bloodstream during a malaria infection and are responsible for some of the host reactions to the disease. We are still far from knowing the precise nature of these substances or understanding the mechanism of their action. Whether any of them can be regarded as a true 'malaria toxin' remains an open question.

African trypanosomiasis

In recent reviews on the pathology of African trypanosomiasis both Ormerod (1970) and Goodwin (1970) stress the fact that growth of trypanosomes in man occurs mainly in the tissues and that this is often overlooked by workers used to studying laboratory-adapted strains in rodents which can produce an overwhelming parasitaemia in the blood-stream and kill a mouse in a few days.

A characteristic of *brucei* group trypanosomes in man and many animals is a succession of peaks of parasitaemia, each representing a population of immunologically distinct trypanosomes. Antibodies develop to variants, each rise in antibody titre being followed by the disappearance of homologous organisms and, when they have been eliminated, a new antigenic type grows up (Gray, 1962). When host defence mechanisms fail the final variant continues to multiply (Weitz, 1962).

Damage to connective tissue during an infection is extensive (Goodwin, 1971) and perivascular oedema and intravascular clotting occur as a result of damage to capillary walls (Hornby, 1949). Much effort has been put into the search for a trypanosome toxin but with little success and there is still no satisfactory explanation for the destructive action of trypanosomes on connective tissue.

Goodwin (1970) suggests that many of the host responses associated with trypanosomiasis may be due to allergic reactions to successive antigenic variants and he draws attention to the fact that antigen–antibody reactions are accompanied by release of a variety of physiologically active substances such as kinins and histamine (Brocklehurst, 1960). No significant changes have been detected in the histamine content of blood or tissues of mice infected with *T. brucei* (Yates, 1970) but peptides of the kinin type have been found in blood of infected mice (Goodwin & Richards, 1960; Richards, 1965) and cattle (Boreham, 1968*a*). In cattle, kinin appearance coincided with a peak of parasitaemia and the appearance of antibodies. Recently kinins have also been detected in human trypanosomiasis (Boreham, 1970).

Studies on the components of the kinin-kininogenase system suggest that trypanosomes may initiate the release of kinin peptides from plasma proteins by activating the precursor of kininogenase (Boreham, 1968*b*); it has been suggested that this activation might result from the absorption of Hageman factor onto the antigen–antibody complex (Boreham & Goodwin, 1969). As already mentioned an important property of kinins is to increase the permeability of blood vessels and it now seems likely from the elegant studies of Goodwin (1970, 1971) that these peptides play a very important role in producing the vascular changes and oedema associated with the chronic form of trypanosomiasis. However, the possibility that a trypanosome toxin contributes to these effects can not yet be ruled out. Recently Seed (1969) has described the preparation of a protein fraction from homogenates of *Trypanosoma gambiense* which increased vascular permeability and caused oedema when injected intradermally into animals. Whether this fraction acts directly or stimulates the release of kinins is not known.

Chagas' disease

In contrast to the African trypanosomes *Trypanosoma (Schizotrypanum) cruzi* occurs in man both as a motile trypomastigote form in the bloodstream and as a non-motile, intracellular amastigote form.

After penetrating through a skin abrasion or mucous membrane trypanosomes eventually leave the site of the primary lesion and reach the bloodstream where they circulate for a period of time, then they invade cells of visceral tissues and transform immediately to amastigote forms. The most commonly parasitised tissues are heart, skeletal and smooth muscle and neuroglia. What factors stimulate trypanosomes to leave the bloodstream and how they select, preferentially, muscle and glial cells is unknown. Trypanosomes appear to leave the bloodstream soon after entering the venous circulation; they do so by actively penetrating the capillary wall, posterior (i.e. non-flagellated) end first; they later penetrate the plasma membrane of the host cell in a similar manner. Electron microscope studies have shown that there is a vesicle at the posterior end of trypomastigote forms of *T. cruzi* and it has been suggested that this may contain enzymes which are used to aid the penetration process (Meyer, Oliveria Musachi & Andrade Mendoza, 1958). Once established in the host cell, amastigote forms continue to divide over a period of about 5 days, forming a cyst-like cavity or pseudocyst (Plate 4) within the host cell. Initially amastigote forms in the pseudocyst appear to be uniform in size, but after a time differences can be seen and at the same time a small amount of fluid accumulates in the pseudocyst and transformation to motile trypomastigote forms begins. The number of amastigotes which begin to transform is variable but only a small number develop into mature trypomastigotes and leave the pseudocyst by penetrating the cell membrane, again, posterior end first. Once the pseudocyst has ruptured all further transformation ceases and remaining amastigotes disintegrate (Romana & Meyer, 1942).

It is now firmly established that cardiomyopathy is the most important and frequent manifestation of acute and chronic Chagas' disease (Köberle, 1968; WHO, 1969). Cardiomegalies are common and are accompanied by thinning of the muscle wall and apical aneurysms (Plate 4). In the acute stages of the disease massive destruction of ganglia occurs in the central and peripheral autonomic systems. This denervation is accompanied by hypertrophy and dilation of hollow organs, particularly the oesophagus and colon, giving rise to the 'mega conditions' which are now recognised as a characteristic of acute Chagas' disease as it occurs in southern South America. Köberle (1968) has made a detailed study of the extent of denervation of a number of organs in the acute form of the disease; he finds that it may be as high as 95 % in the oesophagus and 55 % in the colon. He has also observed that ganglion destruction occurs only in the region of ruptured pseudocysts. He postulates that disintegration of amastigote forms releases a neurotoxic or neurolytic substance but no specific factors have yet been isolated. It is interesting that mega organs apparently do not occur in northern South America or Central America; this has been attributed to strain differences. Lesions of the reticulo-endothelial system appear to predominate in infections with a Columbian strain and lesions of the digestive tract with a Brazilian strain (WHO, 1969). Clearly comparative biochemical studies on these strains would be of the greatest interest and might throw some light on the nature of the pathogenicity of *T. cruzi*.

CONCLUSIONS

If it does nothing else, this survey shows how little we know of the chemical basis of protozoal pathogenicity. In many cases the pathology of protozoal diseases is still incompletely described. What knowledge there is of the biochemistry and physiology of pathogenic protozoa is often restricted to a single stage in a complex developmental cycle or even to a 'culture form' which may be avirulent or non-infective. We still do not understand the mechanism of antigenic variation that enables some protozoa to stay a jump ahead of host defence mechanisms nor how others have adapted themselves to invade and develop in the phagocytes that should be concerned with their removal. Investigation of all these problems is hindered by inadequate culture techniques. Howie & O'Hea (1955) prefaced the 5th Symposium of this Society with the statement 'discussions on how micro-organisms produce disease are very apt to follow a circular course to platitudinous conclusions'. This is still a very real danger in discussing mechanisms of protozoal pathogenicity: it is far too early to draw any conclusions.

I wish to thank: Dr J. R. Baker for his critical reading of the manuscript and for his many helpful comments; Dr R. A. Neal for allowing me to read a typescript before its publication; Professor F. Köberle for providing Plate 3, fig. 2. I am also indebted to the editors or copyright holders of *Advances in Parasitology* (Plate 4, Fig. 1); *Annals of Tropical Medicine and Parasitology* (Plate 2); *Journal of Cell Science* (Plate 1) and the *Journal of Parasitology* (Fig. 2 and Plate 3) for permission to reproduce plates and diagrams.

REFERENCES

AIKAWA, M. (1966). The fine structure of the erythrocytic stages of three avian malaria parasites, *Plasmodium fallax, P. lophurae* and *P. cathemerium. American Journal of Tropical Medicine and Hygiene*, 15, 449–71.

ALLISON, A. C. (1954). Protection afforded by sickle-cell trait against subtertian malaria infection. *British Medical Journal*, 1, 290–4.

ALP, M. H. & HISLOP, I. G. (1969). The effect of *Giardia lamblia* infestation on the gastro-intestinal tract. *Australasian Annals of Medicine*, 18, 232–7.

ANGUS, M. G. N., FLETCHER, K. A. & MAEGRAITH, B. G. (1971). Studies on the lipids of *Plasmodium knowlesi*-infected rhesus monkeys (*Macaca mulatta*). II. Changes in serum non-esterified fatty acids. *Transactions of the Royal Society of Tropical Medicine and Hygiene*, 65, 155–67.

ANGUS, M. G. N., THURNHAM, D. I., FLETCHER, K. A. & MAEGRAITH, B. G. (1967). Biochemical aspects of the pathogenesis of malaria. II. Gas chromatography of serum non-esterified fatty acids in *Plasmodium knowlesi* malaria. *Transactions of the Royal Society of Tropical Medicine and Hygiene*, 61, 4.

BAKER, J. R. (1969). *Parasitic Protozoa*. London: Hutchinson.

BIAGI-F, F. & BELTRAN-H, F. (1969). The challenge of Amoebiasis: understanding pathogenic mechanisms. *International Review of Tropical Medicine*, 3, 219–39.

BIAGI-F, F., ROBLEDO, E., SERVIN, H. & MARTUSCELLI-Q, A. (1962). The effect of cholesterol on the pathogenicity of *Entamoeba histolytica. American Journal of Tropical Medicine and Hygiene*, 11, 333–40.

BIGGS, P. M., LONG, P. L., KENZY, S. G. & ROOTES, D. G. (1968). Relationship between Marek's disease and coccidiosis. II. The effect of Marek's disease on the susceptibility of chickens to coccidial infections. *The Veterinary Record*, 83, 284–9.

BOREHAM, P. F. L. (1968a). Immune reactions and kinin formation in chronic trypanosomiasis. *British Journal of Pharmacology and Chemotherapy*, 32, 493–504.

BOREHAM, P. F. L. (1968b). *In vitro* studies on the mechanism of kinin formation by trypanosomes. *British Journal of Pharmacology and Chemotherapy*, 34, 598–603.

BOREHAM, P. F. L. (1970). Kinin release and the immune reaction in human trypanosomiasis caused by *Trypanosoma rhodesiense. Transactions of the Royal Society of Tropical Medicine and Hygiene*, 64, 394–400.

BOREHAM, P. F. L. & GOODWIN, L. G. (1969). Release of kinins as the result of an antigen–antibody reaction in trypanosomiasis. *Pharmacological Research Communications*, 1, 144–5.

BREWER, G. J. (1967). Genetic and population studies of quantitative levels of adenosine triphosphate in human erythrocytes. *Biochemical Genetics*, 1, 25–34.

BREWER, G. J. & POWELL, R. D. (1965). A study of the relationship between the content of adenosine triphosphate in human red cells and the course of falciparum malaria: a new system that may confer protection against malaria. *Proceedings of the National Academy of Sciences, U.S.A.*, **54**, 741–5.

BROCKLEHURST, W. E. (1960). The release of histamine and the formation of a slow reacting substance (SRS-A) during anaphylactic shock. *Journal of Physiology, London*, **151**, 416–35.

BURKITT, D. P. (1969). Etiology of Burkitt's lymphoma – an alternative hypothesis to a vectored virus. *Journal of the National Cancer Institute*, **42**, 19–28.

CENEDELLA, R. J., JARRELL, J. J. & SAXE, L. (1969). *Plasmodium berghei*: production *in vitro* of free fatty acids. *Experimental Parasitology*, **24**, 130–6.

CLARK, T. B. & WALLACE, F. G. (1960). A comparative study of kinetoplast ultrastructure in the Trypanosomatidae. *Journal of Protozoology*, **7**, 115–24.

COUNCILMAN, W. T. & LAFLEUR, H. A. (1891). Amoebic dysentery. *Johns Hopkins Hospital Reports*, **2**, 395–402.

COX, H. W. (1966). A factor associated with anaemia and immunity in *Plasmodium knowlesi* infection. *Military Medicine*, **131**, 1195–200.

COX, H. W., SCHROEDER, W. F. & RISTIC, M. (1965). Erythrophagocytosis associated with anaemia in rats infected with *Plasmodium berghei*. *Journal of Parasitology*, **51**, 35–6.

CULBERTSON, C. G. (1961). Pathogenic Acanthamoeba (*Hartmanella*). *The American Journal of Clinical Pathology*, **35**, 195–202.

CULBERTSON, C. G. (1970). Pathogenic free-living amoebae. *Industry and Tropical Health*, **VII**, 118–23.

CULBERTSON, C. G. (1971). *Annual Review of Microbiology*, **25**, (In press).

CULBERTSON, J. T. (1941). *Immunity against animal Parasites*. New York: Columbia University Press.

DIAMOND, L. S. (1968). Techniques for axenic cultivation of Entamoeba histolytica-like amoebae. *Journal of Parasitology*, **54**, 1047–56.

DRAPER, G. (1925). The influence of sex upon the constitutional factor in disease. *New York State Journal of Medicine*, **25**, 1065–70.

DUNN, M. J. (1969). Alterations of red blood cell metabolism in simian malaria: evidence for abnormalities of nonparasitised cells. *Military Medicine*, **134**, 1100–5.

EATON, R. D. P., MEEROVITCH, E. & COSTERTON, J. W. (1969). A surface active lysosome in *Entamoeba histolytica*. *Transactions of the Royal Society of Tropical Medicine and Hygiene*, **63**, 678–80.

EATON, R. D. P., MEEROVITCH, E. & COSTERTON, J. W. (1970). The functional morphology of pathogenicity in *Entamoeba histolytica*. *Annals of Tropical Medicine and Parasitology*, **64**, 299–304.

EL-HASHMIT, W. & PITTAM, F. (1970). Ultra-structure of *Entamoeba histolytica* trophozoites obtained from the colon and from *in vitro* cultures. *American Journal of Tropical Medicine and Hygiene*, **19**, 215–26.

FLETCHER, K. A., FIELDING, C. M. & MAEGRAITH, B. G. (1970). Studies on the role of adenosine phosphates in erythrocytes of *Plasmodium knowlesi* infected monkeys. *Annals of Tropical Medicine and Parasitology*, **64**, 487–96.

FLETCHER, K. A. & MAEGRAITH, B. G. (1970). Erythrocyte reduced glutathione in malaria (*Plasmodium berghei* and *Plasmodium knowlesi*). *Annals of Tropical Medicine and Parasitology*, **64**, 481–6.

FULTON, J. D. (1969). Metabolism and pathogenic mechanisms of parasitic protozoa. In *Research in Protozoology*, ed. T. Chen, III, 390–504. London: Pergamon.

FLYNN, I. W. & BOWMAN, I. R. B. (1970). Comparative biochemistry of mono-morphic and pleomorphic strains of *Trypanosoma rhodesiense*. *Transactions of the Royal Society of Tropical Medicine and Hygiene*, **64**, 175–6.

GARNHAM, P. C. C. (1966). *Malaria Parasites and Other Haemosporidia*, 1–1114. Oxford: Blackwell.

GARNHAM, P. C. C., BIRD, R. G. & BAKER, J. R. (1962). Electron microscope studies of motile stages of malaria parasites. III. The ookinetes of *Haemamoeba* and *Plasmodium*. *Transactions of the Royal Society of Tropical Medicine and Hygiene*, **56**, 116–20.

GARNHAM, P. C. C., PIERCE, A. E. & ROITT, I. (1963). *Immunity to Protozoa. Symposium of the British Society for Immunology*. Oxford: Blackwell.

GAVALLER, B. (1951). Enfermedad de Chagas congenita. *Boletin de la Maternidad 'Conception Palacios'*, **4**, 59–64.

GEIMAN, Q. M., SIDDIQUI, W. A. & SCHNELL, J. V. (1969). Biological basis for susceptibility of *Aotus trivirigatus* to species of plasmodia from man. *Military Medicine*, **134**, 780–6.

GILBERTSON, M., MAEGRAITH, B. G. & FLETCHER, K. A. (1970). Resistance to superinfection with *Plasmodium berghei* in mice in which the original infection was suppressed by a milk diet. *Annals of Tropical Medicine and Parasitology*, **64**, 497–512.

GILLES, H. M., FLETCHER, K. A., HENDRICKSE, R. G., LINDNER, R., REDDY, S. & ALLEN, N. (1967). Glucose-6-phosphate dehydrogenase deficiency, sickling and malaria in African children in South Western Nigeria. *The Lancet*, **i**, 138–40.

GINGER, C. D. (1967). Preliminary studies on the lipid constitution of the malaria parasite. *Transactions of the Royal Society of Tropical Medicine and Hygiene*, **61**, 2–3.

GOBLE, F. C. (1951). Studies on experimental Chagas' disease in mice in relation to chemotherapeutic testing. *Journal of Parasitology*, **37**, 408–14.

GOBLE, F. C. (1966). Pathogenesis of blood protozoa. In *Biology of Parasites*, ed. E. J. L. Soulsby. New York: Academic Press.

GODFREY, D. G. & TAYLOR, A. E. R. (1969). Studies of the surface of trypanosomes. *Transactions of the Royal Society of Tropical Medicine and Hygiene*, **63**, 115–16.

GOODWIN, L. G. (1970). The pathology of African trypanosomiasis. *Transactions of the Royal Society of Tropical Medicine and Hygiene*, **64**, 797–812.

GOODWIN, L. G. (1971). Pathological effects of *Trypanosoma brucei* on small blood vessels in rabbit ear chambers. *Transactions of the Royal Society of Tropical Medicine and Hygiene*, **65**, 82–8.

GOODWIN, L. G. & RICHARDS, W. H. G. (1960). Pharmacologically active peptides in the blood and urine of animals infected with *Babesia rodhani* and other pathogenic organisms. *British Journal of Pharmacology and Chemotherapy*, **15**, 152–9.

GORDON, R. M., CREWE, W. & WILLETT, K. C. (1956). Studies on the deposition, migration and development to the blood forms of trypanosomes belonging to the *T. brucei* group. *Annals of Tropical Medicine and Parasitology*, **50**, 426–37.

GRANT, P. T. & SARGENT, J. R. (1960). Properties of L-α-glycerophosphate oxidase and its role in the terminal respiration of *Trypanosome rhodesiense*. *Biochemical Journal*, **76**, 229–37.

GRAY, A. R. (1962). The influence of antibody on serological variation in *Trypanosoma brucei*. *Annals of Tropical Medicine and Parasitology*, **56**, 4–13.

GRIFFITHS, A. J. (1970). Encystment in Amoebae. *Advances in Microbial Physiology*, **4**, 106–29.

HAUSCHKA, T. S. (1947). Sex of host as a factor in Chagas' disease. *Journal of Parasitology*, **33**, 399–404.

HAWKING, F. (1953). Milk diet, p-aminobenzoic acid and malaria (P. berghei). British Medical Journal, 1, 1201–2.

HEGNER, R. W., JOHNSON, C. M. & STABLER, R. M. (1932). Host parasite relations in experimental amoebiasis in Panama. American Journal of Hygiene, 15, 394–443.

HEPLER, P. K., HUFF, C. G. & SPRINZ, H. (1966). The fine structure of the erythrocytic stages of Plasmodium fallax. Journal of Cell Biology, 30, 333–58.

HOARE, C. A. (1950). Amoebiasis in Great Britain, with special reference to carriers. British Medical Journal, 11, 238–41.

HOARE, C. A. (1966). The classification of mammalian trypanosomes. Ergebrüsse der Mikrobiologie, Immunitätsforschung und Experimentellen Therapie, 39, 43–57.

HONIGBERG, B. M. (1967). Chemistry of Parasitism among some protozoa. In Chemical Zoology, ed. M. Florkin, B. T. Scheer and G. W. Kidder, 695–814. New York: Academic Press.

HONIGBERG, B. M. (1970). Effect of cultivation on pathogenicity and antigenicity of trichomonads. Journal of Parasitology, 56, 153–4.

HONIGBERG, B. M., BECKER, R. D., LIVINGSTON, M. C. & McLURE, M. T. (1964). The behaviour and pathogenicity of two strains of Trichomonas gallinae in cell cultures. Journal of Protozoology, 11, 447–65.

HORNBY, H. E. (1949). Animal Trypanosomiasis in Eastern Africa. London: H.M.S.O.

HOWIE, J. W. & O'HEA, A. J. (1955). Mechanisms of Microbial Pathogenicity. Fifth Symposium of Society for General Microbiology. London: Cambridge University Press.

HUFF, C. A., PIPKIN, A., WEATHERSBY, A. & JENSEN, D. (1960). The morphology and behaviour of living exoerythrocytic stages of Plasmodium gallinaceum and P. fallax and their host cells. Journal of Biophysical and Biochemical Cytology, 7, 93–101.

HUTCHISON, W. M., DUNACHIE, J. F. & WORK, K. (1970). Coccidian-like nature of Toxoplasmas gondii. British Medical Journal, 1, 142–4.

HUTCHISON, W. M. & WORK, K. (1969). Observations on the faecal transmission of Toxoplasma gondii. Acta Pathologica et microbiologica scandinavica, 77, 275–82.

JACOBS, R. L. (1964). Role of p-aminobenzoic acid in Plasmodium berghei infection in the mouse. Experimental Parasitology, 15, 213–25.

JACOBS, L. (1967). Toxoplasma and Toxoplasmosis. Advances in Parasitology, 5, 11–145.

JACOBS, L., REMINGTON, J. S. & MELTON, M. L. (1960). The resistance of the encysted form of Toxoplasma gondii. Journal of Parasitology, 46, 11–21.

JAMES, S. P. & TATE, P. (1937). New knowledge of the life cycle of malaria parasites. Nature, London, 139, 545.

JARUMILINTA, R. & KRADOLFER, F. (1964). The toxic effect of Entamoeba histolytica on leucocytes. Annals of Tropical Medicine and Parasitology, 58, 375–81.

JARUMILINTA, R. & MAEGRAITH, B. G. (1969). Enzymes of Entamoeba histolytica. Bulletin of the World Health Organisation, 41, 269–73.

JIROVEC, O. & PETRU, M. (1968). Trichomonas vaginalis and Trichomoniasis. Advances in Parasitology, 6, 117–88.

JONES, E. S., MAEGRAITH, B. G. & SCULTHORPE, H. H. (1951). Pathological processes in disease. III. The oxygen uptake of blood from albino rats infected with Plasmodium berghei. Annals of Tropical Medicine and Parasitology, 45, 244–52.

KÖBERLE, F. (1968). Chagas' disease and Chagas' syndromes: The pathology of American trypanosomiasis. Advances in Parasitology, 6, 63–116.

KREIER, J. P. (1969). Mechanisms of erythrocyte destruction in chickens infected with *Plasmodium gallinaceum*. *Military Medicine*, **134**, 1203–19.

KRETCHMAR, W. (1965). The effects of stress and diet on resistance to *Plasmodium berghei* and malarial immunity in the mouse. *Annales de la Societé belge de médécine tropicale*, **45**, 325–44.

LADDA, R., AIKAWA, M. & SPRINZ, H. (1969). Penetration of erythrocytes by merozoites of mammalian and avian malarial parasites. *Journal of Parasitology*, **55**, 633–44.

LASER, H. (1948). Haemolytic system in the blood of malaria infected monkeys. *Nature, London*, **161**, 560.

LASER, H. (1950). The isolation of a haemolytic substance from animal tissues and its biological properties. *Journal of Physiology*, **110**, 338–55.

LOWE, C. Y. & MAEGRAITH, B. G. (1970). Electron microscopy of an axenic strain of *Entamoeba histolytica*. *Annals of Tropical Medicine and Parasitology*, **64**, 293–8.

LYCKE, E., NORRBY, R. & REMINGTON, J. (1968). Penetration-enhancing factor extracted from *Toxoplasma gondii* which increases its virulence for mice. *Journal of Bacteriology*, **93**, 785–8.

MAEGRAITH, B. G. (1966). Pathogenic processes in malaria. In *The Pathology of Parasitic Diseases. Fourth Symposium. British Society for Parasitology*, ed. A. E. R. Taylor, 15–32. Oxford: Blackwell.

MAEGRAITH, B. G. (1968). Liver involvement in acute mammalian malaria with special reference to *Plasmodium knowlesi* malaria. *Advances in Parasitology*, **6**, 189–231.

MAEGRAITH, B. G., DEEGAN, T. & SHERWOOD-JONES, E. (1952). Suppression of malaria (*P. berghei*) by milk. *British Medical Journal*, **2**, 1382–4.

MAEGRAITH, B. G., JONES, E. S. & ANDREWS, W. H. H. (1951). Pathological processes in malaria. Progress report. *Transactions of the Royal Society of Tropical Medicine and Hygiene*, **45**, 15–33.

MCCONNACHIE, E. W. (1969). The morphology, formation and development of cysts of *Entamoeba*. *Parasitology*, **59**, 41–53.

MCGHEE, R. B. (1964). Autoimmunity in malaria. *American Journal of Tropical Medicine and Hygiene*, **13**, 219–24.

MCGHEE, R. B. (1965). Erythrophagocytosis in ducklings injected with malarious plasma. In *Progress in Protozoology. Proceedings of the 2nd International Congress of Protozoology*, London, 171.

MEYER, H., OLIVEIRA MUSACHI, M. & ANDRADE MENDOZA (1958). Electron microscope study of *Trypanosoma cruzi* in thin sections of infected tissue culture and blood-agar forms. *Parasitology*, **48**, 1–9.

MOTULSKY, A. G. (1964). Hereditary red cell traits and malaria. *American Journal of Tropical Medicine and Hygiene*, **13**, 147–58.

MOULDER, J. W. (1962). *The Biochemistry of Intracellular Parasitism*. Chicago: University Press.

MÜHLPFORDT, H. (1964). Über den kinetoplasten der Flagellaten. *Zeitschrift für Tropenmedizin und Parasitologie*, **15**, 289–323.

NEAL, R. A. (1958). The pathogenicity of *Entamoeba histolytica*. *Proceedings of the 6th International Congress of Tropical Medicine and Malaria*, **III**, 350–9.

NEAL, R. A. (1971). Pathogenesis of Amoebiasis. *Bulletin of the New York Academy of Medicine*, **47**, 462–8.

NEWTON, B. A. (1970). Chemotherapeutic compounds affecting DNA structure and function. *Advances in Pharmacology and Chemotherapy*, **8**, 149–84.

NORRBY, R., LINDHOLM, L. & LYCKE, E. (1968). Lysosomes of *Toxoplasma gondii* and their possible relation to the host-cell penetration of *Toxoplasma gondii*. *Journal of Bacteriology*, **96**, 916–19.

NORRBY, R. & LYCKE, E. (1967). Factors enhancing the host-cell penetration of *Toxoplasma gondii*. *Journal of Bacteriology*, **93**, 53–8.

O'DONOVAN, D. K., MCGRAITH, J. & BOLAND, S. J. (1942). Giardial infestation with steatorrhoea. *The Lancet*, **ii**, 4–6.

OKAMOTO, J. (1954). On the influence of *Clostridium perfringens* upon the experimental infection of *Entamoeba histolytica* in rats. *Keio Journal of Medicine*, **3**, 121–7.

ONABANJO, A. O. & MAEGRAITH, B. G. (1970a). Inflammatory changes in small blood vessels induced by kallikrein (kininogenase) in the blood of *Macaca mulatta* infected with *Plasmodium knowlesi*. *Annals of Tropical Medicine and Parasitology*, **64**, 227–36.

ONABANJO, A. O. & MAEGRAITH, B. G. (1970b). Pathological lesions produced in the brain by kallikrein (kininogenase) in *Macaca mulatta* infected with *Plasmodium knowlesi*. *Annals of Tropical Medicine and Parasitology*, **64**, 237–42.

ORMEROD, W. E. (1970). Pathogenesis and pathology of trypanosomiasis in man. In *The African Trypanosomiases*, ed. H. W. Mulligan and W. H. Potts, 587–601 London: Allen and Unwin.

PETERSON, J. M. (1957). Intestinal changes in *Giardia lamblia* infestation. *American Journal of Roentgenology*, **77**, 670–7.

PHILLIPS, B. P. & BARTGIS, I. L. (1954). Effects of growth *in vitro* with selected microbial associates and of encystation and excystation on the virulence of *Endamoeba histolytica* for guinea-pigs. *American Journal of Tropical Medicine and Hygiene*, **3**, 621–7.

PHILLIPS, B. P., WOLFER, P. A., RUS, C. W., GORDAN, H. A., WRIGHT, W. H. & REYNIERS, J. A. (1955). Studies on the amoeba-bacteria relationship in amoebiasis. Comparative results of the intercaecal inoculation of germ-free, monocontaminated and conventional guinea-pigs with *Entamoeba histolytica*. *American Journal of Tropical Medicine and Hygiene*, **4**, 675–92.

REICHENOW, E. (1921). Untersuchungen neber das verhalten von Trypanosoma gambiense in menschlichen korper. *Zeitschrift für Hygiene und Infectionskrankheiten*, **94**, 266–85.

RICHARDS, W. H. G. (1965). Pharmacologically active substances in the blood, tissues and urine of mice infected with *Trypanosoma cruzi*. *British Journal of Pharmacology and Chemotherapy*, **24**, 124–31.

RILEY, M. V. & DEEGAN, T. (1960). The effect of *Plasmodium berghei* malaria on mouse liver mitochondria. *Biochemical Journal*, **76**, 41–6.

RILEY, M. V. & MAEGRAITH, B. G. (1961). A factor in the serum of malaria infected animals capable of inhibiting the *in vitro* oxidative metabolism of normal liver mitochondria. *Annals of Tropical Medicine and Parasitology*, **55**, 489–97.

RILEY, M. V. & MAEGRAITH, B. G. (1962). Changes in the metabolism of liver mitochondria of mice infected with rapid acute *Plasmodium berghei* malaria. *Annals of Tropical Medicine and Parasitology*, **56**, 473–82.

ROBERTSON, M. (1912). Notes on the polymorphism of *Trypanosoma gambiense* in the blood and its relation to the exogenous cycle in *Glossina palpalis*. *Proceedings of the Royal Society, London*, B, **85**, 527–9.

ROCHA E SILVA, M., BERALDO, W. T. & ROSENFELD, G. (1949). Bradykinin, a hypotensive and smooth muscle stimulating factor released from plasma globulin by snake venom and by trypsin. *American Journal of Physiology*, **156**, 261–73.

ROMANA, C. & MEYER, H. (1942). Estudo do ciclo evolutivo do *Schizotrypanum cruzi* em cultura de tecido de embriao de galinha. *Memorias del Instituto Oswaldo Cruz*, **37**, 19–27.

RONDANELLI, E. G., CAROSI, G., GERNA, G. & DE CARNERI, I. (1968). Sul reperts di corpi filamentosi paranucleari in *Entamoeba histolytica in vitro*. Richerche microelettroniche. *Bollettino dell'Instituto sieroterapico milanese, Milano*, **47**, 401-5.

RUDZINSKA, M. & TRAGER, W. (1959). Phagotrophy and two new structures in the malaria parasite. *Journal of Biophysical and Biochemical Cytology*, **6**, 103-12.

RUDZINSKA, M., TRAGER, W. & BRAY, R. (1965). Pinocytotic uptake and digestion of haemoglobin in malaria parasites. *Journal of Protozoology*, **12**, 563-76.

RUDZINSKA, M. & VICKERMAN, K. (1968). The fine structure. In *Infections Blood Diseases of Man and Animals*, ed. D. Weinman and M. Ristic, 214-73. New York: Academic Press.

RYLEY, J. F. (1962). Studies on the metabolism of the Protozoa. 9. Comparative metabolism of bloodstream and culture forms of *Trypanosoma rhodesiense*. *Biochemical Journal*, **85**, 211-23.

RYLEY, J. F. (1966). Histochemical studies on blood and culture forms of *Trypanosoma rhodesiense*. In *Proceedings of the 1st International Congress of Parasitology*, **1**, 41-2.

SALAMAN, M. H., WEDDERBURN, N. & BRUCE-CHWATT, L. J. (1969). The immunodepressive effect of a murine plasmodium and its interaction with murine oncogenic viruses. *Journal of General Microbiology*, **59**, 383-91.

SASS, M. D. (1967). Glutathione stability and glucose-6-phosphate dehydrogenase activity in red cells of different ages. *Clinica chimica acta*, **15**, 1-6.

SCHNELL, J. V., SIDDIQUI, W. A. & GEIMAN, Q. M. (1969). Analyses on the blood of normal monkeys and owl monkeys infected with *Plasmodium falciparum*. *Military Medicine*, **134**, 1068-73.

SEED, J. R. (1969). *Trypanosoma gambiense* and *T. lewisi:* increased vascular permeability and skin lesions in rabbits. *Experimental Parasitology*, **26**, 214-23.

SEED, T. M. & KREIER, J. P. (1969). Autoimmune reactions in chickens with *Plasmodium gallinaceum* infection: The isolation and characterisation of a lipid from trypsinized erythrocytes which reacts with serum from acutely infected chickens. *Military Medicine*, **134**, 1220-7.

SHUTE, P. G. (1964). Quoted in Garnham, 1966, p. 11.

SINGH, B. N. & DAS, S. R. (1970). Studies on pathogenic and non-pathogenic small free-living amoebae and the bearing of nuclear division in the classification of the order Amoebida. *Philosophical Transactions of the Royal Society, London*, **259**, 435-76.

SINISCALCO, M., BENINI, L., FILIPPI, G., LATTE, B., KAHN, P. M., PIOMELLI, S. & RATTAZZI, M. (1966). Population genetics of haemoglobin variants, thalassaemia and glucose-6-phosphate dehydrogenase deficiency, with particular reference to the malaria hypothesis. *Bulletin of the World Health Organisation*, **34**, 379-93.

SOLOMON, G. B. (1969). Host hormones and parasitic infection. *International Review of Tropical Medicine*, **3**, 101-58.

STABLER, R. M., HONIGBERG, B. M. & KING, V. M. (1964). Effect of certain laboratory procedures on virulence of the Jones' Barm strain of *Trichomonas gallinae* for pigeons. *Journal of Parasitology*, **50**, 36-41.

STEINERT, M. (1960). Mitochondria associated with the kinetonucleus of *Trypanosoma mega*. *Journal of Biophysical and Biochemical Cytology*, **8**, 542-6.

SWARTZWELDER, J. C. (1939). Experimental studies on *Entamoeba histolytica* in the dog. *American Journal of Hygiene*, **29**, 89-109.

TAYLOR, A. E. R. (1968). *Immunity to Parasites. 6th Symposium, British Society for Parasitology*. Oxford: Blackwell.

TAYLOR, A. E. R. & BAKER, J. R. (1968). *The Cultivation of Parasites in vitro.* Oxford: Blackwell.

TAYLOR, D. J., GREENBERG, J. & JOSEPHSON, E. S. (1952). The effect of two different diets on experimental amoebiasis in the guinea-pig and the rat. *American Journal of Tropical Medicine and Hygiene,* 1, 559–66.

TELLA, A. & MAEGRAITH, B. G. (1966). Further studies on bradykinin involvement in *P. knowlesi* malaria. *Transactions of the Royal Society of Tropical Medicine and Hygiene,* 60, 304–17.

TRAGER, W. (1956). The intracellular position of malarial parasites. *Transactions of the Royal Society of Tropical Medicine and Hygiene,* 50, 419–20.

VAN ASSEL, S. & STEINERT, M. (1968). Inhibition selective de la replication du DNA kinetoplastique des trypanosomides par le bromure d'ethidium. *Archives Internationales de Physiologie et de Biochimie,* 76, 388–9.

VAN DER WAAIJ, D. (1964). The transmission of toxoplasmosis before birth. *Tropical and Geographical Medicine,* 16, 327–30.

VICKERMAN, K. (1965). Polymorphism and mitochondrial activity in sleeping sickness trypanosomes. *Nature, London,* 208, 762–6.

VICKERMAN, K. (1969). On the surface coat and flagellar adhesion in trypanosomes. *Journal of Cell Science,* 5, 163–93.

VICKERMAN, K. (1971). Morphological and physiological considerations of extra-cellular blood protozoa. In *Ecology and Physiology of Parasites: a Symposium,* ed. A. M. Fallis, 58–91. Toronto: University Press.

VOLLER, A., RICHARDS, W. H. G., HAWKEY, C. M. & RIDLEY, D. S. (1969). Human malaria (*Plasmodium falciparum*) in owl monkeys (*Aotus trivirgatus*). *Journal of Tropical Medicine and Hygiene,* 72, 153–60.

WANG, S. S. & FELDMAN, H. A. (1967). Isolation of *Hartmanella* species from human throats. *New England Journal of Medicine,* 277, 1174–9.

WEED, R. I. & REED, C. F. (1966). Membrane alterations leading to red cell destruction. *American Journal of Medicine,* 41, 681.

WEISZ, P. B. (1954). *Biology.* New York: McGraw-Hill.

WEITZ, B. (1962). Immunity in trypanosomiasis. In *Drugs, Parasites and Hosts,* ed. L. G. Goodwin and R. H. Nimmo-Smith, 180–90. London: Churchill.

WERBITZKI, F. W. (1910). Über blepharoplastlose Trypanosomen. *Zentralblatt für Bakteriologie, Parasitenkunde, Infecktionskrankheiten und Hygiene (Abteilung 1),* 53, 303–15.

WHO. (1969). Comparative studies of American and African Trypanosomiasis. *World Health Organisation Technical Report Series,* No. 411.

WIJERS, D. J. B. & WILLETT, K. C. (1960).Factors that may influence the infection rate of *Glossina palpalis* with *Trypanosoma gambiense.* II. The number and morphology of the trypanosomes present in the blood of the host at the infected feed. *Annals of Tropical Medicine and Parasitology,* 54, 341–50.

WITTNER, M., ROSENBAUM, R. M. & EINSTEIN, A. (1970). Utilization of axenically-grown strains of *Entamoeba histolytica* in providing further understanding of strain virulence. *Proceedings 2nd International Congress of Parasitology,* 4, 44–5.

WRIGHT, K. A. & HALES, H. (1970). Cytochemistry of the pellicle of bloodstream forms of *Trypanosoma brucei. Journal of Parasitology,* 56, 671–83.

YARDLEY, J. H., TAKANO, J. & HENDRIX, T. B. (1964). Epithelial and other lesions of the jejunum in giardiasis: jejunal biopsy studies. *Bulletin of the Johns Hopkins Hospital,* 115, 389–406.

YATES, D. B. (1970). Pharmacology of trypanosomiasis. *Transactions of the Royal Society of Tropical Medicine and Hygiene,* 64, 167–8.

ZUCKERMAN, A. (1964). Autoimmunization and other types of indirect damage to host cells as factors in certain protozoan diseases. *Experimental Parasitology*, **15**, 138–83.

ZUCKERMAN, A. (1966). Recent studies on factors involved in malarial anaemia. *Military Medicine*, **131**, 1201–16.

EXPLANATION OF PLATES

PLATE 1

Electronmicrographs of *Trypanosoma rhodesiense* (Vickerman, 1969) showing surface differences between bloodstream and culture forms. Preparations fixed in phosphate-buffered glutaraldehyde and stained with uranylacetate. Abbreviations: *far*, flagellum associated reticulum; *ff*, free flagellum; *fl*, attached flagellum; *gr*, granular reticulum; *mit*, mitochondrion; *pm*, pellicular microtubules; *pr*, paraxial rod; *rm*, reduced microtubules; *sm*, surface membrane.

Fig. 1. Bloodstream form. Transverse section of flagellum and adjacent pellicle. The coat overlying the surface membrane (*sm*) is clearly seen.

Fig. 2. Culture form. A comparable section of Fig. 1. Note the absence of the surface coat.

PLATE 2

Entamoeba histolytica showing surface lysosomes (Eaton, Meerovitch & Costerton, 1970).

Fig. 1. Section through a large surface lysosome showing 'trigger' (*t*), ribosomes (*r*), polyribosomes (*p*) and Golgi-like structure (*g*). (*pl* = plasmalemma).

Fig. 2. Scanning electron micrograph of *E. histolytica* trophozoite showing seven lysosomes, three of which have the 'trigger' device in view.

Fig. 3. Close-up of a surface lysosome with protruding 'trigger'.

PLATE 3

Merozoites of *Plasmodium berghei yoeli* (Ladda, Aikawa & Sprinz, 1969). Preparations stained with uranylacetate followed by lead citrate. Abbreviations: *co*, conoid; *db*, dense body; *Dm*, dense inner membrane; *dMb*, double membraned body (primitive mitochondrion); *ER*, endoplasmic reticulum; *Gr*, adherent granular material; *Hm*, host cell membrane; *mLb*, multilaminated membrane body; *N*, nucleus; *Pm*, parasite plasmalemma; *Po*, paired organelles.

Fig. 1. Merozoites oriented with conoid and directed toward a host reticulocyte.

Fig. 2. Merozoite penetrating a mouse erythrocyte. Conoid region is invariably the leading end of the penetrating parasite. Note the tight contact between red cell and parasite at the initial site of penetration. The space between the red cell and parasite is probably a shrinkage artifact. The accumulation of adherent granular material around the extracellular portion of the merozoite indicates the tight contact between the cell surfaces at the site of entrance.

PLATE 4

Fig. 1. Parasitism of human tissues in acute *Trypanosoma cruzi* infection. Striated muscle fibres of oesophagus showing a pseudocyst, packed full of amastigote forms, within the invaded cell (Köberle, 1968).

Fig. 2. Frontal section of a Chagasic heart showing an apical aneurysm. (Reproduced by kind permission of Professor Fritz Köberle).

PLATE 1

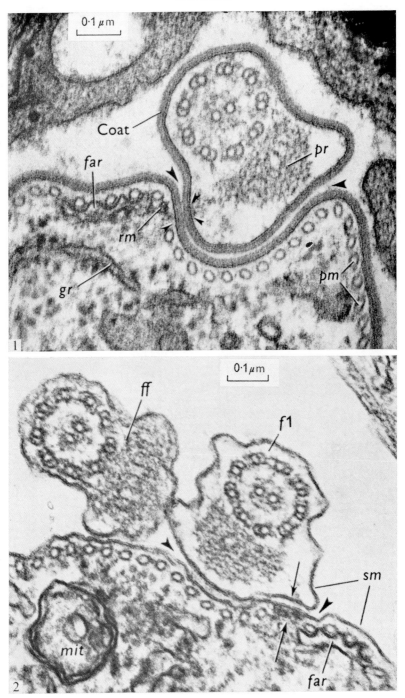

PLATE 2

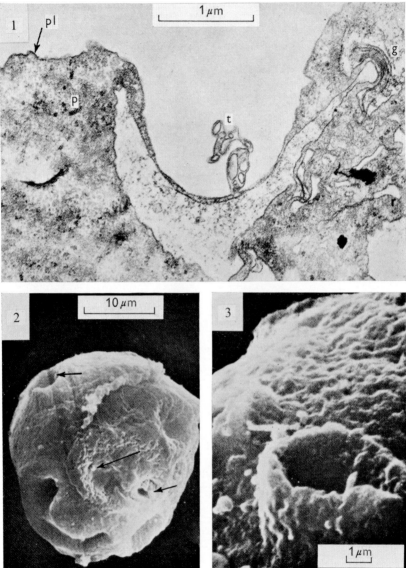

PLATE 3

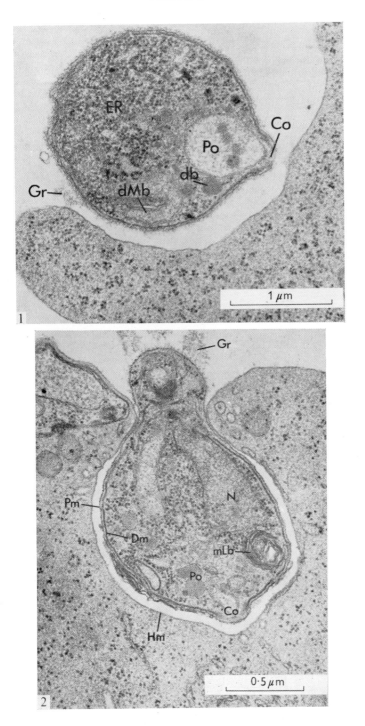

PLATE 4

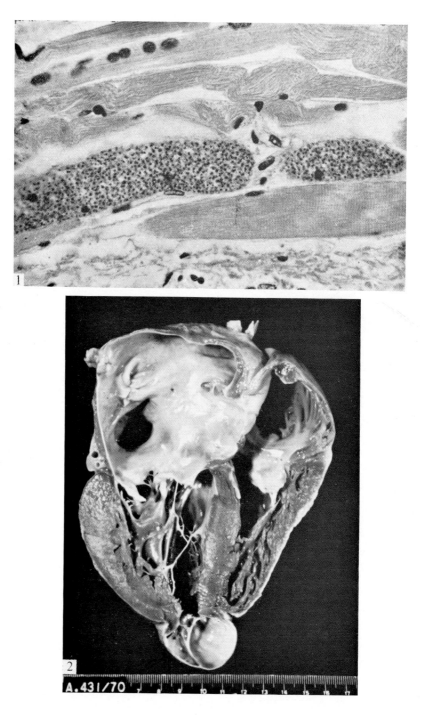

EARLY STAGES OF VIRUS INFECTION: STUDIES *IN VIVO* AND *IN VITRO*

R. BURROWS

The Animal Virus Research Institute, Pirbright, England

INTRODUCTION

In this, the 22nd Symposium, we are asked to look at Microbial Pathogenicity and in particular to consider infection by more natural routes than the usual experimental procedures. My brief was to consider viral pathogenicity in relation to the survival on and penetration of mucosae and of virus growth *in vivo*. I have interpreted my task as; (1) a consideration of those factors which are of importance in determining the initial localisation of virus on the host mucosae and of the susceptibility of the cells to infection; and (2) studies of the early stages of virus infection in the natural host. For brevity, I have referred to review articles in many areas of study.

Much of the information concerning the pathogenicity of viruses for their natural hosts has been acquired from observations of the disease in the field, from surveillance of open and closed communities, from diagnostic records and reports and from the development, production and evaluation of live and inactivated virus vaccines. Although valuable information has been obtained concerning the ability of different virus strains to grow in and to damage host tissues, many of the studies have been made during the middle and later stages of virus infection and detailed knowledge of the very early stages of the pathogenesis of many virus diseases is lacking. We are still ignorant of factors which may decide whether a susceptible host will acquire an infection or not, and the events which occur from the first localisation of virus in the animal to the stage when natural infection is established. Reasons for the apparent neglect of this important aspect of virus research have been both economic and practical. Considerable expense is involved in supplying and housing animals in suitable isolation units and the variation in susceptibility of animals to natural infection requires the use of many animals to establish a pattern of response which can be regarded as typical for any particular virus–host relationship.

In order to infect a new host, the virus must first gain access to susceptible cells, and the probability of this occurring will depend on the occurrence and concentration of virus in the environment, the mode of entry to the body and the physical and other antiviral barriers which may lie between the site of virus deposition or lodgement in the animal and the locality of susceptible cells.

OCCURRENCE OF VIRUS IN THE ENVIRONMENT

Those viruses which multiply predominantly in the respiratory tract are shed primarily in nasal secretions and discharges or suspended in airborne droplets. Excretion by other routes is probably minimal and of little significance. In man the relative importance of coughing, sneezing, talking and breathing in producing infective aerosols apparently differs according to that part of the respiratory tract which is liberating virus. Louden & Roberts (1967) found that one cough produced as many droplets as 30 s of loud talking and that 49 % of droplets produced by coughing remained airborne for at least 30 min, whereas only 6 % of the droplets produced by talking remained airborne for that period. If, as in Coxsackievirus A21 infection, virus is confined to the upper respiratory tract, sneezing produces more infective droplets than coughing. Very little virus is thought to be shed by normal respiration or by talking (Buckland, Bynoe & Tyrrell, 1965; Gerone *et al.* 1966) and, although virus may be recovered from the throat, little virus is thought to be expelled from the mouth. When the lower respiratory tract is involved, as in influenza infections, the clinical evidence suggests that coughs may be a more important source of virus (Tyrrell, 1967).

Many animals have long nasal passages and this probably affords more opportunity for infective aerosols to form during normal respiration than in the shorter nasal passages of man. In the course of experimental foot-and-mouth disease in cattle, sheep and pigs (Hyslop, 1965; Sellers & Parker, 1969; Donaldson, Herniman, Parker & Sellers, 1970) no difficulty was experienced in recovering virus from the air in the immediate vicinity of animals in the early clinical stages of disease, or from the mouth and the upper respiratory tract of people in contact with the animals (Sellers, Donaldson & Herniman, 1970). Similar studies have shown that equine herpesvirus type 1 (equine rhinopneumonitis virus) can be recovered using a multi-stage liquid impinger (May, 1966) from the vicinity of horses experiencing clinical or subclinical infections. Virus was also recovered from the nose of the impinger operator and from pharyngeal samples taken from an in-contact cow (R. Burrows & D. Goodridge, unpublished observation).

Not all the droplets expelled from the upper respiratory tract become or remain airborne and these larger droplets and drips of nasal secretion and saliva fall and contaminate the immediate surroundings. Other possible sources of virus from an infected animal include fluids and cell debris from lesions of the skin and mucous membranes and secretions and excretions from the mammary glands and from the urogenital and alimentary tracts. In some generalised infections such as foot-and-mouth disease, virus may be found in virtually all excretions and secretions.

The decay of virus outside the host depends on the stability of the virus in the environmental conditions which prevail; some viruses, such as the poxviruses and certain picornaviruses, are more resistant than members of other virus groups. Survival of foot-and-mouth disease virus in manure and hay for periods of up to 6 months has been recorded in the favourable conditions of a Russian winter (Cottral, 1969). The survival of viruses in aerosols has been reviewed recently by Akers (1969). Factors such as relative humidity, temperature and suspending fluid, ultraviolet radiation and the 'fresh air' factor (May, Druett & Packman, 1969) have all been shown to be of importance in the survival of airborne virus. The effect of relative humidity on the survival of virus is not uniform; influenza, measles and parainfluenza type 3 viruses survive better when held at low relative humidity, whereas picornaviruses and adenoviruses exhibit greatest survival at high relative humidity values. Most of these studies have been carried out with virus suspended in tissue culture fluids, various salt solutions or homogenates of mouse brain suspensions, and how far these represent the natural aerosols produced by an infected animal is questionable. In studies with foot-and-mouth disease virus, slight differences in virus decay rates were noted in aerosols derived from infected tongue epithelium suspended in bovine saliva, and those produced from infected tissue culture fluids (A. I. Donaldson & D. Barlow, personal communications).

Possibly the most common way in which virus is taken into the body is by inhalation. In the following sections more attention is given to the respiratory tract than to the digestive tract or to the epidermoid mucous membranes in considering factors concerned with the localisation and survival of virus on body surfaces.

VIRUS ENTRY BY INHALATION

The living animal is an extremely efficient air sampling machine with a sampling rate that ranges from 24 ml per minute in the mouse (Guyton, 1947) to over 100 l a minute for a 500 kg horse (Brody, 1945). In addition, animals are inquisitive and use their noses to investigate their companions, their surroundings and possible sources of food and in so doing increase their chances of inhaling potentially infective material from their environment. The lining epithelium of the respiratory tract is mainly composed of ciliated columnar epithelium apart from areas anterior to the turbinates, in the olfactory area, the pharynx, parts of the larynx and the respiratory portions of the lung. Numerous mucus-secreting goblet cells are interposed between the ciliated cells; Rhodin (1966) has estimated that there are about five ciliated cells for every goblet cell in the tracheal mucous membranes. Both serous and mucus-secreting acini occur in the submucous layer throughout most of the respiratory tract and it has been estimated that the total volume of secretion of the submucosal glandular acini is approximately 40 times greater than that of the goblet cells (Kilburn, 1968). The mucous blanket which is formed in the nasal region is directed backwards by ciliary activity towards the pharynx, while that in the conducting portion of the lung is directed forwards to the pharynx.

With the possible exception of those viruses possessing neuraminid-ase activity, viruses may be regarded as inert particles and the probability of their attachment to susceptible cells will depend on the region of deposition and the pathways of clearance in relation to the site of susceptible cells and the presence of protective or inhibitory factors in their immediate environment.

Factors affecting the deposition and retention of virus in the respiratory tract

Viruses inhaled into the respiratory tract are likely to be deposited at sites which are determined by the size and the nature of the droplet to which they are attached. Estimates of nasal retention for man indicate that the majority of particles with a diameter of 15 μm or greater and approximately 50 % of those with a diameter of 6 μm are retained in the nasal region. Particles of about 3 μm may be deposited in both the conducting and respiratory portions of the lung, while those of less than 1 μm are likely to be deposited mainly in the respiratory portion. Immediate total retention has been estimated at more than 80 % for particles of 3 μm or greater and approximately 30 % for particles in the 0·5 μm range (Druett, 1967). Some difference in the deposition and retention of inhaled particles in the respiratory tract of animals could be expected owing to the obvious differences in size, anatomy and posture. The nose of most animals is thought to be a more efficient filter for inhaled particles than the nose of man (Proctor, 1966). Hatch & Gross (1964) found that the retention rates in both the upper and lower respiratory tract for different size particles was similar in man, monkey and guinea-pig. Retention rates of inhaled 1 μm particles in the mouse have been estimated at 30 % (Goldberg & Lief, 1950), 36 % (Harper & Morton, 1953) and 65 % (Rosebury, 1947). Comparable estimates for man range from 25 % to 65 %, depending on the volume and frequency of the breathing cycle (Druett, 1967).

Factors affecting the survival of virus on respiratory mucosae
Normal clearance and the mucous coat

The clearance of particulate matter from the respiratory tract has been the subject of much investigation (Druett, 1967; Tyrrell, 1967; Kilburn, 1968), studies having included direct measurement of mucus transport in the trachea of various species of animals and the rate of clearance of bacteria, viruses and radioactive particles from the nose and lungs of animals and man. Particles deposited on the mucous blanket of the nasal surfaces are swept backwards to the pharynx at varying speeds according to whether they alight on a slow or fast moving ciliary current (Bang & Bang, 1967). In man the mean rate of movement has been estimated at about 7 mm per min (Quinlan *et al.* 1969). However, Bang & Bang (1969*a*) have shown that individual variations in flow rate could be as great as 4-fold in the chicken and 10-fold in the human. Similarly, particles alighting on the mucus overlying the ciliated surfaces of the conducting portions of the lung are moved upwards towards the pharynx. Mean mucociliary clearance rates of 1·6, 4·0, 8·2 and 12·6 mm per min have been measured in the distal, segmental and lobar bronchii and the trachea of dogs (Asmundsson & Kilburn, 1970) and these rates are comparable with those found for man and other animals (Kilburn, 1968). Apart from this velocity gradient from the small to the large airways, the flow of mucus is not uniform owing to such impediments as bronchial orifices and junctions and areas of squamous metaplasia; and it has been suggested that micro-organisms deposited in the regions of such areas may stand a better chance of initiating infection than if deposited in the more rapidly flowing stream (Druett, 1967). The possibility of gaps occurring in the mucous coat was suggested by Platt (1970), recalling the experiments of Ropes (1930) who showed that insufflated carbon particles could subsequently be demonstrated inside tracheal epithelial cells which had apparently phagocytosed them.

Particles deposited in the respiratory portion may be 'floated off' (Kilburn, 1968) on the film of moisture which is contiguous with the mucous layer and so be attracted into the ciliated area or may be engulfed by alveolar macrophages. The majority of macrophages are thought to migrate or to be attracted into the ciliary region, to be cleared on the mucociliary escalator; some, however, may find their way to the drainage lymph node. The importance of the macrophage in the pathogenesis of infection will depend on whether the virus is able to proliferate in the cell or can resist degradation and digestion and escape from the macrophage in due course (Mims, 1964; this volume).

If virus is deposited in the vestibular area of the nose, either on nasal hairs or on a region which is not cleared by the normal mucociliary mechanism, then its survival may be determined more by its inherent stability than by host factors. While it is difficult to measure virus survival in the normal (susceptible) host because of possible confusion between virus persistence and virus growth, it may be determined in an abnormal (insusceptible) host. Sellers *et al.* (1970) studied the persistence of foot-and-mouth disease virus in the upper respiratory tract of humans following exposure to natural aerosols. Nasal swabs were taken immediately after periods of exposure to infected animals and swabs were taken again after the subjects had showered and made two changes of clothing. The rate of disappearance of virus from the nose varied from person to person, but on average a 60-fold loss of virus in 3·5 h was recorded. Virus was recovered from one person after a period of 28 h but not at 48 h after exposure. In these experiments, despite comparable exposure, there was a considerable variation between subjects in the amounts of virus recovered from nasal secretions, saliva and throat swabs and it would appear that minor differences in normal anatomy and breathing pattern influence the deposition of virus in particular sites of the upper respiratory tract.

It is obvious from this brief résumé of the deposition and clearance of inhaled particles that most of the virus trapped in the respiratory tract will be moved fairly quickly towards the pharynx and that in order to reach susceptible cells the virus must first penetrate the mucous blanket. Little is known about the depth and characteristics of mucus found at different levels of the respiratory tract. In the trachea of rats, the total thickness of the blanket is about 5 μm (Dalhamn, 1956) and consists of a stiff, tenacious transport layer of 1–2 μm which overlies a watery sol layer in which the cilia actually beat. The depth of the sol layer is probably controlled by re-absorption of water by the microvilli of the underlying epithelial cells (Kilburn, 1968). As the airways become smaller there is a gradual reduction in the number of goblet cells and submucous acini and it is likely that the deficiency in mucus production in these areas is compensated for by the secretory activity of the alveolar type 2 cells and the Clara cells of the smaller bronchioles. Whether the slower mucus movement observed in the distal parts of the lung is due to a reduced rate of ciliary beat or to a less effectual beat owing to the characteristics of the mucous layer is not known. It is likely, however, that the combination of slow movement and a thinner protective film will increase the possibility of virus making contact with the epithelial cells. Provided that there is no marked difference in susceptibility between the epithelial cells of the upper and lower respiratory tracts, small-particle virus aerosols should prove more infective than large-particle aerosols, as is the case in some bacterial diseases in experimental animals (Druett, 1967).

Factors affecting mucociliary clearance

The probability of virus penetrating the mucous blanket will be influenced by the activity of the mucociliary mechanism. A reduction in the movement of mucus either by depression of ciliary activity or by an increase in mucus secretion should allow more time for virus to penetrate the mucous coat. Factors which affect mucus secretion either in amount or in consistency or which affect ciliary activity have been reviewed by Rivera (1962), Kilburn & Salzano (1966) and Bang & Bang (1969a). *In vitro* studies have shown that desiccation and temperature changes are particularly important but to what extent these factors operate in the living animal is not clear. Normally, the inspired air is humidified during passage through the nasal area and is fully saturated and warmed to body temperature by the time it reaches the trachea (Proctor, 1964). Thus, the effects of humidity and temperature on ciliary activity are more likely to be exerted in the nasal epithelium than in the lower parts of the respiratory tract. Although Dalhamn (1956) found that ciliary activity in the rat trachea was curtailed at relative humidities below 50 %, Quinlan *et al.* (1969) found that some human subjects maintained fast nasomucociliary rates at a relative humidity of 15 %.

Experimentally, the effect of reducing nasomucociliary clearance in chicks, either by subjecting them to low temperatures ($-20°$) immediately before or after a standard exposure to Newcastle disease virus, by paralysing ciliary action with cocaine or by initiating glandular hypersecretion with pilocarpine, resulted in an increase in the number of cells infected in a specialised area of nasal mucosa (Bang & Bang, 1969a; Bang & Foard, 1969). Reduced nasomucociliary clearance was also noted in chicks suffering from the effects of internal dehydration and avitaminosis A (Bang & Bang, 1969a). The importance of vitamin A in the maintenance of the normal distribution of squamous mucus and ciliated cell epithelium in relation to susceptibility to virus infections has been discussed by Bang & Bang (1969b).

Antiviral activity of mucus

The antiviral activity of mucus may be exerted either by virtue of its natural composition in respect of pH and enzyme content or by the presence of antiviral substances such as antibody or inhibitors. The pH of nasal secretions of man ranges from 7·6 to 8·4 (Hilding, 1934) and that of cattle from 6·2 to 8·0 (A. J. M. Garland, personal communication). In cattle the alkalinity of samples taken from the pharynx (mixed mucus and saliva) may approach levels which are above the optimum for virus survival; a mean pH of 8·24 was calculated for diluted samples (Burrows, 1966a) and of 9·0 (range 8·5 to 10·0) for undiluted samples (A. J. M. Garland, personal communication). The inactivation rate of foot-and-mouth disease virus at 37° in such secretions is reduced considerably if the pH is adjusted to 7·6 by the addition of acid.

The antiviral activity of mucus associated with immunoglobulins is now recognised as a major defence against some viruses both in the respiratory tract and in the gastro-intestinal tract and evidence is accumulating that recovery from infections and resistance to re-infections is better correlated with the levels of virus-neutralising antibody in mucus secretions than with those in sera (Tomasi, 1970; Johnson, 1970). There are, however, significant differences in the number and types of immunoglobulin which have been described in the mucus secretions of various animals and difficulties have arisen in defining which immunoglobulin in a particular animal species is analogous to that found in man. In the human it appears that most of the Ig A in such secretions is synthesised locally in the *lamina propria* of the respiratory and gastro-intestinal tracts and interstitially between salivary gland acini; only a small fraction is derived from blood plasma. In other species such as the cow and the sheep there is evidence that the major immunoglobulin found in external secretions is selectively derived from the plasma (Tomasi, 1970) or is produced by lymphoid cells in the *lamina propria* at the base of the villi in the mucosae of the respiratory and gastro-intestinal tracts (Curtain, Clark & Duffy, 1971).

Various non-specific virus inhibitors have been identified in sera, urine and milk of animals and in homogenates of lung and intestinal mucosae (Wasserman, 1968). Some of these are fairly specific for particular viruses – the alpha and gamma inhibitors of myxoviruses, the mucopolysaccharide inhibitors of Theiler's virus (GD7), the mucoid inhibitors of mumps virus and pneumonia virus of mice and the intestinal factor which inhibits murine hepatitis virus (Piazza & Pane, 1967). Other inhibitors are less specific, such as the lipid and non-lipid inhibitors of normal sera. It is possible that glandular secretions from some mucous surfaces may also contain inhibitory substances (Reed, 1969).

Activity of other micro-organisms

The activity of other micro-organisms in the respiratory tract, whether they be commensal or pathogenic, could create a local environment which might favour or inhibit virus survival, cell susceptibility or virus growth. Some mycoplasma species, which are frequently found in the mouth and respiratory tract of mammals and birds, are known to compete with certain viruses for arginine in tissue culture. Inhibitory effects of mycoplasma on the growth of adenoviruses (Rouse, Bonifas & Schlesinger, 1963), herpesviruses (Hargreaves & Leach, 1970) and measles virus (Romano & Brancato, 1970) in tissue culture have been described. Other viruses which have been shown to be dependent on adequate amounts of arginine for the synthesis of mature virus include fowl plague virus (Becht, 1969), polyoma virus (Winters & Consigli, 1971) and SV40 virus (Goldblum, Ravid & Becker, 1968). To what extent this and similar factors which may result in a deficiency of amino acids essential for viral synthesis operate in an animal is not known.

There is ample evidence that viruses, bacteria, fungi and their products may modify certain virus infections by virtue of inducing interferon (Ho, Fantes, Burke & Finter, 1966; Finter, 1966; Vilcek, 1969) and any disturbance in the microflora of mucous membranes may be reflected in changes in the susceptibility of the cells to infection. Miller (1966) found that the susceptibility of pigeons to Venezuelan equine encephalitis virus was increased if the birds were pre-treated for several weeks with oxytetracyclines and he interpreted this effect as due to a depression of gram-negative endotoxin-producing bacteria.

The lower respiratory tract of mammals is virtually sterile despite the continual entry of droplet nuclei containing bacteria. Lees & McNaught (1959) found that bacteria were rarely present in bronchopulmonary secretions unless there was obvious inflammation or exudation. In such circumstances, interference with mucus clearance and alterations in virus cell susceptibility could be expected owing to local changes in pH and the effect of bacterial toxins.

VIRUS ENTRY THROUGH THE ALIMENTARY TRACT

The pathogenesis of virus infections of the alimentary tract of man has been discussed by Downie (1963) and that of animals by Platt (1967).

The relative importance of the upper and lower regions of the digestive tract as sites of initial infection by particular viruses will be influenced by the ability of the virus to survive passage through the acid and bile-containing regions of the stomach and duodenum. Relatively little is known of local factors which may operate in the gastro-intestinal tract to render cells more or less susceptible to virus but the variations in pH, enzyme content and microfloral populations which occur in different regions of the tract must be of some importance.

Cell receptor activity has been shown to be important in determining cell susceptibility to certain enteroviruses (Holland, 1961). However, uptake of virus by mucosal cells may not be entirely dependent on cell receptor activity. Payne et al. (1960) fed milk containing radioactive resin particles (1–5 μm diameter) to a calf and found that small numbers of particles penetrated the mucous membranes of the tonsil, pharynx and small intestine. Although infection is usually associated with virus-contaminated material taken in by mouth, the pharynx, tonsils and the lower alimentary tract are also exposed to virus in the mucus cleared from the nose and lungs and this may be important in determining the alimentary tract as the initial site of infection. Virus taken in by mouth in the form of contaminated food and water will pass rapidly through the pharynx, whereas virus in respiratory tract secretions may have a more leisurely passage over the pharyngeal walls. Experimental studies of some diseases which were thought to be acquired by ingestion have shown that the feeding of massive amounts of virus is frequently less successful in producing infection than the instillation of smaller amounts of virus in the nose, (Dunne, Hokanson & Luedke, 1959 – swine fever; Plowright, 1968 – rinderpest; Sellers, 1971 – foot-and-mouth disease).

VIRUS ENTRY THROUGH THE SKIN AND EPIDERMOID MUCOUS MEMBRANES

Under normal circumstances the stratum corneum of the skin provides an inert physical barrier between the environment and the living cells of the epidermis and so infection through the skin and to some extent through the epidermoid mucosae is only likely to occur through virus contamination at the sites of minor abrasions or injuries or through the bites of virus-carrying insects or animals. Many of the viruses which are pathogenic for the cells of the epidermis tend to localise in those areas which show a degree of hypoplasia due to mechanical, chemical or physical irritation and it is thought that the increase in vascular permeability and the presence of actively multiplying basal cells increases the probability that virus will localise in such areas. Platt (1963a, b) has shown that in areas of skin damage the cells in the deeper layers of the epidermis readily engulf a variety of particle types up to the size of erythrocytes and this phagocytic activity of the epidermal cells is probably an important factor in the initiation of cellular infection by those viruses which are capable of growth in epidermal cells. Factors which may contribute to the resistance or susceptibility of the skin and epidermoid mucous membranes to virus infection in man and animals have been reviewed by Platt (1963b, 1970).

CELL SUSCEPTIBILITY

Factors which are known to affect the susceptibility of animals and cells to virus infection have been reviewed or discussed by Burnet (1960), Bang & Luttrell (1961), Downie (1963), Platt (1967), Fenner (1968) and Mims (1964, 1970). The various factors include the effects of age, sex, genetic constitution, nutrition, body temperature, metabolic activity, season of the year, stress, trauma and intercurrent infection. The influences of many of these factors have been determined by experiment and have been measured by differences in virus growth and dissemination in the host, by the extent and nature of tissue damage and by the antibody response. Studies of epidemics and epizootics indicate that these factors operate in natural infections but whether the effects are due to alterations in the susceptibility of the epithelial surfaces of the animal to initial infection or to differences in virus growth and behaviour after infection has not always been established.

A number of cell types are present in the epithelia lining or communicating with the respiratory and gastro-intestinal tracts and the identification of 'target' cells involved in the initial stage of infection and studies of virus entry into these cells is difficult in the living animal. During the early stages of infection it may not be possible to locate areas in which virus is multiplying and in the later stages of infection vascular, inflammatory and necrotic or proliferative changes may involve a number of cell types. Histological studies using conventional staining procedures or immunofluorescent techniques have shown influenza virus antigen to be present in the ciliated, intermediate and basal cells of human nasal epithelium and possibly also in the goblet cells (Ebisawa, Kitamoto & Takeuchi, 1969). Bang & Foard (1969) found that the earliest lesion in Newcastle disease virus-infected chickens involved the acinar mucous cells and they assumed that these were the first to be infected. Much of our knowledge of comparative cell susceptibility to virus has come from, and is likely to come from, studies in organ culture. In such cultures the cells of mucous membranes and embryonic skin appear to be similar in structure and physiology to those of the intact host and so should be particularly suitable for studies of the susceptibility of the various cell types to different viruses.

Whilst inoculation procedures may be valuable in determining the location and distribution of cells susceptible to haemotogenous infection, they do not represent the usual method of exposure in many infections. Regions of susceptibility of the respiratory tract have been identified by comparing the dose required to infect by small-particle aerosol (0·3–2·5 μm) with that required to infect by the application of virus to the nasal epithelium. Trials with Coxsackievirus A 21 and rhinovirus NIH 1734 showed that the nasal route of exposure required 5 to 20-fold less virus than the aerosol route (Couch et al. 1966a). Similar trials with adenovirus type 4 showed that the lower respiratory tract was the more susceptible (Couch et al. 1966b) and that both were equally easily infected with influenza virus A 2 (Knight, 1970).

A novel approach to the study of the susceptibility and the response of intestinal cells to virus infection has been reported by Pensaert, Haelterman & Burnstein (1970), who carried out a sequential study of transmissible gastro-enteritis virus infection in surgically isolated loops of jejunum in baby pigs. Different phases of the infection were examined by immunofluorescence and histological procedures and by measurements of virus growth and these were found to be similar to those occurring in the intact pig.

The means by which viruses are incorporated into susceptible cells has been the subject of much research and some controversy. Reviews dealing with virus receptors and the attachment and penetration of viruses into cells have been presented by Philipson (1963), Cohen (1963), and Dales (1965). With some viruses such as the influenza viruses and some of the picornaviruses, attachment may depend on the presence of specific receptors but whether receptors are required for other viruses is not known. There appears little specific about the adsorption process of those viruses which will grow in many cell types such as the herpesviruses, the poxviruses and the arboviruses. Dales (1965) holds the belief that the majority of viruses are taken into cells by a process of phagocytosis or 'viropexis' as suggested originally by Fazekas de St Groth (1948). Viruses which have been reported to enter in this way include reovirus, vaccinia, adenovirus, Newcastle disease virus, parainfluenza virus SV5 and influenza virus (Morgan, Rose & Mednis, 1968), vesicular stomatitis virus (Simpson, Hauser & Dales, 1969) and equine herpesvirus type 1 (Abodeely, Lawson & Randall, 1970). Some difference in opinion, however, has arisen concerning the entry of influenza viruses and herpesviruses. Reports by Morgan, Rose & Mednis (1968) and Morgan & Rose (1968) describe the entry of herpes simplex virus into HeLa cells and of influenza virus into chick chorioallantoic membrane cells as a process which entails fusion of the viral envelope with the cell membrane and passage of the virus core or of nucleoprotein directly into the cytoplasm. A similar method of entry has also been described for parainfluenzavirus type 1 (Morgan & Howe, 1968) Recently a third means of virus entry into the cell has been considered by Morgan, Rosenkranz & Mednis (1969), who found that, in addition to the normal means of entry of adenovirus by viropexis, some virus particles could pass directly through the cell membrane without being engulfed in phagocytic vacuoles. The majority of these studies have been carried out in *in vitro* systems and their relevance to the events which occur in the animal has still to be established. Studies of virus attachment and penetration under more natural conditions are possible and Dourmashkin & Tyrrell (1970) have produced some interesting micrographs showing influenza virus particles adsorbed to the cilia and microvilli of epithelial cells in organ cultures and of the fusion and entry of parainfluenza type 1 virus into cilia. They did not, however, express an opinion on the manner by which influenza virus entered the cell.

EARLY STAGES OF VIRUS INFECTION

The methods of study of the early stages of natural infection in the normal host are dictated to a large degree by ethical and economic considerations. Much useful information has been obtained by sequential sampling from accessible mucous membranes and of secretions and excretions during the incubation period; the occurrence and concentration of virus in such samples have indicated likely sites of virus localisation and multiplication. The routine titration approach, in which animals are killed at various times and estimates made of the virus content of different tissues, is useful if the time of infection is known and if a sufficient number of animals is used to reduce the effect of individual variations in susceptibility.

Relatively few detailed studies of infection following natural routes of exposure have been reported since the classical studies of the pathogenesis of mousepox (Fenner, 1948), myxomatosis (Fenner & Woodroofe, 1953) and rabbit pox (Bedson & Duckworth, 1963). Perhaps the most extensive series of investigations of early infection and of the dissemination and growth of virus in the body have been those undertaken with rinderpest, African swine fever, Newcastle disease and infectious laryngotracheitis of chickens.

Studies of African swine fever showed that soon after contact exposure (Greig, 1971), intranasal exposure (Heuschele, 1967; Plowright, Parker & Staple, 1968) and oral exposure (Colgrove, Haelterman & Coggins, 1969), virus was present in the tonsils and/or in the pharyngeal mucosa. Although moderately high concentrations of virus were recovered from the tonsils 16 to 48 h after exposure, specific foci of fluorescence were rarely seen. In some animals virus was present in greater concentration in the drainage lymph node and it was considered possible that virus may have been carried through the mucosa with little or no proliferation at the site of entry. Virus multiplication in the local lymph node was thought to be the source of the cell-associated viraemia and the subsequent dissemination of virus throughout the lymphopoietic tissues of the body (Plowright, Parker & Staple, 1968). In some pigs there was evidence that virus might also enter through the epithelium of both the upper and lower respiratory tract.

Similar findings had been obtained in earlier studies of the pathogenesis of rinderpest in cattle (Plowright, 1968). Virus proliferation was established at an early stage in the pharyngeal lymph nodes and occasionally in other lymph nodes draining the upper respiratory tract. No virus was detected in the mucosae and it was thought that the infecting virus passed through the epithelium without producing a local lesion or proliferating.

Studies of herpesvirus infections of animals, infectious bovine rhinotracheitis (Chennekatu, Gratzek & Ramsey, 1966), pseudorabies virus of pigs (Sabo, Rajcani & Blaskovic, 1968) and herpesvirus of dogs (Appel, Menegus, Parsonson & Carmichael, 1969) showed that virus localised and proliferated in the squamous epithelium of the tonsillar crypts or in pharyngeal epithelium before invading local lymph nodes, nasal mucosa and other susceptible tissues. In both the pseudorabies and African swine fever studies, foci of specific fluorescence were not seen until virus concentrations greater than $10^{4.5}$ to $10^{5.0}$ ID_{50}/g were reached.

Studies of foot-and-mouth disease

For many years it was thought that the normal method of infection by foot-and-mouth disease virus was by ingestion, virus gaining entry to susceptible cells through minor abrasions in the mouth or other areas in the alimentary tract. This view was supported by the known susceptibility of tongue epithelium to inoculation compared with other methods of exposure (Henderson, 1952) and that frequently the first signs of disease were the appearance of vesicles in the mouth cavity. The pathogenesis of natural disease was thought to be similar to that which follows the inoculation of virus into the epithelium of the tongue – a rapid growth of virus, which leads to vesicle formation and infectivity levels of up to $10^{9 \cdot 0}$ ID_{50}/g of vesicular epithelium within 18 to 30 h, depending on

Table 1. *Virus content of samples collected during the early stages of foot-and-mouth disease (infection by indirect contact with inoculated donors; strain O_1 (Burrows, 1968)*

Group	Site	Infectivity			Time (days) before vesicles observed after first detection of virus	
		No. of positive samples	Mean infectivity*	Range	Mean	Range
8 steers	Pharynx	20	3·6	1·9–5·3	2·5	0–5
4 cows	Pharynx	11	3·5	2·5–5·5	2·7	2–5
	Blood	7	2·2†	1·0–4·1	1·8	1–2
	Milk	9	3·0†	1·0–5·2	2·2	1–4
	Rectum	2	1·3	1·0–1·6	1·0	—
	Vagina	3	≥3·2	2·9–>3·3	1·0	—
9 sheep	Pharynx	22	≥2·9	1·2–>3·5	2·5	0–5
10 pigs	Pharynx	41	2·2	0·7–3·3	5·0	2–10
	Rectum	31	1·3	0·6–2·6	4·2	0–7
	Prepuce/vagina	27	1·6	0·6–3·0	3·6	0–7

* Log_{10} p.f.u./sample; mean of samples taken over several days before the appearance of clinical disease.
† Log_{10} p.f.u./ml.

the virus strain and the dose inoculated, the development of viraemia 5 to 15 h after inoculation which usually persists for periods of up to 4 days and in which blood virus levels of up to $10^{6 \cdot 0}$ ID_{50}/ml have been recorded, the onset of generalised disease from 36 to 96 h after inoculation and the first appearance of specific neutralising antibody on the fourth or fifth day. This concept of the pathogenesis of the disease was not questioned until Korn (1957) suggested that the initial site of infection might be in the upper respiratory tract. Hyslop (1965) confirmed that virus was present in aerosol form in the vicinity of infected cattle and that cattle could be infected by artificially produced aerosols of virus.

Some features of recent outbreaks of disease in the United Kingdom had indicated that airborne spread of virus had occurred (Hurst, 1968; Henderson, 1969) and this led to a reappraisal of the pathogenesis of natural disease in farm animals. Initial studies, by sequential sampling procedures, were made of groups of cattle, sheep and pigs which had been exposed by indirect contact to inoculated donor animals (Burrows, 1968a; Burrows et al. 1971). Virus was recovered from the majority of animals for several days before clinical lesions were evident and Table 1 lists the infectivities of series of samples from different tissues and relates the first occurrence of virus in these specimens to the time when clinical lesions were first observed in the individual animal.

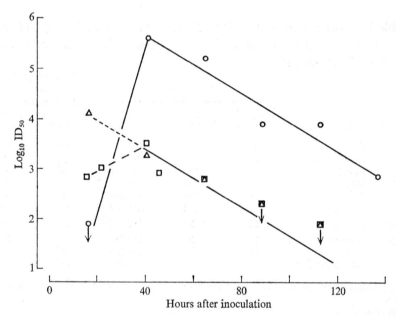

Fig. 1. Mean amounts of virus recovered in a large volume air sampler from loose boxes containing infected animals (60 min sampling at 1000 l/min): Pigs O—O: Cattle □---□: Sheep △- - -△. Drawn from Sellers & Parker, 1969, Tables 1, 2 and 3.

The occurrence of virus in samples taken from the pharynx may or may not have been an indication of virus growth in that region. Subsequent studies by Sellers & Parker (1969) have shown that considerable amounts of virus were likely to have been present in the air of the shed containing infected animals (Fig. 1) and the recovery of virus from the pharyngeal region may have been a measure of the air sampling ability of the animal rather than an indication of virus growth. However, the relationship between the first recovery of virus from this region and the time of appearance of neutralising antibody (Fig. 2) implies that the virus recovery was indicative of infection with this virus strain. The mean time between the appearance of pharyngeal virus and the appearance of antibody was 4·33 days for the twelve cattle, 4·87 for eight of the ten sheep and 4·10 days for ten pigs. An interesting feature was that some pigs developed significant levels of neutralising antibody before clinical disease was recognised.

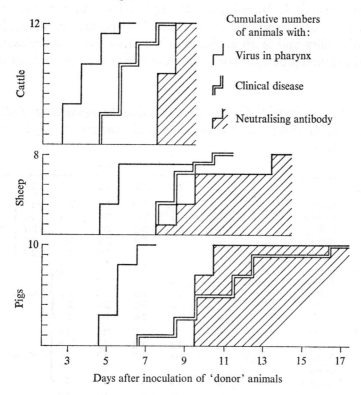

Fig. 2. Relationship between animal size and acquisition of infection and between virus recovery, clinical disease and appearance of neutralising antibody following indirect contact exposure to infected donor animals.

Whilst these observations point to the pharyngeal mucosa as the site of virus growth and entry under these conditions of exposure, they do not provide evidence that the epithelium of the area is more susceptible than that of other regions. In a subsequent experiment in which cattle were infected by the instillation of virus into the mammary gland, some evidence of selective localisation and growth of virus in the pharyngeal area was obtained (Table 2), although it was apparent also that virus growth was occurring in other regions.

Not all our attempts to produce infection by indirect contact exposure have been so successful. Studies with other strains of foot-and-mouth disease virus have resulted in few or no recipient animals acquiring infection, despite the added risk of infection offered by daily sampling procedures. This may have been due to differences in humidity and temperature which may have affected virus survival or to variations in the amounts of virus excreted as aerosols by the donor animals (Donaldson *et al.* 1970). Schulman (1970) has shown that the amount of influenza virus released by donor mice is important in determining transmissibility and is an attribute of the virus strain not necessarily related to lesion production or to differences in growth of the virus in the host.

Table 2. *Virus content of samples taken before the appearance of clinical foot-and-mouth disease lesions* (*udder inoculation: strain* O_1), (Burrows *et al.* 1971)

Animal no.	Sample origin	h after infection							
		4	20	28	44	52	68	76	117
GK 9	Serum	—	T	T	1·1*	1·6	2·1	2·7	⎫
	Pharynx	—	T	1·0	2·6	2·3	2·0	3·7	
	Saliva	—	1·1	2·6	2·2	2·6	—	2·7	⎬ Lesions
	Nose	—	—	—	2·9	2·9	3·0	2·0	
	Vagina	—	—	—	—	1·0	2·5	2·3	⎭
GK 10	Serum	—	T	T	1·9	2·7	2·4		⎫
	Pharynx	1·6	2·6	3·6	3·6	3·2	4·0		
	Saliva	—	—	2·0	1·3	2·2	2·7		⎬ Lesions
	Nose	—	—	1·0	3·2	2·7	2·7		
	Vagina	—	—	—	2·6	1·7	2·5		⎭
GK 11	Serum	—	T	T	2·4	2·7	2·7		⎫
	Pharynx	—	2·4	2·4	3·2	3·2	4·0		
	Saliva	—	—	1·4	1·7	2·7	2·7		⎬ Lesions
	Nose	—	—	—	3·3	3·3	3·3		
	Vagina	—	2·3	2·2	2·5	2·5	2·5		⎭
GK 12	Serum	—	—	T	0·7	1·0	0·4		⎫
	Pharynx	—	—	—	3·0	2·0	3·5		
	Saliva	—	—	—	—	—	—		⎬ Lesions
	Nose	—	0·3	1·7	2·4	2·5	2·8		
	Vagina	—	2·0	1·0	2·0	2·0	2·5		⎭

* Log_{10} p.f.u./ml or sample.
T = Trace amounts of virus detected in BTY cultures only.
— = No virus recovered.
Animals GK 9 and GK 10 each received 10^3 p.f.u. of virus O_1
Animals GK 11 and GK 12 each received 10^6 p.f.u. of virus O_1.

Experiments in which animals were fed high concentrations of virus ($10^{6·7}$ ID_{50}) in food or in water proved unpredictable as previously found (Henderson & Brooksby, 1948). Two animals housed individually did not acquire infection following two feedings at a weekly interval but subsequently proved susceptible to an indirect contact exposure. In a second experiment, in which animals were housed together, virus was recovered from pharyngeal samples of 9 of 16 animals 48 h after exposure to virus in food or water. However, the presence of trace amounts of virus in the nasal passages of some animals indicated that they may have acquired infection by inhalation.

Table 3. *Infection of cattle by various methods with foot-and-mouth disease virus*

Virus strain	Method of exposure	Number of animals					Time of examination: days after exposure
		Exposed	Premature disease	Not infected	Viraemia	No viraemia	
O-1860	Intranasal spray	6	0	3	1	2	2–4
	Intratracheal*	4	0	0	1	3	1–2
	Food	8	1	3	1	3	3–6
	Water	8	3	1	3	1	3–6
A-Eystrup	Indirect contact*	6	0	1	1	4	2–4
A-Iraq 24/64	Indirect contact*	10	2	2	1	5	2–6
		42	6	10	8	18	

* Data provided by A. J. M. Garland and J. A. Mann.

Table 4. *Frequency of virus recovery and mean infectivity of specimens collected from 18 cattle during the early stage of foot-and-mouth disease*

Specimen*	Number examined	Number from which virus was recovered	Mean infectivity of positive samples	Highest infectivity of positive sample
Dorsal surface of soft palate	18	17	2·02†	5·7
Pharynx	18	14	2·16	6·3
Retro-pharyngeal lymph nodes‡	17	8	2·67	5·7
Tonsil	16	6	1·60	4·9
Trachea	13	4	1·62	2·3
Nasal mucosae	18	4	1·02	2·6

* Mucous and superficial epithelium.
† Log_{10} p.f.u./sample or g of tissue.
‡ Tissue.

Other attempts to establish the sites of virus localisation and growth by the 'routine titration approach' were made in cattle exposed to infection by intranasal spray, by intratracheal inoculation or by indirect contact exposure to infected pigs. Of 42 cattle used for this work, 6 developed disease prematurely or were found to have early lesions when examined after slaughter, 10 were found not to be infected and in 8 cattle virus was dispersed widely throughout the body (Table 3). The frequency of virus recovery and the mean infectivity of samples taken from the 18 cattle which were examined before dissemination of virus had taken place indicated that the normal route of entry of foot-and-mouth disease virus in cattle is through the mucosae of the pharyngeal area (Table 4). In this area virus persists for long periods in the convalescent animal (Burrows, 1966a, 1968b) and the ability of a virus strain to establish itself and to persist in the region is probably a measure of its virulence for the animal.

Table 5. *Localisation of virus in the pharynx of cattle exposed to attenuated strains of foot-and-mouth disease virus*

Virus	Route of exposure*	No. of cattle	No. with virus in pharynx
A Kemron Egg	I/M	12	0
	R/P	8	4
C-997 Egg	I/D	12	4
	I/M	56	4
	R/P	4	1
SAT 1 Mouse	I/D	12	0
	I/M	28	0
SAT 2 Mouse	I/M	12	2
	R/P	8	1
SAT 3 Mouse	I/D	12	0
	I/M	20	0

* I/M, intramuscular; I/D, intradermal tongue; R/P, retropharyngeal spray.

Modified strains of foot-and-mouth disease virus are less likely to localise and persist in the pharyngeal area (Table 5) and it is significant that those strains of virus which did succeed in infecting the pharynx of some animals were those which, when used as experimental live vaccines, produced high antibody levels in the majority of animals and, occasionally, mild disease in hypersusceptible animals.

Comparative studies of the virulence of different strains of virus for the normal host have usually been made following inoculation by a particular route which has been found to give a consistent result and to be the most sensitive to minimal amounts of virus. The assessment of virulence may be based on estimates of the concentration of virus necessary to produce a clinical reaction and by the magnitude of that response, by direct measurement of the growth of virus in sequential samples of secretions, exudations or tissues of the animals or, indirectly, by measurement of the antibody response. Thus, the results of the titration of foot-and-mouth disease virus strains in the bovine tongue

(Henderson, 1949) or in the foot of the pig (Burrows, 1966b) and the subsequent clinical response of the animal have formed the basis for comparing the virulence of field strains of virus and of monitoring the modification of virus strains during the process of attenuation. However, it cannot be assumed that estimates of virulence based on the results of inoculation always apply to natural disease. The majority of strains of foot-and-mouth disease virus will produce clinical disease in the pig when inoculated into the foot (Burrows, 1964, 1966b) but not all will spread readily under natural conditions or infect pigs by other routes of inoculation (Brooksby, 1950; Graves & Cunliffe, 1960). In foot-and-

Table 6. *Viraemia patterns in pigs following experimental inoculation of different strains of foot-and-mouth disease virus*

Virus strain	No. of pigs	Days after inoculation			
		1	2	3	4
A-Iraq	4	5·25*	5·80	3·78	0
A-119	7	2·31	4·39	3·76	0·44
A-Kemron	4	NT	3·20	2·30	0
O-Israel	4	2·80	5·07	3·70	0
O-1860	4	2·18	3·80	1·92	0·80
SAT 1	4	3·54	5·62	1·62	0
SAT 3	4	2·70	3·48	3·00	1·10

* Geometric mean, $\log_{10}$ p.f.u./ml.

mouth disease and in other virus diseases it is reasonable to assume that the virus concentrations of the blood are a useful estimate of virus growth in the animal. Table 6 lists the mean virus concentrations in the blood of pigs inoculated with different virus strains and it would appear that the A-Iraq virus was possibly the most virulent virus of this group. However, in a transmission experiment similar to the one illustrated in Fig. 2, none of the pigs acquired infection. In the same experiment 2 pigs housed with 4 inoculated reacting pigs failed to acquire active infection until the 8th and 14th day of contact, despite the fact that virus was recovered from pharyngeal samples taken from the first day of exposure onwards.

RELEVANCE OF *IN VITRO* STUDIES TO VIRUS GROWTH IN THE NORMAL HOST

Many workers have associated, or have attempted to associate, characteristics of virus growth and behaviour *in vitro* with virus behaviour in the normal host. Such characteristics as tissue receptor activity, the production of or susceptibility to interferon, growth and cytopathogenicity in various cultures under differing conditions of temperature and pH and virulence for laboratory animals have all been linked to differences in virulence for the normal host (Holland, 1964). The determinants of virulence, although obviously related to those of virus growth, differ greatly between viruses and between strains of the same virus.

The ability of many viruses to grow in monolayer cultures derived from tissues or hosts which are not normally susceptible to the virus (Chaproniere & Andrewes, 1957), limits the usefulness of such cultures for basic studies of cellular and tissue susceptibilty as they relate to the intact animal. This loss of viral specificity is associated with the rapid de-differentiation of cells which accompanies the multiplication of animal cells *in vitro*. In organ cultures the cells are maintained without active multiplication and retain their normal histological appearance and their functional activity as judged by mucus production and ciliary activity. Most studies of the growth and behaviour of viruses in organ cultures have been made in cultures of respiratory tract epithelia (Hoorn & Tyrrell, 1969) and most human viruses will grow in such cultures. Moreover, the ability of viruses to produce cellular damage in these cultures appears to be related to their ability to produce respiratory disease in man. Comparative studies using cultures prepared from different regions have shown that they retain a similar susceptibility to virus as the corresponding cells in the living animal. Hoorn & Tyrrell (1966) found that smaller concentrations of rhinovirus (HS) were required to infect cultures of nasal epithelium than those of trachea or oesophagus and that greater amounts of virus and a more rapid destruction of cilia were produced in nasal epithelium. A vaccine strain of poliovirus type 1 and a Coxsackievirus B3 grew better in cultures of ileum than in oesophagus or other parts of the gastro-intestinal tract (Rubinstein & Tyrrell, 1970). Shroyer & Easterday (1968) found that, although a variety of bovine tissues would support the growth of infectious bovine rhinotracheitis virus, greatest growth occurred in nasal, tracheal and vaginal epithelia and that the histopathological changes induced in these cultures were similar to those found in natural disease. A good correlation was found between the growth of influenza virus in tissues of experimental ferrets and in organ cultures prepared from those tissues (Basarab & Smith, 1969, 1970).

Few direct comparisons appear to have been made of the growth of virulent and attenuated strains of virus in organ cultures. Banatvala, Best & Kistler (1969) examined a freshly isolated strain and a well-adapted laboratory strain of rubella virus in human embryonic tissues but no comment on possible differences in growth pattern was made.

Campbell, Thompson, Leighton & Penny (1969) found no obvious differences in the growth of two bovine parainfluenza type 3 viruses of different origin. Williams & Burrows (to be published) compared the growth of a number of attenuated strains of foot-and-mouth disease virus with those of their virulent parent strains in cultures of pharyngeal and tongue epithelia, lymph node and lingual muscle. In all experiments the virulent strains multiplied more rapidly and to a higher level than did

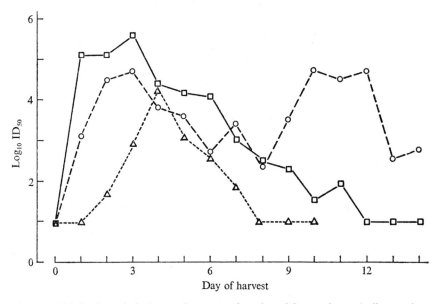

Fig. 3. Multiplication of virulent and attenuated strains of foot-and-mouth disease virus C–997 in organ cultures of pharyngeal epithelium: Virulent □—□: Mouse-attenuated ○- - -○: Egg-attenuated △- - - -△.

the attenuated strain but this was only so for the first 72 h; after this period no consistent pattern of growth was maintained (Fig. 3). Differences were found in the ability of some viruses to grow in particular tissues; both the attenuated and virulent strains of 3 virus types multiplied in cultures of muscle and lymph node, whilst strains of a fourth virus type did not. Foot-and-mouth disease virus has some affinity for muscle *in vivo* and localisation of virus may occur in cardiac and skeletal muscle in some animals. No cellular damage due to viral growth was seen but this is perhaps not surprising as the infectivity levels in the cultures did not approach those found in the animal or in the cells of monolayer cultures.

The use of organ cultures for the study of virus/cell interactions under conditions closely resembling those which obtain in the living animal is now well established and, although infections in organ culture show features which are apparently unlike those which have been recognised in the animal, they should nevertheless extend our knowledge of virus infection.

CONCLUSIONS

Some factors which influence the sites of localisation and the survival of virus in mucous membranes and may affect the susceptibility of the cell to infection have been mentioned. Whilst these are likely to be of importance in determining whether an animal acquires active infection following a brief or intermittent exposure to small amounts of virus, they may be less important under conditions of prolonged exposure to large amounts of virus. It is difficult to escape from the conclusion that the most significant factor in determining susceptibility is the genetic make-up of the host. Studies of natural and experimental virus infection show that marked differences in susceptibility occur between individuals both in the amounts of virus required to initiate infection and in the growth of virus in the body after infection. These differences are barely detectable·in experimental infections with virulent virus; they are more obvious in infections acquired naturally, and they are very obvious in studies with attenuated viruses. Strains of virus which are attenuated for the majority of animals may fail to grow in the hyposusceptible animal and yet may produce disease in the hypersusceptible animal.

REFERENCES

ABODEELY, R. A., LAWSON, L. A. & RANDALL, C. C. (1970). Morphology of enveloped and de-enveloped equine abortion (herpes) virus. *Journal of Virology*, **5**, 513–23.

AKERS, T. G. (1969). Survival of airborne virus, phage and other minute microbes. In *An Introduction to Experimental Aerobiology*, ed. R. L. Dimmick and A. B. Akers. New York: Wiley–Interscience.

APPEL, M. J. G., MENEGUS, M., PARSONSON, I. M. & CARMICHAEL, L. E. (1969). Pathogenesis of canine herpesvirus in specific-pathogen-free dogs: 5- to 12-week old pups. *American Journal of Veterinary Research*, **30**, 2067–73.

ASMUNDSSON, T. & KILBURN, K. H. (1970). Mucociliary clearance rates at various levels in dogs lungs. *American Review of Respiratory Diseases*, **102**, 388–97.

BANATVALA, J. E., BEST, J. M. & KISTLER, G. S. (1969). Studies on the growth and structure of rubella virus grown in human embryonic organ culture. *Symposia Series in Immunobiological Standardization*, **11**, 161–72. Basel: Karger.

BANG, B. G. & BANG, F. B. (1967). Laryngotracheitis virus in chickens. A model for study of acute nonfatal desquamating rhinitis. *Journal of Experimental Medicine*, **125**, 409–27.

BANG, B. G. & BANG, F. B. (1969a). Experimentally induced changes in nasal mucous secretory systems and their effect on virus infection in chickens. I. Effect on mucosal morphology and function. *Journal of Experimental Medicine*, **130**, 105–19.

BANG, B. G. & BANG, F. B. (1969b). Replacement of virus destroyed epithelium by keratinized squamous cells in vitamin A-deprived chickens. *Proceedings of the Society for Experimental Biology and Medicine*, **132**, 50–4.

BANG, F. B. & FOARD, M. A. (1969). Experimentally induced changes in nasal secretory systems and their effect on virus infection in chickens, II. Effects on adsorption of Newcastle disease virus. *Journal of Experimental Medicine*, **130**, 121–40.

BANG, F. B. & LUTTRELL, C. N. (1961). Factors in the pathogenesis of virus diseases. In *Advances in Virus Research*, ed. K. M. Smith and M. A. Lauffer, vol. **8**, 199–244. New York: Academic Press.

BASARAB, O. & SMITH, H. (1969). Quantitative studies on the tissue localization of influenza virus in ferrets after intranasal and intravenous or intracardial inoculation. *British Journal of Experimental Pathology*, **50**, 612–18.

BASARAB, O. & SMITH, H. (1970). Growth patterns of influenza virus in cultures of ferret organs. *British Journal of Experimental Pathology*, **51**, 1–6.

BECHT, H. (1969). Induction of an arginine-rich component during infection with influenza virus. *Journal of General Virology*, **4**, 215–20.

BEDSON, H. S. & DUCKWORTH, M. J. (1963). Rabbitpox: An experimental study of the pathways of infection in rabbits. *Journal of Pathology and Bacteriology*, **85**, 1–20.

BRODY, S. (1945). *Bioenergetics and Growth*. New York: Reinhold Publishing Corporation.

BROOKSBY, J. B. (1950). Strains of the virus of foot-and-mouth disease showing natural adaptation to swine. *Journal of Hygiene*, **47**, 184–95.

BUCKLAND, F. E., BYNOE, M. L. & TYRRELL, D. A. J. (1965). Experiments on the spread of colds. II. Studies in volunteers with Coxsackie A21. *Journal of Hygiene*, **63**, 327–43.

BURNET, F. M. (1960). *Principles of Animal Virology*, 2nd ed. New York: Academic Press.

BURROWS, R. (1964). The behaviour of some modified strains of foot-and-mouth disease virus in the pig. 1. Innocuity. *Bulletin de l'Office International des Epizooties*, **61**, 1251–75.

BURROWS, R. (1966a). Studies on the carrier state of cattle exposed to foot-and-mouth disease virus. *Journal of Hygiene*, **64**, 81–90.

BURROWS, R. (1966b). The infectivity assay of foot-and-mouth disease virus in pigs. *Journal of Hygiene*, **64**, 419–29.

BURROWS, R. (1968a). Excretion of foot-and-mouth disease virus prior to the development of lesions. *Veterinary Record*, **82**, 387–8.

BURROWS, R. (1968b). The persistence of foot-and-mouth disease virus in sheep. *Journal of Hygiene*, **66**, 633–40.

BURROWS, R., MANN, J. A., GREIG, A., CHAPMAN, W. G. & GOODRIDGE, D. (1971). The growth and persistence of foot-and-mouth disease virus in the bovine mammary gland. *Journal of Hygiene* (in press).

CAMPBELL, R. S. F., THOMPSON, H., LEIGHTON, E. & PENNY, W. (1969). Pathogenesis of bovine parainfluenza 3 virus infection in organ cultures. *Journal of Comparative Pathology*, **79**, 347–54.

CHAPRONIERE, D. M. & ANDREWES, C. H. (1957). Cultivation of rabbit myxoma and fibroma viruses in tissues of nonsusceptible hosts. *Virology*, **4**, 351–65.

CHENNEKATU, P. P., GRATZEK, J. B. & RAMSEY, F. K. (1966). Isolation and characterization of a strain of infectious bovine rhinotracheitis virus associated with enteritis in cattle. Pathogenesis studies by fluorescent antibody tracing. *American Journal of Veterinary Research*, **27**, 1583–90.

COHEN, A. (1963). Mechanisms of cell infection. I. Virus attachment and penetration. In *Mechanisms of Virus Infection*, ed. W. Smith, 153–90. London: Academic Press.

COLGROVE, G. S., HAELTERMAN, E. O. & COGGINS, L. (1969). Pathogenesis of African Swine Fever in young pigs. *American Journal of Veterinary Research*, 30, 1343–59.

COTTRAL, G. E. (1969). Persistence of foot-and-mouth disease virus in animals, their products and the environment. *Bulletin de l'Office International des Epizooties*, 71, 549–68.

COUCH, R. B., CATE, T. R., DOUGLAS, R. G., GERONE, P. J. & KNIGHT, V. (1966a). Effect of route of inoculation on experimental respiratory viral disease in volunteers and evidence for airborne transmission. *Bacteriological Reviews*, 30, 517–29.

COUCH, R. B., CATE, T. R., FLEET, W. F., GERONE, P. J. & KNIGHT, V. (1966b). Aerosol-induced adenoviral illness resembling the naturally occurring illness in military recruits. *American Review of Respiratory Disease*, 93, 529–35.

CURTAIN, C. C., CLARK, B. L. & DUFFY, J. H. (1971). The origins of the immunoglobulins in the mucous secretions of cattle. *Clinical and Experimental Immunology*, 8, 335–44.

DALES, S. (1965). Penetration of animal virus into cells. In *Progress in Medical Virology*, ed. J. L. Melnick, vol. 7, 1–43. Basel: Karger.

DALHAMN, T. (1956). Mucous flow and ciliary activity in the trachea of healthy rats and rats exposed to respiratory irritant gases. *Acta Physiologica Scandinavica*, 36, Supplement 123, 1–161.

DONALDSON, A. I., HERNIMAN, K. A. J., PARKER, J. & SELLERS, R. F. (1970). Further investigations on the airborne excretion of foot-and-mouth disease virus. *Journal of Hygiene*, 68, 557–64.

DOURMASHKIN, R. R. & TYRRELL, D. A. J. (1970). Attachment of two myxoviruses to ciliated epithelial cells. *Journal of General Virology*, 9, 77–88.

DOWNIE, A. W. (1963). Pathways of virus infection. In *Mechanisms of Virus Infection*, ed. W. Smith, 101–152. London: Academic Press.

DRUETT, H. A. (1967). The inhalation and retention of particles in the human respiratory system. In *Airborne Microbes, Seventeenth Symposium of the Society for General Microbiology*, ed. P. H. Gregory and J. L. Monteith, 165–202. London: Cambridge University Press.

DUNNE, H. W., HOKANSON, J. F. & LUEDKE, A. J. (1959). The pathogenesis of hog cholera. I. Route of entrance of the virus into the animal body. *American Journal of Veterinary Research*, 20, 615–18.

EBISAWA, I. T., KITAMOTO, O. & TAKEUCHI, Y. (1969). Immunocytologic study of nasal epithelium in influenza. *American Review of Respiratory Diseases*, 99, 507–15.

FAZEKAS DE ST GROTH (1948). Viropexis, the mechanism of influenza virus infection. *Nature, London*, 162, 294–5.

FENNER, F. (1948). The clinical features and pathogenesis of mouse pox (infectious ectromelia of mice). *Journal of Pathology and Bacteriology*, 60, (4), 529–52.

FENNER, F. (1968). *The Biology of Animal Viruses*, vol. II, New York and London: Academic Press.

FENNER, F. & WOODROOFE, G. M. (1953). The pathogenesis of infectious myxomatosis: the mechanism of infection and the immunological response in the European rabbit (*Oryctolagus cuniculus*). *British Journal of Experimental Pathology*, 34, 400–11.

FINTER, N. B. (1966). Interferons in animals and man. In *Interferons*, ed. N. B. Finter, 232–67. Amsterdam: North Holland Publishing Company.

GERONE, P. J., COUCH, R. B., KEEFER, G. V., DOUGLAS, R. G., DERRENBACHER, E. B. & KNIGHT, V. (1966). Assessment of experimental and natural viral aerosols. *Bacteriological Reviews*, 30, 576–84.

GOLDBERG, L. J. & LIEF, W. R. (1950). The use of a radio-active isotope in determining the retention and initial distribution of airborne bacteria in the mouse. *Science*, 112, 299–300.

GOLDBLUM, N., RAVID, Z. & BECKER, Y. (1968). Effect of withdrawal of arginine and other amino acids on the synthesis of tumour and viral antigens of SV 40 virus. *Journal of General Virology*, 3, 143–6.

GRAVES, J. H. & CUNLIFFE, H. R. (1960). The infectivity assay of foot-and-mouth disease virus in swine. *Proceedings of the Sixty-third Annual Meeting of the United States Livestock Sanitary Association*, 340–5.

GREIG, A. (1971). The route of infection in domestic pigs naturally exposed to African Swine Fever. *Journal of Comparative Pathology*, (in press).

GUYTON, A. C. (1947). Measurement of respiratory volumes of laboratory animals. *American Journal of Physiology*, 150, 70–7.

HARGREAVES, F. D. & LEACH, R. H. (1970). The influence of mycoplasma infection on the sensitivity of Hela cells for growth of viruses. *Journal of Medical Microbiology*, 3, 259–65.

HARPER, G. J. & MORTON, J. D. (1953). The respiratory retention of bacterial aerosols; experiments with radio-active spores. *Journal of Hygiene*, 51, 372–85.

HATCH, T. & GROSS, P. (1964). *Pulmonary Deposition and Retention of Inhaled Aerosols*. New York: Academic Press.

HENDERSON, R. J. (1969). The outbreaks of foot-and-mouth disease in Worcestershire. An epidemiological study; with special reference to spread of disease by wind carriage of the virus. *Journal of Hygiene*, 67, 21–33.

HENDERSON, W. M. (1949). The quantitative study of foot-and-mouth disease virus. *Agricultural Research Council Report Series*, No. 8, London: H.M.S.O.

HENDERSON, W. M. (1952). A comparison of different routes of inoculation of cattle for detection of the virus of foot-and-mouth disease. *Journal of Hygiene*, 50, 182–94.

HENDERSON, W. M. & BROOKSBY, J. B. (1948). The survival of foot-and-mouth disease virus in meat and offal. *Journal of Hygiene*, 46, 394–402.

HEUSCHELE, W. P. (1967). Studies on the pathogenesis of African Swine Fever, I. Quantitative studies on the sequential development of virus in pig tissues. *Archiv für die gesamte Virusforschung*, 21, 349–56.

HILDING, A. C. (1934). Notes on some changes in the hydrogen-ion concentration of nasal mucous. *Annals of Otology, Rhinology and Laryngology*, 43, 47–9.

HO, M., FANTES, K. H., BURKE, D. C. & FINTER, N. B. (1966). Interferons or interferon-like inhibitors induced by non-viral substances. In *Interferons*, ed. N. B. Finter, 181–201. Amsterdam: North Holland Publishing Company.

HOLLAND, J. J. (1961). Receptor affinities as major determinants of enterovirus tissue tropism in humans. *Virology*, 15, 312–26.

HOLLAND, J. J. (1964). Viruses in animals and in cell culture. In *Microbial Behaviour 'in vivo' and 'in vitro', Fourteenth Symposium of the Society for General Microbiology*, ed. H. Smith and J. Taylor, 257–86. London: Cambridge University Press.

HOORN, B. & TYRRELL, D. A. J. (1966). A new virus cultivated only in organ cultures of human ciliated epithelium. *Archiv für die gesamte Virusforschung*, 18, 210–25.

HOORN, B. & TYRRELL, D. A. J. (1969). Organ cultures in virology. In *Progress in Medical Virology*. ed. J. L. Melnick, vol. 11, 408–50. Basel: Karger.

HURST, G. W. (1968). Foot-and-mouth disease. The possibility of continental sources of the virus in England in epidemics of October 1967 and several other years. *Veterinary Record*, **82**, 610–14.

HYSLOP, N. ST G. (1965). Airborne infection with the virus of foot-and-mouth disease. *Journal of Comparative Pathology*, **75**, 119–26.

JOHNSON, J. S. (1970). The secretory immune system, a brief review. *Journal of Infectious Diseases*, **121**, (suppl.), 115–17.

KILBURN, K. H. (1968). A hypothesis for pulmonary clearance and its implications. *American Review of Respiratory Diseases*, **98**, 449–63.

KILBURN, K. H. & SALZANO, J. V. (1966). Symposium on structure, function and measurement of respiratory cilia. *American Review of Respiratory Diseases*, **93**, (suppl.), 1–184.

KNIGHT, V. (1970). The importance of particle size in airborne viral respiratory infections. In *Aerobiology. Proceedings of the Third International Symposium*, ed. I. H. Silver, 273. London: Academic Press.

KORN, G. (1957). Experimentelle Untersuchungen zum Virusnachweis im Inkubationsstatium der Maul- und Klauenseuche und zu ihrer Pathogenese. *Archiv für Experimentelle Veterinarmedizin*, **11**, 637.

LEES, A. W. & MCNAUGHT, W. (1959). Bacteriology of lower respiratory secretions, sputum, and upper respiratory tract secretions in 'normals' and chronic bronchitics. *The Lancet*, **2**, 1112–15.

LOUDON, R. G. & ROBERTS, R. M. (1967). Droplet expulsion from the respiratory tract. *American Review of Respiratory Diseases*, **95**, 435–42.

MAY, K. R. (1966). Multistage liquid impinger. *Bacteriological Reviews*, **30**, 559–70.

MAY, K. R., DRUETT, H. A. & PACKMAN, L. P. (1969). Toxicity of open air to a variety of microorganisms. *Nature, London*, **221**, 1146–7.

MILLER, W. S. (1966). Infection of pigeons by airborne Venezuelan equine encephalitis virus. *Bacteriological Reviews*, **30**, 589–95.

MIMS, C. A. (1964). Aspects of the pathogenesis of virus diseases. *Bacteriological Reviews*, **28**, 30–71.

MIMS, C. A. (1970). The pathogenesis of viral infections of the respiratory tract. In *Aerobiology. Proceedings of the Third International Symposium*, ed. I. H. Silver, 243–7. London: Academic Press.

MORGAN, C. & HOWE, C. (1968). Structure and development of viruses as observed in the electron microscope. IX. Entry of parainfluenza 1 (Sendai) virus. *Journal of Virology*, **2**, 1122–32.

MORGAN, C. & ROSE, H. M. (1968). Structure and development of viruses in the electron microscope. VIII. Entry of influenza Virus. *Journal of Virology*, **2**, 925–36.

MORGAN, C., ROSE, H. M. & MEDNIS, B. (1968). Electron microscopy of herpes simplex virus. I. Entry. *Journal of Virology*, **2**, 507–16.

MORGAN, C., ROSENKRANZ, H. S. & MEDNIS, B. (1969). Structure and development of viruses as observed in the electron microscope. X. Entry and uncoating of adenovirus. *Journal of Virology*, **4**, 777–96.

PAYNE, J. M., SANSON, B. F., GARNER, R. J., THOMSON, A. R. & MILES, B. J. (1960). Uptake of small resin particles (1–5 μ diameter) by the alimentary tract of the calf. *Nature, London*, **188**, 586–7.

PENSAERT, M., HAELTERMAN, E. O. & BURNSTEIN, T. (1970). Transmissible gastroenteritis of swine: virus intestinal cell reactions. I. Immunofluorescence, histopathology and virus production in the small intestine through the course of infection. *Archiv für die gesamte Virusforschung*, **31**, 321–34.

PHILIPSON, L. (1963). The early interaction of animal viruses and cells. In *Progress in Medical Virology*, ed. E. Berger and J. L. Melnick, vol. 5, 43–78. New York: Karger.

PIAZZA, M. & PANE, E. (1967). Effect on the infectivity of various viruses by the intestinal factor of normal mice which inactivates murine hepatitis virus. *Nature, London*, **213**, 293–4.

PLATT, H. (1963*a*). The engulfment of colloidal and particulate materials by epidermal cells. *Journal of Pathology and Bacteriology*, **86**, 113–22.

PLATT, H. (1963*b*). The susceptibility of the skin and epidermoid mucous membranes to virus infections in man and other animals. *Guy's Hospital Reports*, **112**, 479–97.

PLATT, H. (1967). Pathogenesis. In *Viral and Rickettsial Diseases of Animals*, ed. A. O. Betts and C. J. York, 167–210. New York: Academic Press.

PLATT, H. (1970). Factors contributing to resistance and susceptibility of stratified squamous and respiratory epithelia. In *Resistance in Infectious Diseases*, ed. R. H. Dunlop and H. W. Moon, 173–83. Saskatoon: Saskatoon Modern Press.

PLOWRIGHT, W. (1968). Rinderpest virus. *Virology Monographs*, **3**, 25–110. Wien: Springer Verlag.

PLOWRIGHT, W., PARKER, J. & STAPLE, R. F. (1968). The growth of a virulent strain of African swine Fever virus in domestic pigs. *Journal of Hygiene*, **66**, 117–34.

PROCTOR, D. F. (1964). Respiration. I. *Handbook of Physiology*, 306–41. Washington, DC: American Physiological Society.

PROCTOR, D. F. (1966). Airborne disease and the upper respiratory tract. *Bacteriological Reviews*, **30**, 498–513.

QUINLAN, M. F., SALMON, S. D., SWIFT, L., WAGNER, H. N. & PROCTOR, D. F. (1969). Measurement of mucociliary function in man. *American Review of Respiratory Diseases*, **99**, 13–23.

REED, S. E. (1969). Persistent respiratory virus infection in tracheal organ cultures. *British Journal of Experimental Pathology*, **50**, 378–88.

RHODIN, J. A. G. (1966). Ultrastructure and function of the human tracheal mucosa. *American Review of Respiratory Diseases*, **93**, (suppl.) 1–15.

RIVERA, J. A. (1962). *Cilia, Ciliated Epithelium and Ciliary Activity*. London: Pergamon Press.

ROMANO, N. & BRANCATO, P. (1970). Inhibition of growth of measles virus by mycoplasma in cell cultures and the restoring effect of arginine. *Archiv für die gesamte Virusforschung*, **29**, 39–43.

ROPES, M. W. (1930). Phagocytic activity and morphological variations of the ciliated epithelial cells of the trachea and bronchii in rabbits. *Contribution to Embryology*, **22**, 79–90.

ROSEBURY, T. (1947). *Experimental Airborne Infection*. Baltimore: The Williams and Wilkins Company.

ROUSE, H. C., BONIFAS, V. H. & SCHLESINGER, R. W. (1963). Dependence of adenovirus replication of arginine and inhibition of plaque formation by pleuropneumonia like organism. *Virology*, **20**, 357–65.

RUBINSTEIN, D. & TYRRELL, D. A. J. (1970). Growth of viruses in organ cultures of intestine. *British Journal of Experimental Pathology*, **51**, 210–16.

SABO, A., RAJCANI, J. & BLASKOVIC, D. (1968). Studies on the pathogenesis of Aujeszky's disease. I. Distribution of the virulent virus in piglets after peroral infection. *Acta Virologica*, **12**, 214–21.

SCHULMAN, J. L. (1970). Transmissibility as a separate genetic attribute of influenza viruses. In *Aerobiology. Proceedings of the Third International Symposium*, ed. I. H. Silver, 248–59. London: Academic Press.

SELLERS, R. F. (1971). Quantitative aspects of the spread of foot-and-mouth disease. *Veterinary Bulletin* (in press).

SELLERS, R. F., DONALDSON, A. I. & HERNIMAN, K. A. J. (1970). Inhalation, persistence and dispersal of foot-and-mouth disease virus by man. *Journal of Hygiene*, **68**, 565–73.

SELLERS, R. F. & PARKER, J. (1969). Airborne excretion of foot-and-mouth disease virus. *Journal of Hygiene*, **67**, 671–7.

SHROYER, E. L. & EASTERDAY, B. C. (1968). Growth of infectious bovine rhino-tracheitis virus in organ cultures. *American Journal of Veterinary Research*, **29**, 1355–62.

SIMPSON, R. W., HAUSER, R. E. & DALES, S. (1969). Viropexis of vesicular stomatitis virus by L cells. *Virology*, **37**, 285–90.

TOMASI, T. B. (1970). Structure and function of mucosal antibodies. *Annual Review of Medicine*, **21**, 281–98.

TYRRELL, D. A. J. (1967). The spread of viruses of the respiratory tract by the airborne route. In *Airborne Microbes. Seventeenth Symposium of the Society for General Microbiology*, ed. P. H. Gregory and J. L. Monteith, 286–306. London: Cambridge University Press.

VILCEK, J. (1969). Interferon. *Virology Monographs*, **6**, Wien: Springer Verlag.

WASSERMAN, F. E. (1968). Methods for the study of viral inhibitors. In *Methods in Virology*, ed. K. Maramorosch and H. Koprowski. New York: Academic Press.

WINTERS, A. L. & CONSIGLI, R. A. (1971). Effects of arginine deprivation on polyoma virus infection of mouse embryo cultures. *Journal of General Virology*, **10**, 53–63.

HOST DEFENCES AGAINST VIRUSES AND THE LATTER'S ABILITY TO COUNTERACT THEM

C. A. MIMS

Department of Microbiology, John Curtin School of Medical Research, Australian National University, Canberra, Australia

INTRODUCTION

Viruses, considered as infectious agents, are in many ways similar to other infectious agents. When we consider the initiation of infection through epithelial surfaces, the spread of infection in the body, the role of immune responses, phagocytic cells, inflammation, or the process of recovery and repair, the general principles are the same be it a virus, rickettsium, bacterium, fungus, or protozoa. There are close parallels, for instance between the response of a mouse to infection with ectromelia virus, the leprosy bacillus, and the malaria protozoan. Nevertheless, viruses are different enough from other micro-organisms for there to be a number of unique host–parasite principles and possibilities.

It will be the purpose of this paper to outline the nature of host defences against viruses and then to consider, with free use of teleology, the ways in which viruses as successful parasites overcome these defences. This they do firstly by methods available to other infectious micro-organisms such as by growing in macrophages, by becoming latent, or by antigenic variation. Sometimes, however, viruses overcome host defences by persisting intracellularly in a defective, non-infectious form, or even insinuating themselves into the host genome and reproductive system so that they are transmitted vertically. In the end it may become difficult to determine what is virus and what is host.

HOST DEFENCES AGAINST VIRUSES

Barriers at body surfaces

Viruses are immobile parasites, incapable of replication until inside cells, and might be expected to have problems in infecting the host across body surfaces. Yet they infect the epithelium of the respiratory and alimentary tracts with surprising efficiency, and it is unfortunate that so little is known about the way in which this is accomplished.

The skin is an effective barrier to most viruses; infection requires a break in the skin, as with smallpox vaccination, or the direct introduction of virus through the skin by a biting arthropod or the heroin addict's needle.

The respiratory and intestinal tracts, being lined by live, naked cells are the principle sites for virus entry into the body. The upper and lower respiratory tract is continually cleansed by a mucociliary mechanism which traps inhaled particles, whether viral, bacterial or inert, and then transports them to the back of the throat where they are swallowed. Virus particles which pass beyond this cleansing system to the alveoli encounter alveolar macrophages and may be phagocytosed. The infectious process is then halted unless virus persists or multiplies in macrophages rather than suffers the usual fate of phagocytosed material by being digested. The intestinal tract would seem to pose equally formidable barriers to infection. Epithelial cells are covered with a film of mucus, and virus particles in the lumen are exposed to acids, alkalis, enzymes, bile salts, and the products of a host of other microorganisms.

Phagocytic cells

Macrophages monitor the principal body compartments and phagocytose virus particles (Mims, 1964). If viruses are then digested and degraded, the macrophages can be said to have functioned well. If however, virus grows in macrophages, these cells have if anything merely hastened the infectious process by their enthusiastic uptake of the particle. By infecting macrophages, the virus has breached one of the body's major defence mechanisms. Thus, the Hampstead mouse strain of ectromelia virus has been shown to be virulent for mice compared with the Hampstead egg strain because of an increased ability to grow in macrophages. The liver is a target organ and virus growth in liver macrophages leads to extensive infection of hepatic cells (Roberts, 1963a). Increased susceptibility of immature animals has been correlated with a difference in macrophage response in the case of herpes simplex virus infection in mice (Johnson, 1964; Hirsch, Zisman & Allison, 1970). Immature mouse macrophages are infected no more readily than adult macrophages, but unlike adult macrophages are able to pass on infection to other cell types. There are also other examples (Bang & Warwick, 1960) where increased macrophage susceptibility is correlated with increased host susceptibility.

Occasionally there is a failure of macrophages to take up a particular type of virus, and this may grossly alter the course of infection. Thus the Armstrong strain of lymphocytic choriomeningitis virus, when injected intravenously into mice, is not cleared from the blood by liver and spleen macrophages, and therefore fails to establish significant infection in these organs. The WE₃ strain of virus, in contrast, is taken up by these macrophages, grows in them, and thus produces extensive infection in both organs (Tosolini & Mims, 1971).

Polymorphonuclear cells, although they do not feature prominently in the host response to virus infections, are nevertheless present. They appear early in inflammatory sites, phagocytosing and digesting extracellular debris. They phagocytose virus particles (Hanson, Kempf & Boand, 1957) and may contain phagocytosed virus particles in infected tissues. Polymorphonuclear cells as transportable sources of digestive enzymes may therefore have a significant virus-killing function in tissues. There is, so far, no evidence that they are successfully infected by viruses. Even if infected, their inevitable early disruption ensures eventual phagocytosis of contents by macrophages.

Other cell types, including epithelial and fibroblastic cells, actively engage in pinocytosis or phagocytosis, but this is on a small scale compared with polymorphonuclear cells and macrophages, and may either increase infection rate or kill ingested viruses, depending on cell susceptibility. It may be noted that an important type of barrier to the spread of viruses in a host is a layer of insusceptible cells barring entry to a target organ. Thus, circulating viruses may fail to negotiate vascular endothelium in the brain, placenta, or skin, and therefore fail to infect these tissues. There is a blood–brain, blood–placenta, or blood–skin barrier.

Fever

Thomas Sydenham (1624–89) wrote that 'Fever is a mighty engine which Nature brings into the world for the conquest of her enemies'. There has been much experimental work, especially with Coxsackie B 1 virus, myxoma virus, Semliki Forest virus, and poliovirus, indicating that artificially raised body temperatures are associated with decreased virus growth in target tissues (Bennett & Nicastri, 1960). For instance, strains of poliovirus which are less virulent for monkeys grow poorly in tissue cultures at febrile temperatures, and mice infected intracerebrally with virulent strains of poliovirus survive longer when kept at 36°, a treatment which raises their rectal temperature by 2–3° (Lwoff, 1959). Animals with raised body temperatures, however, show a variety of secondary changes in metabolism, in hormone secretion, even immune responses. It is therefore difficult to assign a beneficial role to increased body temperature *per se* in resistance to virus diseases. Also, fever is generally seen at the end of the incubation period, when viral replication is largely completed. For this reason, it seems possible that fever often occurs as a response to antigen–antibody complexes (Mott & Wolff, 1966) or to a generalised cell-mediated immune response (Atkins, 1960) rather than playing some teleologically satisfactory role in limiting virus growth.

Interferon

Most cells produce and liberate interferon after infection with viruses. Interferon is a protein which acts on uninfected cells and, with no detectable effect on their metabolism, renders them resistant to infection with a wide range of viruses. Viruses differ both in the amount of interferon they induce and in their susceptibility to interferon. Those that induce an early inhibition of RNA and protein synthesis in the infected cell can thereby depress interferon production.

Theoretically the first infected cell in an animal could liberate interferon and protect neighbouring cells from infection. In this way, interferon can operate without the delay characteristic of the immune response. Immune mechanisms cannot operate until the response itself has been generated and sensitised cells or antibody subsequently delivered to the site of infection. It seems inevitable that interferon, a host resistance mechanism which seems to have particular relevance in relation to viral and rickettsial infections, should be important in resistance to virus diseases. The difficulty in assessing its importance is that its antiviral activity can seldom be dissociated from the activities of antibodies and sensitised immune cells. Indeed, there is evidence that sensitised lymphocytes produce increased amounts of interferon on re-exposure to the original antigen, viral or non-viral (Green, Coopergand & Kibrick, 1969). Also, interferon may have important local activity without being plentiful enough to be detectable in tissue extracts. There are many instances where strains of virus which are less susceptible to interferon or which induce lesser amounts of interferon, are also less virulent. There are also many instances where this is not the case (Baron, 1970). It would be nice to have a clear experimental situation such as an interferon-defective mutant mouse or man, with no accompanying defect in immune responsiveness.

Antibody

Twenty-five years ago antibody was accepted as the major influence in recovery from virus infections and resistance to re-infection. Since then, interferon has been discovered and advances in immunology have led to a clearer distinction between humoral and cellular components of the immune response. Today, although antibody is still considered of critical importance in resistance to re-infection with most viruses, there has been a re-assessment of its role in recovery from primary infection. This has been partly because of observations on the apparently normal recovery from virus infections in patients with agammaglobulinaemia, and also because of experimental work on the significance of immune cells in recovery from certain virus infections (Allison, 1967; Blanden, 1971).

Antibodies are produced as a result of most virus infections, often for long periods, even when persistent infection can almost certainly be excluded (Sawyer, 1931). A large, complex virus such as a poxvirus may induce the formation of 10–20 different types of antibodies (Westwood, Zwartouw, Appleyard & Titmuss, 1965), but most of these will be directed against internal components of the virion and against non-virion proteins appearing during the course of infection. The important antibodies are those that react with antigenic determinants on the surface of virus particles and thereby neutralise infectivity. Other types of antibody are irrelevant from the point of view of the host, although by reacting with antigens or by forming immune complexes they sometimes play an important pathogenic role. Surprisingly, there is still no clear understanding of the mechanism of neutralisation of virus infectivity by antibody. The virus–antibody combination can be reversed and infectious virus recovered from neutralised mixtures, for instance by dilution, treatment with ultrasonics, acid, proteolytic enzymes or genetron. Complement potentiates some virus neutralisation systems in an unspecified fashion, and there have been reports of antibodies which do not neutralise except in the presence of complement (Yoshino & Taniguchi, 1964), but there is no evidence that complement has a killing action comparable to that seen with mycoplasma (Brunner, Razin, Kalica & Chanock, 1971) and with certain bacteria. In some infections non-neutralising antibody combines with virus particles in the blood to form infectious virus-antibody complexes which can be precipitated by antiglobulin serum (Oldstone & Dixon, 1970). Probably, virus which is combined with neutralising antibody is non-infectious because it is prevented from attaching to, and entering susceptible cells, the complex remaining extracellular until it is phagocytosed or the virus is thermally inactivated. Neutralised viruses may be phagocytosed and digested by macrophages, even viruses which would otherwise grow in macrophages (Mims, 1964), and this is another mechanism of virus neutralisation.

There are biological differences between the different immunoglobulin classes, some of which reflect differences in function. Ig G antibodies are probably the most important, particularly in resistance to re-infection with systemic viruses, since they persist for long periods. Ig M antibodies are larger, do not enter the extravascular tissues spaces to any great extent (Hall, Smith, Edwards & Shooter, 1969), and are not transmitted to the foetus via the placenta. But they do neutralise viruses. They are considered a more primitive type of immunoglobulin, but a purely vestigial significance for them is hard to accept. From the narrow point of view of resistance to virus infections, Ig M antibodies are important because they appear a day or two earlier than Ig G antibodies. This may be of critical importance in resistance to infection, where there is a race between virus replication and the production of specific virus inhibitory factors. A type of antibody appearing a day earlier may turn an otherwise severe infection into a mild one (Schell, 1960).

Ig A (secretory) antibodies have an obvious role to play in resistance to virus infections of mucosal surfaces, whether respiratory, alimentary or urogenital. These antibodies are secreted onto mucosal surfaces, neutralise viruses, and are more resistant than other antibodies to degradation by enzymes. Little is known about the persistance of this type of antibody, or about secondary-immune responses. There are indications that their production is localised to infected regions of the alimentary canal or lung (Ogra & Karzon, 1969).

There is at present no evidence about the possible significance of Ig E (reaginic) antibodies in resistance to virus infections of mucosal surfaces.

Cell-mediated immunity

It has been recognised for some time that 'cellular immunity', as opposed to antibody, may be of some consequence in resistance to virus infections, but it is only recently that the roles of immune macro-phages, cytophilic antibodies, and immune lymphocytes have begun to be disentangled.

There is no good evidence that macrophages from immune animals are themselves more resistant to virus infection. Roberts, indeed, found that peritoneal macrophages from mice immune to ectromelia virus were, after thorough washing, more readily infected than normal macrophages, probably because of their greater phagocytic activity (Roberts, 1963b). Experiments on immune macrophages must take into account the possible presence of cytophilic antibody on their surfaces. Macrophages become 'activated' in the presence of immune lymphocytes reacting with antigen. They develop enhanced phagocytic and digestive powers, but there is no clear evidence that this by itself affects their susceptibility to virus infection.

There have been many general indications that immune lymphocytes are important in recovery from some virus infections. Patients with dysplasia or absence of the thymus often show a striking inability to control poxvirus and other infections, in spite of normal antibody responses (Rosen, 1968), and experimental work with antilymphocyte serum and other immunosuppressives has sometimes implicated a cellular immune response rather than antibody or interferon as critical in recovery from infection (Allison, 1967). Blanden, however, has obtained clear evidence that thymus-derived lymphocytes are of key significance in the recovery of mice from ectromelia virus infection (Blanden, 1971). Sensitised lymphocytes enter infectious lesions and react with viral antigens to liberate lymphokinins, which elicit an infiltration of bone-marrow-derived monocytes. The phagocytic activity of these infiltrating cells, perhaps with the help of locally pro-duced antibody and interferon, sterilises the infected focus. In this model antibody and interferon by themselves are relatively ineffective. Although evidence such as this implicates cell-mediated immunity in host resistance to certain virus infections, it is equally clear that in other instances cell-mediated immunity is comparatively unimportant. For instance, antilymphocyte serum does not depress the resistance of mice to intranasally administered influenza virus or Sendai virus (Hirsch & Murphy, 1968; C. A. Mims, unpublished). Perhaps it is a general feature of virus infections of epithelial surfaces that interferon and anti-body are more important than cell-mediated immunity in recovery from infection. However, a sharp distinction between antibody, cell-mediated immunity and interferon may in the end be difficult. Sensitised lympho-cytes for instance, produce interferon when exposed to antigen. It is possible that in resistance to all virus infections antibody, cell-mediated immunity and interferon each play a part, although one of these factors is often much more important than the others.

Cell-mediated immunity probably helps to keep latent infections in the latent state. Convincing evidence for this is not at present available. It is possible that the occurrence of zoster in patients with Hodgkin's disease (Sokal & Firat, 1965) is attributable to defective cell-mediated immunity, and that zoster occurs in older people for a similar reason. Transplant patients treated with steroids, blood transfusions and various immunosuppressives may develop generalised cytomegalovirus infection (Craighead, 1969) or severe herpes simplex lesions (Montgomerie et al. 1969), and depressed cell-mediated immunity may be important but it is not clear how much is attributable to depressed antibody formation or transfusion with blood containing infected leucocytes. Surprisingly, there have been few studies of the role of cell-mediated immunity in maintaining latency in model animal systems.

Cell-mediated immunity also functions in immune surveillance for tumour cells by eliminating cells bearing foreign antigens on their surfaces (Burnet, 1970). Viruses can transform infected cells into tumour cells, causing virus specific, but non-virion antigens, to appear on the cell surface (Klein, 1966). Cell-mediated immunity is the mechanism for the control and elimination of such dividing transformed cells. In many instances viral antigens appear on the surface of infected cells and are necessary for the maturation of virus, which buds from the cell through a portion of the cell membrane containing viral antigens. When the infected cell is destroyed by virus there is little opportunity for immune reactions to take place with the antigenically changed cell membrane, but in non-cytocidal infections cell-mediated immunity can play an important part in resistance as well as in pathogenesis. In lymphocytic choriomeningitis (LCM) virus infection of mice the immune cellular response has an antiviral function in infectious foci, but in the course of carrying out this function it causes a cellular infiltration and inflammation which, in the brain, may be lethal (C. A. Mims, unpublished).

The host cell membrane is of supreme importance for viruses simply because it is the only route in and out of cells. Entry into susceptible cells may require specific receptors on the cell membrane, and in some cases, at least, the viral core enters the cytoplasm as a consequence of fusion of the viral envelope with the cell membrane (Morgan & Rose, 1969; Zee & Talens, 1971). The cell membrane is often necessary for viral maturation as discussed above, and release of virus from the cell by budding rather than cell-bursting can help ensure a non-cytocidal yet productive infection. It is possible that the induction of new antigens on the cell membrane, which forces the host to respond to the infected cell as if it were a foreign cell, is a unique feature of virus infections. But there have been few if any studies of rickettsial, protozoal, or bacterial infections from this point of view.

THE ABILITY OF VIRUSES TO COUNTERACT HOST DEFENCES

Barriers at the body surface

In spite of the cleansing mechanisms and the alveolar macrophages, many viruses infect the respiratory tract with some efficiency. Those which grow in alveolar macrophages after phagocytosis, such as ectromelia virus in mice (Roberts, 1962), can extend the infection to adjacent tissues as discussed below. Infection of ciliated epithelial cells becomes easier to understand when there is a receptor to bind the virus particle to the cell surface, as in the case of influenza virus. The haemagglutinin subunits on the surface of virus particles unite with N-acetyl neuraminic acid receptor on the cell surface, virus and cell membranes fuse, and the virus core (nucleocapsid) subsequently enters the cell. When we see an electron microscope picture (Dourmashkin & Tyrrell, 1970) of an influenza virus particle attached to a susceptible respiratory epithelial cell, we can understand how important receptors may be for a virus which is being steadily borne past susceptible cells in a film of mucus. The enzyme neuraminidase seems important for the virus in freeing itself from the infected cell rather than in getting into the uninfected cell (Schulman, Khakpour & Kilbourne, 1968). Many other viruses, such as the rhinoviruses or measles virus, infect the respiratory tract without great difficulty but so far specific receptors have not been described. A virus whose surface binds to receptors on the host cell can be regarded as having become adapted for this purpose. Without such viral surface properties infection would be difficult or impossible.

Viruses have long been known to differ in their transmissibility, and Schulman (1970) has carefully studied the transmissibility of influenza virus between mice. Transmissibility behaves independently of virulence (disease or lesion production), but may be related to the neuraminidase activity of virus which affects the extent of virus shedding from infected animals. It may also involve marked differences in the ability of viruses to pass host barriers at the body surface and initiate infection. In societies where viral contamination of air, food, fingers and water is becoming less profuse, there would be particularly strong selection favouring strains of virus which established infection with greater efficiency, that is, with fewer virus particles. The important problem of transmissibility requires further study.

The intestinal tract also, with its acid, bile salts, mucus etc. is regularly infected. In most cases nothing is known about mechanisms, about the relative importance of receptors (Holland, 1961) and passive uptake of virus by cells. Certainly the enteroviruses, which enter the body and are shed into the alimentary canal, tend to be resistant to inactivation by acids and bile salts. Even when virus receptors are known to be necessary it is possible that the absorptive functions of the intestinal tract might further facilitate infection. Electron microscopic evidence shows that polystyrene latex particles enter intestinal epithelial cells of adult rats by phagocytosis (Sanders & Ashworth, 1961). There is also evidence that bacteriophage particles pass through the normal intestine of mice and enter lymphatics (Mims, 1964). Further discussion is not warranted because of uncertainties about the way in which virus particles enter susceptible cells *in vivo*. Phagocytosis is important for some viruses (perhaps poxviruses) and for some cells (macrophages), viral cores entering the cell cytoplasm through the wall of a phagocytic vacuole (Dales, 1965). In other instances there is good electron microscopic evidence suggesting fusion of viral coat with the outer cell membrane, as a result of which the viral core enters the cell (Morgan & Rose, 1969; Zee & Talens, 1971). Both mechanisms, and indeed any mechanism, requires that the virus genome at least, passes across the cell membrane, but the phagocytic entry would also expose the particle to lysosomal enzymes.

Production of toxins and similar substances

Viruses, in contrast to bacteria, do not appear to produce toxins. All toxic effects so far described apply to highly artificial situations in experimental animals; there is no evidence that viral toxins operate in natural infections, although of course toxic non-viral materials can be formed in infected tissues. Certain phenomena, such as pyrexia, malaise, or post-infectious depression could conceivably be due to viral toxins, but this seems unlikely. The pyrexia seen in rabbits when large doses of influenza virus are injected intravenously is probably caused by liberation of endogenous toxin from polymorphs and macrophages after phagocytosis of virus.

Bacterial toxins often have a significance in enabling bacteria to spread in tissues, preventing phagocytosis, or damaging host cells. From this last point of view viral toxins would seem less advantageous because damage to host cells would merely reduce the number of available growth sites. If however, there were a selective action on host defence cells (lymphocytes or macrophages) a toxin could be more useful. Some viruses are phagocytosed poorly or not at all by macrophages, (Mims, 1964) but nothing is known about mechanisms. Viruses which are not phagocytosed by macrophages, such as the Armstrong strain of

LCM virus (Tosolini & Mims, 1971), may thereby be prevented from establishing infection in target organs. On the other hand, it would be an advantage to avoid phagocytosis if this merely led to intracellular digestion. Also, defective phagocytosis ('clearing') of virus from the blood would make a viraemia easier to maintain, and thereby improve conditions for invasion of target organs.

One useful strategy by which viruses could counteract host defences would be by producing interferon antagonists, or in some other way inhibiting the action of interferon. Although there have been reports of stimulons, enhancers and anti-interferons recovered from infected tissue cultures, (Chany & Brailovsky, 1967; Truden, Sigel & Dietrich, 1967; Kato, Okada & Ota, 1965; Ghendon, 1965), similar substances have also been found in uninfected tissues (Kato & Eggers, 1969; Fournier, Rousset & Chany, 1968). Further work on these possibly important factors is needed before their significance can be assessed. A few viruses, such as parainfluenza 3, have been shown to block the action of interferon *in vitro*, as evidenced by decreased action of added interferon and increased growth of a second infecting virus (Hermodsson, 1963). The blocking action appears to be associated with infectious virus particles. It is not clear whether cells are made more susceptible to the interferon-blocking virus itself, and the relevance for the infected host cannot at present be assessed.

Growth in macrophages or lymphocytes

Macrophages are strategically placed near the sites where initial infection takes place and also in a position to monitor the major bodily compartments and circulating fluids. Often, as described above, an infection goes further and is more virulent because virus can replicate in these cells rather than suffer intracellular digestion. We do not know what it is that makes a virus able to replicate in macrophages rather than be digested, but even when viral replication does occur, yields are very low compared with yields from other cell types. When mouse macrophages are infected with herpes simplex (Stevens & Cook, 1971), Sendai, ectromelia or lymphocytic choriomeningitis virus for instance, large amounts of virus antigens are produced as detected by fluorescent antibody staining, but yields of infectious virus are low (Stevens & Cook, 1971; C. A. Mims, unpublished). If macrophages are infected and these cells then move through blood, lymph, or tissues, virus spread to target tissues may be accelerated (Mims, 1964) while virus is protected from the action of antibody.

So far we have been considering ways in which a virus by *directly confronting* host defence mechanisms, manages to overcome them. Before going on to the subtle, more devious ways in which defence mechanisms can be overcome, I shall give one striking though artificial example of the overwhelming of the host defences by direct confrontation. When mice are injected intravenously with $10^{9 \cdot 0}$ LD_{50} of Rift Valley Fever virus, the injected virus passes straight through Kupffer cells in the liver to infect nearly all hepatic cells. Hepatic cells show nuclear changes within an hour, necrosis by four hours, and as the

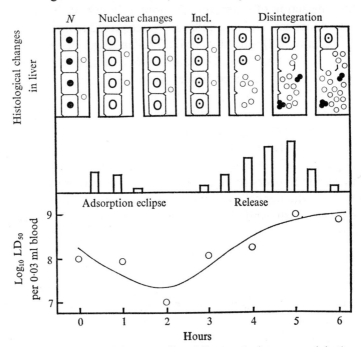

Fig. 1. Single cycle of growth in mouse liver following the intravenous injection of $10^{9 \cdot 0}$ LD_{50} Rift Valley Fever virus. Massive liver necrosis and death occurs by 6 hours. Histological changes in the liver represented diagrammatically. Virus adsorption and liberation calculated from the blood growth curve. N = normal appearances; Incl. = nuclear inclusions; small open circles are erythrocytes (adapted from Mims, 1957).

progeny of the single cycle of growth in hepatic cells is released, mice die with severe liver necrosis, only six hours after initial infection (Fig. 1 adapted from Mims, 1957). Host defences are completely overcome by the intravenous route of injection, by the inability of Kupffer cells to prevent infection of hepatic cells, and by the almost universal infection and destruction of hepatic cells which killed the host long before there was time for an immune response. Even interferon, if produced in infected cells, would have been ineffective because only a single cycle of growth took place.

Viruses can overcome the defensive role of lymphocytes by growing in them and either destroying them or otherwise depressing the subsequent immune response. Virulent strains of ectromelia virus, for instance, grow in lymphocytes and the immune response occurs against the background of lymphoid tissue necrosis (Mims, 1964). Mice which are destined to die show an ineffectual immune response as indicated by the absence of primary footpad swelling (a reflection of humoral and cellular immunity) after infection into the footpad. In the spleens of those mice that survive, reactive follicles and immunoblasts appear in the midst of necrotic follicles. Growth of ectromelia virus in lymphoid tissue also enables virus to spread from the site of infection more readily by 'growing through' the local lymph node defences. Instead of entering lymph nodes to be taken up by macrophages and processed as non-replicating antigen, as occurs with avirulent poxviruses, ectromelia virus grows in the node, appears in efferent lymph and eventually enters the blood. Viruses which grow in lymphoid tissue and thus spread via lymphatic pathways include distemper (Liu & Coffin, 1957) tick-borne encephalitis virus (Malkova, 1968), and LCM virus (Mims & Tosolini, 1969). Instead of acting as a filter and disposal mechanism for viruses drained from tissues, the lymphatic system in such instances spreads the infection. When large enough quantities of virus arrive at a lymph node draining infected tissues, uptake by macrophages is incomplete. Virus appears in efferent lymph even when there is no replication in the node, so that the node has failed in its function as a filter (Wallnerova & Mims, 1970).

There are a number of viruses which infect lymphoid tissues with little or no cell damage and cause immunosuppression in the infected host (Notkins, 1970). Immune responses to unrelated antigens are suppressed, sometimes cell-mediated immunity as well as antibody production. Mechanisms are not understood, and the same cell which is non-cytolytically infected with murine leukaemia virus, for instance, can be involved in an immune response to sheep red blood cells and actually produce antibody (Celada, Asjo & Klein, 1970). Immune responses are generally reduced rather than absent, and antigenic competition is probably important. In spite of its interest, immunosuppression cannot be regarded as a means by which viruses counteract host defences unless the response to the infecting virus itself is also suppressed, as in tuberculoid leprosy in man (Bullock, 1968). Such 'autoimmunosuppression', distinct from tolerance (see below) would by its nature be difficult to demonstrate, and is not known to occur.

Induction of tolerance in the host

Tolerance means an immunologically specific reduction in host response to a given antigen, and is thus distinct from immunosuppression. From a virus' point of view there are two ways in which tolerance can be induced. First, a considerable degree of tolerance can exist in association with vertical transmission from the mother through the egg or early embryo. This is generally explained as resulting from the presence of viral antigens during the ontogenesis of the immune system, so that immune cells reacting with viral antigens are eliminated, as are those reacting with host antigens. Examples include LCM and leukaemia viruses in mice, where there is infection of the egg and foetus, absence of significant immune response, with persistent widespread infection during the entire life of the host (Mims, 1968). The immune tolerance in these two examples is not complete, because antibodies are in fact produced in small amounts, and the complexes formed with circulating viral antigen are deposited in the glomeruli (Oldstone & Dixon, 1969; Hirsch, Allison & Harvey, 1969). The degree of tolerance, however is sufficient to enable virus to persist in tissues, including ovaries, and thus infect offspring. There is also an accompanying failure to produce interferon in response to the infecting virus, and this raises an interesting question about the recognition mechanism in cells for the foreign double-stranded RNA molecules which induce interferon. Many cell types other than lymphocytes produce interferon, but there have been almost no studies of recognition mechanisms, nor of interferon type tolerance, where there is a specific reduction in host interferon response to a given virus.

The second way in which viruses can induce tolerance, is by being poor antigens. There are genetically determined differences in the ability of a given animal to respond to different antigens and an antigenic mutant virus which induced a poor immune response in its host would experience strong selection pressure in its favour. Alternatively, a virus could induce some degree of tolerance by possessing antigens which cross-reacted with those of the host. The mimicking of host antigens by infectious agents has been discussed (Jenkin, 1963) and there are examples especially in the case of bacteria, but it is not known whether this also occurs with viruses. Techniques are available to test for this interesting possibility. Viral proteins can be separated in pure form and there are sophisticated immunological methods for detecting antigenic similarities between virus and host.

At first sight there would appear to be no better way of viruses acquiring host antigens than to mature by budding from the cell membrane, thus acquiring a complete covering of host antigens. Many viruses mature in this fashion, including myxoviruses, paramyxoviruses, leukaemia viruses, rhabdoviruses, arenaviruses and many arboviruses. Although virus antigens are present on the cell membrane, sometimes localised to those particular areas where budding takes place (Morgan *et al.* 1961), the modified membrane may still contain host cell components. Thus, purified Newcastle disease virus propagated in eggs contains Forssman and blood group A antigens derived from the host (Rott, Drzenick, Saber & Reichert, 1966). However, careful studies with SV 5 (a budding paramyxovirus) have shown that all the proteins of the SV 5 envelope appear to be virus-specific, although the glycosphingolipids and other lipids of the host cell membrane are incorporated quantitatively into the envelope (Klenk & Choppin, 1970). The carbohydrate moieties of the glycosphingolipids could serve as antigenic determinants, and indeed such compounds are associated with Forssman and blood group antigenicity. At present there is no evidence that host antigens present in the envelope of budding viruses make them more tolerogenic and less immunogenic. A more important immune consequence of the budding phenomenon is that when infection is not acutely cytocidal there is an immune response to the foreign antigens on host cells, leading to the destruction of infected but healthy cells, as the case of LCM virus in mice. Viruses whose antigens were poorly immunogenic would be at an advantage in such circumstances.

It is relevant at this stage to refer to the so-called 'slow viruses', because they replicate in the host, and apparently avoid inducing any immune response whatsoever even in those infected as adults. These agents include scrapie, kuru, and transmissible mink encephalopathy. Careful tests for immune responses to the scrapie agent in infected mice have so far given completely negative results. If these agents have indeed totally avoided inducing an immune response, perhaps it is by becoming basically different from ordinary infectious agents. They can perhaps be regarded as non-antigenic substances associated with host cell membranes, which somehow induce the cell to replicate them.

Antigenic variation

Animals recovering from virus infections typically show specific immunity to re-infection with the same virus. In the case of infections limited to mucosal surfaces this immunity is largely mediated by secretory Ig A antibodies, but in systemic infections circulating Ig G antibodies are more important. With infections limited to mucosal surfaces resistance to re-infection is sometimes of limited duration, and the same strain of virus may reinfect later in life (Chanock, 1970). Perhaps this is indicative of the limited memory of the Ig A system. As immunological memory fades, there would be strong selective advantage for virus strains with altered antigenic characters. The time between initial infection and virus shedding is short for infections limited to epithelial surfaces, and an antigenically altered mutant could infect, replicate, and be shed from the body before a significant local secondary response was generated, especially if it were a weak one. In contrast to this, an antigenic variant of a virus producing a systemic infection would experience greater difficulty. Pre-existing circulating antibodies are likely to be present, and these, even though they neutralised the variant virus less effectively, are strategically placed to interfere with the spread of virus through the body. Also the secondary immune response, even if it did not hinder the initial establishment of infection on the epithelial surface, would have time to come into action and prevent the spread of infection through the body. Incubation periods for systemic infections such as measles, rubella or smallpox are 2–3 times as long as in the case of respiratory virus diseases. Thus, people immune to measles may show antibody boosts following re-exposure to infection (Stokes, Reilly, Buynak & Hilleman, 1961), but rarely shed virus.

For reasons such as these, antigenic variation is a feature of virus infections restricted to the respiratory and perhaps the alimentary tract. It is a mechanism by which viruses overcome host immune defences. Influenza B virus, for instance, shows constant small changes in its haemagglutinin and neuraminidase over the years, and this has been called immunological drift. Established strains of influenza A show similar drift, but at intervals new pandemic strains of influenza A virus emerge to infect man, probably after the production of more drastic antigenic changes in a bird reservoir. Human enteroviruses perhaps show immunological drift, but evidence is negligible compared with influenza virus. Human rhinoviruses are almost certainly evolving rapidly in this way, and foot-and-mouth disease virus, another rhinovirus, certainly shows immunological drift. Hyslop (1965) showed that antigenic variants of foot-and-mouth disease virus could be produced experimentally by serial passage of virus in cell cultures in the presence of antibody. Foot-and-mouth disease, it should be noted, is a systemic infection, so that systemic infections do not inevitably exclude immunological drift.

As a viral adaptation for the overcoming of host immunity, antigenic variation is more likely to be important in longer-lived species such as the horse or man, when there is a need for multiple re-infections during an individual's life time if the virus does not have the ability to become latent (see below) and is to remain in circulation. In the case of mice, on the other hand, populations renew themselves rapidly and uninfected susceptibles could appear fast enough to maintain a virus cycle.

Human respiratory viruses are among the most successful animal viruses in the world. Many show regular antigenic variation, and because of assured increases in human numbers and density, and the unlikelihood in most cases of successful chemoprophylaxis, this group of viruses are perhaps entering their golden age, with an almost unlimited supply of susceptible hosts in the foreseeable future.

Persistent infection and latency

As discussed above, viruses can overcome host resistance mechanisms by antigenic variation, but they only have the opportunity to do this when there are enough susceptible hosts for infection to be continuously transmitted by contagion. It is only recently, in the evolutionary sense, that human populations have become large enough to support continuously transmitted infections. Measles, for instance, needs a minimum population of 500000 if it is to maintain itself by successive infection of susceptible individuals without re-introduction of the virus from outside (Black, 1966). In paleolithic times, when men lived in small isolated groups of 30–50, such infections could not have existed. Under such circumstances viruses could be maintained in a community either by having alternative hosts, perhaps with vectors, or by persistent infection. In recent years evidence has accumulated to show that sterile immunity is *not* the rule after primary virus infection. Many viruses persist in the body in spite of an immune response and the continued production of antibody.

Latency can be defined as a type of persistent infection. After recovery from initial infection virus persists in the body, and later in life is once again shed, often in association with further clinical and pathological signs. Latency reflects a striking ability of viruses to overcome host defences. Chicken pox, for instance, can maintain itself indefinitely in a population of 1000 or less (Black, 1966) because after infecting all susceptibles it does not disappear until re-introduced like measles or smallpox, but remains latent in the body. Many years later, when a fresh group of young susceptibles have appeared, an older person develops zoster, and the virus shed from these vesicles can once again produce chicken pox. It is not known whether virus persists in the body in infectious form, perhaps replicating in a limited local fashion in the presence of antibody and interferon, or whether a non-infectious viral component persists in cells later to be activated, with virus release. There are several tissue culture models for viral persistence, but their relevance for the infected host is not clear. The two common latent virus infections of man, herpes simplex and varicella, are probably maintained in dorsal root ganglion cells after primary infection, and subsequently activated and liberated from these sites to produce cold sores or zoster (Paine, 1964; Hope-Simpson, 1965).

Persistent viruses may be shed in urine or saliva and others are recoverable from apparently normal tissues. Often virus is not isolated by mere addition of tissue suspensions to susceptible cells. There is usually a need to cultivate cells *in vitro*, giving time for the disappearance of antibody and perhaps an alteration in virus–cell balance before infectious virus is released. Since the introduction of such methods a variety of viruses have been isolated from tissues of healthy hosts. Adenoviruses were first isolated from the adenoids of healthy children after growth of explanted tissues *in vitro*, and Israel (1962) recovered adenoviruses from 26 % of the tonsils and 58 % of the adenoids of 390 children. From normal human kidneys cytomegaloviruses, reoviruses, herpes simplex, varicella etc. have been isolated (Hsiung, 1968). Persistent viruses like cytomegalovirus and adenoviruses would no doubt maintain themselves in small isolated communities as does varicella (see above). In attempts to isolate the agent of kuru from explanted tissues of affected chimpanzees 47 different virus strains, including adenoviruses, reoviruses and foamy viruses, were recovered from brain, lymph nodes, thymus, kidney, etc. of 9 chimpanzees (Rogers, Basnight, Gibbs & Gajdusek, 1967). Mice have been studied as thoroughly as men, and persistent murine viruses include pneumonia virus of mice, lactic dehydrogenase virus, cytomegalovirus, Theilers virus, lymphocytic choriomeningitis virus, reovirus etc.

From the virus' point of view, it is no use becoming persistent unless virus can later infect new hosts. This is accomplished by shedding into urine (polyoma virus in mice, Rowe, 1961), saliva (herpes simplex in man), intestine, respiratory tract (psittacosis) or blood (equine infectious anaemia in horses, serum hepatitis in man). Patients infected with serum hepatitis have the infectious agent associated with Australian antigen in their plasma, where it may persist for years. Transmission by contaminated needles is common in developed countries, but the natural history of the infection in undeveloped countries is less clear. It would seem inevitable that biting arthropods, acting as 'flying pins' would play some role in the spread of infection, but there have been no experiments to test for this possibility.

We still have no understanding of the mechanism by which a virus like cytomegalovirus persists in the salivary gland, or polyoma virus in the mouse kidney, in spite of circulating neutralising antibody. One interesting possibility is that since one surface of the infected cell faces the external world (salivary gland lumen or kidney tubule) a non-cytopathic liberation of virus confined to this surface could persist, because even if this surface of the cell bears viral antigens, it is not exposed to circulating antibodies or sensitised lymphocytes.

Rubella virus persists in congenitally infected infants in the presence of circulating antibodies, and is shed in the urine to infect susceptible nurses in hospitals. Persistence is not really understood. In the kidney it could be by the mechanism suggested above, but defective cell-mediated immunity may also be involved.

Integration of viral genome into host cells

Viruses can overcome immune responses and the interferon response by staying in cells without producing infectious virus, or even viral antigens. Viral nucleic acids must be replicated as cells divide, and the most effective way of doing this is by integration of viral nucleic acid into the host cell genome. Such viruses (eminently plausible, but at present hypothetical) have given up regular mechanisms of shedding and transmission to fresh hosts and depend on uniform infection of a cell clone. In practice, this means vertical transmission in the host species, with infection of the ovum and all derived cells. There are still opportunities for spread to uninfected lineages because of sexual reproduction in the host species.

LCM virus and the leukaemia viruses of mice might be said to have gone some way towards achieving this end. Infectious virus and viral antigen, however, are still produced in infected cells, and this in fact is why such infections have been identified. LCM virus antigen can be detected in the cells of congenitally infected mice by the fluorescent antibody method, even in ova in the ovary (Mims, 1966). Ovum transfer experiments indicate that similar infection occurs with murine leukaemia virus. Both infections are basically non-cytopathic in mouse cells, and most cells in the body are infected throughout life. Leukaemia (leukaemia viruses) or kidney lesions (LCM virus) occur as a late, and from the virus point of view, irrelevant consequence. Infectious virus is still produced in mice congenitally infected with LCM virus, although significantly less per cell than in mice infected for the first time as adults (Mims, 1970), and there are large amounts of antigen. Transmission is vertical in infected colonies, but shedding of infectious virus in urine, faeces and respiratory secretions makes horizontal transmission a possibility. In the case of murine leukaemia virus, which appears to be present in all mouse colonies, very little infectious virus is produced in some strains of mice, although viral antigens may be present in cells. Accordingly there is no horizontal transmission, vertical transmission being adequate to maintain uniform infection of the species. Final integration of such a virus into the host would involve stable incorporation of virus-induced DNA into the host genome, with no infectious virus and probably no viral antigens.

DNA viruses are even more obvious candidates for integration into the host cell genome with consequent vertical transmission via germ cells. Integration of viral DNA into the host cell genome *in vitro* has been shown to occur, for instance, with SV_{40} and polyoma viruses, but there is no evidence about vertical transmission. Work has centred on the tumour-producing capacities of cells transformed by these viruses, transformation involving unrestricted cell division in association with the appearance of tumour-specific transplantation antigens on the cell surface. There seems no reason why viral integration should necessarily involve transformation of this sort with changes in surface antigens and mitotic activity. Diphtheria bacilli infected by temperate bacteriophage β give no signs of infection except the production under certain circumstances of the diphtheria toxin. Also, from the virus' point of view there would seem to be little significance or advantage in being able to infect cells, become integrated into the cell genome, and produce tumours. It is not inconceivable that this ability reflects the ability of similar, as yet unidentified viruses, to become integrated into cells without such gross transformational effects. If this involved germ line cells, vertical transmission would be achieved. Only a small portion of the viral genome need be integrated into host cells. From the evolutionary point of view, survival is what is successful, and this certainly applies to a restricted portion of a viral genome associated with and replicating with the host genome.

If these at present hypothetical events occur, it would be difficult to distinguish the virus or virus fragment from the genetic constitution of the host. To become, in effect, part of the host and the same as the host, is the final device used by viruses to counteract host defence mechanisms. It is the logical, the ultimate, viral subterfuge, and is at the same time a total evolutionary commitment to a given host species. Viruses are unique among micro-organisms in their ability to insinuate themselves into the host and become this type of parasite. It calls to mind the reverse process, suggested for the origin of viruses, where host cell genome fragments become, in the end, independent infectious viruses. Even the word parasite is inappropriate if there are no disadvantages to the host species; indeed, careful studies might reveal advantages to the host, so that the word symbiosis became a better one.

ADDENDUM

It is a general feature of long-established virus–host associations that there is minimal damage to the host. Host defences need to be overcome only enough to enable virus to infect, replicate, and be shed to infect fresh hosts. Serious damage to the host before the end of reproductive life would mean fewer descendents and an eventual shortage of hosts. Chicken pox can be taken as an example of a long-established infection in man. There is a very mild disease in young people (chicken means child, from Middle English) followed by latency, and then zoster, more severe but a fresh source of virus, in older people. Rhinoviruses in man represent a comparatively recent but ecologically sound type of virus–host association. There is a rapid infection of the upper respiratory tract and efficient virus shedding by the induction of a profuse watery secretion, with minimal host damage.

In the above discussion prominence has been given to viruses which short circuit host defence mechanisms by becoming latent, persistent, or integrated into host cells in a defective form. These are usually successful adaptations from the virus' point of view, with a balance established between host and virus. However, even if a virus–host association is balanced for the population as a whole, at the individual level an infection may still lead to serious illness. Medicine today deals more and more with the unusual host response to common virus infections. Moreover, as human society evolves, there is an even more rapid evolution of viruses in search of hosts. Consequently there are always many unbalanced virus–host associations, where infection often leads to significant illness. Such infections are the concern of modern medicine, and although viruses causing significant illness are now being vaccinated out of existence, new viruses may still emerge. Indeed, they are more likely to emerge in a world where ecological patterns are changing so rapidly with one species of animal over-running the earth. A new virus–host association would be represented by an influenza virus pandemic caused by a strain of virus which invaded cardiac muscle to produce acute myocarditis and death in three-quarters of those infected. The human population would be decimated before a more stable type of virus–host balance evolved, analogous to that seen with myxoma virus in the Australian rabbit (Fenner & Ratcliffe, 1965). The mechanism by which this virus overcame host defences would become of great import-ance, however limited the eventual success of the virus strain. The mechanism would presumably be by direct confrontation and over-whelming of regular host defence measures as discussed earlier. Although a less interesting viral strategy than that involving latency, persistence, or integration, this is the type of infection which may one day plague the earth.

REFERENCES

ALLISON, A. C. (1967). Cell-mediated immune responses to virus infections and virus-induced tumours. *British Medical Bulletin*, **23**, 60.

ATKINS, E. (1960). Pathogenesis of fever. *Physiological Reviews*, **40**, 580.

BANG, F. B. & WARWICK, A. (1960). Mouse macrophages as host cells for the mouse hepatitis virus and the genetic basis of their susceptibility. *Proceedings of the National Academy of Sciences of the U.S.A.*, **46**, 1065.

BARON, S. (1970). The biological significance of the interferon system. *Archives of Internal Medicine*, **126**, 84.

BENNETT, I. L. & NICASTRI, A. (1960). Fever as a mechanism of resistance. *Bacteriological Reviews*, **24**, 16.

BLACK, F. L. (1966). Measles endemicity in insular populations: Critical community size and its evolutionary implications. *Journal of Theoretical Biology*, **11**, 207.

BLANDEN, R. V. (1971). Mechanisms of recovery from a generalised viral infection: Mousepox. II. Passive transfer of recovery mechanisms with immune lymphoid cells. *The Journal of Experimental Medicine*, **133**, 1074.

BRUNNER, H., RAZIN, S., KALICA, A. R. & CHANOCK, R. M. (1971). Lysis and death of *Mycoplasma pneumoniae* by antibody and complement. *The Journal of Immunology*, **106**, 907.

BULLOCK, W. E. (1968). Studies of immune mechanisms in leprosy. I. Depression of delayed allergic response to skin test antigens. *The New England Journal of Medicine*, **278**, 298.

BURNET, F. M. (1970). The concept of immunological surveillance. *Progress in Experimental Tumor Research*, **13**, 1.

CELADA, F., ASJO, B. & KLEIN, G. (1970). The presence of Moloney virus induced antigen on antibody-producing cells. *Clinical and Experimental Immunology*, **6**, 317.

CHANOCK, R. M. (1970). Control of acute Mycoplasmal and viral respiratory disease. *Science*, **169**, 248.

CHANY, C. & BRAILOVSKY, C. (1967). Stimulating interactions between viruses (stimulons). *Proceedings of the National Academy of Sciences of the U.S.A.*, **57**, 87.

CRAIGHEAD, J. E. (1969). Immunological response to cytomegalovirus infection in renal allograft recipients. *American Journal of Epidemiology*, **90**, 506.

DALES, S. (1965). Penetration of animal viruses into cells. *Progress in Medical Virology*, **7**, 1.

DOURMASHKIN, R. R. & TYRRELL, D. A. J. (1970). Attachment of two myxoviruses to ciliated epithelial cells. *The Journal of General Virology*, **9**, 77.

FENNER, F. & RATCLIFFE, F. N. (1965). *Myxomatosis*. London and New York: Cambridge University Press.

FOURNIER, F., ROUSSET, S. & CHANY, C. (1968). Un facteur tissulaire antagoniste de l'action de l'interferon. *Compte Rendus Académie de Science*, **266**, 2306.

GHENDON, Y. Z. (1965). On the ability of certain viruses to block the effect of interferon. *Acta Virologica*, **9**, 186.

GREEN, J. A., COOPERGAND, S. R. & KIBRICK, S. (1969). Immune specific induction of interferon production in cultures of human blood leucocytes. *Science*, **164**, 1415.

HALL, J. G., SMITH, M. E., EDWARDS, P. A. & SHOOTER, K. V. (1969). The low concentration of macroglobulin antibodies in peripheral lymph. *Immunology*, **16**, 773.

HANSON, R. J., KEMPF, J. E. & BOAND, A. V. (1957). Phagocytosis of influenza virus. II. Its occurrence in normal and immune mice. *The Journal of Immunology*, **79**, 422.

HERMODSSON, S. (1963). Inhibition of interferon by an infection with parainfluenza virus type 3(PIV-3). *Virology*, **20**, 333.

HIRSCH, M. S. & MURPHY, F. A. (1968). Effects of antilymphoid sera on viral infections. *The Lancet*, **ii**, 37.

HIRSCH, M. S., ALLISON, A. C. & HARVEY, J. J. (1969). Immune complexes in mice infected neonatally with Moloney leukaemogenic and murine sarcoma viruses. *Nature, London*, **223**, 739.

HIRSCH, M. S., ZISMAN, B. & ALLISON, A. C. (1970). Macrophages and age-dependent resistance to herpes simplex virus in mice. *The Journal of Immunology*, **104**, 1160.

HOLLAND, J. J. (1961). Receptor affinities as major determinants of enterovirus tissue tropisms in humans. *Virology*, **15**, 312.

HOPE-SIMPSON, R. E. (1965). The nature of herpes zoster: a long-term study and a new hypothesis. *Proceedings of the Royal Society of Medicine*, **58**, 9.

HSIUNG, G. D. (1968). Latent virus infections in primate tissues with special reference to simian viruses. *Bacteriological Reviews*, **32**, 185.

HYSLOP, N. ST G. (1965). Isolation of variant strains of foot-and-mouth disease virus propagated in cell cultures containing antiviral sera. *The Journal of General Microbiology*, **41**, 135.

ISRAEL, M. S. (1962). The viral flora of enlarged tonsils and adenoids. *The Journal of Pathology*, **84**, 169.

JENKIN, C. R. (1963). Heterophile antigens and their significance in the host–parasite relationship. *Advances in Immunology*, **3**, 351.

JOHNSON, R. T. (1964). The pathogenesis of herpes virus encephalitis. II. A cellular basis for the development of resistance with age. *The Journal of Experimental Medicine*, **120**, 359.

KATO, N., OKADA, A. & OTA, F. (1965). A factor capable of enhancing virus replication appearing in parainfluenza virus type I (HVJ) infected allantoic fluid. *Virology*, **26**, 630.

356   C. A. MIMS

KATO, N. & EGGERS, H. J. (1969). A factor capable of enhancing interferon synthesis. *Virology*, 37, 545.

KLEIN, G. (1966). Tumor antigens. *Annual Review of Microbiology*, 20, 223.

KLENK, H. D. & CHOPPIN, P. W. (1970). Glycosphingolipids of plasma membranes of cultured cells and an enveloped virus (SV₅) grown in these cells. *Proceedings of the National Academy of Sciences of the United States of America*, 66, 57.

LIU, C. & COFFIN, D. L. (1957). Studies on canine distemper infection by means of fluorescein labelled antibody. I. The pathogenesis, pathology and diagnosis of the disease in experimentally infected ferrets. *Virology*, 3, 115.

LWOFF, A. (1959). Factors influencing the evolution of viral disease at the cellular level and in the organism. *Bacteriological Reviews*, 23, 109.

MALKOVA, D. (1968). The significance of the skin and the regional lymph nodes in the penetration and multiplication of tick-borne encephalitis virus after subcutaneous infection of mice. *Acta Virologica*, 12, 222.

MIMS, C. A. (1957). Rift Valley Fever Virus in mice. VI. Histological changes in the liver in relation to virus multiplication. *The Australian Journal of Experimental Biology and Medical Science*, 35, 595.

MIMS, C. A. (1964). Aspects of the pathogenesis of virus diseases. *Bacteriological Reviews*, 28, 30.

MIMS, C. A. (1966). Immunofluorescent study of the carrier state and mechanism of vertical transmission in lymphocytic choriomeningitis virus infection in mice. *The Journal of Pathology and Bacteriology*, 91, 395.

MIMS, C. A. (1968). The pathogenesis of viral infections of the fetus. *Progress in Medical Virology*, 10, 194.

MIMS, C. A. (1970). Observations on mice infected congenitally or neonatally with lymphocytic choriomeningitis (LCM) virus. *Archiv für die gesamte Virusforschung*, 30, 67.

MIMS, C. A. & TOSOLINI, F. A. (1969). Pathogenesis of lesions in lymphoid tissue of mice infected with lymphocytic choriomeningitis (LCM) virus. *The British Journal of Experimental Pathology*, 50, 584.

MONTGOMERIE, J. Z., BECROFT, D. M. O., CROXSON, M. C., DOAK, P. B. & NORTH, J. D. K. (1969). Herpes simplex virus infection after renal transplantation. *The Lancet*, ii, 867.

MORGAN, C. & ROSE, H. M. (1969). Structure and development of viruses as observed in the electron microscope. IX. Entry of parainfluenza I (Sendai) virus. *The Journal of Virology*, 2, 1122.

MORGAN, C., HSU, K. S., RIFKIND, R. A., KNOX, A. W. & ROSE, H. M. (1961). The application of ferritin-conjugated antibody to electron microscope studies of influenza virus in infected cells. I. The cellular surface. *The Journal of Experimental Medicine*, 114, 825.

MOTT, P. M. & WOLFF, S. M. (1966). The association of fever and antibody response in rabbits immunized with human serum albumin. *The Journal of Clinical Investigation*, 45, 372.

NOTKINS, A. L. (1970). Effect of virus infections on the function of the immune system. *Annual Review of Microbiology*, 24, 525.

OGRA, P. L. & KARZON, D. T. (1969). Distribution of poliovirus antibody in serum, nasopharynx and alimentary tract following sequential immunization of lower alimentary tract with poliovaccine. *The Journal of Immunology*, 102, 1423.

OLDSTONE, M. B. A. & DIXON, F. J. (1969). Pathogenesis of chronic disease associated with persistent lymphocytic choriomeningitis viral infection. I. Relationship of antibody production to disease in neonatally infected mice. *The Journal of Experimental Medicine*, 129, 483

OLDSTONE, M. B. A. & DIXON, F. J. (1970). Persistent lymphocytic choriomeningitis viral infection. III. Virus–antiviral–antibody complexes and associated chronic disease following transplacental infection. *The Journal of Immunology*, **105**, 829.

PAINE, T. F. (1964). Latent herpes simplex infection in man. *Bacteriological Reviews*, **28**, 472.

ROBERTS, J. A. (1962). Histopathogenesis of mousepox. I. Respiratory infection. *The British Journal of Experimental Pathology*, **43**, 451.

ROBERTS, J. A. (1963a). Histopathogenesis of mousepox. III. Ectromelia virulence. *The British Journal of Experimental Pathology*, **44**, 465.

ROBERTS, J. A. (1963b). Growth of virulent and attenuated ectromelia virus in cultured macrophages from normal and ectromelia-immune mice. *The Journal of Immunology*, **92**, 837.

ROGERS, N. G., BASNIGHT, M., GIBBS, C. J. & GAJDUSEK, D. C. (1967). Latent viruses in chimpanzees with experimental Kuru. *Nature, London*, **216**, 446.

ROSEN, F. S. (1968). The lymphocyte and the thymus gland. Congenital abnormalities. *The New England Journal of Medicine*, **279**, 643.

ROTT, R., DRZENICK, R., SABER, M. S. & REICHERT, E. (1966). Blood group substances, Forssman and mononucleosis antigens in lipid-containing RNA viruses. *Archiv für die gesamte Virusforschung*, **19**, 273.

ROWE, W. P. (1961). The epidemiology of mouse polyoma virus infection. *Bacteriological Reviews*, **25**, 18.

SANDERS, E. & ASHWORTH, C. T. (1961). A study of particulate intestinal absorption and hepatocellular uptake. *Experimental Cell Research*, **22**, 137.

SAWYER, W. A. (1931). Persistence of yellow fever immunity. *The Journal of Preventative Medicine*, **5**, 413.

SCHELL, K. (1960). Studies on the innate resistance of mice to infection with mousepox. II. Route of inoculation and resistance, and some observations on the inheritance of resistance. *The Australian Journal of Experimental Biology and Medical Science*, **38**, 289.

SCHULMAN, J. L. (1970). Effects of immunity on transmission of influenza: Experimental studies. *Progress in Medical Virology*, **12**, 128.

SCHULMAN, J. L., KHAKPOUR, M. & KILBOURNE, E. D. (1968). Protective effects of specific immunity to viral neuraminidase on influenza virus infection of mice. *The Journal of Virology*, **2**, 778.

SOKAL, J. E. & FIRAT, D. (1965). Varicella-zoster infection in Hodgkin's disease. *The American Journal of Medicine*, **39**, 452.

STEVENS, J. G. & COOK, M. L. (1971). Restriction of herpes simplex virus by macrophages. *The Journal of Experimental Medicine*, **133**, 19.

STOKES, J., REILLY, C. M., BUYNAK, E. B. & HILLEMAN, M. R. (1961). Immunologic studies of measles. *The American Journal of Hygiene*, **74**, 293.

TOSOLINI, F. A. & MIMS, C. A. (1971). Effect of murine strain and viral strain on the pathogenesis of lymphocytic choriomeningitis infection, and a study of footpad responses. *The Journal of Infectious Diseases*, **123**, 134.

TRUDEN, J. L., SIGEL, M. M. & DIETRICH, L. S. (1967). An interferon antagonist: Its effect on interferon action in mengo-infected Ehrlich ascites tumor cells. *Virology*, **33**, 95.

WALLNEROVA, Z. & MIMS, C. A. (1970). Thoracic duct cannulation and haemal node formation in mice infected with cowpox virus. *The British Journal of Experimental Pathology*, **51**, 118.

WESTWOOD, J. C. N., ZWARTOUW, H. T., APPLEYARD, G. & TITMUSS, D. H. J. (1965). Comparison of the soluble antigens and virus particle antigens of vaccinia virus. *The Journal of General Microbiology*, **38**, 47.

YOSHINO, K. & TANIGUCHI, S. (1964). The appearance of complement-requiring neutralizing antibodies by immunization and infection with herpes simplex virus. *Virology*, **22**, 193.

ZEE, Y. C. & TALENS, L. (1971). Entry of infectious Bovine Rhinotracheitis virus into cells. *The Journal of General Virology*, **11**, 59.

MECHANISMS OF VIRUS CYTOPATHIC EFFECTS

R. BABLANIAN

State University of New York, Downstate Medical Centre, Brooklyn, N.Y.

INTRODUCTION

Virus cytopathic effects have been recognised for well over two decades; yet the mechanisms by which viruses cause cell damage remain largely unknown both in cell cultures and in the animal host. There are indications that the underlying mechanisms of virus-induced cell damage are fundamentally similar in both systems. Similarities can be found between the development of pathological changes in animals and the development of cell damage in culture.

Multiplication of a virus in a host is a prerequisite for the development of a disease but virus multiplication does not always lead to disease. Furthermore, pathological changes can also be caused by the virus in the absence of demonstrable production of new infectious virus particles (Schlesinger, 1950; Ginsberg, 1951). At the cell level, virus multiplication usually leads to cell damage, although in some virus–cell systems virus multiplication occurs without apparent morphological damage to cells (Choppin, 1964). Pathological changes can occur in cells even if only some products of virus multiplication are made and new infective particles are not produced (Henle, Girardi & Henle, 1955; Ledinko, 1958; Appleyard, Westwood & Zwartouw, 1962). Thus, both in animals and in cultured cells, viruses can cause cell damage with or without the production of infectious virus.

Enders (1954) coined the term 'virus cytopathic effect', or CPE, and defined it as any virus-induced cell damage, whether morphological or biochemical. This definition would include not only the effects produced by cytocidal viruses, which during the course of their multiplication kill cells, but also such non-lethal changes as cell transformation. However, the term 'cytopathic effect', or CPE, has been used by many investigators to denote exclusively morphological damage that may accompany virus–cell interaction. The reason for this restricted usage is probably because there is abundant information concerning virus-induced morphological damage, when contrasted to the paucity of knowledge concerning metabolic derangements in infected cells.

Recently however, as a result of increasing studies of the biochemical alterations in virus-infected cells, a picture of the various patterns of biochemical changes in cells infected with different viruses is beginning to emerge. This article will deal with the mechanisms of virus-induced cell damage caused by cytocidal viruses. Both the morphological lesions and the biochemical alterations of infected cells will be discussed.

MECHANISMS OF VIRUS-INDUCED CYTOPATHOLOGY

Under the usual conditions of infection, cytocidal viruses cause cell damage during the course of their active multiplication in susceptible cells. The overall time relationship between viral replication and the development of morphological abnormalities in cells has been studied with many cytocidal viruses (Pereira, 1961). Such studies were useful in establishing similarities and differences in morphological changes caused by different viruses. In fact, the similarities in the cytopathic effects within a group were so great that Barski (1962) proposed a classification of viruses on this basis. In many instances, however, virus infection leads to cell degeneration without the production of new infectious virus. This type of acute and non-transmissible virus-inflicted morphological damage has been referred to as the cytotoxic effect. In order to show the toxic effect of viruses it has always been necessary to infect cells at a high multiplicity.

Virus-induced cytopathology can occur under the following conditions: (1) when the virus is inherently incapable of multiplying in a given cell type (Henle et al. 1955) or multiplication is restricted (Sturman & Tamm, 1966); (2) when chemical inhibitors are used to prevent the replication of infectious virus (Tamm & Eggers, 1963); (3) when the virus is inactivated by ultraviolet irradiation (Brown, Mayyasi & Officer, 1959; Hanafusa, 1960; Appleyard et al. 1962); and (4) by pre-formed viral components (Pereira & Kelly, 1957; Everett & Ginsberg, 1958; Rowe, Hartley, Roizman & Levy, 1958).

Virus-induced cell damage under restricted or abortive infections

The cytotoxic effect was first observed in cultured cells by Henle et al. (1955), who showed that HeLa cells, which could not sustain influenza virus reproduction when exposed to large quantities of the virus, underwent degeneration, yielding viral haemagglutinins and complement-fixing antigens but no infectious virus. The proportion of affected cells in the cultures was directly dependent on the concentration of the input virus. Newcastle disease virus (NDV) (Prince & Ginsberg,

1957) and fowl plague virus (Franklin & Breitenfeld, 1959) have also been shown to be cytotoxic; in both cases viral products were formed in the absence of production of infectious virus (Walker, 1960; Westwood, 1963). Sturman & Tamm (1966) demonstrated that the murine picornavirus, GDVII, caused cytopathology in several restrictive cell strains with little infectious virus production. Later it was found (Sturman & Tamm, 1969) that the maximum virus yields in restrictive or permissive cells were reached at the same time. Thus, they concluded that the differences in the rate of development of virus cytopathic effects in the productive and restricted cycles may reside in the different rates of accumulation of cytotoxic viral products in the two host cell types. Buck, Granger, Taylor & Holland (1967) and Wall & Taylor (1969) also showed that cells of a restrictive host infected with mengovirus produced viral antigens and a limited number of infectious virus particles, yet were killed. Holland (1968) showed that there are no qualitative differences in mengovirus-directed protein synthesis in productively or abortively infected cells as determined by acrylamide gel electrophoresis. However, the total amount of viral proteins in non-permissive host cells was reduced. Thus, virus-induced cytopathology can occur when, (1) some viral products are produced in the absence of production of infectious virus, or (2) when a much reduced number of infectious virions are made.

Virus-induced cell damage in the presence of chemical inhibitors of virus reproduction

Using chemical inhibitors, Ackermann, Rabson & Kurtz (1954) and Ledinko (1958) have shown that although infectious poliovirus is not formed, infected cells degenerate. Wecker, Hummeler & Goetz (1962) have shown that in the absence of production of infectious poliovirus, viral antigens are formed in the presence of *p*-fluorophenylalanine (FPA). Compounds such as 5-fluoro-2'-deoxyuridine (Shatkin, 1963), *p*-fluorophenylalanine (Loh & Payne, 1965) and isatin-β-thiosemicarbazone (Appleyard, Hume & Westwood, 1965) inhibit infectious vaccinia virus production but allow some virus-induced protein synthesis to occur. But these three compounds did not prevent virus cytopathic effects (Bablanian, 1968). On the other hand, known inhibitors of protein synthesis, such as puromycin, cycloheximide and streptovitacin A prevented both infectious virus production and virus-induced morphological damage (Bablanian, 1968, 1970). The removal of streptovitacin A from infected cultures is followed by a delay in the development of virus cytopathic effects This 'cytopathic lag' of about

Table 1. *Incorporation of L-leucine-4,5-³[H] by vaccinia virus-infected or uninfected LLC-MK2 cells in the presence or after the removal of streptovitacin A (300 µg/ml)*

Time of label after infection*	Untreated cells		Virus and drug		Virus and drug†		Cells and drug		Cells and drug†	
	counts/min	% of control	counts/min	% of control	counts/min	% of control	counts/min	% of control	counts/min	% of control
5 h	6539	100	53	0·8	213	3·3	42	0·6	146	2·2
6 h	5643	100	47	0·8	475	8·4	30	0·5	661	11·7
7 h	6733	100	40	0·6	307	4·6	28	0·4	978	14·5

* Duration of pulse-label 20 min.
† Removed 4 h after treatment.

Table 2. *Incorporation of L-leucine-4,5-³[H] by vaccinia virus-infected or uninfected LLC-MK2 cells in the presence or after the removal of cycloheximide (300 µg/ml)*

Time of label after infection*	Untreated cells		Virus and drug		Virus and drug†		Cells and drug		Cells and drug†	
	counts/min	% of control	counts/min	% of control	counts/min	% of control	counts/min	% of control	counts/min	% of control
4 h	5423·6	100	44·5	0·8	2722·7	50·2	35·9	0·7	3207·0	59·1
4 h 15 min	6646·1	100	55·1	0·8	2176·5	32·7	34·8	0·5	3360·5	50·6
4 h 30 min	7350·6	100	42·1	0·6	2245·3	30·5	40·1	0·5	3004·3	40·9
5 h	7670·7	100	65·6	0·8	1971·1	25·7	46·1	0·6	3937·6	51·3

* Duration of pulse-label 20 min.
† Removed 4 h after treatment.

Table 3. *Incorporation of L-leucine-4,-5-³[H] by vaccinia virus-infected or uninfected LLC-MK2 cells in the presence or after the removal of puromycin (330 μg/ml)*

Time of label after infection*	Untreated cells		Virus and drug		Virus and drug†		Cells and drug		Cells and drug†	
	counts/ min	% of control	counts/ min	% of control	counts/ min	% of control	counts/ min	% of control	counts/ min	% of control
4 h	3869·8	100	88·1	2·3	1887·3	48·8	94·1	2·4	2464·2	63·7
4 h 15 min	2725·2	100	81·3	3·0	1109·5	40·7	63·5	2·3	1745·0	64·0
4 h 30 min	2555·2	100	64·9	2·5	735·3	28·8	70·9	2·8	2316·1	90·6
5 h	2213·1	100	80·0	3·6	494·5	22·3	62·3	2·8	2025·6	91·5

* Duration of pulse-label 20 min.
† Removed 4 h after treatment.

2 h after the removal of this inhibitor coincided with a similar lag in the resumption of protein synthesis (Table 1). On the other hand, when cycloheximide was removed from infected cultures, protein synthesis resumed without a lag period (Table 2). In such cultures virus cytopathic effects became evident within 20 min and were striking by 1 to $1\frac{1}{2}$ h after reversal (Bablanian & Baxt, in preparation). Similar results were also obtained with puromycin (Table 3). These results indicate that the appearance of virus-induced morphological damage in the vaccinia virus system is intimately related to the resumption of protein synthesis.

Bablanian, Eggers & Tamm (1965b) demonstrated that the virus-induced morphological changes in poliovirus-infected cells are also directly related to the active synthesis and accumulation of viral proteins. The addition of guanidine as late as 2 h after infection, before the onset of any detectable amounts of viral RNA polymerase, viral RNA, and proteins, completely prevented the development of virus-induced cytopathological changes in a single cycle infection. When guanidine was added at later times its inhibitory effect on cytopathological changes progressively decreased, and when it was added during the exponential increase of virus, it had virtually no effect on virus-induced morphological changes. However, if puromycin (a known inhibitor of protein synthesis) was added during the exponential rise of virus, cytopathological changes were avoided.

The addition of guanidine to infected cells during the exponential phase of poliovirus replication stopped the ongoing synthesis of RNA polymerase (Baltimore, Eggers, Franklin & Tamm, 1963), viral RNA and infectious virus (Eggers, Reich & Tamm, 1963) but a limited amount of viral protein was synthesised (Halperen, Eggers & Tamm, 1964). Thus, it was proposed that the development of virus-morphological changes were dependent on the synthesis of viral proteins. Recently, studies in our laboratory (Bablanian & Shepler, in preparation) have confirmed and strengthened these conclusions. We took advantage of the fact that in the presence of guanidine, poliovirus suppressed cellular protein synthesis. We studied the development of virus cytopathology in relation to the onset of viral protein synthesis after the removal of the drug from infected cells. Under conditions where cellular protein synthesis was essentially arrested, a more meaningful relationship could be sought between the development of virus-induced cytopathology and virus-induced protein synthesis. When LLC-MK2 cells were infected with poliovirus, in the presence of guanidine, protein synthesis was inhibited by over 90 % when measured 6 to 7 h after infection and treatment. On removal of guanidine from infected-treated cultures (reversal) virus-induced morphological damage became evident $3\frac{1}{2}$ h later and developed exponentially until about 6 h when cell damage became maximal. The re-addition of guanidine 2 h after reversal prevented virus-induced morphological damage but the re-addition of the drug at later times could no longer prevent virus-induced cytopathology (Fig. 1). A comparison of the rates of protein synthesis from reversed cultures to those of unreversed cultures showed that 3 h after

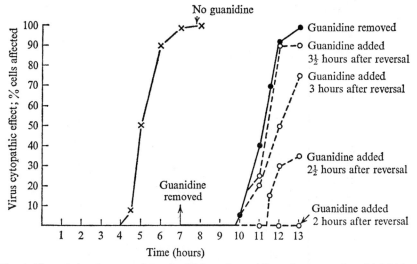

Fig. 1. The relation between time of addition of guanidine after reversal and inhibition of the development of virus-induced cell damage in poliovirus-infected LLC-MK2 cells.

the removal of the drug, the rate of virus-induced protein synthesis went up, whereas that of the unreversed cultures fell slowly (Fig. 2). Thus, the time when a visible change in the rate of viral protein synthesis took place (3 h) closely coincided with the time when the re-addition of guanidine to reversed cultures could no longer prevent virus-induced morphological damage. These results indicate that once active peptide polymerisation begins in reversed cultures, guanidine can no longer prevent cell damage. However, when streptovitacin A (a powerful inhibitor of protein synthesis) was added to cultures 3 h after the removal of guanidine, virus-induced cytopathology was prevented at a time when cell damage in infected reversed cultures was greatest (6 h after the removal of guanidine from infected cultures). These results confirm and further support our previous conclusions (Bablanian, Eggers & Tamm, 1965a, b) that poliovirus-induced cytopathology is directly related to the synthesis and accumulation of virus-induced proteins.

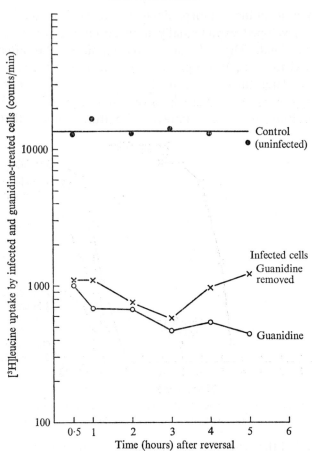

Fig. 2. The rate of protein synthesis in reversed and unreversed cultures
of poliovirus-infected LLC-MK2 cells.

Virus-induced cell damage by ultraviolet-irradiated virus

Some viruses that have lost their infectivity after ultraviolet irradiation
can still cause cytotoxic effects. Ultraviolet-irradiated vaccinia virus
cause cytotoxic effects in a variety of cells (Scherer, 1952; Brown *et al.*
1959; Hanafusa, 1960; Nishmi & Bernkopf, 1958). Ultraviolet-
irradiated rabbitpox virus which has lost its infectivity still can cause
morphological damage to cells with the production of some viral
antigens. (Appleyard *et al.* 1962). These authors have also demonstrated
a cytotoxic effect when multiplication of infectious rabbitpox virus was
prevented by chemical inhibitors (Appleyard *et al.* 1962). The time
course of the appearance of morphologic damage was the same as that
in untreated infected cells which produced virus. Furthermore, under

these conditions some viral antigens were formed. Loh and his co-workers (Loh & Oie, 1969) discovered that ultraviolet-irradiated reovirus, which had lost its infectivity, acquired a 'toxic' characteristic which was absent when unirradiated virus was used to infect cells. This curious phenomenon could not be associated with any significant changes in such viral properties as the production of haemagglutinins, RNA polymerase activity, interferon-inducing ability, virus architecture, buoyant density in CsCl, capsid protein components, etc. (Subasinghe & Loh, 1971). Since both transcription and translation of reovirus message is under strict control (Zweerink & Joklik, 1970) this control mechanism may be deranged by irradiation causing the 'untimely' synthesis of some virus-induced proteins, resulting in early death of the cells.

Virus-induced cytopathology can also vary with the same virus and the same cell line but under different cultural conditions. Ensminger & Tamm (1969) studied reovirus-induced morphological and biochemical changes in L-cell monolayers and L-cell suspension cultures. They observed that 14 h after infection, 90 % of cells in suspension were damaged, whereas in monolayer cultures morphological damage was not evident at this time. Similar yields of infectious virus were obtained in both systems.

Virus-induced cell damage by preformed viral products

Another well-known form of morphological damage induced by a number of viruses is the formation of polykaryocytes or syncytia. Herpesviruses and paramyxoviruses cause giant-cell formation. With several of the paramyxoviruses such as Sendai virus (Okada & Tadokoro, 1963), Newcastle disease virus (NDV) (Kohn, 1965), measles virus (Cascardo & Karzon, 1965) and Simian virus 5 (SV 5) (Holmes & Choppin, 1966), an early polykaryocytosis occurs but only when high virus multiplicates are used. Large doses of some NDV strains can cause giant-cell formation within 2–3 h after infection even after ultraviolet irradiation. This phenomenon was termed 'fusion from without' implying that there was a pre-formed viral product that caused the formation of syncytia and *de novo* protein synthesis was not necessary to cause cell damage (Bratt & Gallaher, 1969). However, protein synthesis was required for fusion to occur with other strains of NDV (Bratt & Gallaher, 1969; Reeve & Poste, 1971). The formation of early syncytia by high multiplicities of SV 5 in the presence of puromycin and actinomycin D also indicated that a pre-formed viral product was the cause of the giant cell formation in BHK 21 cells (Holmes & Choppin, 1966). Tokumaru (1968) reported that the syncytium-producing components of herpesvirus could be found both on the virus particles as well as free. These components had the physiochemical properties of lipoproteins and appeared in infected cytoplasms before the appearance of infectious virus.

Work on adenoviruses have also produced well-documented evidence that a virus-directed separable, soluble antigen can cause morphological changes to cells in culture. Pereira & Kelly (1957), Everett & Ginsberg (1958) and Rowe et al. (1958) found a toxin-like viral component in crude extracts of adenovirus-infected cells. It was separable from the virion and when added to cell cultures caused cell rounding and cell detachment. This viral antigen has been termed the penton (Ginsberg, Pereira, Valentine & Wilcox, 1966). The relationship of the antigen to the virion capsid was shown by Valentine & Pereira (1965). Maizel, White & Scharff (1968) showed that the penton (vertex capsomere) isolated from infected cells was serologically identical to peptide III, one of the major virion components, as revealed by SDS-gel electrophoresis. Interestingly, the penton does not cause depression of macromolecular synthesis. However, the fibre (vertex projection) inhibits cellular RNA, DNA and protein synthesis (Levine & Ginsberg, 1967).

Recently, Cordell-Stewart & Taylor (1971) have reported that double-stranded RNA from bovine enterovirus causes rapid cell death without the production of infectious virus. The effect of this double-stranded RNA was non-specific since it killed cells that are normally not susceptible to this enterovirus. Synthetic double-stranded polymers could also cause cell death but to a lesser extent. Whether or not this is a mechanism of enterovirus-induced cytopathology remains to be seen. Although these authors found no infectious virus production by the double-stranded RNA of this virus, encephalomyocarditis (EMC) virus and poliovirus double-stranded RNA is known to be infectious (Montagnier & Saunders, 1963; Pons, 1964). It is quite conceivable that an incomplete cycle of virus replication might have occurred in these cells treated with double-stranded RNA with the production of some viral proteins which in turn could have been responsible for cell death. It would also be desirable to test the effect of double-stranded RNA on monolayer cultures since it is known that virus-induced cell damage can vary under different cultural conditions (Ensminger & Tamm, 1969).

In conclusion, the underlying mechanism which brings about morphological damage to cells in the presence or absence of productions of infectious virus appears to be the same, namely, the synthesis of some virus-induced proteins which may or may not be incorporated into the fully infectious virus particle.

MECHANISMS OF VIRUS-INDUCED METABOLIC
ALTERATIONS IN CELL CULTURE

Virus cytopathic effects also include metabolic alterations of infected cells. Ackermann (1958); Ackermann, Loh & Payne 1959; Salzman, Lockart & Sebring (1959) were among the first to study the metabolic alterations in cultured cells resulting from poliovirus infection. Later work with poliovirus (Holland, 1962; Darnell, 1962; Fenwick, 1963; Zimmerman, Heeter & Darnel, 1963) and other picornaviruses such as mengovirus (Baltimore & Franklin, 1962; Franklin & Baltimore, 1962), EMC virus (Martin, Malec, Sved & Work, 1961) and ME-virus (Scholtissek *et al.* 1962) confirmed Salzman's and his coworkers' findings by showing that infection with these picornaviruses resulted in a severe depression of both RNA and protein synthesis of the infected cells. Baltimore & Franklin (1962) and Franklin & Baltimore (1962) showed that the DNA synthesis of mengovirus-infected cells was inhibited later than the RNA and protein synthesis. Similar results were also obtained with another picornavirus, ME-virus (Scholtissek *et al.* 1962) and poliovirus (Ackermann *et al.* 1966). There are presently many RNA- and DNA-containing viruses belonging to several major groups which have been shown to cause depression of cellular macro-molecular synthesis (Martin & Kerr, 1968).

The mechanisms by which cytocidal viruses cause inhibitions of cellular macromolecular synthesis, the so-called cut-off phenomenon, is still not well understood. It has been suggested that inhibition of cellular protein synthesis in cells infected with poliovirus, mengovirus, and EMC virus is due to the breakdown of polysomes (Penman, Scherrer, Becker & Darnell, 1963; Joklik & Merigan, 1966; Dalgarno, Cox & Martin, 1967). The virus-induced inhibition of cellular RNA synthesis is thought to be due to the inability of cellular messenger RNA in infected cells to reassociate with ribosomes (Willems & Penman, 1966). Recently, Ensminger & Tamm (1970) have presented evidence that NDV and mengovirus-induced inhibition of cellular protein synthesis affected a process required for the initiation of DNA synthesis on regulatory units of DNA replication.

The metabolic alterations in infected cells in the absence of production of infectious virus have also been studied. Hanafusa (1960) reported that L-cells infected with inactivated vaccinia virus showed the same degree of cellular DNA depression as when infected with active virus. Holland (1964a) demonstrated that the RNA and protein synthesis of poliovirus-infected HeLa cells treated with guanidine in the absence of infectious virus multiplication was depressed. Vaccinia virus is known to cause depression of cellular RNA and protein synthesis (Joklik, 1966). Moss (1968) has reported that HeLa cells infected with vaccinia virus, in the presence of actinomycin D and cycloheximide, are incapable of resuming protein synthesis after the removal of cycloheximide. On this basis, he concluded that the input virions have a pre-formed substance that can suppress cellular protein synthesis without the need of *de novo* protein synthesis. Our results (Bablanian & Baxt, in preparation) are not in agreement with those of Moss. We find that LLC-MK2 cells infected with vaccinia virus and treated either with streptovitacin A (Table 1), cycloheximide (Table 2) or puromycin (Table 3), showed a resumption of protein synthesis upon removal of the drugs. Furthermore, up to $\frac{1}{2}$ h after reversal, the amounts of proteins synthesised in infected reversed cultures and uninfected reversed cultures were almost equal for all compounds (Tables 1–3). Virus-induced depression of cellular protein synthesis became evident only after 45 min to 1 h post-reversal (Tables 2 & 3). In some experiments we also included actinomycin D in addition to cycloheximide to mimic Moss' conditions but again, on removal of cycloheximide there was a resumption of an equal amount of protein synthesis both in the infected reversed and uninfected reversed cultures. The discrepancy between our results and those of Moss could be due to the differences in the cell line used or due to cultural conditions, since he used spinner cultures of HeLa cells and we used monolayers of LLC-MK2 cells. Furthermore, cycloheximide may be less efficient in inhibiting protein synthesis in HeLa cells than it is in LLC-MK2 cells, thus allowing some virus-induced proteins to be made in the presence of the drug. Our results with streptovitacin A indicate that virus-induced depressions can occur even if close to 90 % of the cells' protein synthesis is inhibited (compare columns 5 and 6 to columns 9 and 10 in Table 1).

Ultraviolet irradiation has different effects on the ability of viruses to cause inhibition of cellular macromolecular synthesis. Ultraviolet-irradiated poliovirus loses its ability to cause depression of cellular RNA and protein synthesis (Penman & Summers, 1965). On the other hand, vesicular stomatitis virus (VSV) which has completely lost its infectivity through ultraviolet irradiation, can still inhibit RNA and protein synthesis of cells (Huang & Wagner, 1965; Wagner & Huang, 1966) indicating that a pre-formed virus component may be the cause of these depressions. With adenoviruses it has been shown that the fibre antigen, part of the virus capsid, inhibits cellular biosynthetic processes (Levine & Ginsberg, 1967; Ginsberg, Bello & Levine, 1967). Thus, it appears that there are at least two separate and distinct modes of viral inhibition of cellular macromolecular synthesis: (1) inhibition caused by pre-formed viral proteins, e.g., adenovirus fibre antigen; and (2) inhibition caused by virus-induced products in the course of either productive or abortive virus replication.

Recently, Ehrenfeld & Hunt (1971) have reported that double-stranded poliovirus RNA inhibits the initiation of protein synthesis in lysates of rabbit reticulocytes. Whether inhibition of host-cell protein synthesis is also caused by double-stranded RNA is still unanswered. The big question that remains is why does poliovirus inhibit cellular RNA and protein synthesis in the presence of guanidine when under these conditions RNA-dependent RNA polymerase is not made (Baltimore et al. 1963) and without the polymerase, viral double-stranded RNA cannot be formed.

The effect of interferons on the cut-off phenomenon of infected cells has also been studied. Levy (1964) showed that pre-treatment with interferon had no effect on the mengovirus-induced inhibition of RNA and protein synthesis. Gauntt & Lockart (1966, 1968) confirmed these findings. However, Levy, Snellbacker & Baron (1963) showed that Sindbis virus-induced cellular depression was abolished by interferon. A similar conclusion was reached by Lockart (1963) using Western equine encephalitis (WEE) virus. Recently, Haase, Baron, Levy & Kasel (1969) have reported that the cut-off of host-cell protein synthesis by mengovirus can be prevented when high concentrations of interferon are used. On the other hand, Yamazaki & Wagner (1970) reported that VSV-induced inhibition of host-cell protein synthesis is not affected by high concentrations of interferon. These results taken together suggest that viruses respond to the action of interferons, as far as the cut-off phenomenon is concerned, in three different ways: (1) interferon can completely prevent virus-induced depression of host-cell macromolecular synthesis, e.g. for WEE (Lockart, 1963) and Sindbis virus (Levy et al. 1963), (2) high doses of interferon are required to prevent inhibition, e.g. for mengovirus (Haase et al. 1969) and (3) interferon cannot prevent the cut-off phenomenon, e.g. for VSV (Yamazaki & Wagner, 1970).

RELATIONSHIP BETWEEN VIRUS-INDUCED METABOLIC ALTERATIONS AND VIRUS-INDUCED MORPHOLOGICAL DAMAGE TO CELLS

Bablanian et al. (1965a) were the first to point out that virus-induced morphological damage was not directly related to the cut-off phenomenon. These authors showed that in the presence of guanidine, poliovirus depressed both the RNA and protein synthesis of the cells without the appearance of virus-induced cytopathological changes. Amako & Dales (1967) using mengovirus came to a similar conclusion. Haase *et al.* (1969) demonstrated that the addition of cycloheximide, up to 4 h after infection with mengovirus, prevents virus-induced cytopathic effects. By 4 h after infection severe depression of macromolecular synthesis had occurred. These authors therefore, concluded that the 'cut-off' phenomenon was unrelated to morphological cell damage. Bablanian (to be published), using high multiplicities of the guanidine dependent strain of poliovirus type 1 (Loddo, Ferrari, Botzu & Spanedda, 1962) showed that under restrictive conditions cellular RNA and protein synthesis were depressed at the same rate and to the same extent as under permissive conditions. However, virus cytopathology was evident only under the permissive conditions.

A lack of parallelism between virus-induced metabolic alterations and virus-induced cytopathology can also be found with viruses other than the picornavirus group. Ensminger & Tamm (1969) demonstrated that reovirus infection markedly inhibited cellular DNA synthesis by 14 h after infection, yet morphological damage was not present at that time in monolayer cultures of L-cells. Scholtissek *et al.* (1962) demonstrated that in spite of the lack of inhibition of cellular metabolic functions, influenza virus-infected cells were killed. Holmes & Choppin (1966) showed that fusion of BHK 21 cells by SV 5 took place much before depression of cellular RNA, DNA and protein synthesis was evident. Falke & Peterknecht (1968) have shown that giant-cell formation by herpesvirus was not related to the virus-induced depression of cellular protein synthesis but rather was associated with the synthesis of virus-induced early proteins. The experiments with adenoviruses provide the best evidence that the virus-induced morphological lesions to cells are not the result of virus-induced metabolic alterations because the penton antigen caused cell rounding only, whereas it was the fibre antigen which brought about the depression of cellular RNA, DNA and protein synthesis.

Even though the evidence presented indicated a lack of parallelism between virus-induced cell damage and virus-induced metabolic alterations, the evidence cited above does not cover all the major virus groups. Therefore, it is possible that in some virus-cell systems metabolic alterations could occur at the same time as virus-induced cell damage.

CURRENT CONCEPTS OF THE MECHANISM OF VIRUS-INDUCED CELL DAMAGE

Having considered the central facts of virus cytopathic effects, some of the current thoughts on the mechanisms of virus-induced cell damage can now be examined. There is no single hypothesis that can explain the mechanism of virus cytopathic effects. Ginsberg (1961) proposed that, in general, viruses can cause cell injury by two basic mechanisms: (1) by the actual process of viral replication that would either cause depletion of cellular components essential for cell life or cause damage to cells by viral components made in excess: or (2) by producing toxin-like substances which cause cell death without viral multiplication. Martin & Work (1961) were the first to suggest that the cell-killing property of EMC virus may reside in its ability to inhibit normal cellular RNA synthesis. Holland (1963, 1964b) also postulated that the major cause of poliovirus-induced cell damage may reside in its ability to depress cellular RNA and protein synthesis. Undoubtedly, if cellular macromolecular synthesis is inhibited cells will eventually die but whether virus-induced inhibition is the immediate and direct cause of cell death is the point in question. Baltimore & Franklin (1962) suggested that although the inhibition of RNA in mengovirus-infected L-cells may eventually lead to cell death, it could not entirely explain the advanced virus-induced morphological changes, since cells treated with actinomycin D, which also inhibited cellular RNA synthesis, underwent much less drastic morphological changes than those which were virus-infected. Bablanian (1968, 1970) also observed that puromycin, cycloheximide and streptovitacin A at doses which inhibit over 99 % of the protein synthesis in LLC-MK2 cells caused no morphological changes up to 24 h after treatment. These results indicate that the mere switch-off of cellular RNA and protein synthesis does not bring about cytopathological changes of the type seen with cytocidal viruses. Moreover, in some virus cell interactions, virus-induced cytopathology was not directly related to virus-induced inhibition of macromolecular synthesis.

It has been suggested that the activation of lysosomal enzymes may be a mechanism of virus-induced cytopathology (Allison, 1967). The release of lysosomal enzymes might play a late role in causing cell death but the primary injury that leads to cell death requires virus-induced protein synthesis (Bubel, 1967; Blackman & Bubel, 1969). In fact, Guskey, Smith & Wolff (1970) suggested that a poliovirus-induced protein produced during the third hour after infection is responsible for lysosomal enzyme release. However, Wolff & Bubel (1964) have shown that with some cytocidal viruses, such as vaccinia and vesicular stomatitis virus, cytopathological changes can occur without any significant enzyme release.

The work of Bablanian *et al.* (1965*b*) and Bablanian (1968, 1970) has shown that the synthesis and accumulation of virus-induced proteins is essential for the occurrence of poliovirus and vaccinia virus cytopathology. Evidence obtained with several other cytocidal viruses has supported this view (Gauntt & Lockart, 1966; La Placa, 1966; Scholtissek, Becht & Drzeniek, 1967; Amako & Dales, 1967; Bubel, 1967; Gauntt & Lockart, 1968; Blackman & Bubel, 1969; Haase *et al.* 1969; Guskey, Smith & Wolff, 1970). The evidence obtained with these cytocidal viruses indicated that virus 'toxicity' is manifested only after the virus enters the cells and induces some protein synthesis. The time of appearance of cytopathological lesions varies with these different viruses. Adenoviruses, however, are known to have a separable viral capsid protein (penton) which is capable of inducing cell rounding (Pereira & Kelly, 1957; Everett & Ginsberg, 1958; Rowe *et al.* 1958). Herpesviruses also seem to have separable virus-directed components, lipoproteins in nature (Tokumaru, 1968). These are the only known instances where viral products can be directly related to the cause of cell rounding and giant-cell formation. It is quite conceivable that there are other such viral components that await discovery. In this respect the work of Birkbeck & Stephen (1970) should be mentioned. These authors are isolating and purifying vaccinia virus-specific components and testing these for the reproduction of toxic effects in uninfected cells.

All the evidence presented so far suggests that viruses cause cell destruction by inducing the synthesis of proteins that may or may not be part of the structural viral proteins or lipoproteins.

CONCLUDING REMARKS

The evidence presented in this article shows that virus cytopathic effects are a combination of virus-induced morphological changes and virus-induced metabolic alterations in cells. Evidence is also presented to show that these two virus-induced lesions are not directly related, although both the cut-off phenomena and virus cytopathology are brought about by some virus-induced proteins which may or may not be part of the virion.

There are certain common findings that are important to keep in mind for future studies of virus cytopathic effects. First, the manifestations of virus cytopathic effects depend on the cultural conditions

(Ensminger & Tamm, 1969), the types of cells used (Bablanian, 1965a) and the virus strains employed (Wilson, 1968). Second, the amount and the types of inhibitors used also determine whether or not some viral products will be made (Bablanian, 1968, 1970; Friedman, 1968; Long & Burke, 1970). Low concentrations of inhibitors prevent the production of infectious virus but usually allow the synthesis of some viral products, whereas larger doses of inhibitors are needed to inhibit most virus-induced alterations (Wecker, Hummeler & Goetz, 1962; Bablanian, 1968, 1970; Long & Burke, 1970; Gauntt & Lockart, 1968; Haase et al. 1969). Third, interferons also seem to act the same way as the chemical inhibitors of virus replication. Low doses of interferon usually inhibit infectious virus production but allow some viral antigens to be made (Gauntt & Lockart, 1968). Much higher doses of interferon are required to prevent virus-induced metabolic alterations and cytopathological changes (Haase et al. 1969).

The virus-induced metabolic alterations and the cytopathological changes at the cell level should be used as a guide-line in the study of the pathological changes in the organism. For example, Wagner & Huang (1966) have suggested that the reason VSV does not induce the production of interferon in Krebs-2 cells is because the virus inhibits host cell RNA synthesis. Recently Wertz & Younger (1970) have shown that a small-plaque mutant of VSV does not shut off RNA and protein synthesis completely. In such cultures the production of interferon is evident. On the other hand, the wild-type VSV inhibits RNA and protein synthesis completely by 8 h, and in these cultures there is no production of interferon. Mutants that have a variable effect in cell culture should also be tested in animals. One such study (Amako & Dales, 1967) shows that a small-plaque variant of mengovirus which causes less severe morphological changes in culture than the large-plaque variant, is also less virulent to mice. Clearly, more studies are needed in the future to seek similarities between virus-induced virulence in the organism and virus cytopathic effects.

The pathological changes in most viral diseases are ultimately related to the alteration or destruction of cells by the virus. Delineation of the mechanisms by which viruses cause cytopathic effects will therefore provide insight into the pathogenesis of viral diseases.

REFERENCES

ACKERMANN, W. W. (1958). Cellular aspects of the cell-virus relationship. *Bacteriological Reviews*, **22**, 223.

ACKERMANN, W. W., COX, D. C., KURTZ, H., POWERS, C. D. & DAVIES, S. J. (1966). Effect of poliovirus on deoxyribonucleic acid synthesis in HeLa cells. *Journal of Bacteriology*, **91**, 1943.

ACKERMANN, W. W., LOH, P. C. & PAYNE, F. E. (1959). Studies of the biosynthesis of protein and ribonucleic acid in HeLa cells infected with poliovirus. *Virology*, **7**, 170.

ACKERMANN, W. W., RABSON, A. & KURTZ, H. (1954). Growth characteristics of poliovirus HeLa cell cultures: Lack of parallelism in cellular injury and virus increase. *Journal of Experimental Medicine*, **100**, 437.

ALLISON, A. (1967). Lysosomes in virus-infected cells. In *Virus-Directed Host Response. Perspectives in Virology*, **5**, 29.

AMAKO, K. & DALES, S. (1967). Cytopathology of mengovirus infection. I. Relationship between cellular disintegration and virulence. *Virology*, **32**, 201.

APPLEYARD, G., HUME, V. B. M. & WESTWOOD, J. C. N. (1965). The effect of thiosemicarbazones on the growth of rabbitpox virus in tissue culture. *Annals of the New York Academy of Sciences*, **130**, 92.

APPLEYARD, G., WESTWOOD, J. C. N. & ZWARTOUW, H. T. (1962). The toxic effect of rabbitpox virus in tissue culture. *Virology*, **18**, 159.

BABLANIAN, R. (1968). The prevention of early vaccinia virus-induced cytopathic effects by inhibition of protein synthesis. *Journal of General Virology*, **3**, 51.

BABLANIAN, R. (1970). Studies on the mechanism of vaccinia virus cytopathic effects: effect of inhibitors of RNA and protein synthesis on early virus-induced cell damage. *Journal of General Virology*, **6**, 221.

BABLANIAN, R., EGGERS, H. J. & TAMM, I. (1965a). Studies on the mechanism of poliovirus-induced cell-damage. I. The relation between poliovirus-induced metabolic and morphological alterations in cultured cells. *Virology*, **26**, 100.

BABLANIAN, R., EGGERS, H. J. & TAMM, I. (1965b). Studies on the mechanism of poliovirus-induced cell damage. II. The relation between poliovirus growth and virus induced morphological changes in cells. *Virology*, **26**, 114.

BALTIMORE, D., EGGERS, H. J., FRANKLIN, R. M. & TAMM, I. (1963). Poliovirus-induced RNA polymerase and the effects of virus specific inhibitors on its production. *Proceedings of the National Academy of Sciences of the U.S.A.*, **49**, 843.

BALTIMORE, D. & FRANKLIN, R. M. (1962). The effect of mengovirus infection on the activity of the DNA-dependent RNA polymerase of L-cells. *Proceedings of the National Academy of Sciences of the U.S.A.*, **48**, 1383.

BARSKI, G. (1962). The significance of *in vitro* cellular lesions for classification of viruses. *Virology*, **18**, 152.

BIRKBECK, T. H. & STEPHEN, J. (1970). Specific removal of host-cell or vaccinia-virus antigens from extracts of infected cells by polyvalent disulphide immunosorbents. *Journal of General Virology*, **8**, 133.

BLACKMAN, K. E. & BUBEL, H. C. (1969). Poliovirus-induced cellular injury. *Journal of Virology*, **4**, 203.

BRATT, M. A. & GALLAHER, W. R. (1969). Preliminary analysis of the requirements for fusion from within and fusion from without by Newcastle disease virus. *Proceedings of the National Academy of Sciences of the U.S.A.*, **64**, 536.

BROWN, A., MAYYASI, S. A. & OFFICER, J. E. (1959). The 'toxic' activity of vaccinia virus in tissue culture. *Journal of Infectious Diseases*, **104**, 193.

BUBEL, H. C. (1967). Protein leakage from mengovirus-infected cells. *Proceedings of the Society for Experimental Biology and Medicine*, **125**, 783.

BUCK, C. A., GRANGER, G. A., TAYLOR, M. W. & HOLLAND, J. J. (1967). Efficient, inefficient and abortive infection of different mammalian cells by small RNA viruses. *Virology*, **33**, 36.

CASCARDO, M. R. & KARZON, D. T. (1965). Measles virus giant-cell inducing factor (fusion factor). *Virology*, **26**, 311.

CHOPPIN, P. W. (1964). Multiplication of myxovirus (SV5) with minimal cytopathic effects and without interference. *Virology*, **23**, 224.

CORDELL-STEWART, B. & TAYLOR, M. W. (1971). Effect of double stranded viral RNA on mammalian cells in culture. *Proceedings of the National Academy of Sciences of the U.S.A.*, **68**, 1326.

DALGARNO, L., COX, R. A. & MARTIN, E. M. (1967). Polyribosomes in normal Krebs-2 ascites tumour cells and in cells infected with encephalomyocarditis virus. *Biochimica et biophysica acta*, **138**, 316.

DARNELL, J. E. (1962). Early events in poliovirus infection. *Cold Spring Harbor Symposium on Quantitative Biology*, **27**, 149.

EGGERS, H. J., REICH, E. & TAMM, I. (1963). The drug-requiring phase in the growth of drug-dependent anteroviruses. *Proceedings of the National Academy of Sciences of the U.S.A.*, **50**, 183.

EHRENFELD, E. J. & HUNT, T. (1971). Double-stranded poliovirus RNA inhibits initiation of protein synthesis by reticulocyte lysates. *Proceedings of the National Academy of Sciences of the U.S.A.*, **68**, 1075.

ENDERS, J. F. (1954). Cytopathology of virus infections. *Annual Reviews of Microbiology*, **8**, 473.

ENSMINGER, W. & TAMM, I. (1969). Cellular DNA and protein synthesis in reovirus-infected L-cells. *Virology*, **39**, 357.

ENSMINGER, W. D. & TAMM, I. (1970). The step in cellular DNA synthesis blocked by Newcastle disease or mengovirus infection. *Virology*, **40**, 152.

EVERETT, S. F. & GINSBERG, H. S. (1958). A toxin-like material separable from type 5 adenovirus particles. *Virology*, **6**, 770.

FALKE, D. & PETERKNECHT, W. (1968). DNS-, RNS-, und Proteinsynthese und ihre Relation zur Riesenzellibildung *in vitro* nach Infektion mit *Herpesvirus hominis*. *Archiv für die gesamte Virusforschung*, **24**, 267.

FENWICK, M. L. (1963). The influence of poliovirus infection on RNA synthesis in mammalian cells. *Virology*, **19**, 241.

FRANKLIN, R. M. & BALTIMORE, D. (1962). Patterns of macromolecular synthesis in normal and virus-infected mammalian cells. *Cold Spring Harbor Symposium On Quantitative Biology*, **27**, 175.

FRANKLIN, R. M. & BREITENFELD, P. M. (1959). The abortive infection of Earle's L-cells by fowl plague virus. *Virology*, **8**, 293.

FRIEDMAN, R. M. (1968). Protein synthesis directed by an arbovirus. *Journal o, Virology*, **2**, 26.

GAUNTT, C. J. & LOCKART, R. Z., JR (1966). Inhibition of mengovirus by interferon. *Journal of Bacteriology*, **91**, 176.

GAUNTT, C. J. & LOCKART, R. Z., JR (1968). Destruction of L-cells by mengovirus: use of interferon to study the mechanism. *Journal of Virology*, **2**, 567.

GINSBERG, H. S. (1951). Mechanism of production of pulmonary lesions in mice by Newcastle disease virus (NDV). *Journal of Experimental Medicine*, **101**, 25.

GINSBERG, H. S. (1961). Biological and biochemical basis for cell injury by animal viruses. *Federation Proceedings*, **20**, 656.

GINSBERG, H. S., BELLO, L. J. & LEVINE, A. J. (1967). Control of biosynthesis of host macromolecules in cells infected with adenoviruses. In *Symposium of the Molecular Biology of Viruses*, ed. J. S. Colter, 547. New York: Academic Press.

GINSBERG, H. S., PEREIRA, H. S., VALENTINE, R. C. & WILCOX, W. C. (1966). A proposed terminology for the adenovirus antigens and virion morphological subunits. *Virology*, **28**, 782.

GUSKEY, L. E., SMITH, P. C. & WOLFF, A. (1970). Patterns of cytopathology and lysosomal enzyme release in poliovirus-infected HEP-2 cells treated with either 2-(α-hydroxybenzyl)-benzimidazole or guanidine HCl. *Journal of General Virology*, **6**, 151.

HAASE, A. T., BARON, S., LEVY, H. & KASEL, J. A. (1969). Mengovirus-induced cytopathic effect in L-cells: protective effect of interferon. *Journal of Virology*, **4**, 490.

HALPEREN, S., EGGERS, H. J. & TAMM, I. (1964). Evidence for uncoupled synthesis of viral RNA and viral capsids. *Virology*, **24**, 36.

HANAFUSA, H. (1960). Killing of L-cells by heat and UV-inactivated vaccinia virus. *Biken's Journal*, **3**, 191.

HENLE, G., GIRARDI, A. & HENLE, W. (1955). A non-transmissible cytopathogenic effect in influenza virus in tissue culture accompanied by formation of non-infectious hemagglutinins. *Journal of Experimental Medicine*, **101**, 25.

HOLLAND, J. J. (1962). Inhibition of DNA-primed RNA synthesis during poliovirus infection of human cells. *Biochemical and Biophysical Research Communications*, **9**, 556.

HOLLAND, J. J. (1963). Depression of host-controlled RNA synthesis in human cells during poliovirus infection. *Proceedings of the National Academy of Sciences of the U.S.A.*, **49**, 23.

HOLLAND, J. J. (1964a). Inhibition of host cell macromolecular synthesis by high multiplicities of poliovirus under conditions preventing virus synthesis. *Journal of Molecular Biology*, **8**, 574.

HOLLAND, J. J. (1964b). Enterovirus entrance into specific host cells and subsequent alterations of cell protein and nucleic acid synthesis. *Bacteriological Reviews*, **28**, 3.

HOLLAND, J. J. (1968). Virus-directed protein synthesis in different animal and human cells. *Science*, **160**, 1346.

HOLMES, K. V. & CHOPPIN, P. W. (1966). On the role of the response of the cell membrane in determining virus virulence. Contrasting effects of the parainfluenza virus SV 5 in two cell types. *Journal of Experimental Medicine*, **124**, 501.

HUANG, A. S. & WAGNER, R. R. (1965). Inhibition of cellular RNA synthesis by non-replicating vesicular stomatitis virus. *Proceedings of the National Academy of Sciences of the U.S.A.*, **54**, 1579.

JOKLIK, W. K. (1966). The pox viruses. *Bacteriological Reviews*, **30**, 33.

JOKLIK, W. K. & MERIGAN, T. C. (1966). Concerning the mechanism of action of interferon. *Proceedings of the National Academy of Sciences of the U.S.A.*, **56**, 558.

KOHN, A. (1965). Polykaryocytosis induced by Newcastle disease virus in monolayers of animal cells. *Virology*, **26**, 228.

LA PLACA, M. (1966). On the mechanism of the cytopathic changes produced in human amnion cell cultures by the molluscum contagiosum virus. *Archiv für die gesante Virusforschung*, **18**, 374.

LEDINKO, N. (1958). Production of noninfectious complement-fixing poliovirus particles in HeLa cells treated with proflavine. *Virology*, **6**, 512.

LEVINE, A. J. & GINSBERG, H. S. (1967). Mechanism by which fiber antigen inhibits multiplication of type 5 adenovirus. *Journal of Virology*, **1**, 747.

LEVY, H. B. (1964). Studies on the mechanism of interferon action. II. The effect of interferon on some early events in mengovirus infection in L-cells. *Virology*, **22**, 575.

LEVY, H. B., SNELLBACKER, L. F. & BARON, S. (1963). Studies on the mechanism of action of interferon. *Virology*, **21**, 48.

LOCKART, R. Z. (1963). Production of an interferon by L-cells infected with Western equine encephalomyelitis virus. *Journal of Bacteriology*, **85**, 556.

LODDO, B., FERRARI, W., BROTZU, G. & SPANEDDA, A. (1962). *In vitro* inhibition of infectivity of polio viruses by guanidine. *Nature, London*, **193**, 97.

LOH, P. C. & OIE, H. K. (1969). Growth characteristics of reovirus type 2. Ultraviolet light inactivated virion preparations and cell death. *Archiv für die gesamte Virusforschung*, **26**, 197.

LOH, P. C. & PAYNE, F. E. (1965). Effect of *p*-fluorophenylalanine on the synthesis of vaccinia virus. *Virology*, **25**, 560.

LONG, W. F. & BURKE, D. C. (1970). The effect of infection with fowl plague virus on protein synthesis in chick embryo cells. *Journal of General Virology*, **6**, 1.

MAIZEL, J. V., JR, WHITE, D. O. & SCHARFF, M. D. (1968). The polypeptides of adenovirus. II. Soluble proteins, cores, top components and the structure of the virion. *Virology*, **36**, 126.

MARTIN, E. M. & WORK, T. S. (1961). The localization of metabolic changes within subcellular fractions of Krebs II mouse-ascites-tumour cells infected with encephalomyocarditis virus. *Biochemical Journal*, **81**, 514.

MARTIN, E. M. & KERR, I. M. (1968). Virus-induced changes in host-cell macromolecular synthesis. In *The Molecular Biology of Viruses*, ed. L. V. Crawford and M. G. P. Stoker, 15. London: Cambridge University Press.

MARTIN, E. M., MALEC, J., SVED, S. & WORK, T. S. (1961). Studies on protein and nucleic acid metabolism in virus infected mammalian cells. I. Encephalomyocarditis virus in Krebs II mouse-ascites-tumour cells. *Biochemical Journal*, **80**, 585.

MONTAGNIER, L. & SAUNDERS, F. K. (1963). Replicative form of encephalomyocarditis virus ribonucleic acid. *Nature, London*, **199**, 664.

MOSS, B. (1968). Inhibition of HeLa cell protein synthesis by the vaccinia virion. *Journal of Virology*, **2**, 1028.

NISHMI, M. & BERNKOPF, H. (1958). The toxic effect of vaccinia virus on leucocytes *in vitro*. *Journal of Immunology*, **81**, 460.

OKADA, Y. & TADOKORO, J. (1963). The distribution of cell fusion capacity among several cell strains of cells caused by HVJ. *Experimental Cell Research*, **32**, 417.

PENMAN, S., SCHERRER, K., BECKER, Y. & DARNELL, J. E. (1963). Polyribosomes in normal and poliovirus-infected HeLa cells and their relationship to messenger RNA. *Proceedings of the National Academy of Sciences of the U.S.A.*, **49**, 654.

PENMAN, S. & SUMMERS, D. (1965). Effects on host cell metabolism following synchronous infection with poliovirus. *Virology*, **27**, 614.

PEREIRA, H. G. (1961). The cytopathic effect of animal virus. *Advances in Virus Research*, **8**, 245.

PEREIRA, H. G. & KELLY, B. (1957). Dose-response curves of toxic and infective actions of adenovirus in HeLa cell cultures. *Journal of General Microbiology*, **17**, 517.

PONS, M. W. (1964). Infectious double-stranded poliovirus RNA. *Virology*, **24**, 467.

PRINCE, A. M. & GINSBERG, H. S. (1957). Immunohistochemical studies on the interaction between ascites tumor cells and Newcastle disease virus. *Journal of Experimental Medicine*, **105**, 177.

REEVE, P. & POSTE, G. (1971). Cell fusion by Newcastle disease virus in the absence of RNA synthesis. *Nature, London*, **229**, 157.

ROWE, W. P., HARTLEY, J. W., ROIZMAN, B. & LEVY, H. B. (1958). Characterization of a factor formed in the course of adenovirus infection of tissue cultures causing detachment of cells from glass. *Journal of Experimental Medicine*, **108**, 713.

SALZMAN, N. P., LOCKART, R. Z. & SEBRING, E. D. (1959). Alterations in HeLa cell metabolism resulting from poliovirus infection. *Virology*, **9**, 244.

SCHERER, W. F. (1952). Agglutination of a pure strain of mammalian cells (L strain, Earle) by suspensions of vaccinia virus. *Proceedings of the Society for Experimental Biology and Medicine*, **80**, 598.

SCHLESINGER, R. W. (1950). Incomplete growth cycle of influenza virus in mouse brain. *Proceedings of the Society for Experimental Biology and Medicine*, **74**, 541.

SCHOLTISSEK, C., BECHT, H. & DRZENIEK, R. (1967). Biochemical studies on the cytopathic effect of influenza viruses. *Journal of General Virology*, **1**, 219.

SCHOLTISSEK, C., ROTT, R., HOUSEN, P., HOUSEN, H. & SCÄFER, W. (1962). Comparative studies of RNA and protein synthesis with a myxovirus and a small polyhedral virus. *Cold Spring Harbor Symposium on Quantitative Biology*, **27**, 245.

SHATKIN, A. J. (1963). Actinomycin D and vaccinia virus infection of HeLa cells. *Nature, London*, **199**, 357.

STURMAN, L. S. & TAMM, I. (1966). Host dependence of G-DVII virus: complete or abortive multiplication in various cell types. *Journal of Immunology*, **97**, 885.

STURMAN, L. S. & TAMM, I. (1969). Formation of viral ribonucleic acid and virus in cells that are permissive or nonpermissive for murine encephalomyelitis virus (G-DVII). *Journal of Virology*, **3**, 8.

SUBASINGHE, H. A. & LOH, P. C. (1971). Some properties of UV-inactivated reovirus type 2. *Bacteriological Proceedings*, 178.

TAMM, I. & EGGERS, H. J. (1963). Specific inhibition of replication of animal viruses. *Science*, **142**, 24.

TOKUMARU, T. (1968). The nature of toxins of herpes virus. I. Syneytial giant cell producing components in tissue culture. *Archiv für die gesamte Virusforschung*, **24**, 104.

VALENTINE, R. C. & PEREIRA, H. G. (1965). Antigens and structure of the adenovirus. *Journal of Molecular Biology*, **13**, 13.

WAGNER, R. R. & HUANG, A. S. (1966). Inhibition of RNA and interferon synthesis in Krebs-2 cells infected with vesicular stomatitis virus. *Virology*, **28**, 1.

WALKER, D. L. (1960). *In vitro* cell–virus relationships resulting in cell death. *Annual Reviews of Microbiology*, **14**, 177.

WALL, R. & TAYLOR, M. W. (1969). Host-dependent restriction of mengovirus replication. *Journal of Virology*, **4**, 681.

WECKER, E., HUMMELER, K. & GOETZ, O. (1962). Relationship between viral RNA and viral protein synthesis. *Virology*, **17**, 110.

WERTZ, G. W. & YOUNGER, J. S. (1970). Interferon production and inhibition of host synthesis in cells infected with vesicular stomatitis virus. *Journal of Virology*, **6**, 476.

WESTWOOD, J. C. N. (1963). Virus pathogenicity. In *Mechanisms of Virus Infection*, ed. W. Smith, 299. New York: Academic Press.

WILLEMS, M. & PENMAN, S. (1966). The mechanism of host cell protein synthesis inhibition by poliovirus. *Virology*, **30**, 335.

WILSON, D. E. (1968). Inhibition of host-cell protein synthesis and ribonucleic acid synthesis by Newcastle disease virus. *Journal of Virology*, **2**, 1.

WOLFF, D. A. & BUBEL, H. C. (1964). The disposition of lysosomal enzymes as related to specific viral cytopathic effects. *Virology*, **24**, 502.

YAMAZAKI, S. & WAGNER, R. R. (1970). Action of interferon: kinetics and differential effects on viral functions. *Journal of Virology*, **6**, 421.

ZIMMERMAN, E. F., HESTER, M. & DARNELL, J. E. (1963). RNA synthesis in poliovirus infected cells. *Virology*, **19**, 400.

ZWEERINK, H. J. & JOKLIK, W. K. (1970). Studies on the intracellular synthesis of reovirus-specified proteins. *Virology*, **41**, 501.

AN ASSESSMENT OF THE ROLE OF THE ALLERGIC RESPONSE IN THE PATHOGENESIS OF VIRAL DISEASES

H. E. WEBB AND J. G. HALL

St Thomas' Hospital, London SE1 and Department of Experimental Pathology, The Medical School, University of Birmingham

INTRODUCTION

In this paper we examine the proposition that the allergic response to infection by viruses may sometimes be more damaging to the host than the cytopathic effects of the virus itself. This idea must at first seem ridiculous to a society that is committed to the practice of demonstrably effective programmes of preventive vaccination and where the life-long immunity that follows the virus diseases of childhood is apparent to the least observant individual. Certainly, it is not part of our argument to suggest that the occasional deleterious effects of the allergic response detract from its usual protective role. The allergic response, like all products of evolution is to some extent a compromise; if it cures 999 individuals and kills one, it still has an enormous value. However, the clinician is not in the position to abandon the treatment of the occasional patient who is paying the price for the minor imperfections of the evolutionary process; he must try and do something to prevent a fatal or crippling outcome. Every day multitudes of people are exposed to potentially pathogenic viruses and only few succumb to clinical disease. It is amongst this latter highly selected group, that represents only a minority of the total number that have had contact with the virus, that occasional individuals may be found whose illness owes more to their allergic response than the primary cytopathic effect of the virus. In other words, the immediate problem is clinical and not necessarily of central importance to the academic biologist or immunologist. Nevertheless, there exist already virus diseases of experimental animals where the morbidity and mortality are almost certainly the consequence of an allergic response.

CLASSIFICATION OF ALLERGIC RESPONSES

Coombs (1968) has stressed that many infecting micro-organisms would show little pathogenicity on their own account in an animal whose allergic responses were completely suppressed. Their pathogenicity, Coombs suggested, is due to the antigenicity of the organisms and their products and consequent tissue-damaging allergic reactions which may occur wherever the surviving organisms or their products happen to be. We propose to examine this concept in relation to virus diseases; and the classification of 'allergic' reactions proposed by Gell & Coombs (1968), Coombs & Smith (1968) and Coombs (1968) will be used. Allergic reactions which afford protection from disease may be said to be 'immune reactions' whereas those allergic reactions which are responsible for disease are the reactions of clinical hypersensitivity. However, the response to viruses is complex; while part of the response may be protective and thus an 'immune' response other, damaging 'allergic' components may co-exist. For 'antigen' or 'allergen' we would like to substitute the virus to see if some of the syndromes met in both clinical and experimental virology can be accounted for by these 'allergic' reactions. Coombs' classification is as follows.

The Type I reaction ('anaphylactic', 'reagin-dependent') is initiated by allergen reacting with tissue mast cells which have become allergised passively by reaginic antibody (probably of the Ig E type), produced elsewhere, leading to the release of pharmacologically active substances such as histamine. The local release of histamine and other vaso-active amines is followed by an increase in capillary permeability with consequent oedema and also by the contraction of smooth muscle. These phenomena are responsible for the familiar signs and symptoms of hay-fever and allergic asthma.

More severe reactions lead to general anaphylactic shock; a profound and sometimes fatal circulatory collapse associated with broncho-spasm, pulmonary oedema and urticaria. Such generalised reactions result from antigen reaching the circulation of a highly sensitised person.

The Type II reaction (cytotoxic) is initiated by humoral IgG or IgM antibody reacting either with an antigenic component of a cell membrane or basement membrane or with an antigen or hapten which has become intimately associated with these. Complement is usually necessary to effect the cellular damage. Examples from clinical medicine, other than those provided by virus diseases, include blood transfusion reactions, haemolytic disease of the newborn, nephrotoxic activity in certain forms of glomerulonephritis and drug reactions such as that seen in Sedormid purpura. A recent development in this category concerns a stimulating activity rather than the cytotoxic effect that is produced usually by complement and antibody to cell membrane antigens. Though the reaction itself is not lethal to the cells the consequences are nonetheless damaging to the tissue and could well underlie certain disease processes. In these studies on chick-bone rudiments in organ culture, this type of allergic reaction leaves the target cells fully viable but stimulates their metabolic activity excessively, the lysosomal system is activated and the enzymes released from the cells completely degrade the cartilage matrix (Fell, Coombs & Dingle, 1966; Dingle, Fell & Coombs, 1967).

The Type III reaction (Arthus-type damage by antigen–antibody complexes) is initiated by antigen reacting in the tissue spaces with antibody to form microprecipitates in and around blood vessels, causing damage to cells and tissues secondarily by a variety of means. This is the situation of the classical Arthus reaction. Alternatively, antigen in excess may react with antibody in the bloodstream to form soluble circulating complexes which are deposited in the blood vessel walls or in basement membranes, causing local inflammation. This occurs in classical serum sickness.

Besides the Arthus reaction and serum sickness, Type III reactions are involved in allergic pulmonary interstitial alveolitis, allergic vasculitis, polyarteritis, certain other forms of glomerulonephritis and in the formation of the tissue lesions in systemic lupus erythematosus and other connective tissue diseases. Obviously the precise histological site where the complexes lodge determines the overall pathology and clinical picture.

An anaphylactic-like syndrome can also be produced by this type of reaction after activation of complement by antigen–antibody complexes and the consequent production of anaphylotoxin and other products of complement activation.

The Type IV reactions are mediated by actively allergised mononuclear leucocytes which infiltrate the site where antigen has become localised.

Clinically, Type IV reactions manifest themselves in contact dermatitis and in delayed skin reactions such as the Mantoux test. They play an important part in rejection of allogeneic grafted tissue and pathologically their involvement can be seen in allergic granulomatous conditions (e.g. tuberculosis) and as a prominent feature of the lesions of some auto-allergic diseases.

The Type IV reaction is clearly one of major importance in a consideration of the pathogenesis of virus disease. Unfortunately, it is also one of the most complex. The basic phenomena surround the destruction or damage of target cells (e.g. tumour cells, allografted normal cells or host cells that have acquired a new, e.g. viral, antigen) by mononuclear cells of the host. For experimental convenience the process has been investigated usually in terms of the destruction of tissue culture or tumour cells *in vitro* by mononuclear cells obtained from the lymphoreticular tissue of actively and specifically allergised donors. Three main types of attacking cells exist. First, there are antibody-producing lymphoid blast cells or 'immunoblasts'. These cells arise in lymphoid tissue within 3 to 4 days of the arrival of an allergen and are discharged into the lymphatic and blood vascular circulations (Hall, 1969). They achieve their cytotoxic effect by virtue of the lytic, complement-dependent antibody which they secrete. Because they are mobile and motile they can enter the extravascular compartment and, by local antibody production, carry out, in a micro-environment, the Type I, II and III reactions already described (Hall, 1969; Grant, Denham, Hall & Alexander, 1970). However, these antibody-producing blasts have a somewhat ephemeral existence and they are soon replaced by a second generation of cytotoxic lymphoid cells (Denham, Grant, Hall & Alexander, 1970) which appear 1 to 3 weeks after the primary arrival of the allergen. These cells do not depend on antibody and complement for their cytotoxic activity, the molecular basis of which is quite unknown. Even their morphological structure is uncertain; probably they are small lymphocytes which can rapidly transform into cytotoxic blast cells on coming in contact with the antigen and there is compelling evidence that, unlike the antibody-producing blasts, they are derived from thymic lymphoid tissue (Cerottini, Nordin & Brunner, 1970). These thymus-derived lymphocytes represent the quintessence of what is usually denoted by the term 'cell-mediated immunity' and it follows that congenital or acquired diseases which damage the thymic lymphoid system must also depress this type of allergic response.

Finally, the third cell that may mediate cytotoxic reactions is the macrophage. Macrophages are the final effectors of that component of cell-mediated immunity that is responsible for overcoming infections by facultative intracellular parasites, such as the chronic infective granulomata in man (Mackaness, 1967). Macrophages may also be the prime movers in the rejection of certain types of tumour (Evans & Alexander, 1970) and it therefore seems likely that actively allergised macrophages may play an important part in the rejection of homografts and the development of delayed hypersensitivity reactions. Again, the molecular basis of the cytotoxicity of macrophages is obscure.

Briefly, Type IV reactions express themselves as localised lesions which are characterised by a cellular exudate (often initially of peri-vascular localisation) composed predominantly of mononuclear cells, although occasional neutrophil, basophil and eosinophil granulocytes may be present also. Since some of the mononuclear cells will be anti-body-producing blast cells it follows that, in the microcosm of the Type IV reaction, elements of the Types I, II and III reactions may also be present and thus the whole 'immunological orchestra' may be assembled at one site.

SPECIAL FEATURES OF ALLERGIC RESPONSES TO VIRUSES

Before considering the interaction of particular viruses with the immune apparatus it will be profitable to consider some general features of this host–parasite relationship and its associated terminology.

Latency

Some virus infections are not eradicated but persist in a so-called latent state (e.g. herpes in humans). A latent infection is characterised by relatively high titres of specific antibody in the blood together with smaller or undetectable amounts of virus. A detectable viraemia, when it occurs at all, is usually associated with some intercurrent episode, such as immuno-suppression with drugs or X-irradiation or another infection.

Tolerance

In animals that are tolerant to a given virus (usually because of infection during the perinatal period before proper immunocompetence has developed) much infective virus may be present in the blood but detectable antibody is usually absent. In fact, contrary to early im-pressions, antibody may indeed be formed but since it is immediately converted to antigen–antibody complexes it is difficult to detect by conventional methods. In spite of the large amounts of virus present the infected animal may be free from obvious disease.

Immunosuppression by viruses

Blumhardt, Pappano & Moyer (1968) showed that following live measles vaccination there was a reduction, for 3 to 6 weeks, of cutaneous sensi-tivity to poison ivy. Mims & Wainwright (1968) using lymphocytic choriomeningitis virus in mice have shown a decreased susceptibility to anaphylaxis with ovalbumen. Mouse leukaemia viruses cause similar effects. These effects may be caused by viruses which are able to cause long term cytopathic infections of the lymphoid tissue. Hanaoka, Suzuki & Hotchin (1969) described apparently selective destruction of thymus-derived lymphocytes and thymocytes in mice with lymphocytic choriomeningitis infections. Olson *et al.* (1968) stated that a number of viruses caused impaired responsiveness of lymphocytes to transformation by phytohaemagglutinin. In fact many experimental animals undergo a period of demonstrable immunodepression following infection with viruses (Notkins, Mergenhagen & Howard, 1970). This is particularly true of oncogenic viruses in mice but the phenomenon is not restricted to these situations. Defects in cellular immunity in man, as judged by the ability of peripheral blood lymphocytes to respond to mitogens (see below), occur in several common virus diseases. The mechanism of such immunosuppression is uncertain and may vary in different situations. Certainly some viruses are known to parasitise preferentially lymphoid blast cells and this could directly reduce the number of antigen-reactive cells. Also, fluctuations of the blood lymphocyte count and sometimes a marked lymphopoenia, are commonplace features of virus infections. Since most circulating blood lymphocytes are believed to be of thymic origin and to initiate cell-mediated immunity, their destruction or sequestration could reduce significantly the cell-mediated allergic response. Hard data on the kinetics of lymphocyte recirculation in virus diseases is in short supply but a single dose of influenza virus can halt the re-circulation of lymphocytes through a single lymph node for 24 h (Smith & Morris, 1970). It is unlikely that the immunosuppression associated with these phenomena is either complete or particularly long-lasting but it may allow the virus a few more hours of unopposed replication and could alter the exact nature and magnitude of the resurgent allergic state.

Nature of immunity to viruses

Both humoral and cell-mediated immunity are needed to eradicate naturally occurring virus diseases (Smith, 1970). The necessity of cell-mediated immunity, especially in the early stages of the disease, has been stressed (Hall, 1969). Although serum antibody can ultimately become a potent anti-viral agent it is unlikely to reach an effective titre for several days after infection and even then physical factors may prevent its transudation into the tissue spaces. The Bruton type of agammaglobulinaemia does not seem to impair recovery from some virus infections, although in the Swiss type there appears to be an inability to limit spread. A cellular immunity, deficiency syndrome has been described which includes normal incidence of bacterial infection and normal levels of antibody to both viruses and bacteria but abnormal susceptibility to virus infections (Di George, 1965).

Certainly, the accumulation of lymphocytes and lymphoid blast cells at the site of virus infection is characteristic of many virus diseases. The protective role of the cell response in several experimental systems has been reviewed (Allison, 1967). As stated earlier, many of the lymphoid cells present during the relatively early stages are concerned with local antibody production and thus are almost by definition excluded from the mysteries of 'pure' cell-mediated immunity. The function of the thymus-derived lymphoid cells which also appear and which do not make antibody is hard to understand in terms of an immunologically protective function. Such cells are concerned with delayed hypersensitivity reactions which may promote the penetration of serum antibody. For example, Brown, Glynn & Holborow (1967) have shown that humoral antibody and delayed-type cellular reactions have a synergistic action in bringing about a localised immune reaction. However, this is not an adequate explanation since there is compelling evidence that even in the absence of humoral antibody, cell-mediated mechanisms may defeat a virus infection, whereas in the absence of cell-mediated mechanisms such an infection may prove fatal even if antibody is present (Burnet, 1968).

The apparent necessity for some elements of cell-mediated immunity must be taken into account also in the immunoprophylaxis of virus disease. Parenteral immunisation with vaccines of killed virus induces titres of IgG and IgM antibodies in the blood which are often high enough to prevent a viraemia and the clinically recognisable disease which would otherwise follow. However, viruses such as poliomyelitis which gain entry by first infecting the epithelium of mucous surfaces can probably be completely excluded from the body by providing the mucous membrane with IgA antibodies in the local secretions. This local antibody production by the plasma cells which lie beneath the mucosa can only be fully stimulated by living attenuated virus which is administered by mouth and which later replicates in the intestinal epithelium (Ogra, Karzon, Righthand & MacGillivray, 1968; Ogra & Karzon, 1969). IgA antibodies would thus seem to be the first line of defence against many viruses and although information on their neutralising ability and their specificity is incomplete their appearance is a desirable sequel of an immunisation procedure. However, they do not appear in the secretions of the respiratory or intestinal tracts following parenteral immunisation. The allegedly beneficial effects of live oral vaccines may go beyond the local production of IgA antibodies. Resistance to facultative intracellular bacterial parasites may only be acquired following immunisation with the living replicating organism and is ascribed to cell-mediated phenomena rather than to humoral antibodies (Mackaness, 1967). Likewise, immunisation with a living replicating virus may be effective by producing a more effective state of cellular immunity than a killed vaccine. However, although satisfying, there is little hard data to support this idea. Nonetheless, it seems proper to consider whether or not a high titre of serum antibody is, on its own, always completely beneficial. Although such antibody can prevent viraemia it may not always be able to prevent initial replication at the site of entry. Thus, in the absence of local immunity, a situation may be created in which a considerable amount of locally produced virus suddenly impinges on a circulatory system that is replete with specific antibody. As detailed later, this can have unfortunate consequences.

The anatomical site of the infection

Many viruses are dangerous because they replicate in the cells of vital organs. Neurotropic viruses provide an example in which it is easy to see how phenomena other than the primary cytopathic effect of the virus may be responsible for the loss of function. The brain and much of the spinal cord are confined by the bones of the skull and vertebral column and they are particularly vulnerable to anoxia and ischaemia. Gross inflammatory oedema, even if it were caused by a potentially beneficial immune reaction, might increase the tissue tension enough to produce a circulatory insufficiency. This may well account for the reversible nature of some of the paretic manifestations of poliomyelitis.

Another situation where the site of virus disease may lead to unusual sequelae is that of 'immunologically privileged' sites. These areas, such as the brain, the anterior chamber of the eye and perhaps the testis, seem normally to be devoid of a detectable immune surveillance mechanism, so that transplantation-type antigens that are confined to these sites do not sensitise the host. However, once sensitisation has occurred, by, e.g. the same antigen being presented to the host via a conventional route, a powerful allergic response may develop in the hitherto protected area. Once this has happened the incoming host lymphocytes may be confronted not only with the foreign antigen but also with normal host antigens which they have never previously encountered and which may thus not be recognised as 'self' components.

This situation may form the basis for some of the apparently 'auto-immune' phenomena which follow some virus disease. However, this question is associated intimately with the behaviour of the so-called 'budding' viruses which not only present their own antigens on the surfaces of host cells but which on leaving a cell carry with them a lipoprotein coat derived from the cytoplasmic membrane of the host cell. In this way a virus may not only induce the formation of new antigens on the surfaces of host cells but may also itself acquire host antigens which, because of their association with the virus, are more immunogenic than they are in their native condition. Because this property may often be crucial to the nature of the allergic response it is considered in some detail.

The budding viruses

Tissue damage can stimulate the production of tissue-specific immunoglobulins. This has been noted in myocardial infarction (Ehrenfeld, Gery & Davies, 1961; Van der Geld, 1964), following burns (Pavkova, 1962), irradiation to the thyroid (Irvine, 1964) and various liver diseases (Mackay & Gadjusek, 1958). It seems even more likely to occur in tissue damage from virus infection because so many viruses form their outer coat either from the cell membrane or from internal constituents of the cell in which they have multiplied. In this way the new virions from the damaged cell might carry antigens which could stimulate the production of anti-cell antibody. This property of viruses and their affinity to grow in neoplastic tissue prompted the use of viruses as oncolytic agents (Webb & Smith, 1970).

Hoyle (1950, 1954) observed filamentous microvilli bulging from cells infected with influenza virus; fluorescent antibody and ferritin-conjugated antibody techniques made it clear that these protrusions contained virus-specific material and that the cell surface was altered. This happens with all the viruses with lipoprotein coats (the lipoviruses) including the arboviruses, myxoviruses and lymphocytic-choriomeningitis virus; see the pictures of Murphy, Webb, Johnson & Whitfield (1969) for machupo virus. Harboe, Schoyen & Bye-Hansen (1966) showed that the haemagglutinin of fowl plague and influenza virus which had grown in the endodermal cells of the chick chorioallantoic membrane was inhibited by antibody to normal chick allantoic material. Similarly, in cells infected with avian leucosis sarcoma virus, the budding virus particles incorporate part of the cell into their own new structure (Morgan, 1968) and herpes virus contains antigenic determinants of the host cell (Watson & Wildy, 1963). In short, many workers have concluded that immunological alterations specified by the viral genome take place at the surface of infected cells (Rose & Morgan, 1960; Marcus, 1962; Duc-Nguyen, Rose & Morgan, 1966).

Almeida & Waterson (1969) review the morphology of virus–antibody interaction at the ultrastructural level, particularly their own experiments with avian infectious bronchitis virus, a pleomorphic virus which displays distinctive projections. They made complexes of virus with two types of antisera; (1) homotypic antiserum prepared by immunising chickens, and (2) heterotypic antiserum prepared in rabbits (Berry & Almeida, 1968). Homotypic antibody was attached only to the projections whereas the heterotypic antibody was attached to the projections and also to the envelope of the virus. This suggested that, whereas the projections contained the virus antigen, the envelope had chicken material in it and therefore did not elicit a host response and would only react with the heterotypic antiserum. In other words, the chicken-derived envelope of the virus was antigenic to the rabbit and therefore the heterotypic rabbit serum not only reacted with the projections but also the envelope. Finally serum from rabbits immunised against normal chick embryo fibroblasts neutralised the avian infectious bronchitis virus and the antibody actually attached to the envelope of the virus.

An electron microscope study of the destruction of virus by antibody has shown the initial event to be a 'pitting' of the envelope, but this occurred only if complement was present. In other words, the event is similar to the complement-dependent action of haemolytic antibodies on red cells that has been demonstrated at an ultrastructural level by Borsos, Dourmashkin & Humphrey (1964). It is mainly the C3 component of complement that is visualised in this way (Coombs & Lachmann, 1968) i.e. the component that has been implicated in the pathogenesis of lymphocytic choriomeningitis in mice (see below).

Obviously this type of reaction can protect the host insofar as it destroys the virus, but it may also have a damaging effect on the cells of the host. There are two similar ways in which an auto-allergic reaction may be induced. Firstly host cell membrane may become incorporated into new virions and, secondly, the process of viral infection may induce the formation of neo-antigens on the surface of host cell. In either case the association of viral antigens with normal membrane may cause normal membrane determinants to become immunogenic and, later, the target of immune attack so that the resulting allergic response may damage normal uninfected cells. This phenomenon has been discussed by Rook & Webb (1970) who showed that lymphocytes from animals that had been sensitised to Langat virus destroyed non-neuronal cells of brain that were infected with the virus but also damaged non-infected cells as well. It follows that lymphocytes from an animal that has recovered from an infection with an encephalitogenic virus that behaves in this way, may be sensitised against normal brain cells of a syngeneic animal. Wiktor, Kuwert and Kaprowski (1968) found that rabies-infected cells were lysed by antibody even before virus budding could be observed and Smith (1970) suggests that this may be an important factor in limiting virus spread. All these phenomena tend to incriminate the Type II reaction which is usually cytotoxic and mediated by antibody (either derived from the serum or locally produced by an infiltrate of lymphoid cells) reacting either with an antigenic component of a cell membrane or basement membrane or with a virus antigen which has become intimately associated with these. Complement is usually necessary. However, in this category Fell *et al.* (1966) and Dingle *et al.* (1967) have found the *stimulating* rather than a cytotoxic activity produced by complement and antibody to cell membrane antigens. The consequences, though not damaging to the cells themselves may be destructive to adjoining tissue. This mechanism might apply to the destruction of liver cells that is seen in the proliferative cirrhosis of partially immune dogs infected with canine hepatitis virus (Gocke *et al.* 1967). Illavia & Webb (1970) and Zlotnik (1968) have noted the proliferative effect certain encephalitogenic viruses appear to have on glia *in vitro* and *in vivo*. In the situation *in vivo* a Type II allergic response may be contributory to the proliferative gliosis seen. This proliferation is a feature of many chronic central nervous system diseases which are thought to have a viral aetiology, e.g. sub-acute sclerosing panencephalitis, progressive multifocal leuco-encephalopathy, Creutzfeldt–Jakob's disease, kuru, scrapie and disseminated sclerosis. Many of the neurotropic viruses are 'budding' viruses and could well initiate a Type II response of the proliferative nature. Nonetheless, proliferation can occur in the absence of an allergic response. Illavia & Webb (1970, 1971) and Webb, Illavia & Laurence (1971) found a strongly proliferative effect on the non-neuronal cells of the central nervous system with many different potentially neurotropic viruses in experimental situations where the allergic response could be excluded.

Viruses which multiply specifically in, or attach themselves to, lymphocytes in baby animals probably create a situation in which the lymphocytes cannot mount a cell mediated response to that virus, i.e. they become tolerant. This seems to apply to the visceral strain of lymphocytic-choriomeningitis virus but not to the more aggressive cerebral strain. The latter has arisen from serial passage through cerebral cells and may not be adapted for growth in lymphocytes. Lymphocytic-choriomeningitis is a budding virus (Murphy *et al.* 1969), and after many passages through cerebral cells its coat will probably have in it specific components of the cytoplasmic membrane of brain cells. The antigenic determinants of these components will be distinct from that of the virus itself and so may initiate a specific allergic response that is directed against autochthonous brain tissue. In this way a cell-mediated response could be mounted against cerebral cells whether they contained virus or not. However, antigens (particularly of the transplantation type) which are confined to the brain are sometimes less immunogenic than they would be in other parts of the body which have a conventional lymphatic drainage. For example, tissue grafts placed in the brain take much longer to be rejected than in other sites, indicating how foreign cells in the brain fail to sensitise the host effectively so that little allergic reaction is generated. However, when the host is sensitised to the graft antigens following challenge via a conventional route, an allergic response develops, the allergised lymphoid cells are conveyed to the brain via the blood, and attack the graft that had not previously been molested.

Simons (1968) has suggested that rubella virus persists in congenitally infected babies as a tolerated infection of long-lived lymphocytes. Although humoral antibody is present in these infants they do not seem to be able to mount a cell-mediated response and although they do not fall ill as a result of their infection they continually excrete virus so that unfortunately those who look after them frequently become infected. True tolerance appears to occur in at least one natural infection (machupo virus) of wild animals (Justines & Johnson, 1969). Certainly, in this infection the virus has been shown to multiply in and have a close relationship to the lymphocyte (Murphy *et al.* 1969) and it is tempting to believe that this induces tolerance.

There is no doubt that viruses themselves destroy cells in many tissue culture systems *in vitro* (Bablanian, this volume). That they can do it *in vivo* also is certain and considerable destruction probably occurs by this process in many virus infections. However, with the examples of lymphocytic-choriomeningitis, machupo and junin viruses, it is clear that allergic reactions which may give rise to clinical hypersensitivity also play an important part in causing tissue damage. It is generally accepted that measles and vaccinia encephalitis occurring 1–2 weeks after infection accords well with the concept that allergic reactions directed against the virus are responsible for much of the disease process. We cannot distinguish between these and other neurotropic infections in which encephalitis occurs. The time intervals are just the same, the only difference is that the viraemia and infective phase goes undetected because of no obvious clinical exanthemata. Moreover, it seems reasonable to suggest that the orchitis, öophoritis, pancreatitis and thyroiditis which occur 1–4 weeks after a virus infection may have a similar cause, i.e. an allergic reaction induced by a virus infection.

EXAMPLES FROM VIRUS DISEASES IN WHICH THE ALLERGIC RESPONSES APPEAR TO BE HARMFUL

In man
Measles

The administration of dead measles vaccine to children may be followed by an atypical clinical response when the children later experience either naturally occurring measles or the live measles vaccine (Fulginiti, Eller, Downie & Kempe, 1967; Buser, 1967; McNair Scott & Bonanno, 1967). Alteration in the rash, pain in the head and abdomen, peripheral oedema, pneumonia, and occasionally pleural effusions were noted, as also were signs of central nervous system involvement. The usual viraemia was conceivably prevented to some extent because of the level of serum antibody, but the absence of an optimal cell-mediated immunity response allowed the virus to multiply in the target tissue. Later, when virus was released into the circulation it combined with the pre-existing serum antibody and caused anaphylactic/arthus/serum sickness reactions.

Smallpox

Dixon (1962) drew attention to a 'pulmonary allergy', often very crippling in nature, with severe bronchospasm, which tends to affect persons vaccinated against smallpox when they later come into contact with cases of that disease.

Virus pneumonia from respiratory syncitial virus

This pneumonia tends to be more severe in the first two months of life and this has been attributed to an Arthus-type reaction occurring between a large amount of virus and a low level of circulating antibody possibly transferred from the mother (Chanock *et al.* 1967). This would be a Type III reaction and could produce considerable bronchial constriction with all the symptoms of bronchospasm and obstructive air entry into the lungs. Lennon & Isacson (1967) have observed a delayed-type hypersensitivity in similar circumstances.

Parrott *et al.* (1967) found that children immunised with a poorly immunogenic respiratory syncytial virus (RSV) vaccine developed a more severe disease when infected with a live wild RSV at a future date. This also suggests a Type I and III reaction. But here, because a similar antigen has been given originally, a Type IV response may also be involved as the cell-mediated immune response will have been activated also. Other factors may also be involved. Blandford (1970) has suggested that cellular debris resulting from a widespread Arthus reaction is inhaled and causes an obstruction of the bronchioles which compounds the pneumonic process. Gardner, McQuillin & Court (1970) suggest that the bronchiolitis is due to a second infection with the virus which causes an anaphylactic reaction; because, although in RSV pneumonia the virus is abundant in the tissues, it is scanty in bronchiolitis where immunoglobin can be detected in the tissues with approximately the same distribution as the virus. In this situation the anaphylactic reaction could arise from a Type I and/or a Type III reaction and/or Type IV reaction.

Yellow fever and dengue

In yellow fever the virus antigen present in the circulation (Hughes, 1933) causes anaphylaxis in passively sensitised animals (Davis, 1931), a Type I reaction again. There is no reason why similar conditions should not be present in other virus infections. Situations have certainly been recorded in which a second exposure to a given virus (or an antigenically similar virus) has been followed by severe clinical disease with haemorrhages and hypotensive shock. Halstead (1966) reported a severe disease of this nature following infection with Dengue virus and which has caused many deaths in children in south-east Asia since 1958. Dengue is usually unpleasant but not dangerous. Seriously affected children appear to have an accelerated immunological response and during the clinical crisis, complement (β_1 c/a) is consumed and disappears from the circulation. The secondarily infected patients had an IgG but not an IgM or IgA response to the infection (Russell, Intavivat & Kanchanapilant, 1969). These features are consistent with the idea that a previous infection with a related virus caused the patients either to have pre-existing antibodies or to mount a rapid secondary allergic response specific enough to provide cross-reacting antibodies which caused Types I–III reactions.

Mumps

This infection is common, and though usually mild it may occasionally be very severe with complications such as pancreatitis, orchitis, öophoritis and meningo-encephalitis. Many people with clinical mumps have abnormal cerebrospinal fluids without any other suggestion of meningeal involvement except possibly a headache. Johnson (1968) found no association between the severity of the mumps encephalitis and the titre of specific antibody in the cerebrospinal fluid and suggested that a type of delayed hypersensitivity was responsible for the inflammation. Possibly a similar immunological reaction accounts for the orchitis, öophoritis and pancreatitis which sometimes occur in this disease. Similarly, Volpe, Row & Ezrin (1967) have clearly shown the relationship between thyroid damage caused by virus infection and the subsequent development of thyroid auto-immunity. Sub-acute thyroiditis in man was found to be associated with high titres of antibodies to influenza, mumps, coxsackie, adeno- and echoviruses and these titres fell in the months that followed the disappearance of clinical disease. However, it was during this later phase that anti-thyroid auto-antibodies were detected. In one patient the titre of anti-thyroid auto-antibody rose to a high level, at the same time as clinical hypothyroidism occurred. The occasional occurrence of thyrotoxicosis in epidemic form (Iversen, 1948) suggests that infective agents sometimes contain an antigen capable of stimulating the formation of thyroid stimulating auto-antibody. All these reactions suggest that a Type II reaction takes place. It may also be stimulatory (see above, p. 385).

Infections with coxsackie B *virus*

The importance of these viruses has recently been recognised in the pathogenesis of myocarditis (Sainani, Krompotic & Slodki, 1968). In some patients prolonged heart disease occurred beginning 1–2 weeks after onset of the infection. Adams (1969) suggested that in this condition the breakdown of heart muscle releases antigens which are normally concealed so that an auto-allergic response is mounted against the myocardium. Coxsackie virus may also induce a new antigen in the heart muscle or take some of the host cell lipo-protein into its coat and thus present the host's allergic mechanisms with an auto-antigen so that an anti-heart-cell response is induced, again another Type II response.

Skin rashes in virus infections

Mims (1966) stated that skin rashes in virus diseases depend on both humoral and cell-mediated responses. In pox virus infection, the pock itself is caused by virus multiplication in the skin. At first there is cellular proliferation and then lysis. The cellular proliferation may be a direct stimulatory effect of virus on the cell, as seen with some of those viruses when they cause a tumour in a suitable host. On the other hand, the proliferative effect could be from the immunological proliferative effect of the Type II reaction described by Fell *et al.* (1966) and Dingle *et al.* (1967). This follows the reaction of complement and antibody with cell membrane antigens. The lysis seen in pox infections could result from a similar effect. In man the pox take several days to develop, and the surrounding erythema does not become apparent until the 7th day, so there is time after experiencing this type of virus infection for an allergic response to be mounted. This view is supported by the experiments of Flick & Pincus (1963) who showed that in immunologically tolerant rabbits, in which no detectable antibody developed, the skin lesions were either absent or abortive; nonetheless, Pincus & Flick (1963*a*) showed that 4 days after infection a cutaneous delayed-type hypersensitivity could be elicited. They also found that the papule and vesicle, but not the erythema, were prevented by the intradermal administration of an antiserum against mononuclear cells (Pincus & Flick, 1963*b*).

In experimental animals

Canine infective hepatitis

Gocke *et al.* (1967) have found that non-immune dogs given this virus developed acute hepatic necrosis. On the other hand, dogs which had a low level of immunity before infection developed a slow and insidious disease with intense infiltration of the perivascular spaces of the liver with mononuclear cells and finally an intense cirrhosis. The acute damage was done by primary multiplication of virus in the liver cells before an allergic reaction would occur, whereas the chronic process which ultimately was just as lethal, was the consequence of a chronic allergic reaction certainly involving Type IV and probably Types I–III reactions.

Disease of New Zealand B mice

These mice were thought to suffer from an hereditary auto-immune disorder which led to the development of a haemolytic anaemia and glomerulonephritis (Bielschowsky, Helyer & Howie, 1959; Howie & Helyer, 1968; Russell, Hicks & Burnet, 1966). It now appears that this condition is caused by a virus (Holborow & Denman, 1968). East, de Sousa, Prosser & Jaquet (1967) have demonstrated virus particles, and Mellors & Huang (1967) transmitted the disease to Swiss mice by the injection of cell-free virus particles obtained from the spleens of affected New Zealand B mice. There may be many such latent viruses in apparently normal tissue which are sterile by conventional techniques of isolation (Rogers, Basnight, Gibbs & Gajdusek, 1967). If these viruses were 'budding' viruses then a continuous shower of cell membrane-coated virus particles would be presented to the body's allergic systems throughout life. The normally weak host antigens may be strengthened by their association with the virus particle and thus antibodies may be developed against them, provoking an auto-allergic phenomenon.

Kyasanur Forest disease

Webb & Burston (1966) first tried to explain the shock-like state in monkeys infected with the virus of Kyasanur Forest disease by a disturbance of the autonomic nervous system. These animals became hypotensive with a bradycardia, diarrhoea and peripheral vascular collapse 24 h before death, which occurred about the 10th to 12th day just as the viraemia was terminating. At the height of their viraemia the blood of these animals contained up to 10^9 virions/ml. Early in the infection, when antigen is grossly in excess of antibody the complexes are probably small and soluble. As antibody increases, larger 'insoluble' complexes are probably formed, so that a Type III reaction is possible and damage may be caused by microprecipitates from antigen–antibody reactions occurring in and around vessel walls (the Arthus type reaction). Later, the deposition of complexes in vessel walls and basement membranes may be followed by local inflammation and serum sickness. Also after activation of complement by antigen–antibody complexes anaphylatoxin can be produced. These allergic reactions may account for the clinical signs seen in susceptible monkeys infected with Kyasanur Forest disease virus rather than a primary effect on the autonomic nervous system. This view is supported by the observation that these monkeys develop agglutinins against their own white blood cells, platelets and red cells (Webb & Chatterjea, 1962). McKay & Margaretten (1967) suggested that the deposition of immune complexes on vessel walls probably stimulates acute disseminated intravascular coagulation. In acute herpetic encephalitis there is a necrotising arteriolitis which may be caused by a similar type of reaction.

Blue tongue virus of sheep

Richards & Cordy (1967) have shown that the foetuses of pregnant ewes which had been infected with the virus before the 9th to 10th week of gestation developed either necrotic lesions or congenital abnormalities. When the ewes were infected later in pregnancy their foetuses developed predominantly inflammatory lesions. In sheep the ability to mount an allergic response develops about the 11th week of gestation. These findings suggest that before the foetus develops immune competence the damage is done primarily by the direct cytopathic effect of the virus. After the foetus has developed immune competence, the lesions resulting from virus infection are a combination of this direct cytopathic effect and allergic reactions to the virus antigen.

IMMUNOSUPPRESSANTS AND THE IMMUNOLOGICAL ROLE IN THE PATHOGENESIS OF VIRUS DISEASE

The use of immunodepressants in clinical practice for the therapy of cancer and transplant surgery can cause the activation of latent bacterial and virus infections. Montgomerie *et al.* (1969) considers 4 fatal cases of generalised herpes simplex in renal transplant patients who had received azathioprine and prednisone. All the patients had pre-existing, complement-fixing antibody to the virus, suggesting that the herpes simplex virus was present in a latent state. The effects of immunosuppression in experimentally-induced virus disease of the central nervous system have been reviewed (Nathanson & Cole, 1970). The effects of the more common agents of immunosuppression such as X-irradiation, cyclophosphamide, cortisone, anti-lymphocytic serum, thymectomy and 6-thioguanine are summed up as follows.

(1) The levels of virus in blood, brain and other tissues was elevated and virus appeared earlier and persisted for longer periods as compared to controls.

(2) More cells were eventually infected before death and a greater number appeared destroyed. Sub-clinical infections became symptomatic and often fatal.

(3) The appearance of antibody in serum or tissues is retarded and death may intervene before antibody is detected by the techniques they use.

(4) Interferon levels are often directly related to virus titre and may be higher in tissues of suppressed animals than in infected but unsuppressed controls.

(5) The physiological derangements produced by many immunosuppressive techniques usually, but not always, reduce the interferon response.

Many workers feel that immunodepressants allow viruses to multiply longer and in more cells, thereby causing a primary destruction of cells by the virus in its process of multiplication. This may well be, but there is enough conflicting evidence to question whether a primary cytopathic effect or an allergic response plays the more important part. With a primary cytopathic effect it is particularly difficult to explain why, if a lethal dose of virus is given to control and immunosuppressed mice, the immunosuppressed animals usually survive significantly longer than the controls, although more virus is present for longer in brain and blood (Webb et al. 1968b; Rook & Webb, 1970). Also, when animals are infected with a dose of virus which only kills some controls, those controls which die, frequently do so before the immunosuppressed animals, even though in the immunosuppressed animals there is eventually a higher death rate. The effect of methotrexate on lympho-cytic-choriomeningitis-infected mice is also of some interest. Severe histological involvement of the central nervous system was found in methotrexate-treated mice surviving after infection with an otherwise lethal dose of lymphocytic-choriomeningitis virus. Mice given this immunosuppressive drug remained clinically healthy, despite the presence of extensive acute or chronic lesions comparable to those seen in untreated animals (Lerner & Haas, 1958). Stone, Lerner & Goode (1969) make a critical distinction between the histological and clinical aspects of auto-immune encephalitis. They produced in a strain of guinea-pig (2/N) a histological pattern of severe, acute or chronic auto-immune encephalitis, though the animals remained overtly healthy. They attribute the failure of these animals to show clinical disease to a 'deficiency' at a late stage in the pathogenesis of auto-immune encephal-itis. They postulated, in most strains of the animal some 'paralysing' material directly released from lymphoid cells as the final effector agent, i.e. a 'lymphotoxin' similar to that described by Granger & Kolb (1968) and Williams & Granger (1969) which is directly damaging to neuronal cells. This stresses again the possible role of the lymphocyte, suppressed in the above case by methotrexate, in the production of the clinical disease.

IMMUNOLOGICAL RESPONSE TO LYMPHOCYTIC-CHORIOMENINGITIS VIRUS IN THE PATHOGENESIS OF DISEASE OF THE CENTRAL NERVOUS SYSTEM

This virus produces acute encephalitis but it may persist in a latent state and can produce tolerance. The distinction is as follows. In latency the virus titre is low, antibody is present and, although the virus is not readily apparent, it can be activated by immunosuppressant drugs, ultraviolet light, X-rays and anti-lymphocytic serum. This is similar to herpes simplex virus and to other viruses. In a persistent tolerant in-fection the virus titre is high and fully infective for other animals; free antibody against virus appears to be absent but in fact antibody bound to circulating infective virus can be demonstrated (Oldstone & Dixon, 1969). This persistent lymphocytic-choriomeningitis infection with antigen excess can result in immune complex disease in the form of glomerulonephritis. This was originally called 'late' disease by Hotchin, Benson & Seamer (1962) and Hotchin (1965), as it may take many months to develop. Hirsch, Murphy & Hicklin (1968) found it could be accelerated under circumstances which favour the formation of high levels of circulating antigen–antibody complexes (Hirsch & Murphy, 1968). Oldstone & Dixon (1967, 1969) have shown that the glomerulo-nephritis, the cardinal feature of this disease, is associated with deposits of complexes of lymphocytic-choriomeningitis virus, antibody and complement fractions just outside the basement membrane of the glomerular capillary. The study of this tolerant state has considerable importance and relevance to the role which the immune response may play in the pathogenesis of viral disease. Animals with neonatal 'tolerant' infections can be 'cured' to a large extent by grafting immune isologous lymphoid cells. However, traces of virus can usually be detected after this procedure, in spite of high titres of circulating neutralising and complement-fixing antibody (Volkert & Larsen, 1964, 1965). The 'balance' of antigen–antibody reactions seems important; it is possible to create a situation in which both infectious virus and cerebrospinal fluid antibody can be detected in the animal over long periods of time (Hirsch, Murphy & Hicklin, 1968).

In mice that are more than one week old the intracerebral inoculation of virus causes an acute disease, a severe encephalitis which appears about one week later. Histological examination shows a severe chorio-meningitis consisting primarily of lymphocytes and other mononuclear cells (Findlay & Stern, 1936; Lillie & Armstrong, 1945). Cole, Gilden, Monjan & Nathanson (1971) have shown that adult mice immuno-suppressed with cyclophosphamide on the 3rd day of infection develop a persistent life-long infection which resembles the lymphocytic-chorio-meningitis tolerant state. These mice will die of a fatal encephalitis if they are given spleen cells ($\pm 10^8$) from syngeneic adult mice immunised with lymphocytic-choriomeningitis virus. This encephalitis is very similar on clinical and histological grounds to that caused acutely by lymphocytic-choriomeningitis virus inoculated intracerebrally into normal adult mice. Comparative studies of lymphocytic-choriomeningitis-virus infection in the central nervous systems of suckling mice and rats, infected at various times after birth, showed that both species develop an age-related, immune-mediated granule cell necrosis of the cerebellum.

Different strains of virus have different virulence (Hotchin *et al.* 1962; Hotchin, 1965). Virus passaged by intracerebral inoculation kills much more acutely than a viscerally passaged virus which is more likely to produce a latent or tolerant infection. The strain of mice and their age plays a large role in determining the outcome of the infection. However, as a rule, virus replicates in both fatally affected and surviving animals, but in groups with a high survival ratio the brain titres are relatively lower and choriomeningitis, if present, is mild.

In relation to the role of the immune response in pathogenesis, the important points are as follows. In the acute disease the titre of virus in the brain correlates with the incidence of disease – suggesting that the more target cells are infected, the more likely it is that clinical disease will occur. Lymphocytic-choriomeningitis undoubtedly depresses the cell-mediated immunity response, and with the late onset of cell-mediated immunity more target cells will be infected by the time the cell-mediated immunity reaction is mounted. Although the cell-mediated immunity response may be protective if mounted early, when few target cells are infected, it may be extremely damaging if mounted late at a time when many cells are involved. Billingham (1969) has stated that only a few cells of the allergic response are needed to cause destruction in a cell-mediated response. In general, however, it seems clear that in virus infections a viraemia of high titre is associated with an increased infection and risk of damage to the target organ although there are some exceptions to this as seen with the pathogenic and non-pathogenic strains of Semliki Forest virus (Bradish, Allner & Maber, 1971; Pusztai, Gould & Smith, 1971). With the definitely neurotropic viruses the high viraemia not only increases the chance of invasion but also increases the number of infected target cells and thus to more chance of a damaging anti-target cell immunological response being initiated by the virions produced.

In persistent tolerant infection of mice, large quantities of virus are continually present in the brain but this does not correlate with clinical disease, and little cell-mediated or humoral antibody response can be detected. Moreover, it is difficult to produce tolerance with a neurotropic strain of lymphocytic-choriomeningitis virus, possibly because it has a strong antigenic determinant from brain tissue in its coat which would set up an anti-brain allergic reaction resulting in the encephalitic picture seen in the acute disease. In the visceral-adapted strain the virus coat would have cell membrane from host cells already recognised as 'self', so no anti-cell action would be set up. The immunodepressant effect of the virus itself acting on the lymphocytes would aid the production of tolerance. Immunological responses may develop to the brain cell antigen in virus coats, because brain cells are in an immuno-logically privileged site, and they are cells of greater longevity than most other cells of the body (see above); the breakdown products of the cells, therefore, will not be continually 'reminding' the hosts' immunological mechanisms that they are 'self' and perhaps they are more likely to be treated as foreign material. Therefore we suggest that tolerance to lymphocytic-choriomeningitis is initiated in the newborn, because at that age if brain cell membrane in the coat of disseminated virus is presented to the allergic mechanisms, it will be considered as 'self'. This auto-antigen will continue to be produced as the virus goes on multiplying in the brain and elsewhere, and so tolerance to brain will be reinforced. Similarly, immunodepressants, X-irradiation, anti-lymphocytic serum and drugs will suppress adequately the anti-brain immunological response so that a persistent tolerant state can be achieved.

Clearly, an allergic process plays an important part in the patho-genesis of lymphocytic-choriomeningitis infections. There is further evidence to support this. Skin grafts from mice chronically infected with lymphocytic choriomeningitis virus are rejected by uninfected syngeneic recipients (Holtermann & Majde, 1969). This suggests that a virus-specific surface antigen on the cells of the skin graft is initiating an immunological response which is sufficient to cause the rejection of the graft.

In culture, infected cells can remain viable while supporting virus replication over a long time (Benson & Hotchin, 1960; Benda & Cinatl, 1962; Seamer, 1965; Lehmann-Grube, 1967). Such cells contain virus-specific antigenic determinants in their plasma membranes which can bind anti-viral antibody (Dalton *et al.* 1968), thereby activating com-plement and causing cellular injury (Oldstone & Dixon, 1970). They suffer cytotoxic effects with lymphoid cells from animals which have been immunised against lymphocytic-choriomeningitis virus (Volkert & Lundstedt, 1968; Oldstone & Dixon, 1970).

CONCLUSION

The evidence quoted in this paper clearly suggests that the allergic response can and does add to the morbidity and mortality in virus diseases. It seems to us that the relative roles of the primary effect of virus on the cell and the secondary destruction caused by the virus pro-voking an allergic response cannot yet be assessed. However, certain major points require explanation, perhaps by giving thought to these some useful ideas may be gained.

Firstly, how can we explain the fact that when a lethal dose of virus is given to immunosuppressed animals they survive longer than controls (Rook & Webb, 1970; Webb, Wight, Platt & Smith, 1968a; Webb et al. 1968b; Berlin, 1964; Goldberg, Brodie & Stanley, 1935; Dubin, Baylin & Goebel, 1946; Nathanson & Cole, 1970)? These authors used either X-irradiation, anti-lymphocytic serum or cyclophosphamide and provided evidence that the suppression of the cell-mediated response plays a large part in the survival mechanism. Berlin (1964) showed the increased survival of immunosuppressed mice inoculated intranasally with a lethal strain of influenza virus and commented specifically on the quantitatively greater presence of virus in the lungs, but a marked reduction in the lung lesions, of the immunosuppressed, infected animals. The controls showed pneumonic consolidation due to an inflammatory reaction. The irradiated virus-infected mice which died had no evidence of lung pathology. In contrast to these results, the same immunosuppressants can increase the mortality particularly when a relatively harmless virus is given to the mice (Robinson, Cureton & Heath, 1969; Nathanson & Cole, 1970). A hypothesis can be put forward to explain these apparently opposite effects. When a virus of low virulence is given to immunosuppressed animals there is considerable delay in mounting the allergic response, virus can multiply uninhibited in cells, possibly damaging them by a primary action for a much longer time. Such a damaging effect is not, in fact, borne out in most instances by a quantitative/histological study. But the useful functioning of infected cells can be destroyed without histological changes being apparent. Before any allergic response appears a vast number of 'target' cells are infected and a considerable amount of antigen is present. If at this stage there is even a limited cell-mediated response and humoral antibody release, the type of allergic reaction with excess antigen–little antibody can occur. This can be severe, and in itself productive of anaphylaxis. Without the immunosuppression the quick response of the allergic mechanisms, both cell-mediated and humoral, will have prevented many 'target' cells becoming involved and an excess antigen–little antibody situation will not develop. The prolongation by immunosuppressants of survival from a lethal dose of virus is likely to be due to a delay in mounting the allergic response. There is the same amount of virus, or more, present for longer in the target organs of the immunosuppressed animals than control animals. This suggests that the extent of virus multiplication in the cells is not the sole cause of their disruption and loss of function. The good clinical state observed in these mice before death is not being affected by the virus multiplication in the cells. This is supported by the state of tolerance observed with viruses such as lymphocytic-choriomeningitis and machupo. Here large amounts of virus can be produced from the target organ for great lengths of time without the development of obvious histological lesions. We do not believe that lymphocytic-choriomeningitis virus is different from others, except in that it has a more severe effect on the cells of the allergic response and therefore is more immunodepressant. It is, perhaps, this action that makes it such a potent 'tolerance' producer. The initial phase of virus-induced immuno-suppression or the presentation of conjugated host-virus antigens deviate and pervert the immune response and there may be many other factors of which we are ignorant.

Even if this concept of a deleterious immune response proves un-acceptable, it does not follow that all the lesions of virus diseases can be attributed logically to the primary cytopathic effect. There are many examples of virus diseases in animals where, for example, major paresis may be present in the absence of any detectable histological abnormality in the infected nerve cells. Thus, virus infections may induce non-lethal reversible biochemical lesions that can abrogate function without affecting structure. Conversely, some animals whose nervous tissue presents the stigmata of severe cytopathic effects are yet without any detectable loss of function (Lerner & Haas, 1958). A consideration of this problem is beyond the scope of this paper, but neurophysiological techniques should be applied to the study of the effects of neurotrophic viruses on the function of nerve cells and similar techniques for other virus infections.

At the beginning we suggested that the allergic responses to virus infections in man could lead to a crisis in the disease that might have a serious or even fatal outcome. Naturally, the question arises as to whether the use of immunosuppressive agents as a therapeutic pro-cedure is justified in such situations. This question must be approached with caution. It can be argued that the slightly extended survival of immunosuppressed experimental animals that occurs after the adminis-tration of a *lethal* dose of virus is irrelevant; both the test animals and the controls are doomed and their death is only a matter of time. On the other hand where a normally sub-lethal dose of a virus is given, immuno-suppression increases the morbidity and is frequently fatal. At first sight these are compelling reasons for dismissing immunosuppressives as therapeutic agents. However, it must be remembered that the immuno-suppressive agents are given to experimental animals in large doses, early in the disease or even before infection with the virus has occurred. Once an allergic or immune response has been established it is much less susceptible to immunosuppressive agents than in this early, inductive phase. The types of clinical crisis with which we are concerned occur when a full-blown allergic state has already been established. At such a time moderate immunosuppressive therapy can do little to alter the fundamental structure of the response but the judicious use of, for example, steroids can ameliorate the more extreme manifestations of the inflammatory processes which are triggered by allergic mechanisms. The mechanism of action of steroids in these situations is ill-understood and in the relevant dose range their anti-inflammatory action is probably more important than immunosuppression *per se*. The investigation of such a therapeutic protocol is difficult in experimental animals; the situation is analagous to that of organ homotransplants. In man, for ethical and logistic reasons, it is often impossible to give the high doses of immunosuppressives that are needed to suppress graft rejection in experimental animals; even so, treatment which in animal terms is homeopathic is nonetheless sometimes beneficial in man. Perhaps a similar situation can be anticipated in the context of virus disease.

Although our simple view of the allergic response in the pathogenesis of viral disease may have something to recommend it at the superficial level where generalisations are appropriate, it must necessarily become inadequate when particular cases are considered in depth. In any situation where the known is heavily outweighed by the unknown it is simple to propose a theory that will fit the small number of available facts. It does not follow that any real progress has been made. Although we have inclined to the view that the destructive effects of an allergic response may be sometimes more damaging than the primary cytopathic effect of the virus itself, we are well aware that in either case the details of the process are largely unknown. Thus it is hardly feasible to attempt to predict in detail what will happen in an outbred large primate when a replicating, pathogenic virus gains entry by any one of several possible routes, parasitises the cells of the allergic apparatus and incorporates host antigens into its envelope. The answer can only come from direct experiment. Where antibody production is concerned, such an approach is feasible, albeit laborious. However, in the case of cell-mediated immunity the situation is confused and, for technical reasons, less susceptible to investigation by precise, quantitative experiments. All we can say is that cell-mediated responses seem to be crucial in allergic reactions in some virus diseases and to be more susceptible to either endogenous or exogenous immunosuppression than the production of humoral antibodies. A real understanding of the problem will remain impossible until the mysteries of cell-mediated immunity have been unravelled.

Despite the large areas of uncertainty we feel that a number of lesions associated with virus diseases result from an allergic response. In feeling this, we do not wish to suggest that the response of the allergic apparatus is necessarily either 'good' or 'bad'. One would like to think of it always being 'good' but clearly it sometimes gets out of step and this results in a very complex and rapid sequence of events that can be damaging. Not only each species of animal, but each individual animal of that species, will develop a different allergic response. Even the immune response of an inbred rodent to a non-replicating, non-pathogenic antigen is a mystery. A whole spectrum of specific immunoglobulin molecules of differing avidities, structure and function will appear at different times, in different concentrations in different tissue fluids. These inter-relationships will alter with the dose of the antigen, the route of its administration, the species or even the particular genotype of the recipient. It is an allergic response to virus infection which is either quantitatively or qualitatively inappropriate to the situation occurring in an individual that appears to determine whether a pathogenic reaction occurs. It may vary from the too rapid response following pre-sensitisation – as for example in the children who had received formalinised measles vaccine – to the delay in the response which allows such a target organ as the brain to become excessively infected with virus with the consequences of an increased primary virus effect on the cells and a subsequent damaging allergic reaction. When a member of an orchestra gets a few bars out of time, his continued efforts are likely to result in a chaotic disharmony and he would serve the cause of music better if he kept quiet.

REFERENCES

ADAMS, D. D. (1969). A theory of the pathogenesis of rheumatic fever, glomerulo-nephritis and other autoimmune diseases triggered by infection. *Clinical Experimental Immunology*, **5**, 105.

ALLISON, A. C. (1967). Cell-mediated immune response to virus infections and virus-induced tumours. *British Medical Bulletin*, **23**, 60.

ALMEIDA, D. & WATERSON, A. P. (1969). The morphology of virus-antibody interaction. In *Advances in Virus Research*, **15**, 307. New York: Academic Press.

BENDA, R. & CINATL, J. (1962). Multiplication of lymphocytic choriomeningitis virus in bottle cell cultures. Experimental data for the preparation of highly infectious fluids. *Acta Virologica*, **6**, 159.

BENSON, L. & HOTCHIN, J. E. (1960). Cytopathogenicity and plaque formation with lymphocytic choriomeningitis virus. *Proceedings of the Society of Experimental Biology and Medicine*, **103**, 623.

BERLIN, B. S. (1964). Sparing effect of X-rays for mice inoculated intranasally with egg-adapted influenza virus, CAM strain (29720). *Proceedings of the Society of Experimental Biology and Medicine*, **117**, 864.

BERRY, D. M. & ALMEIDA, J. D. (1968). The morphological and biological effects of various antisera on avian infectious bronchitis virus. *Journal of General Virology*, **3**, 97.

BIELSCHOWSKY, M., HELYER, B. J. & HOWIE, J. B. (1959). Spontaneous haemolytic anaemia in mice of the NZB/BL strain. *Proceedings of the University of Otago Medical School*, **37**, 9.

BILLINGHAM, R. E. (1969). The role of the lymphocyte in transplantation immunity. Symposium on the Life History of Lymphocytes. *The Anatomical Record*, **165**, no. 1, 121.

BLANDFORD, G. (1970). Arthus reaction and pneumonia. *British Medical Journal*, **1**, 758.

BLUMHARDT, R., PAPPANO, J. E. & MOYER, D. G. (1968). Depression of poison ivy skin tests by measles vaccine. *Journal of the American Medical Association*, **206**, 2739.

BORSOS, T., DOURMASHKIN, R. R. & HUMPHREY, J. H. (1964). Lesions in erythrocyte membranes caused by immune haemolysis. *Nature, London*, **202**, 251.

BRADISH, C. J., ALLNER, K. & MABER, H. B. (1971). The virulence of original and derived strains of Semliki Forest virus for mice, guinea-pigs and rabbits. *Journal of General Virology*, **12**, (in press).

BROWN, P. C., GLYNN, L. E. & HOLBOROW, E. J. (1967). The dual necessity for delayed hypersensitivity and circulating antibody in the pathogenesis of experimental allergic orchitis in guinea pigs. *Immunology*, **13**, 307.

BURNET, F. M. (1968). Measles as an index of immunological function. *The Lancet*, **ii**, 610.

BUSER, F. (1967). Side reaction to measles vaccination suggesting the Arthus phenomenon. *New England Journal of Medicine*, **277**, 250.

CEROTTINI, J. C., NORDIN, A. A. & BRUNNER, K. T. (1970). Specific *in vitro* cytotoxicity of thymus-derived lymphocytes sensitised to allo-antigens. *Nature, London*, **228**, 1308.

CHANOCK, R. M., SMITH, C. B., FRIEDWALD, W. T., PARROTT, R. H., FORSYTH, B. R., COATES, H. V., KAPIKIAN, A. Z. & CHAPURE, M. A. (1967). In *Vaccines against Viral and Rickettsial Diseases of Man*, **53**, Washington DC: Pan American Health Organization Scientific Publications, no. 147.

COLE, G. A., GILDEN, D. H., MONJAN, A. A. & NATHANSON, N. (1971). Lymphocytic choriomeningitis virus: pathogenesis of acute central nervous system disease. *Proceedings of the Federation of American Societies for Experimental Biology*, **31**, (in press).

COOMBS, R. R. A. (1968). Immunopathology. *British Medical Journal*, **1**, 597.

COOMBS, R. R. A. & LACHMANN, P. J. (1968). Immunological reactions at the cell surface. *British Medical Bulletin*, **24**, 113.

COOMBS, R. R. A. & SMITH, H. (1968). The allergic response in immunity. In *Clinical Aspects of Immunology*, ed. P. G. H. Gell and R. R. A. Coombs, 2nd edition, 423. Oxford: Blackwell Scientific Publications.

DALTON, A. J., ROWE, W. P., SMITH, G. H., WILSNACK, R. E. & PUGH, W. E. (1968). Morphological and cytochemical studies on lymphocytic choriomeningitis virus. *Jounral of Virology*, **2**, 1465.

DAVIS, G. E. (1931). Complement fixation in yellow fever in monkey and in man. *American Journal of Hygiene*, **13**, 79.

DENHAM, S., GRANT, C. K., HALL, J. G. & ALEXANDER, P. (1970). The occurrence of two types of cytotoxic lymphoid cells in mice immunised with allogeneic tumour cells. *Transplantation*, **9**, 366.

DI GEORGE, A. M. (1965). Discussion on 'A new concept of the cellular basis of immunity'. *Journal of Pediatrics*, **67**, 907.

DINGLE, J. T., FELL, H. B. & COOMBS, R. R. A. (1967). The breakdown of embryonic cartilage and bone cultivated in the presence of complement-sufficient antiserum. 2. Biochemical changes and the role of the lysosomal system. *International Archives of Allergy*, **31**, 282.

DIXON, C. W. (1962). *Smallpox*, 41. London: Longmans.

DUBIN, I. N., BAYLIN, G. J. & GOEBEL, W. G. (1946). Effect of roentgen therapy in experimental virus pneumonia; on pneumonia produced in white mice by swine influenza virus. *American Journal of Roentgenology*, **55**, 478.

DUC-NGUYEN, H., ROSE, H. M. & MORGAN, C. (1966). An electron microscope study of changes at the surface of influenza-infected cells as revealed by ferritin-conjugated antibodies. *Virology*, **28**, 404.

EAST, J., DE SOUSA, M. A. B., PROSSER, P. A. & JAQUET, H. (1967). Malignant changes in New Zealand black mice. *Clinical and Experimental Immunology*, **2**, 427.

EHRENFELD, E. N., GERY, I. & DAVIES, A. M. (1961). Specific antibodies in heart-disease. *The Lancet*, **i**, 1138.

EVANS, R. & ALEXANDER, P. (1970). Co-operation of immune lymphoid cells with macrophages in tumour immunity. *Nature, London*, **228**, 620.

FELL, H. B., COOMBS, R. R. A. & DINGLE, J. T. (1966). The breakdown of embryonic (chick) cartilage and bone cultivated in the presence of complement-sufficient antiserum. 1. Morphological changes, their reversibility and inhibition. *International Archives of Allergy*, **30**, 146.

FINDLAY, G. M. & STERN, R. O. (1936). Pathological changes due to infection with virus of lymphocytic choriomeningitis. *Journal of Pathology and Bacteriology*, **43**, 327.

FLICK, J. A. & PINCUS, W. B. (1963). Inhibition of the lesions of primary vaccinia and of delayed hypersensitivity through immunological tolerance in rabbits. *Journal of Experimental Medicine*, **117**, II, 633.

FULGINITI, V. A., ELLER, J. J., DOWNIE, A. W. & KEMPE, C. H. (1967). Altered reactivity to measles virus. Atypical measles in children previously immunised with inactivated measles virus vaccines. *Journal of the American Medical Association*, **202**, 1075.

GARDNER, P. S., McQUILLIN, J. & COURT, S. D. M. (1970). Speculation on pathogenesis in death from respiratory syncytial virus infection. *British Medical Journal*, **1**, 327.

GELL, P. G. H. & COOMBS, R. R. A. (1968). Classification of allergic reactions responsible for clinical hypersensitivity and disease. In *Clinical Aspects of Immunology*, ed. P. G. H. Gell and R. R. A. Coombs, 2nd edition, 575. Oxford: Blackwell Scientific Publications.

GOCKE, D. J., PREISIG, R., MORRIS, T. Q., McKAY, D. G. & BRADLEY, S. E. (1967). Experimental viral hepatitis in the dog: production of persistent disease in partially immune animals. *Journal of Clinical Investigation*, **46**, II, 1506.

GOLDBERG, S. A., BRODIE, M. & STANLEY, P. (1935). Effect of X-rays on experimental encephalitis in mice inoculated with the St Louis strain. *Proceedings of the Society of Experimental Biology and Medicine*, **32**, 587.

GRANGER, G. A. & KOLB, W. P. (1968). Lymphocyte *in vitro* cytotoxicity mechanisms of immune and non-immune small lymphocyte mediated target L-cell destruction. *Journal of Immunology*, **101**, 111.

GRANT, C. K., DENHAM, S., HALL, J. G. & ALEXANDER, P. (1970). Antibody and complement-like factors in the cytotoxic action of immune lymphocytes *Nature, London*, **227**, 509.

HALL, J. G. (1969). Effector mechanisms in immunity. *The Lancet*, **i**, 25.

HALSTEAD, S. B. (1966). Mosquito-borne haemorrhagic fevers of South and South-East Asia. *Bulletin of the World Health Organisation*, **35**, 3.

HANAOKA, M., SUZUKI, S. & HOTCHIN, J. (1969). Thymus-dependent lymphocytes: destruction by lymphocytic choriomeningitis virus. *Science*, **163**, 1216.

HARBOE, A., SCHOYEN, R. & BYE-HANSEN, A. (1966). Haemagglutination inhibition by antibody to host material of fowl plague virus given in different tissues of chick chorioallantoic membranes. *Acta pathologica et microbiologica scandinavica*, **67**, 573.

HIRSCH, M. S. & MURPHY, F. A. (1968). Effects of anti-lymphoid sera on viral infections. *The Lancet*, **ii**, 37.

HIRSCH, M. S., MURPHY, F. A. & HICKLIN, M. D. (1968). Immunopathology of lymphocytic choriomeningitis virus infection of newborn mice. Anti-thymocyte serum effects on glomerulonephritis and wasting disease. *Journal of Experimental Medicine*, **127**, 757.

HOLBOROW, E. J. & DENMAN, A. M. (1968). Lessons from studies of spontaneous auto-immunity in New Zealand mouse strains. In *Proceedings of the International Symposium on Gammapathies, Infections, Cancer and Immunity*, ed. V. Chini, L. Bonomo and C. Sirtori, 64. Milan: Carlo Erba Foundation.

HOLTERMANN, O. A. & MAJDE, J. A. (1969). Rejection of skin grafts from mice chronically infected with lymphocytic choriomeningitis virus by non-infected syngeneic recipients. *Nature, London*, **223**, 624.

HOTCHIN, J. (1965). Chronic disease following lymphocytic choriomeningitis virus inoculation and possible mechanisms of slow virus pathogenesis. In *Slow, Latent and Temperate Virus Infections*, ed. D. G. Gajdusek, C. J. Gibbs and M. Alpers, NINDB Monogr. No. 2, 341. Washington, DC: U.S. Government Print. Off.

HOTCHIN, J., BENSON, L. M. & SEAMER, J. (1962). Factors affecting the induction of persistent tolerant infection of newborn mice with lymphocytic choriomeningitis. *Virology*, **18**, 71.

HOWIE, J. B. & HELYER, B. J. (1968). The immunology and pathology of New Zealand B mice. *Advances in Immunology*, **9**, 215.

HOYLE, L. (1950). Multiplication of influenza viruses in fertile egg; report to Medical Research Council, *Journal of Hygiene*, **48**, 277.

HOYLE, L. (1954). Release of influenza virus from infected cells. *Journal of Hygiene*, **52**, 180.

HUGHES, T. P. (1933). Precipitin reaction in yellow fever. *Journal of Immunology*, **25**, 275.

ILLAVIA, S. J. & WEBB, H. E. (1970). An encephalitogenic virus (Langat) in mice: isolation and persistence in cultures of brains after intraperitoneal infection with the virus. *The Lancet*, **ii**, 284.

ILLAVIA, S. J. & WEBB, H. E. (1971). The effect of encephalitogenic viruses on tissue culture of non-neuronal cells of mouse and human brain. *Neurology* (in press).

IRVINE, W. J. (1964). Thyroid auto-immunity as a disorder of immunological tolerance. *Quarterly Journal of Experimental Physiology*, **49**, 324.

IVERSEN, K. (1948). *Temporary Rise in Frequency of Thyrotoxicosis in Denmark*, 1941–45. Copenhagen: Rosenkilde & Bagger.

JOHNSON, R. T. (1968). Mumps virus encephalitis in the hamster. Studies of the inflammatory response in non-cytopathic infections of neurons. *Journal of Neuropathology and Experimental Neurology*, **27**, 80.

JUSTINES, G. & JOHNSON, K. M. (1969). Immune tolerance in *Calomys callosus* infected with machupo virus, *Nature, London*, **222**, 1090.

LEHMANN-GRUBE, F. (1967). A carrier state of lymphocytic choriomeningitis virus in L-cell cultures. *Nature, London*, **213**, 770.

LERNER, E. M. & HAAS, V. H. (1958). Histopathology of lymphocytic choriomeningitis in mice spared by amethopterin. *Proceedings of the Society of Experimental Biology and Medicine*, **98**, 395.

LENNON, R. G. & ISACSON, P. (1967). Delayed dermal hypersensitivity following killed measles vaccine. *Journal of Pediatrics*, **71**, 525.

LILLIE, R. D. & ARMSTRONG, G. (1945). Pathology of lymphocytic choriomeningitis in mice. *Archives of Pathology and Laboratory Medicine*, **40**, 141.

MACKANESS, G. B. (1967). The relationship of delayed hypersensitivity to acquired cellular resistance. *British Medical Bulletin*, **23**, 52.

MACKAY, I. R. & GADJUSEK, D. C. (1958). An auto-immune reaction against human tissue antigens in certain acute and chronic diseases. II. Clinical correlations. *Archives of Internal Medicine*, **101**, 30.

MCKAY, D. G. & MARGARETTEN, W. (1967). Disseminated intravascular coagulation in virus diseases. *Archives of Internal Medicine*, **120**, 129.

MCNAIR SCOTT, T. F. & BONANNO, D. E. (1967). Reactions to live measles virus vaccine in children previously inoculated with killed virus vaccine. *New England Journal of Medicine*, **277**, 248.

MARCUS, P. I. (1962). Dynamics of surface modification in myxovirus infected cells. *Symposium of Quantitative Biology*, **27**, 351.

MELLORS, R. C. & HUANG, C. Y. (1967). Immunopathology of NZB/BL mice VI, virus separable from spleen and pathogenic for Swiss mice. *Journal of Experimental Medicine*, **126**, 53.

MIMS, C. A. (1966). Pathogenesis of rashes in virus diseases. *Bacteriological Reviews*, **30**, 739.

MIMS, C. A. & WAINWRIGHT, S. (1968). The immunodepressive action of lymphocytic choriomeningitis virus in mice. *Journal of Immunology*, **101**, 717.

MONTGOMERIE, J. Z., BECROFT, D. M. O., CROXSON, M. C., DOAK, P. N. & NORTH, J. D. K. (1969). Herpes-simplex virus infection after renal transplantation. *The Lancet*, **ii**, 867.

MORGAN, H. R. (1968). Ultrastructure of the surfaces of cells infected with avian leukosis-sarcoma viruses. *Journal of Virology*, **2**, 1133.

MURPHY, F. A., WEBB, P. A., JOHNSON, K. M. & WHITFIELD, S. G. (1969). Morphological comparison of machupo with lymphocytic choriomeningitis virus: basis for a new taxonomic group. *Journal of Virology*, **4**, 535.

NATHANSON, N. & COLE, G. A. (1970). Immunosuppression and experimental virus infection of the nervous system. In *Advances in Virus Research*, **16**, 397. New York: Academic Press.

NOTKINS, A. L., MERGENHAGEN, S. E. & HOWARD, R. J. (1970). Effect of virus infections on the function of the immune system. *Annual Review of Microbiology*, **24**, 525.

OGRA, P. L., KARZON, D. T., RIGHTHAND, F. & MACGILLIVRAY, M. (1968). Immunoglobulin response in serum and secretions after immunisation with live and inactivated poliovaccine and natural infection. *New England Journal of Medicine*, **279**, 893.

OGRA, P. L. & KARZON, D. T. (1969). Distribution of poliovirus antibody in serum, nasopharynx and alimentary tract following segmental immunisation of the lower alimentary tract with polio vaccine. *Journal of Immunology*, **102**, 1423.

OLDSTONE, M. B. A. & DIXON, F. J. (1967). Lymphocytic choriomeningitis: production of antibody by 'tolerant' infected mice. *Science*, **158**, 1193.

OLDSTONE, M. B. A. & DIXON, F. J. (1969). Pathogenesis of chronic disease associated with persistent lymphocytic choriomeningitis viral infection. I. Relationship of antibody production to disease in neo-natally infected mice. *Journal of Experimental Medicine*, 129, 483.

OLDSTONE, M. B. A. & DIXON, F. J. (1970). Pathogenesis of chronic disease associated with persistent lymphocytic choriomeningitis viral infection. II. Relationship of the anti-lymphocytic choriomeningitis immune response to tissue injury in chronic lymphocytic choriomeningitis disease. *Journal of Experimental Medicine*, 131, 1.

OLSON, G. B., DENT, P. B., RACOLS, W. E., SOUTH, M. A., MONTGOMERY, J. P., MELNICK, J. L. & GOOD, R. A. (1968). Abnormalities of *in vitro* lymphocyte responses during rubella virus infections. *Journal of Experimental Medicine*, 128, 47.

PARROTT, R. H., KIM, H. W., ARROBIO, J., CHANCHOLA, J. G., BRANDT, C. D., DEMEIO, J. L., JENSEN, K. E. & CHANOCK, R. M. (1967). In *Vaccines against Viral and Rickettsial Diseases of Man*, 35. Washington, DC: Pan American Health Organisation Scientific Publication, no. 147.

PAVKOVA, L. (1962). Formation and significance of auto antibodies in burned patients. *Vox Sanguinis*, 7, 116.

PINCUS, W. B. & FLICK, J. A. (1963*a*). The role of hypersensitivity in the pathogenesis of vaccinia virus infection in humans. *Journal of Pediatrics*, 62, 57.

PINCUS, W. B. & FLICK, J. A. (1963*b*). Inhibition of the primary vaccinial lesion and of delayed hypersensitivity by an antimononuclear cell serum. *Journal of Infectious Diseases*, 113, 15.

PUSZTAI, R., GOULD, E. A. & SMITH, H. (1971). Infection patterns in mice of a virulent and avirulent strain of Semliki Forest virus. *British Journal of Experimental Pathology*, (in press).

RICHARDS, W. P. C. & CORDY, D. R. (1967). Blue tongue virus infection: pathologic responses of nervous system in sheep and mice. *Science*, 156, 530.

ROBINSON, T. W. E., CURETON, R. J. R. & HEATH, R. B. (1969). The effect of cyclophosphamide on Sendai virus infection of mice. *Journal of Medical Microbiology*, 2, 137.

ROGERS, N. G., BASNIGHT, M., GIBBS, C. J. & GAJDUSEK, D. C. (1967). Latent viruses in chimpanzees with experimental kuru. *Nature, London*, 216, 446.

ROOK, G. A. W. & WEBB, H. E. (1970). Antilymphocyte serum and tissue culture used to investigate role of cell-mediated response in viral encephalitis in mice. *British Medical Journal*, 4, 210.

ROSE, H. M. & MORGAN, C. (1960). Fine structure of virus infected cells. *Annual Review of Microbiology*, 14, 217.

RUSSELL, P. J., HICKS, J. D. & BURNET, F. M. (1966). Cyclophosphamide treatment of kidney disease in (NZB XNZW)F1 mice. *The Lancet*, i, 1279.

RUSSELL, P. K., INTAVIVAT, A. & KANCHANAPILANT, S. (1969). Anti-dengue immunoglobulins and serum beta 1 c-a globulin levels in dengue shock syndrome. *Journal of Immunology*, 102, 412.

SAINANI, G. S., KROMPOTIC, E. & SLODKI, S. J. (1968). Adult heart disease due to the coxsackie virus B infection. *Medicine (Baltimore)*, 47, 133.

SEAMER, J. (1965). Mouse macrophages as host cells for murine viruses. *Archiv für die gesamte Virusforschung*, 17, 654.

SIMONS, M. J. (1968). Congenital rubella: an immunological paradox? *The Lancet*, ii, 1275.

SMITH, C. E. G. (1970). Immunology and virus diseases. *Journal of the Royal College of Physicians, London*, 5, 31.

SMITH, J. B. & MORRIS, B. (1970). The response of the popliteal node of sheep to swine influenza virus. *Australian Journal of Experimental Biology and Medicinal Science*, **48**, 33.

STONE, S. H., LERNER, E. M. & GOODE, J. H. (1969). Acute and chronic auto-immune encephalomyelitis: age, strain and sex dependency. The importance of the source of antigen (34210). *Proceedings of the Society of Experimental Biology and Medicine*, **132**, 341.

VAN DER GELD, H. (1964). Anti-heart antibodies in the post pericardiotomy and the post myocardial-infarction syndromes. *The Lancet*, **ii**, 617.

VOLKERT, M. & LARSEN, J. H. (1964). Studies on immunological tolerance to lymphocytic choriomeningitis virus. 3. Duration and maximal effect of adoptive immunization of virus carriers. *Acta Pathologica et Microbiologica Scandinavica*, **60**, 577.

VOLKERT, M. & LARSEN, J. H. (1965). Immunological tolerance to viruses. *Progress in Medical Virology*, **7**, 160.

VOLKERT, M. & LUNDSTEDT, C. (1968). The provocation of latent lymphocytic choriomeningitis virus infections in mice by treatment with antilymphocyte serum. *Journal of Experimental Medicine*, **127**, 327.

VOLPE, R., ROW, V. V. & EZRIN, C. (1967). Circulating viral and thyroid antibodies in subacute thyroiditis. *Journal of Clinical Endocrinology and Metabolism*, **27**, 1275.

WATSON, D. H. & WILDY, P. (1963). Some serological properties of herpes virus particles studied with the electron microscope. *Virology*, **21**, 100.

WEBB, H. E. & BURSTON, J. (1966). Clinical and pathological observations with special reference to the nervous system in *Macaca radiata* infected with Kyasanur Forest disease virus. *Transactions of the Royal Society of Tropical Medicine and Hygiene*, **60**, 3.

WEBB, H. E. & CHATTERJEA, J. B. (1962). Clinico-pathological observations on monkeys infected with Kyasanur Forest disease virus, with special reference to the haemopoietic system. *British Journal of Haematology*, **8**, 401.

WEBB, H. E., ILLAVIA, S. J. & LAURENCE, G. D. (1971). Measles vaccine viruses in tissue culture of non-neuronal cells of human foetal brain. *The Lancet*, (in press).

WEBB, H. E. & SMITH, C. E. G. (1970). Viruses in the treatment of cancer. *The Lancet*, **i**, 1206.

WEBB, H. E., WIGHT, D. G. D., PLATT, G. S. & SMITH, C. E. G. (1968a). Langat virus encephalitis in mice. I. The effect of the administration of specific anti-serum. *Journal of Hygiene*, **66**, 343.

WEBB, H. E., WIGHT, D. G. D., WIERNIK, G., PLATT, G. S. & SMITH, C. E. G. (1968b). Langat virus encephalitis in mice. II. The effect of irradiation. *Journal of Hygiene*, **66**, 355.

WIKTOR, T. J., KUWERT, E. & KOPROWSKI, H. (1968). Immune lysis of rabies virus-infected cells. *Journal of Immunology*, **101**, 1271.

WILLIAMS, T. W. & GRANGER, G. A. (1969). Lymphocyte *in vitro* cytotoxicity: correlation of derepression with release of lymphotoxin from human lympho-cytes. *Journal of Immunology*, **103**, 170.

ZLOTNIK, I. (1968). The reaction of astrocytes to acute virus infections of the central nervous system. *British Journal of Experimental Pathology*, **49**, 555.

SPECIFICITY OF VIRUSES FOR TISSUES AND HOSTS

F. B. BANG

Department of Pathobiology, The Johns Hopkins University,
School of Hygiene and Public Health, Baltimore, Maryland

INTRODUCTION

The unique specificity of localisation and effect which is a hallmark of many virus infections allowed Rhazes (see Major, 1932) to differentiate smallpox from measles. However, the way in which the two agents differentially affect the same tissues is not yet understood. Tissue specificity, whereby influenza is usually limited to the respiratory tract, while vaccinia grows nicely in the skin following Jennerian inoculation, has been studied in organ cultures, but as yet there is no molecular explanation for these differences.

Specificity of viruses for certain host species, and resistance of other hosts, have been analysed by tissue culture methods. Specificity of virus growth in one genetic variety of inbred animal and failure of growth in another inbred line of the same species (mouse or chicken) have also been analysed by tissue culture; the beginnings of an understanding of the molecular events are at hand. It is clear then that specificity of virus growth and effect is a term covering many different adaptations on the part of the virus to its host, or certain tissues of the host.

It may be useful to outline the problem from an evolutionary point of view. If we examine one of the latest classifications of viruses (Andrewes & Pereira, 1967), examples of specificity of effect on target organ and differential susceptibility of different hosts are found within each class. This would suggest that specificity of viral effect, like other adaptations of biological populations to particular ecological niches, arises in each species of parasite in its own way. In the evolution of viruses, adaptation involves survival of both host and parasite. The evolution of an RNA virus such as a member of the chicken leukaemia complex may have progressed so far, and the transmission become so 'vertical', that the presence of the virus is detectable only as an antigen in certain inbred lines of chickens in which it seems to appear as a genetic character rather than as an infectious agent (Payne & Chubb, 1968). Smallpox on the other hand persists as a highly infectious and virulent agent because transmission is associated with virulence, i.e. the more ill the patient the greater the transmission to others (Rao, Jacob & Kamalakski, 1969; Thomas *et al.* 1971).

This suggests that, as with macro-parasites, there are multiple characteristics to be studied. At times the virus seems to have acquired new characteristics, such as the reverse transcriptases of the sarcoma RNA viruses (Temin & Mizutani, 1970; Baltimore, 1970) or the different polymerases necessary for growth of RNA bacteriophages *in vitro* (Spiegelman *et al.* 1965). More frequently, however, we study the loss of those characteristics which make the viruses dependent upon specific cells, or hosts.

Discussion of the problems of specificity will be undertaken in two sections. The first will deal with specificity of tissue effect. In this, because of available material, emphasis will be on pathological changes produced in various organs and tissues. The second will deal with specificity for the host itself; that is, resistance and susceptibility of different species and different genetic strains of host. In the latter the beginnings of an understanding of the virological and molecular events will be discussed.

TISSUE SUSCEPTIBILITY

Release of virus from the cell and from the host is the last step in the sequence of events of infection. Since a large number of virus particles is required to overcome the hazards of transmission, it is not surprising that cell destruction often accompanies this terminal stage in which each cell produces great numbers of virions. The destructive effects of pox viruses on the epithelium of the host are well known, whether smallpox, mouse pox or myxomatosis. In one of the most complete studies of pathogenesis – the sequences in which different organs are involved – Fenner (1948) and his colleagues have analysed factors contributing to virulence and susceptibility. Yet even this study, which will no doubt be further analysed in this Symposium, is more a description of the sequences than an understanding of why specific tissues become involved.

One of the most interesting studies of pathogenesis was that of Goodpasture (1925) who showed that inoculation of a neurotropic strain of herpes onto the scarified eye of a rabbit was followed by a definite sequence of growth of the virus, first in the fifth nerve, then progression by way of the trigeminal ganglion to the fifth nucleus, then to the brain proper. This pathological study, based solely on a study of sequential changes – with particular emphasis on the presence of typical eosinophilic intranuclear inclusions in the affected cells – is a beautiful example of specificity. A more extreme example is that of a diabetes mellitus-like syndrome induced in mice by selective destruction of the islets of Langerhans by encephalomyocarditis virus (Craighead & Steinke, 1971). However, most of these pathological studies which have established specificity as a frequent phenomenon only led to the idea of tropisms, but did not yield explanations.

It was again Goodpasture (1941) and his collaborators, especially Buddingh (1950), who showed in chick embryos that much of the tissue specificity which is characteristic of a number of viruses is reproducible in the embryo. First fowl pox, with its specific skin lesions, was shown to grow admirably on the chorio-allantoic membrane where it produced a thick grey layer of swollen cells containing huge irregular lipid inclusions which in turn contained great numbers of virus particles (Woodruff & Goodpasture, 1931). Next it was shown that vaccinia grew on the chorio-allantoic membrane in sufficient quantities for the production of vaccine (Goodpasture, Woodruff & Buddingh, 1931); then that vaccinia spread to the embryo in which it produced scattered pocks (Buddingh, 1936) which in every way resembled disseminated vaccinia.

A number of viruses were subsequently grown in the chick embryo, and at one time no virus laboratory was without an incubator for this prime experimental host. Perhaps, as Buddingh (1970) recently suggested, it was this very specificity of tissue effect that was the most important contribution of the chick embryo to virology. For example: (1) both herpes and pseudorabies (a highly destructive porcine herpes virus) grow nicely on the chorio-allantoic membranes of embryos of certain ages, but herpes disseminated into scattered foci in the embryo (Anderson, 1940) while pseudorabies destroys the embryo's entire nervous system (Bang, 1942); (2) influenza virus grows nicely on the chorionic surface of the membrane but does much better on the allantoic side, yet, when combined with the bacterium *Haemophilus influenzae* on the chorio-allantoic membrane it disseminates through the fluids and causes pulmonary destruction (Bang, 1943); (3) Rous sarcoma virus, first shown by Keogh (1938) to produce small pocks or hyperplastic foci on the chorio-allantoic membrane, can also disseminate by way of the vascular system and cause haemangiomas throughout the embryo (Lo & Bang, 1955; Coates, Borsos, Foard & Bang, 1968). The latter are particularly prominent following intravenous inoculation, and in line with the original interest of the Goodpasture–Buddingh school it seems that cells derived from the original mesoderm are selectively altered by chicken leukemia agents (Coates *et al.* 1968); (4) the virulent strain of Newcastle disease virus grows throughout the embryo, but a vaccine

strain which grows particularly in the allantoic sac has a selective
destructive effect on the pulmonary tissues of 16–17 day embryos (Liu
& Bang, 1953).

Thus, as Buddingh (1970) pointed out in a recent editorial concerning
the continuing potential of the chick embryo for research: 'This orderly
progression of embryonic metabolic activities from reaction patterns,
which characterise more simple bacterial cells in culture, to the highly
complex activities of the specialised, organised and differentiated cells
of specific organs and tissues should provide present cellular and "viral
molecular" biologists with a system that can be explored at well-
defined intervals and in which the knowledge now available at the level
of cells in tissue culture can be applied in a more meaningful manner
to the eventual development of the infectious process in the intact host.'

Specificity of lesions in other developing animals

Age-dependent resistance to viral infection is a well known phenomenon.
It is important to the context of specificity in that specific viral lesions
may be produced only at particular ages. For instance the characteristic
destruction of muscle produced in mice by Coxsackie virus occurs
primarily in newborn mice. Kantoch (Kantoch & Kuczkowska, 1964;
Kantoch & Sieminska, 1965) has suggested that this is due to the
particular susceptibility of muscle cells prior to development of cross-
striations. Tissue cultures of muscle cells from young mice, grown on
collagen, showed a great predominance of non-striated cells which were
readily destroyed by the virus, while cultures of muscle cells from older
mice on the same substrate were not destroyed.

Even more striking examples of selective effects are the studies on
the destruction of the cells of the external granular layer of the cerebellum by a parvovirus, including particularly the rat virus, discovered
by Kilham. This virus also produces a haemorrhagic encephalopathy,
apparently due to destruction of the developing vascular endothelium,
and therefore Margolis & Kilham (1970) have suggested that the agent
selectively destroys rapidly proliferating cells in the neonatal brain, such
as the external germinal layer of the cerebellum and the sub-ependymal
cell plate. This suggestion is in line with autoradiographic evidence of
cell proliferation and migration in the primitive ependymal zone of the
nervous system (Sidman, Miale & Feder, 1959; Miale & Sidman, 1961)
and it would be of further interest to see whether these two different
studies, one on pathogenesis (Margolis & Kilham, 1970) and one on
tissue proliferation (Sidman, Miale & Feder, 1959; Miale & Sidman,
1961), can be more closely correlated. In this connection Monjan,

Gelden, Cole & Nathanson (1971) have shown that lymphocytic chorio-meningitis, a member of an entirely different group of viruses, caused cerebellar hypoplasia and produced maximum neurological signs of disease in animals infected at 4 days old, while animals infected at 1 or 7 days developed only transient ataxia. In a test of the Margolis–Kilham hypothesis (Margolis & Kilham, 1970) that the rat parvovirus preferentially infects tissues with actively dividing cells, Tennant, Layman & Hand (1969) showed in rat embryo cultures that there was good correlation between thymidine uptake (indicating active proliferation) and the presence of rat virus protein as demonstrated by fluorescent antiviral antibody.

Johnson & Johnson (1968) have shown that hamsters develop an internal hydrocephalus following intracerebral inoculation of mumps virus into newborn animals, apparently due to destruction of the ependymal lining of the aqueduct of Sylvius. Again it was necessary to inoculate at specific ages so that the appropriate cells could be attacked to produce a 'congenital defect'. These recent demonstrations of high degrees of specificity of viral lesions within the central nervous system, plus other well established instances such as the effect of poliomyelitis virus on the anterior horn cells or avian encephalomyelitis on Purkinje cells of the chicken (Olitsky, 1939), make it probable that other specific lesions of the central nervous system will in the future be found to be related to virus infection. (I am indebted to Dr Y. C. Hsu for several interesting suggestions in this particular section.)

Organ cultures

Many questions concerning tissue specificity of viruses can be approached by using cultures of differentiated cell systems. In general, differentiation is maintained in cultures by keeping these cells in close association with each other. This may be done in a variety of ways: (a) by maintaining relatively small fragments of intact tissue (often epithelium) on a nutrient surface such as a plasma clot, a method developed by Honor Fell (Fell, 1951); (b) by maintaining the explant on a reconstituted collagen substrate, as developed by Gey (Ehrmann & Gey, 1956); or (c) as Murray (1966) did with nerve cells, by carefully preserving the original tissue on cover slips. All three methods are applicable to the study of the effect of differentiation on tissue susceptibility to viruses.

The first method has been used a little more than the others for virological studies, perhaps because a greater preservation of the interrelationships of different cells is maintained. It was first used in virological studies by Bang & Niven (1958) who maintained cultures of respiratory epithelium with the specific purpose of trying to grow viruses of the 'common cold'. This was later done successfully by Hoorn (1963) and Hoorn & Tyrrell (1965). It is important then to evaluate the extent to which a culture of differentiated cells succeeds in pinpointing problems of tissue specificity in viral infections.

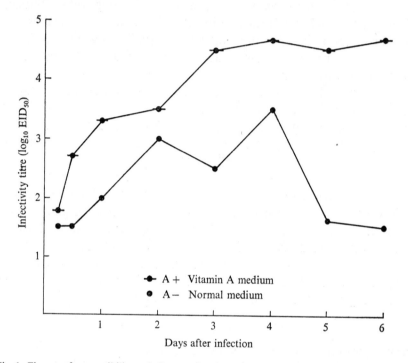

Fig. 1. Change of susceptibility to influenza virus in organ culture of 10-day chick embryonic epidermis after 11 days' cultivation with low dose (13 μg) of added vitamin A.

A direct question concerning specificity was raised by Huang & Bang (1964): would the conversion of squamous epithelium of chick embryo skin into mucous-secretory epithelium by the use of excess vitamin A (Fell, 1957) change the susceptibility of the cells to influenza virus? This virus was chosen because of its well-known destructive effect on mucociliated epithelium. Increased yield of influenza virus was obtained from the vitamin A-treated cultures with increased time of exposure, and in general the greatest yield of virus was obtained when the greatest amount of reordering of differentiation had occurred (Figs. 1, 2). Since the effect of vitamin A does not work by directly converting squamous epithelium to mucous epithelium, but rather by re-ordering the process of differentiation from the basal layer of epithelial cells, one would expect (and it was found) that excess vitamin A would have no immediate effect upon, nor would it increase the susceptibility of, chick embryo chorio-allantoic cells. It appeared that the increased amount of mucus available in the differentiated skin cells was responsible for their greater

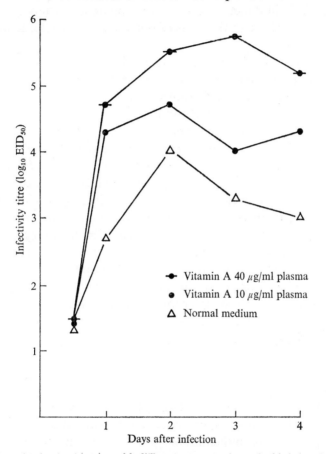

Fig. 2. Effect of 8 days' cultivation with different concentrations of added vitamin A on the change of susceptibility to influenza virus in organ culture of 12-day chick embryonic epidermis.

susceptibility and higher yield of virus, but how this changed metabolic activity had its effect on virus yield is not yet known. That the effect was at least somewhat specific for influenza virus was apparent since there was a decreased yield of vaccinia virus in the same cultures of embryonic skin treated with excess vitamin A. Thus specific susceptibility of differentiated tissues in this case seemed to follow morphological characteristics.

A different test of reproducibility in organ culture of specificity is the ability of explants of human embryonic trachea to support the growth of various rhinoviruses, particularly those that have not been successfully grown on other types of cells (Hoorn & Tyrrell, 1965; McIntosh *et al.* 1967). The susceptibility of tracheal ciliated cells in organ culture to these viruses is in direct contrast to the apparent resistance of ciliated cells in organ culture to the virus of poliomyelitis (Barski, Kourilski & Cornefert, 1957).

A third test of whether the specific lesions produced by a virus *in vivo* are being reproduced *in vitro* is to determine whether the characteristic morphological lesions of the virus (herpes, for example) may also be seen in the organ culture. Herpes virus grown on human embryonic skin did produce typical microvesicles within the affected cell layers, with typical intranuclear inclusions in the cells (Bang & Niven, 1958; McGowan & Bang, 1960; McGowan, 1963). Organ cultures have now been used in the study of a great number of different viruses, and have even been used in studies of chemotherapy of virus infection (Herbst-Laier, 1970), but it is important to point out that in some cases (mouse hepatitis and more especially human hepatitis) they have so far failed to support virus growth. Finally, Herbst-Laier (1970) reports that vervet monkey trachea cultures are unusual in that they are resistant to influenza A virus, even to strains of virus which grow well in monkey kidney cultures. Thus although problems of specificity can be studied in organ cultures, these cultures have as yet not reproduced all of the phenomena of specificity seen in the intact animal.

Tissue cultures

Although the original Li–Rivers minced cultures of tissues which were used for the growth of viruses (Rivers, 1928) undoubtedly had different cell types within them, the appearance, and possible selective destruction, of cell types such as epithelium or fibroblasts was not studied. Present day preparations of trypsinised layers of cells also do not allow one to distinguish what cell type is destroyed.

Primary non-trypsinised explants of tissue onto glass or collagen do allow many cell differences to be clearly distinguished. In a study of the comparative susceptibility of epithelium, fibroblasts and macrophages to fowl pox (Bang, Levy & Gey, 1951), the high susceptibility and rapid destruction of the epithelium was clear. Primary explants of mouse liver grown on collagen produced great numbers of macrophages as well as hepatocytes and these cultures showed selective destruction of macrophages when exposed to mouse hepatitis virus (Bang & Warwick, 1960).

Findings by others using strains of virus more adapted to tissue culture, or using trypsinised cells, do not contradict these results, for trypsinisation changes the surface coating of cells in a variety of ways and may thus offer an opportunity to study the nature of differential susceptibility.

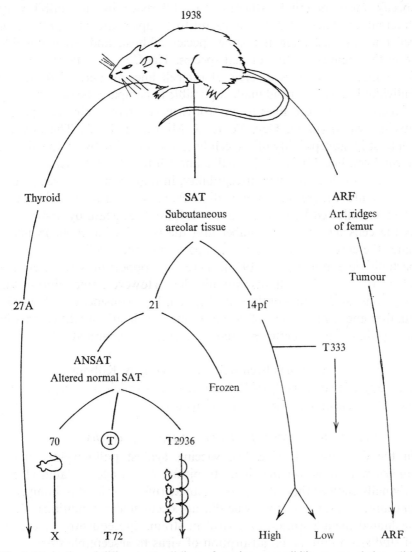

Fig. 3. Derivation of different rat cell lines of varying susceptibility to encephalomyelitis virus (constructed from information furnished by G. O. and M. K. Gey).

Cell cultures

Now that tissue-derived cell culture has almost replaced the use of chick embryos in the cultivation of viruses, the differing susceptibility of cell lines to different viruses is well established. In order, however, to specify more definitely wherein the difference in susceptibility of different cell lines resides, it is useful to compare the susceptibility of cell lines derived from the same species of host, and when possible from the same individual of that species. The Geys maintained *in vitro* for a number of years a series of rat cell lines all derived from one individual rat (Fig. 3) and made it possible to compare the susceptibility of these different cell lines to an arbovirus (Eastern equine encephalomyelitis virus) (Bang, Gey, Foard & Minnegan, 1957). The greatest contrast in susceptibility of the cell lines was found between the altered normal cell line 14pf and the malignant cell line T333. Other cell lines were intermediate in their susceptibility, in that when virus was placed on the culture there was an immediate extensive destruction, which was however followed by recovery of the culture, then again by destruction as the cells increased in number and the colonies increased in mass. Since the chromosome number and pattern of these cell lines has been partially determined (Gey, 1955), there is an opportunity to determine where the gene for cell susceptibility lies. However, since during the long-continued culture of the cells there has been considerable change in number and arrangement of chromosomes, such striking differences in susceptibility might better be studied in recently separated cell lines or types.

Such a possibility has been offered by Wong & Kilbourne (1961) in the study of two variants of HeLa, one line of which is susceptible and the other resistant to the growth of myxovirus.

Molecular explanations of tissue specificity

In the development of a live vaccine against poliomyelitis, many studies were done on the tissue tropisms of this agent. Sabin (1957) originally showed that adsorption of an avirulent strain of poliomyelitis to central nervous system tissue did not occur under conditions which facilitated the adsorption of a virulent strain. Subsequently, in a series of studies on differential adsorption of virus to susceptible tissues and cells, Holland and his coworkers (summarised in Holland & Hoyer, 1962) showed that microsome fractions containing plasma membrane material selectively adsorbed and uncoated both poliomyelitis and Coxsackie viruses. Preparations of cells from insusceptible tissues (except liver) did not have receptor sites; cells from tissues such as kidney, which did not adsorb in the intact animal and was not susceptible, did develop receptor sites as they became susceptible in tissue cultures. Finally, tissue cultures of susceptible cells *in vitro* were shown to have receptor sites, but insusceptible cells lacked adsorption sites. Much of this work was done before 1962 and there is little work since which has defined the nature of these receptor components.

HOST (SPECIES) SPECIFICITY AND CELLULAR INSUSCEPTIBILITY

This part of the discussion is concerned with why one host is susceptible to a virus and another resistant. If one ignores the known variation in capacity to produce antibody, or the rapidity with which interferon is formed, the question may be more directly examined by studying susceptibility of cells from different hosts to the same virus in tissue culture.

Host differences in susceptibility may be used to investigate the sequential factors involved in the growth of viruses. When differences in susceptibility between different species of animals, such as primates *vis-à-vis* other warm-blooded vertebrates, are studied, a much broader category of differences at the molecular level may be involved than for differences within species. The important finding of Holland & Hoyer (1962) that a specific receptor substance is responsible for the adsorption of virus to susceptible cells, and that this is absent in resistant cells, may include within it a complex of differences. On the other hand, when differences of resistance are brought about by a single gene difference within a species, the results may demonstrate extremely discrete differences; these discrete differences may yield information on specific sequences in the viral growth cycle. Clearly both inter- and intra-species studies are important.

Returning to differences in susceptibility of unrelated host species to the same virus, a clear demonstration that this can be studied *in vitro* was the work of Chaproniere & Andrewes (1957). They showed that although rabbit myxoma and fibroma viruses are highly host-specific in naturally occurring infections, tissues of several unrelated species which are not susceptible to infections as whole animals readily supported virus growth. Thus, tissue susceptibility did not reflect host susceptibility.

It was the use of human tissue – sometimes derived from such improbable sources as preputial and uterine tissues – that allowed Enders and his colleagues (Robbins, Enders & Weller, 1950) to grow poliomyelitis virus in culture. The unexpected susceptibility of these non-nervous tissues, derived subsequently from a variety of different primates and tissues, led to an explosion of interest in tissue culture. The reason for the difference in susceptibility of primates and non-primates was then investigated by Holland, McLaren, Hoyer & Syverton (1960), who demonstrated brilliantly that specific receptor substances are present in the tissues of susceptible hosts and are absent from the tissues of insusceptible animals. Susceptible cells were fractionated and the microsomal portion containing a great deal of membrane structure was shown to adsorb and inactivate the virus – a finding which led to the concept of uncoating of viruses. This in turn suggested experiments that demonstrated that if this first step of adsorption and uncoating could be sidestepped, then cells coming from resistant animals such as rabbits, chicks, guinea-pigs and hamsters could be infected with virus RNA alone (Holland *et al.* 1960; Alexander, Koch, Mountain & Van Damme, 1958); and that intact virions which were infectious only for the originally susceptible primate host cells (primates) developed from the 'resistant' cells.

Recently Chan & Black (1970) have shown again that two susceptible tissue culture lines yielded active preparations of receptors which uncoated poliomyelitis virus whereas two insusceptible strains lacked these receptors.

Host specificity as reflected in cellular susceptibility

About ten years ago when we had stumbled onto the fact that macrophages of a strain of mice (PRI) which was susceptible to mouse hepatitis were themselves susceptible to the virus, we tested the susceptibility of cells from another strain of mice (C3H), and found these to be insusceptible. These tests were done with macrophages as they migrated from explants of liver of newborn mice; the differential susceptibility of macrophages (see above) and the failure of both the parenchymal liver cells and the fibroblasts to succumb to the virus (Bang & Warwick, 1960) also were apparent in these cultures. These findings have led us on a continuous study which grows in interest as it proceeds. However, to review it chronologically would be confusing, so the data have been rearranged. First an important warning is in order: viruses change as well as, or perhaps better than, hosts and therefore the major portion of the work has been concerned with the susceptibility of tissues and cells to a strain of mouse hepatitis *as it is grown in PRI mice or cells.* Thus work on resistance to mouse hepatitis in the past was really work on resistance to the virus taken from PRI mice. Later, Shif (Shif &

Bang, 1970b) showed that another 'strain' of virus could be developed which was capable of destroying resistant cells and killing resistant mice. So we now designate the two strains of virus as MHV (PRI) and MHV (C3H). The situation is summarised below.

Mouse strain	Susceptibility to virus	
	MHV (PRI)	MHV (C3H)
PRI	+	+
C3H	−	+

The change of the virus from MHV (PRI) to MHV (C3H) does not seem to be simply the selection of a mutant, for it is greatly influenced by the medium in which the resistant C3H cells grow (Lavelle & Bang, 1971). This change, which can be readily induced, may represent a very small change in viral genetic information and yet be important in the adaptation of a virus to new tissues or to closely related hosts.

The relationship of cell susceptibility to host susceptibility

There are two sets of evidence for the idea that resistance on the part of C3H mice to MHV (PRI) resides in the cells, particularly the macrophages. First, in a series of genetic crosses and segregation experiments, each time a comparison was made between the susceptibility of cells from the adult mouse and a genetic analysis made of the susceptibility of its offspring, a clear correlation was obtained between the behaviour of the mouse and the behaviour of the macrophages. These are now routinely obtained from peritoneal washings of adult mice and thus the same mouse may be used for both tests (Kantoch, Warwick & Bang, 1964). Heterozygous mice obtained from a cross of PRI and C3H mice were themselves susceptible, yielded susceptible cells, and F_2 generations obtained from these showed a 3 to 1 ratio of mice with susceptible cells to mice with resistant cells. (Macrophages were considered resistant if when inoculated with 10^4 ID_{50} doses of virus they showed no change during a period of at least 7 days; macrophages from susceptible mice succumbed in 2 to 3 days to about 1 ID_{50} of virus, i.e. the same concentration required to kill susceptible mice.)

Secondly, a strain of C3H mice which however has the gene for susceptibility, was recently developed by a series of back-crosses of susceptible hybrids to the resistant C3H mice. Susceptibility was determined in the back-crosses solely on the basis of tests of in vitro susceptibility of the macrophages. A new strain of C3H mice obtained by crossing the susceptible 20th back-crosses and eliminating all resistant mice when they segregated, has been shown not only to yield susceptible cells but also to be susceptible itself when inoculated intra-peritoneally with MHV (PRI) (Bang, Shif & Vellisto, 1971). Thus selection of a susceptible hybrid on the basis of the susceptibility of the cells was successful in producing a susceptible mouse. These data, including the persistent production of 50 % susceptible hybrids by back-crossing, indicate that the genetic factor for susceptibility is unifactorial and of course dominant. This suggests in turn that only one protein–enzyme sequence is involved in the difference between resistance and susceptibility *in this particular example*.

Comparisons of the appropriate crosses for susceptibility to West Nile virus (H. Koprowski & I. Shif, unpublished) and to mouse leukaemia virus (T. Pincus & I. Shif, unpublished) show that these resistance factors are on different chromosomes, and thus we cannot talk about *a* genetic factor for resistance to *viruses* in mice, but must consider different genetic factors for at least different classes of viruses.

In an earlier section of this review we mentioned the effect of age on the susceptibility of the host to viruses. Sabin (1952) in his classic study of the genetics of resistance of mice to yellow fever pointed out that the clear evidence of Mendelian inheritance which he presented could be obtained only in grown mice; baby mice were not resistant. Thus there is an ontogeny of resistance. In the case of MHV (PRI), baby C3H mice are initially susceptible, but resistance develops gradually so that at three weeks they are resistant to 10^7 ID_{50} of virus. Liver cultures prepared from newborn C3H mice yielded susceptible cells, whereas cells from similar preparations from older mice became resistant. This acquired cell resistance paralleled mouse resistance at the time the cultures were made (Gallily, Warwick & Bang, 1967). There was however no apparent further development of resistance of the cultured macrophages with time. Thus some factor involved in the maturation of resistance *in vivo* was not operative *in vitro*.

Mechanism of resistance

The evidence for differential receptor sites for poliomyelitis virus in cells of primates, compared to non-primates, made it essential to test the adsorption of MHV (PRI) to resistant and susceptible cells. Virus was readily adsorbed to cultures of both cell types (Shif & Bang, 1970a). Similar studies have been done by others on Rous sarcoma virus (Crittenden, 1968); this complex of RNA viruses, which is made up of several different strains each with its appropriate susceptible and resis-tant cells, shows equal adsorption by such cells. Thus the broad species differences in susceptibility to poliomyelitis between vertebrates seem to be related to adsorption, but the differences of susceptibility within a species (mouse or chicken) are apparently unrelated to the capacity to adsorb virus. This does not describe the entire situation, for Holland (Holland & Hoyer, 1962) has shown that susceptibility of nervous and enteric cells, in contrast to the resistance of other types of primate cells, is related to the presence or absence of receptor sites. Perhaps, then, the variations in results are due to different viruses rather than to different hosts. We conclude that resistance to MHV within the mouse, and resistance to Rous sarcoma virus within the chicken, occurs at some stage past the initial adsorption. It is not known whether this is at the stage of uncoating, or later.

Virus adaptation and the development of a C3H-virulent strain of virus: MHV (C3H)

When cultures of resistant (C3H) macrophages are inoculated with *large* doses of the standard virus, MHV (PRI), destruction of cells often takes place. This sometimes occurs with as little as $10^5\,\mathrm{ID}_{50}$ of virus, and was at first thought to be due to a toxic effect. Investigation showed that a new strain of virus had emerged from the destroyed cells which on passage was able to kill C3H cells at considerable dilution, and which also killed adult C3H mice. On continued passage in C3H cells it maintained this ability, and also retained the original capacity to kill PRI cells. There was a complete correlation between the appearance of the new virus and destruction in C3H cells inoculated with large doses of MHN (PRI). In other words, whenever destruction occurred, new virus (termed MHV (C3H)) was found, and at inoculum dilutions above those at which destruction occurred no new virus was found. This suggests that at no time is the original MHV (PRI) virus able to destroy C3H (resistant) cells, but that when it is given in a sufficiently high multiplicity some reaction occurs which facilitates the formation of a new virus adapted to C3H cells (Shif & Bang, 1970*b*). This could occur during the long period of persistence of MHV (PRI) which occurs within C3H cells when they are infected with lower dilutions of the unadapted virus.

In early studies, we had seen that the amount of MHV (PRI) necessary to cause destruction of C3H cells seemed to vary significantly from one experiment to another. A fairly consistent increase in destruction was obtained when the C3H cells were grown in diluted horse serum instead of the usual 90 % horse serum medium. Comparisons of different sera were then made by Lavelle (Lavelle & Bang, 1971), using a standard concentration of virus. Depending on the serum used, very large differences in 'conversion' of MHV (PRI) to MHV (C3H) were obtained (Table 1). This dependence on serum concentration and source supports the view that the change from MHV (PRI) to MHV (C3H) is not due to a simple selection of mutants. It also means that our previous experiments (Kantoch & Bang, 1962; Gallily, Warwick & Bang, 1964) in which resistant cells appeared susceptible after treatment with extracts of susceptible cells or with cortisone, must be reinvestigated. It is entirely possible that the rather limited 'increase in susceptibility' of these cells *in vitro* was actually due to increased conversion of MHV (PRI) to MHV (C3H).

Table 1. *Influence of type and concentration of serum on titration of MHV (PRI) virus in cultures of C3H and PRI macrophages*

% concentration (v/v) and type of serum in medium	Infectivity of virus (log TC ID_{50}/0·05 ml)*	
	On C3H cells	On PRI cells
90; horse	1·3, 2·5, <1·0	7·3, 7·6
20; horse	3·8, 4·8, 4·5	8·3, 7·8
10; C3H mouse	<1·0, <1·0	7·8, 8·0
20; horse + C3H mouse	<1·0, <1·0	Not done
10; PRI mouse	<1·0	6·9
10; foetal calf	5·0	Not done
10; rat	3·3	Not done

* Results of 3 different experiments.

One strain of mouse hepatitis virus was found by Gledhill, Dick & Andrewes (1952) to be significantly pathogenic for mice when it grew in the presence of eperythrozoon, but caused a mild insignificant infection in the absence of this red cell parasite. It was not surprising, then, that G. C. Lavelle (unpublished) was able to show that MHV (PRI) readily killed C3H mice if these genetically resistant mice were given eperythrozoon one day before they were given MHV. Analysis of the growth of the virus in these eperythrozoon-converted susceptible mice, made by determining the capacity of the virus growing in different tissues – particularly the liver – to kill both C3H and PRI cells, showed that most of the virus early in growth was actually MHV (PRI). Only late in virus growth did the titres of virus, as measured on C3H cells, equal those which were obtained on PRI macrophages. Very similar results were obtained by Willenborg (1971) when he followed the growth of MHV (PRI) in C3H mice made susceptible by cortisone treatment or by cytoxan. These results suggest that all three factors, eperythrozoon, cortisone and cytoxan, make the mouse more susceptible to the original virus. Attempts to make C3H macrophages highly susceptible to MHV (PRI) virus *in vitro* by use of eperythrozoon, cytoxan or cortisone have failed.

SUMMARY

Specificity of effect on certain tissues, and specificity for growth in certain hosts, are characteristics of viruses. Tissue specificity has been recognised in the terms 'neurotropic', 'dermatropic' and so on. The tendency to localise in and destroy certain tissues and the tendency to be parasitic upon specific hosts are now recognised as characteristics present in many different groups of viruses and therefore cannot be used in general classification schemes although shown by many different diseases. These tissue tropisms and adaptations to different hosts appear to have evolved separately among the different classes of viruses and therefore may have occurred in a number of different ways. The fact that different genetic mechanisms are responsible for resistance to different viruses supports this multiphyletic point of view.

High degrees of tissue specificity are seen commonly in the course of experimental infection, as well as in natural disease. Specificity has been mirrored in studies of virus pathogenesis in developing chick embryos, and analysis has commenced of some examples in cultures of differentiated cells (organ cultures). Individual cell lines obtained from the same host have varied in their susceptibility.

Specificity for host involves on the one hand large species differences (poliomyelitis in primates as against non-primates), and on the other hand minute but important differences in susceptibility of strains of mice or chickens. Differences between primates and non-primates in susceptibility to poliomyelitis is clearly related to specific adsorption sites, as Holland showed, and further proven by the production of infection of resistant cells by virus RNA.

Individual genes control susceptibility to arboviruses, mouse hepatitis and mouse sarcoma. These genetic differences are reflected in the case of MHV in cultures of macrophages from the appropriate hosts. A mechanism which does not involve adsorption of virus to the cells is responsible for this genetic specificity and its elucidation should shed light on the sequence of events involved in animal virus growth. However, continuous adaptation of virus to genetically different hosts (and probably tissues) does occur *in vitro*.

REFERENCES

ALEXANDER, H. E., KOCH, G., MOUNTAIN, I. M. & VAN DAMME, O. (1958). Infectivity of ribonucleic acid from poliovirus in human cell monolayers. *Journal of Experimental Medicine*, **108**, 493.

ANDERSON, K. (1940). Pathogenesis of herpes simplex virus infection in chick embryos. *American Journal of Pathology*, **16**, 137.

ANDREWES, C. & PEREIRA, H. G. (1967). *Viruses of Vertebrates*, 4th edition. Baltimore: Williams and Wilkins.

BALTIMORE, D. (1970). RNA-dependent DNA polymerase in virions of RNA tumor viruses. *Nature, London*, **226**, 1209.

BANG, F. B. (1942). Experimental infection of the chick embryo with the virus of pseudorabies. *Journal of Experimental Medicine*, **76**, 263.

BANG, F. B. (1943). Synergistic action of *Hemophilus influenza suis* and the swine influenza virus on the chick embryo. *Journal of Experimental Medicine*, **77**, 7.

BANG, F. B., GEY, G. O., FOARD, M. A. & MINNEGAN, D. (1957). Chronic infections produced in cultured cell strains by the virus of Eastern equine encephalitis. *Virology*, **4**, 404.

BANG, F. B., LEVY, E. & GEY, G. O. (1951). Some observations on host-cell virus relationships in fowl pox. I. Growth in tissue culture. II. The inclusion produced by the virus on the chick chorio-allantoic membrane. *Journal of Immunology*, **66**, 329.

BANG, F. B. & NIVEN, J. S. F. (1958). Study of infection in organized organ cultures. *British Journal of Experimental Pathology*, **39**, 317.

BANG, F. B., SHIF, I. & VELLISTO, I. (1971). Transfer of the gene for susceptibility to mouse hepatitis virus into a genetically resistant strain of mice (further data establishing the role of macrophages in mouse susceptibility). *Bacteriological Proceedings*, Abst. **209**, 202.

BANG, F. B. & WARWICK, A. (1960). Mouse macrophages as host cells for the mouse hepatitis virus and the genetic basis of their susceptibility. *Proceedings of the National Academy of Sciences of the U.S.A.*, **46**, 1065.

BARSKI, G., KOURILSKI, R. & CORNEFERT, F. (1957). Resistance of respiratory ciliated epithelium to action of polio and adenoviruses in vitro. *Proceedings of the Society for Experimental Biology and Medicine*, **96**, 386.

BUDDINGH, G. J. (1936). A study of generalized vaccinia in the chick embryo. *Journal of Experimental Medicine*, **63**, 227.

BUDDINGH, G. J. (1950). The culture and effects of viruses in chick embryo cells. In *The Pathogenesis and Pathology of Virus Diseases*, ed. J. G. Kidd, 19. New York: Columbia University Press.

BUDDINGH, G. J. (1970). Editorial: The chick embryo for the study of infection and immunity. *Journal of Infectious Diseases*, **121**, 660.

CHAN, V. F. & BLACK, F. L. (1970). Uncoating of poliovirus by isolated plasma membranes. *Journal of Virology*, **5**, 309.

CHAPRONIERE, D. M. & ANDREWES, C. H. (1957). Cultivation of rabbit myxoma and fibroma viruses in tissues of nonsusceptible hosts. *Virology*, **4**, 351.

COATES, H., BORSOS, T., FOARD, M. & BANG, F. B. (1968). Pathogenesis of Rous sarcoma virus in the chick embryo with particular reference to vascular lesions. *International Journal of Cancer*, **3**, 424.

CRAIGHEAD, J. E. & STEINKE, J. (1971). Diabetes mellitus-like syndrome in mice infected with encephalomyocarditis virus. *American Journal of Pathology*, **63**, 119.

CRITTENDEN, L. B. (1968). Observations on the nature of a genetic cellular resistance to avian tumor viruses. *Journal of the National Cancer Institute*, **41**, 145.

EHRMANN, R. L. & GEY, G. O. (1956). The growth of cells on a transparent gel of reconstituted rat-tail collagen. *Journal of the National Cancer Institute*, 16, 1375.

FELL, H. B. (1951). Histogenesis in tissue culture. In *Cytology and Cell Physiology*, ed. G. H. Bourne, 419. London: Oxford University Press.

FELL, H. B. (1957). The effect of excess vitamin A on cultures of embryonic chicken skin explanted at different stages of differentiation. *Proceedings of the Royal Society of London*, B, 146, 242.

FENNER, F. (1948). The pathogenesis of the acute exanthems. An interpretation based on experimental investigations with mousepox (infectious ectomelia of mice). *The Lancet*, ii, 915.

GALLILY, R., WARWICK, A. & BANG, F. B. (1964). Effect of cortisone on genetic resistance to mouse hepatitis virus *in vivo* and *in vitro*. *Proceedings of the National Academy of Sciences of the U.S.A.*, 51, 1158.

GALLILY, R., WARWICK, A. & BANG, F. B. (1967). Ontogeny of macrophage resistance to mouse hepatitis *in vivo* and *in vitro*. *Journal of Experimental Medicine*, 125, 537.

GEY, G. O. (1955). Some aspects of the constitution and behaviour of normal and malignant cells maintained in continuous culture. In *The Harvey Lectures*, *Series L*, 1954–5, 154. New York: Academic Press.

GLEDHILL, A. W., DICK, G. M. A. & ANDREWES, C. H. (1952). Production of hepatitis in mice by the combined action of two filtrable agents. *The Lancet*, ii, 509.

GOODPASTURE, E. W. (1925). The axis-cylinders of peripheral nerves as portals of entry to the central nervous system for the virus of Herpes simplex in experimentally infected rabbits. *American Journal of Pathology*, 1, 11.

GOODPASTURE, E. W. (1941). The cell–parasite relationship in bacterial and virus infections. *Transactions of the College of Physicians of Philadelphia*, 9, 11.

GOODPASTURE, E. W., WOODRUFF, A. M. & BUDDINGH, G. J. (1931). The cultivation of vaccinia and other viruses in the chorio-allantoic membrane of chick embryos. *Science*, 74, 371.

HERBST-LAIER, R. (1970). Organ cultures in evaluation of anti-viral drugs. I. Virus infection in trachea explants as potential model systems for evaluation of anti-viral drugs against respiratory viruses. *Archiv für die gesamte Virusforschung*, 30, 379.

HOLLAND, J. J. & HOYER, B. H. (1962). Early stages of enterovirus infections. *Cold Spring Harbor Symposium on Quantitative Biology*, 27, 101.

HOLLAND, J. J., McLAREN, L. C., HOYER, B. H. & SYVERTON, J. T. (1960). Enteroviral ribonucleic acid. II. Biological, physical and chemical studies. *Journal of Experimental Medicine*, 112, 841.

HOORN, B. (1963). Respiratory viruses in model experiments. *Acta Otolaryngolica Supplementum*, 188, 138.

HOORN, B. & TYRRELL, D. A. (1965). On the growth of certain 'newer' respiratory viruses in organ cultures. *British Journal of Experimental Pathology*, 46, 109.

HUANG, J. S. & BANG, F. B. (1964). The susceptibility of chick embryo skin organ cultures to influenza virus following excess vitamin A. *Journal of Experimental Medicine*, 120, 129.

JOHNSON, R. T. & JOHNSON, K. P. (1968). Hydrocephalus following viral infection: The pathology of aqueductal stenosis developing after experimental mumps virus infection. *Journal of Neuropathology and Experimental Neurology*, 27, 591.

KANTOCH, M. & BANG, F. B. (1962). Conversion of genetic resistance of mammalian cells to susceptibility to a virus infection. *Proceedings of the National Academy of Sciences of the U.S.A.*, 48, 1153.

KANTOCH, M. & KUCZKOWSKA, B. (1964). Studies on the susceptibility of newborn and adult mice to infection with Coxsackie viruses. *Archivum Immunologiae et Therapiae Experimentalis (Warszawa)*, **12**, 497.

KANTOCH, M. & SIEMINSKA, A. (1965). Studies on the susceptibility of mouse muscle cultures to Coxsackie A-4 viruses *in vitro*. *Archivum Immunologiae et Therapiae Experimentalis (Warszawa)*, **13**, 413.

KANTOCH, M, WARWICK, A., & BANG, F. B. (1964). The cellular nature of genetic susceptibility to a virus. *Journal of Experimental Medicine*, **117**, 781.

KEOGH, E. V. (1938). Ectodermal lesions produced by the virus of Rous sarcoma. *British Journal of Experimental Pathology*, **19**, 1.

LAVELLE, G. C. & BANG, F. B. (1971). Influence of type and concentration of sera *in vitro* on susceptibility of genetically resistant cells to mouse hepatitis virus. *Journal of General Virology*, **12**, 233.

LIU, C. & BANG, F. B. (1953). An analysis of the difference between a destructive and a vaccine strain of Newcastle disease virus in the chick embryo. *Journal of Immunology*, **70**, 538.

LO, W. H. Y. & BANG, F. B. (1955). Rous sarcoma virus infection in the chick embryo. I. Pathogenesis of the Rous virus infection in the chick embryo with particular emphasis on the hemorrhagic disease. *Bulletin of the Johns Hopkins Hospital*, **97**, 227.

MAJOR, R. H. (1932). *Classic Descriptions of Disease*. Baltimore: C. C. Thomas.

MARGOLIS, G. & KILHAM, L. (1970). Parvovirus infections, vascular endothelium and hemorrhagic encephalopathy. *Laboratory Investigation*, **22**, 478.

McGOWAN, T. R. (1963). Long-term support of intract skin in organ cultures, with application to the study of virus infection. *Journal of the National Cancer Institute*, Monograph 11, 95.

McGOWAN, T. R. & BANG, F. B. (1960). Organ cultures of human fetal skin for the study of skin diseases. *Bulletin of the Johns Hopkins Hospital*, **107**, 63.

McINTOSH, K., DEES, J. H., BECKER, W. B., KAPIKIAN, A. Z. & CHANOCK, R. M. (1967). Recovery in tracheal organ cultures of novel viruses from patients with respiratory disease. *Proceedings of the National Academy of Sciences of the U.S.A.*, **57**, 933

MIALE, I. L. & SIDMAN, R. L. (1961). An autoradiographic analysis of histogenesis in the mouse cerebellum. *Experimental Neurology*, **4**, 277.

MONJAN, A. A., GELDEN, D. H., COLE, G. A. & NATHANSON, N. (1971). Cerebellar hypoplasia in neonatal rats caused by lymphocytic choriomeningitis virus. *Science*, **171**, 194.

MURRAY, M. R. (1966). Nervous tissue *in vitro*. In *Cells and Tissues in Culture*, ed. E. N. Willmer, 373. New York: Academic Press.

OLITSKY, P. K. (1939). Experimental studies of the virus of infectious avian encephalomyelitis. *Journal of Experimental Medicine*, **70**, 565.

PAYNE, L. N. & CHUBB, R. C. (1968). Studies on the nature and genetic control of an antigen in normal chick embryos which reacts in the CORFAL test. *Journal of General Virology*, **3**, 379.

RAO, A. R., JACOB, E. S. & KAMALAKSKI, S. (1969). Experimental studies in small-pox. A study of intrafamilial transmission in a series of 254 infected families. *Indian Journal of Medical Research*, **56**, 1826.

RIVERS, T. M. (1928). *Filterable Viruses*. Baltimore: Williams and Wilkins.

ROBBINS, F. C., ENDERS, J. F. & WELLER, T. H. (1950). Cytopathogenic effect of poliomyelitis viruses *in vitro* on human embryonic tissues. *Proceedings of the Society for Experimental Biology and Medicine*, **75**, 370.

SABIN, A. B. (1952). Nature of inherited resistance to viruses affecting the nervous system. *Proceedings of the National Academy of Sciences of the U.S.A.*, **38**, 540.

SABIN, A. B. (1957). Properties of attenuated poliovirus and their behaviour in human beings. In *Cellular Biology, Nucleic Acids and Viruses*, ed. O. V. Whitlock, 113. New York: New York Academy of Science.

SHIF, I. & BANG, F. B. (1970*a*). *In vitro* interaction of mouse hepatitis virus and macrophages from genetically resistant mice. I. Adsorption of virus and growth curves. *Journal of Experimental Medicine*, **131**, 843.

SHIF, I. & BANG, F. B. (1970*b*). *In vitro* interaction of mouse hepatitis virus and macrophages from genetically resistant mice. II. Biological characterization of a variant virus (MHV (C3H) isolated from stocks of MHV (PRI). *Journal of Experimental Medicine*, **131**, 851.

SIDMAN, R. L., MIALE, I. L. & FEDER, N. (1959). Cell proliferation and migration in the primitive ependymal zone; an autoradiographic study of histogenesis in the nervous system. *Experimental Neurology*, **1**, 322.

SPIEGELMAN, S., HARUNA, I., HOLLAND, I. B., BEAUDREAU, G. & MILLS, D. (1965). The synthesis of a self-propagating and infectious acid with a purified enzyme. *Proceedings of the National Academy of Sciences of the U.S.A.*, **54**, 919.

TEMIN, H. M. & MIZUTANI, S. (1970). RNA-dependent DNA polymerase in virions of Rous sarcoma virus. *Nature, London*, **226**, 1211.

TENNANT, R. W., LAYMAN, K. R. & HAND, R. E., JR. (1969). Effect of cell physiological state on infection by rat virus. *Journal of Virology*, **4**, 872.

THOMAS, D. B., ARITA, I., McCORMACK, W. M., KHAN, M. M., ILAM, S. & MACK, T. M. (1971). Endemic smallpox in rural East Pakistan. II. Intravillage transmission and infectiousness. *American Journal of Epidemiology*, **93**, 373.

WILLENBORG, D. O. (1971). *The effect of chemical and physical agents on the genetic resistance of mice to mouse hepatitis virus (MHV-PRI)*. Thesis, The Johns Hopkins University School of Hygiene and Public Health, Baltimore, Maryland.

WONG, S. C. & KILBOURNE, E. D. (1961). Changing viral susceptibility of a human cell line in continuous cultivation. I. Production of infective virus in a variant of the Chang conjunctival cell following infection with swine or N-WS influenza virus. *Journal of Experimental Medicine*, **113**, 95.

WOODRUFF, A. M. & GOODPASTURE, E. W. (1931). The susceptibility of the chorioallantoic membrane of chick embryos to infection with the fowl-pox virus. *American Journal of Pathology*, **7**, 209.

INDEX

abortion: bovine mycotic, 258, 262; infections causing, 201
Absidia ramosa, mastitis caused by, 259
Absidia sp., ulcers caused by, 258
acetylcholine: inhibition of release of, in botulism, 114
N-acetyl-galactosamine uronic acid, Vi antigen of *S. typhi* a polymer of, 87
N-acetyl neuraminic (sialic) acid, as receptor: for influenza virus, 341; for mycoplasma neurotoxin? 233, 234; for mycoplasmas, 237, 240
Acholeplasma laidlawii (mycoplasma), 228, 236; virus infection of, 241–2
acid hydrolases: in macrophages, 61, 68; in neutrophils, 60, 67
acid polysaccharides, K antigens of *E. coli* as, 87, 103
acids, volatile organic: produced by indigenous microbes, 48, 49
actinomyces, thermophilic: cause of farmers' lung, 257
actinomycin D, inhibitor of protein synthesis, 370, 373; does not affect giant-cell formation caused by virus products, 367
actinophage, endotoxin and formation of antibody to, 100
adenosine, in malaria, 17
adenosine-3'-5'-cyclic monophosphate (cyclic AMP), cholera toxin and, 134–5
adenoviruses: cells lacking 'replication factors' for, 8; entry of, 314, 315; fibre antigen of, 370, 372; humidity and survival of, 305; inhibitory effects of mycoplasmas on, 311; penton (separable capsid protein) of, 15, 368, 372, 374; persistence of, 350
adenyl cyclase, activated by enterotoxins, 135, 137, 145, 147
adiaspiromycosis, 256
adrenalin: effect of suppression of inflammation by, on infections by staphylococci, 77
aflatoxins (fungal), 16, 265
African swine fever virus, 316
agalactia, contagious, 221
agammaglobulinaemia: recovery from virus diseases in, 337, 388–9; serum antibacterial activity in, 99
age, and susceptibility: to fungus infections, 253; to virus infections, 11, 418–19, 428
agglutination of erythrocytes: inhibition of, by K antigens of *E. coli*, 85, 86, 87, 88
agglutinins, bacterial, 28, 32, 33

aggressins: bacterial, 11, 35, 75, 102, 196; fungal, 12; protozoal, 13
alkaline phosphatase: in intestinal epithelium, 29–30; in neutrophils, 60, 67
allergic responses: in bacterial infections, 185–7; classification of, 384–7; in fungal infections, 252, 253, 255, 267; in pathogenesis of virus diseases, 383–408
allergy, 157; contact, 256
Allescheria boydii, mastitis caused by, 259
alveolitis, allergic pulmonary interstitial, 385
p-aminobenzoic acid, growth factor for *Plasmodium berghei*, 9–10, 283
ammonia, and bacterial infection of kidney, 205
amoebic dysentery, 270; *see also Entamoeba histolytica*
anaemia: induced by soluble antigens of *Plasmodium*, 288; in infections, 98; and susceptibility to infections, 98
anaphylactic shock, 384; in different species, 120
anaphylatoxin, 165, 186, 385
anaphylaxis, 160, 384; from bacterial antigens, 159, 160–3; in piglet bowel oedema, 163–6; reversed passive, 164
anopheline mosquitoes, introduction of *Plasmodium* by, 272
anthrax, 116–17; *see also Bacillus anthracis*
antibiotics: and indigenous microbes, 6, 25, 27, 40, 44, 46, 47, 50; in treatment of *Cl. welchii* infections, 118
antibodies, 13, 69, 158; to bacterial antigens adsorbed on to tissue cells, 179–80; circulating (IgG, IgM), 160, 167, 288, 348, 384, 389; cytophilic, 338; to fungi, 257, 264; to gonococci, 82; harmful complexes of antigens and, *see* Arthus responses; of immunoblasts, 386, after immunosuppressants, 400; to impedin factor of staphylococci, 78; interaction of virus and, 392; iron and, 97; to K antigens of *E. coli*, 85, 86–7, 91; in mucus, 310; to mycoplasmas, 229–30; to *P. septica*, 95; in presence of latent virus, 387, 402; reagin (IgE), 160, 384; receptors for, on phagocytes, 65, 66; recovery of viruses from combination with, 337; secretory (IgA), 338, 348, 389; in syphilis, 184–5, 186; to viruses, 318–19, 336–8, 388; *see also* auto-antibodies
anti-cardiolipins, detected by diagnostic tests for syphilis, 184